Molecular and Cellular Aspects of Calcium in Plant Development

NATO ASI Series

Advanced Science Institutes Series

A series presenting the results of activities sponsored by the NATO Science Committee, which aims at the dissemination of advanced scientific and technological knowledge, with a view to strengthening links between scientific communities.

The series is published by an international board of publishers in conjunction with the NATO Scientific Affairs Division

A	**Life Sciences**	Plenum Publishing Corporation
B	**Physics**	New York and London
C	**Mathematical and Physical Sciences**	D. Reidel Publishing Company Dordrecht, Boston, and Lancaster
D	**Behavioral and Social Sciences**	Martinus Nijhoff Publishers
E	**Engineering and Materials Sciences**	The Hague, Boston, and Lancaster
F	**Computer and Systems Sciences**	Springer-Verlag
G	**Ecological Sciences**	Berlin, Heidelberg, New York, and Tokyo

Recent Volumes in this Series

Volume 97—Interactions between Electromagnetic Fields and Cells
edited by A. Chiabrera, C. Nicolini, and H. P. Schwan

Volume 98—Structure and Function of the Genetic Apparatus
edited by Claudio Nicolini and Paul O. P. Ts'o

Volume 99—Cellular and Molecular Control of Direct Cell Interactions
edited by H.-J. Marthy

Volume 100—Recent Advances in Nervous System Toxicology
edited by Corrado L. Galli, Luigi Manzo, and Peter S. Spencer

Volume 101—Chromosomal Proteins and Gene Expression
edited by Gerald R. Reeck, Graham A. Goodwin,
and Pedro Puigdomènech

Volume 102—Advances in Penicillium and Aspergillus Systematics
edited by Robert A. Samson and John I. Pitt

Volume 103—Evolutionary Biology of Primitive Fishes
edited by R. E. Foreman, A. Gorbman, J. M. Dodd,
and R. Olsson

Volume 104—Molecular and Cellular Aspects of Calcium in Plant Development
edited by A. J. Trewavas

Series A: Life Sciences

Molecular and Cellular Aspects of Calcium in Plant Development

Edited by
A. J. Trewavas

University of Edinburgh
Edinburgh, Scotland

Plenum Press
New York and London
Published in cooperation with NATO Scientific Affairs Divison

Proceedings of a NATO Advanced Research Workshop on
Molecular and Cellular Aspects of Calcium in Plants,
held July 15–20, 1985,
in Edinburgh, Scotland

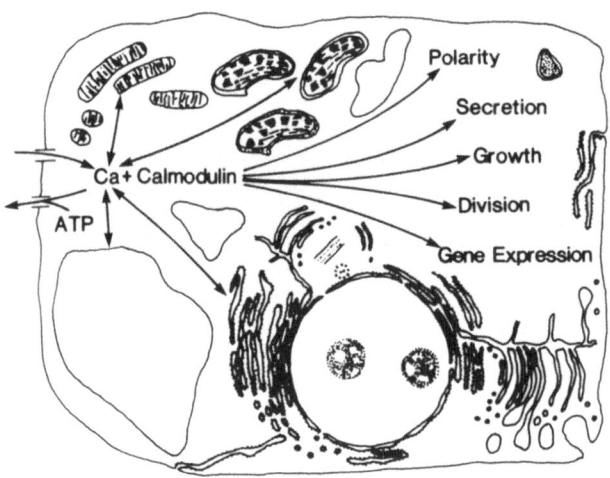

Library of Congress Cataloging in Publication Data

NATO Advanced Research Workshop on Molecular and Cellular Aspects of
Calcium in Plants (1985: Edinburgh, Lothian)
    Molecular and cellular aspects of calcium in plant development.

    (NATO ASI series. Series A, Life sciences; v. 104)
    "Proceedings of a NATO Advanced Research Workshop on Molecular and
Cellular Aspects of Calcium in Plants, held July 15–20, 1985, in Edinburgh,
Scotland"—T.p. verso.
    "Published in cooperation with NATO Scientific Affairs Division."
    Includes bibliographies and index.
    1. Calcium—Congresses. 2. Plants—Development—Congresses. 3. Plants—
Metabolism—Congresses. 4. Plant cells and tissues—Congresses. 5. Cellular
control mechanisms—Congresses. 6. Molecular biology—Congresses. I.
Trewavas, A. J. II. Title. III. Series
QK898.C2N37 1985   581.3   85-31169
ISBN-13:978-1-4612-9282-1   e-ISBN-13:978-1-4613-2177-4
DOI:10.1007/978-1-4613-2177-4

PREFACE

This volume summarises the lecture and poster sessions of a NATO advanced
workshop held in Edinburgh, July 15th-19th, 1985. The workshop was held to
bring together plant scientists of many different disciplines but who share
a common interest in the regulatory role of calcium in plant development.
Although this volume covers the formal proceedings, an equal length of time
was devoted to discussion both in large and small groups. A little of the
flavour of the directions and character of the discussions will be found in
the final article by David Clarkson which was written to cover this otherwise uncovered area of the workshop.

The volume reflects much of the current excitement in the field of
plant calcium research. Many of the participants are pioneers in their respective areas and the extent to which the last five years has seen a dramatic unfolding, a complete inversion of the role of calcium from simple macro-
nutrient to major metabolic and developmental controller is recounted here.
The material is new and much of it unpublished. In plant physiology, the
eighties may yet be designated the decade of calcium.

The formal part of the workshop consisted of half-hour lectures, the
summaries of which are found in the longer articles. There were five sessions. These were: calcium receptors and sub-cellular calcium-regulated
processes; calcium regulated enzymes; methods of calcium estimation; physiological and cellular responses to calcium and regulation of free cytosol
calcium entry. The poster abstracts which follow the main articles are
arranged in the same approximate order.

In addition to NATO who provided most of the funding of the workshop,
thanks are also due to the British Council, Shell Research, Zeiss and
'Vat 69' who provided additional help. I was also greatly helped with the
day-to-day conference running by Jane Longland, David Drakeford, David
Blowers and Simon Gilroy and by the International Organising Committee of
Jack Hanson, Stan Roux, Deiter Marmé, Andreas Sievers, Raoul Ranjeva and
Alistair Hetherington. Three prizes were awarded for outstanding posters
and these went to the posters presented by A.G. Sandelius and J. Morré;
M. Vantard, H. Stoeckel, P. Picquot, L. Van Eldik and A.M. Lambert; and
finally, M.J. Saunders. In the organising committee's judgement these
posters were representative of the outstanding advances presently being
made in plant calcium research.

A.J. Trewavas
Botany Department
University of Edinburgh
Edinburgh EH9 3JH
1985

CONTENTS

The Role of Calcium in the Regulation of Plant Metabolism   1
    D. Marmé

SUB-CELLULAR RESPONSES TO CALCIUM. CALCIUM RECEPTORS
AND SUB-CELLULAR CALCIUM REGULATED PROCESSES

Molecular Mechanisms of Calmodulin Action   11
    D.M. Roberts, T.J. Lukas, H.M. Harrington
        and D.M. Watterson

Aluminum-Induced Changes in Calmodulin   19
    A. Haug and C. Weis

Ca^{2+} Binding to Calmodulin and Interactions with Enzymes   27
    J. Haiech, N. Vidal, J. Sallantin and J.-C. Cavadore

Calcium and Calcium-Binding Proteins in Phloem   33
    D.D. Sabnis and A.R. McEuen

A Novel Plant Calciprotein as Transient Subunit of Enzymes   41
    R. Ranjeva, A. Graziana, M. Dillenschneider,
        M. Charpenteau and A.M. Boudet

Role of Ca^{2+} in the Regulation of α-Amylase Synthesis and
        Secretion in Barley Aleurone   49
    R.L. Jones, J. Deikman and D. Melroy

Cellulose Microfibril Synthesis and Orientation in Oocystis
        Solitaria: Evidence for the Involvement of Calcium   57
    H. Quader

Effect of Calcium on Lipid Metabolism During the Growth of a
        Calcifuge Plant (Lupinus luteus L.) or a Calcicolous
        One (Vicia faba)   65
    A. Oursel and B. Citharel

SUB-CELLULAR RESPONSES TO CALCIUM: CALCIUM-REGULATED ENZYMES

Plant Leaf Calcium-Dependent Protein Kinases   75
    G.M. Polya, V. Micucci, S. Basiliadis,
        T. Lithgow and A. Lucantoni

The Role of Calcium and Calmodulin in Hormone Action in
    Plants: Importance of Protein Phosphorylation   83
    B.W. Poovaiah and K. Veluthambi

Plant NAD Kinase: Regulation by Calcium and Calmodulin   91
    P. Dieter

Calcium-Dependent Protein Phosphorylation in Suspension-
    Cultured Soybean Cells   99
    C.L. Putnam-Evans, A.C. Harmon and M.J. Cormier

Roles of Calmodulin Dependent and Independent NAD Kinases in
    Regulation of Nicotinamide Coenzyme Levels of Green
    Plant Cells   107
    S. Muto and S. Miyachi

Modulation of Enzyme Activities in Isolated Pea Nuclei
    by Phytochrome, Calcium and Calmodulin   115
    N. Datta, Y.-R. Chen and S.J. Roux

Calcium/Calmodulin Dependent Membrane Bound Protein Kinase   123
    A.M. Hetherington, D.Blowers and A. Trewavas

Ca^{2+}-Dependence of Callose Synthesis and the Role of Polyamines
    in the Activation of 1,3-β-Glucan Synthase by Ca^{2+}   131
    H. Kauss

ESTIMATION OF CYTOPLASMIC CALCIUM

The Definition and Measurement of Intracellular Free Ca   141
    M.V. Thomas

Measurement of Cytoplasmic Calcium Activity with
    Ion-Selective Microelectrodes   149
    D. Sanders and A.J. Miller

NMR and X-ray Microanalysis Methods for Measurement
    of Calcium in Plant Cells   157
    W.A. Hughes

PHYSIOLOGICAL AND CELLULAR RESPONSES TO CALCIUM:
LONG DISTANCE TRANSPORT OF CALCIUM

Calcium Regulation of Mitosis: The Metaphase/Anaphase Transition . .   167
    P.K. Hepler

Calcium and Calmodulin as Regulators of Chromosome Movement
    During Mitosis in Higher Plants   175
    A.-M. Lambert and M. Vantard

Calcium Mediation of Cytokinin-Induced Cell Division   185
    M.J. Saunders

Uptake and Release of Ca^{2+} in the Green Algae
    Mougeotia and Mesotaenium   193
    M.H. Weisenseel

Calcium Pools, Calmodulin and Light-Regulated Chloroplast
    Movements in Mougeotia and Mesotaenium 201
    S. Jacobshagen, D. Altmüller, F. Grolig and G. Wagner

Calcium and Polarity in Tip Growing Plant Cells 211
    H.-D. Reiss, W. Herth and E. Schnepf

Transduction of the Gravity Stimulus in Cress Roots:
    A Possible Role of an ER-Localized Ca^{2+} Pump 219
    A. Sievers

Electrical Events in Growing Lepidium Root Tips Seem to be
    Correlated with Gravitropic Dynamics 225
    H. Lühring, H.M. Behrens and A. Sievers

Localization of Calcium in the Sensory Cells of the
    Dionaea Trigger Hair by Laser Micro-Mass Analysis (LAMMA) . . 233
    B. Buchen and W.H. Schröder

Long Distance Transport of Calcium 241
    J.A. Raven

REGULATION OF FREE CYTOSOL CALCIUM ENTRY

Controls on Calcium Influx in Corn Root Cells 253
    J.B. Hanson, M. Rincon and S.A. Rogers

ATPases and Membrane Properties in Relation to
    Ecological Differences 261
    A. Kylin and M. Sommarin

Function of Calcium-Calmodulin in Chloroplasts 269
    F.L. Crane and R. Barr

The Role of Intracellular Organelles in the
    Regulation of Cytosolic Calcium Levels 277
    A.L. Moore, M.O. Proudlove and K.E.O. Åkerman

Calcium Involvement in Plant Hormone Action 285
    D.C. Elliott

Calcium Modulation of Auxin-Membrane Interactions
    in Plant Cell Elongation 293
    D.J. Morré

Evidence for a Mechanism by which Auxins and Fusicoccin
    May Induce Elongation Growth 301
    R.W. Parish, H. Felle and B. Brummer

POSTER ABSTRACTS

The Concentration of Calmodulin Alters During Early Stages
    of Cell Development in the Root Apex of Pisum sativum 311
    E.E.F. Allan and A.J. Trewavas

Intracellular Localization of Calmodulin on Embryonic
    Axes of Cicer arietinum L. 313
    J. Hernández-Nistal, J.J. Aldasaro, D. Rodriguez,
    J. Babiano and G. Nicolás

Activation of Plant and Animal Calmodulin-Dependent Enzymes
    by Plant, Animal, Protist and Fungal Calmodulins 317
    A.C. Harmon, H.W. Jarrett and M.J. Cormier

The Control of Calmodulin Synthesis and Localisation
    in Pea Root Tissue 319
    R. Butcher and D.E. Evans

Immunocytochemical Localization of Calmodulin in Plant Tissue . . . 321
    B.S. Serlin, M. Dauwalder and S.J. Roux

Aluminium-Induced Deformations and Malfunctions of Calmodulin . . . 323
    A. Haug and C. Weis

Calcium and Calcium-Binding Proteins in Phloem 325
    D.D. Sabnis and A.R. McEuen

Calcium Nutrition, Calmodulin Content and Susceptibility
    to TMV in Tobacco (Nicotiana tabacum L. c.v. Xanthi) 327
    S. Ferrario, A. Poupet and D. Blanc

Role of Ca^{2+} in the Induction of the Phytoalexin
    Defense Response in Soybean Cells 331
    M.R. Stäb and J. Ebel

Ca-Stimulated α-Amylase Secretion from Barley
    Aleurone Protoplasts 333
    D.S. Bush, M.-J. Cornejo, C. Huang and R.L. Jones

Calcium Uptake and Exchange in Barley Aleurone Layers
    and Protoplasts during Ca-Stimulated Amylase Secretion 335
    D.S. Bush and R.L. Jones

The Control of α-Amylase Synthesis by Calcium in the
    Barley Aleurone 337
    J. Deikman and R.L. Jones

Tyrosine Specific Protein Kinases in Plant Tissues 339
    D. Blowers and A.J. Trewavas

Protein Phosphorylation in Normal and Transformed Cells:
    Effect of Cytokinin on Calcium Regulation In Vivo and In Vitro 341
    D.C. Elliott

Properties of a Calcium-Dependent Protein Kinase
    from Silver Beet Leaves 343
    G.M. Polya and V. Micucci

Calcium-Dependent Protein Phosphorylation in Germinated Pollen . . . 345
    G.M. Polya, V. Micucci, A.L. Rae, P.J. Harris and A.E. Clarke

In Vitro and In Vivo Protein Phosphorylation in Oat Coleoptiles:
    Effects of Calcium, Calmodulin Antagonists and Auxin 347
    K. Veluthambi and B.W. Poovaiah

Phosphorylation of Tonoplast Proteins in Acer pseudoplatanus
    Cell Suspension Cultures 349
    C. Teulières, G. Alibert and R. Ranjeva

Calcium-Calmodulin Requirements of Phosphatidyl Inositol
    Turnover Stimulated by Auxin 351
    A.S. Sandelius and D.J. Morré

Isolation of Plasma Membrane and Tonoplast from the Same
    Homogenates of Plant Cells by Free-Flow Electrophoresis . . . 353
    A. Sandelius, C. Penel, G. Auderset, A. Brightman,
    K. Safranski, H. Greppin and D.J. Morré

Phosphorylation of Membrane-Located Proteins of Soybean: In Vitro
    Response of Purified Plasma Membranes to Auxin and Calcium . . 355
    R.L. Varnold, D.J. Morré and A.S. Sandelius

Distribution of Calmodulin-Dependent and Calmodulin-Independent
    NAD Kinase during Early Cell Development in the Root Apex
    of Pisum sativum 357
    E.E.F. Allan and A.J. Trewavas

Ca^{2+}, Calmodulin Dependent NAD Kinase: Regulation by Light 359
    P. Dieter and D. Marmé

Ca^{2+}-Dependent Generation of a Membrane-Borne Proteinase
    Involved in Volume Regulation of Poterioochromonas 363
    H. Kauss

The Regulation of Plant Peroxidases by Calcium 365
    C. Penel, F.J. Castillo, S. Kiefer and H. Greppin

The In Vitro Reversible Association of the Regulatory and
    Catalytic Subunits of Quinate: NAD^+ Oxidoreductase 367
    A. Graziana, M. Dillenschneider, M. Charpenteau
    and R. Ranjeva

Plant Aspartate Kinase Is Not Activated by Calmodulin or Calcium . . 369
    P.L.R. Bonner, A. Hetherington and P.J. Lea

An Assessment of the Usefulness of Quin-2-AM for Cytoplasmic
    Calcium Measurements in Plant Cells 371
    R.J. Cork

Some Practical Aspects of the Application of Quin-2
    to Plant Systems 373
    S. Gilroy, W. Hughes and A.J. Trewavas

Calcium Ions and the Dynamics of the Microtubular Cytoskeleton
    During the Initiation and the Progression of Mitosis
    in Endosperm Cells 375
    M. Vantard, H. Stoeckel, P. Picquot, L. Van Eldik
    and A.M. Lambert

Mitosis and Cytokinesis in Funaria Are Accompanied by
    Inward Current Localized at the Nuclear Zone 379
    M.J. Saunders

Quantification of <u>Mesotaenium</u> Calmodulin by Improved Cyclic
        Nucleotide Phosphodiesterase Test   381
    S. Jacobshagen, F. Grolig and G. Wagner

Calcium Effects on Stomatal Guard Cells   383
    E.A.C. MacRobbie

Calcium and the Current-Voltage Relations of Stomatal Guard Cells .   385
    M.R. Blatt

Abscisic Acid, Calcium Ions and Stomatal Function   387
    A.M. Hetherington, D.L.R. De Silva, R.C. Cox and T.A. Mansfield

Calcium Regulation in Apple Fruit   389
    M. Fukumoto and M.A. Venis

Ca^{2+} Contributes to the Signal Transduction Chain in
        Phytochrome-Mediated Spore Germination   391
    R. Wayne and P.K. Hepler

A Change in Intracellular pH Does Not Contribute to the
        Signal Transduction Chain in Phytochrome-Mediated
        Spore Germination   393
    R. Wayne and P.K. Hepler

Localization of Potential Ca^{2+}-Binding Sites in
        Lily Pollen Tubes and Maize Calyptra Cells   395
    W. Herth, H.-D. Reiss, B. Hertler, R. Bauer and K. Traxel

Ca^{2+} Regulation of Myosin Sliding Along Characeae Actin Bundles . .   397
    T. Shimmen, M. Yano and K. Kohama

Mechanism of Ca^{2+}-Control of Cytoplasmic Streaming in Characeae . .   399
    Y. Tominaga and M. Tazawa

The Effect of Segment Orientation and Cell Growth on
        the Acropetal Flux of Calcium   401
    R.K. dela Fuente and C.C. de Guzman

The Role of Calcium in the Phototactic Response of
        <u>Chlamydomonas</u> <u>reinhardtii</u>   403
    N. Morel-Laurens

Calmodulin Antagonists Inhibit Adventitious Root Growth
        of <u>Tradescantia</u>   405
    S. Muto and T. Hirosawa

Direct Estimation of Plasmodesmatal Conductivity   407
    B.R. Terry and A.W. Robards

Current-Voltage Analysis as a Means to In Vivo "Separation"
        of Primary Electrogenic and Coupled Secondary Transport   . . .   409
    M.R. Blatt and C.L. Slayman

The Role of Ca^{++} in <u>Chara</u> K^+ State   411
    M.J. Beilby

The Role of Ca^{++} in the Excitation of the Single Membrane
     Samples of <u>Chara</u> 413
  M.J. Beilby

Control of Plasma Membrane Permeability of <u>Chara</u>
     by Cytoplasmic Calcium 415
  D. Sanders, U.-P. Hansen and D. Gradmann

Properties and Distribution of ATPases in Oat and Wheat
     with Special Reference to the Plasmalemma 417
  M. Sommarin, T. Lundborg and A. Kylin

A Calcium Antagonist Binding Site in Plants 419
  E. Andrejauskas, R. Hertel and D. Marmé

Characterization of Calcium Channels in Carrot Cells 421
  M.-Jose Gillery and R. Ranjeva

Binding of Verapamil to Maize Root Membranes 423
  D. Drakeford and A.J. Trewavas

The Influence of Calcium on Mitochondrial Membrane
     Transport and Fluidity 425
  A. Cooke, D. Collison, F.E. Mabbs and M.J. Earnshaw

Effect of Photoperiod on the Calcium Uptake by
     Microsomes and Mitochondria from Green Leaves 427
  V. Stosic, C. Penel and H. Greppin

The Calcium Content of Chloroplast, Mitochondrial and
     Cytosolic Fractions of Pea Leaf Cells 429
  M.O. Proudlove and A.L. Moore

Limits on a Role for Calmodulin in Chloroplast Metabolism 431
  A.R. Ashton

Effects of Calmodulin Antagonists on Transmembrane Auxin
     Transport in <u>Cucurbita pepo</u> L. Hypocotyl Segments 433
  M.C. Astle

Growth Control by Transplasmalemma Redox:
     Evidence for a Calcium Role 435
  F.L. Crane, R. Barr and T.A. Craig

The Effects of Fusicoccin and Indole-3-Acetic Acid
     on the Cytosolic pH of <u>Zea mays</u> Cells 437
  B. Brummer, A. Bertl, I. Portrykus, H. Felle and R.W. Parish

Calcium in the Agriculture of Turkey 439
  S. Terzioglu

Calcium Complexing Ligands Derived from Cystine: Exploratory Work . 441
  L. Cazaux, N. Leygue, C. Picard and P. Tisnès

Future Development of Calcium Studies: Evidence in the
     Support of Concepts 443
  D.T. Clarkson

Index 449

THE ROLE OF CALCIUM IN THE REGULATION OF PLANT METABOLISM

Dieter Marmé

Institute of Biology III
University of Freiburg
7800 Freiburg, FRG

INTRODUCTION

The free calcium ion is now recognized as a major intracellular regulator of numerous biochemical and physiological processes in plants. Over the last six or seven years a large body of evidence has been accumulated which allows us to propose a working hypothesis for the mode of action of calcium-dependent mechanisms. This hypothesis consists of essentially three parts: (1) The free cytoplasmic calcium concentration is low (less than 1 µM) and under metabolic control; (2) The cytoplasmic calcium concentration can be regulated by various extra- (or intra-) cellular signals; (3) The cytoplasmic calcium binds to receptor proteins (calmodulin beeing the most important one) which become activated and capable to modify enzyme or other activities. It is the aim of this review to put together all the essential information which supports our working hypothesis.

CALMODULIN

Calmodulin is one member of a family of low relative molecular mass acidic calcium binding proteins. Many of these proteins, such as parvalbumin, troponin C, intestinal calcium-binding protein are found only in vertebrates and, even in these animals, are restricted to a small number of tissues or cells. In addition, the function, when known, is specific and limited to one metabolic process or pathway. Calmodulin is the unique member of this protein class, as it is present in all eukaryotic cells and serves many functions (Eldik van and Watterson, 1985). Calmodulin is a single 148 aminoacid polypeptide chain containing four calcium-binding sites and is highly conserved throughout evolution. Even in the absence of calcium calmodulin exhibits a significant α-helicity. Calcium binding results in an increase in the α-helical content. This calcium-induced conformational change allows the protein to interact with appropriate enzymes and also with pharmacological agents such as phenothiazines which are known as calmodulin antagonists.

Recently the three-dimensional structure of rat testis calmodulin has been presented as determined crystallographically at 3.0 Å resolution (Babu et al., 1985). The molecule consists of two globular lobes connected by a long exposed α-helix. Each lobe binds two calcium ions through helix-loop-helix domains. The long helix between the lobes may be involved in inter-

1

actions with enzymes and drugs. An approach that can now be used to study
structure/function relationships is to alter specifically appropriate regions
of the molecule and to see how these modifications affect enzyme binding
and activation. Recombinant DNA techniques provide the tools for such
studies: the calmodulin gene has been cloned and expressed in bacteria
(Means et al., 1985).

The amino acid sequence for spinach has recently become available
(Lukas et al., 1984). The data show only minor differences as compared to
vertebrate calmodulin (Fig. 1). Out of the 148 amino acids only 13 are
exchanged. Nine of these differences between the two calmodulins are found
in the carboxyl-terminal half of the molecule. Two of the differences are
novel: a cysteine at residence 26 and a glutamine at residue 96. At the
DNA level these changes require a minimum of two base changes from the
vertebrate codons and both changes are in the proposed calcium binding
loops. It is not yet completely clear what alterations in calcium-binding
properties would result from such a change.

So far we could not detect a significant functional difference between
calmodulins from such different sources as bovine brain, drosophila heads and
zucchini (Fig. 2). Furthermore the calcium sensitivity of the activation
process is similar for calmodulin from bovine brain and zucchini. Both
calmodulins need about 1 μM free calcium for half-maximal stimulation of
the cAMP-phosphodiesterase at a calmodulin concentration of 0.12 μM (Marmé
and Dieter, 1983).

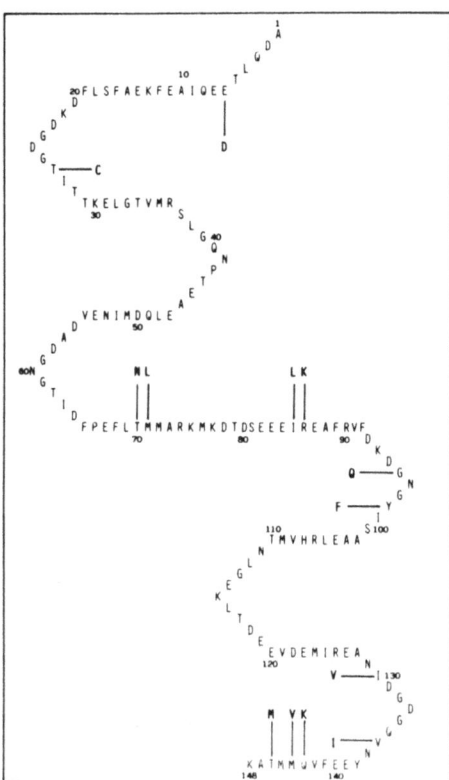

Fig. 1. Amino acid sequence of rat testis calmodulin. The sequence is arranged to show
schematically the topology of the molecule.
Note the long (helix) part between the
lobes. The 13 exchanged amino acids from
spinach are marked by bars.

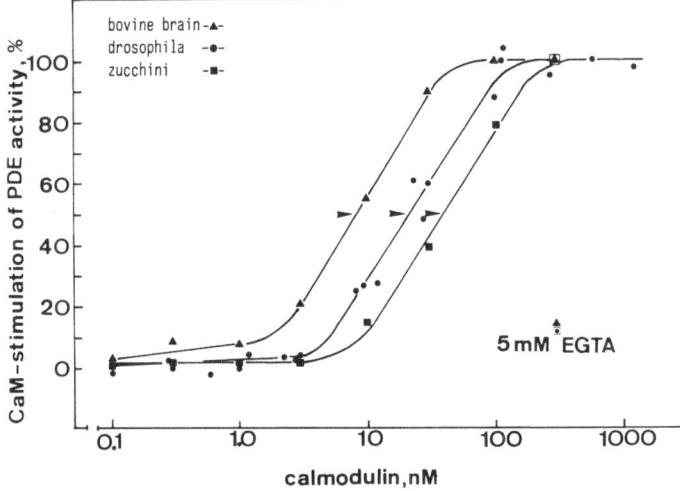

Fig. 2. Stimulation of cAMP-phosphodiesterase by calmodulin from various sources in the presence of 10 µM free calcium. In the presence of 5 mM EGTA almost no activation can be observed.

All this information suggests that animal and plant calmodulin are very similar as far as structure and biochemical function are concerned.

CALCIUM TRANSPORT MECHANISMS

The control and the regulation of free cytoplasmic calcium are pre-requisites for the calcium-mediated modulation of intracellular processes. Direct measurements of the free calcium concentration and its changes upon cellular stimulation in small plant cells have not yet been achieved because of technical problems. There is only one report by Williamson (1981) that the calcium-specific photoprotein aequorin has been successfully used to determine the free calcium concentration in large algal cells to be about 1 µM. The very elegant technique proposed by Tsien et al. (1982) using the fluorescent EGTA derivative Quin-2 apparently cannot be applied to plants. Our own and others experience with Quin-2 suggests that the compound does not penetrate the plasma membrane or does not get hydrolysed in the cytoplasm.

We therefore concentrated our efforts on the active and passive mechanisms which are thought to be important for the control and regulation of cytoplasmic free calcium. It has been reported that the control of cyto-plasmic calcium in plant cells is achieved as in animal cells by ATP-dependent extrusion of calcium out of the cell (Dieter and Marmé, 1983), by accumulation of calcium into cellular organelles like mitochondria (Hodges and Hanson, 1965), endoplasmic reticulum and vaccuoles (Gross, 1982). We have been able to demonstrate that the calcium uptake into a microsomal, plasma membrane-enriched fraction, from dark-grown plant tissues is stimulated by the addition of calmodulin (Dieter and Marmé, 1980d). Fig. 3 (panel A) shows the ATP dependent accumulation of calcium into a microsomal fraction and indicates that the saturation level of calcium accumulation is due to an equilibrium between active uptake and passive efflux. Panel B

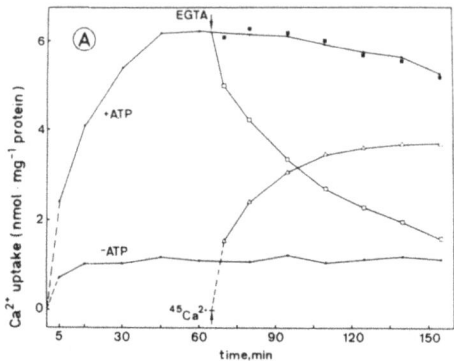

Fig. 3. Kinetics of calcium accumulation by a plant microsomal
fraction. Panel A, calcium uptake in the absence and
presence of ATP. After 70 min either EGTA or $^{45}Ca^{2+}$ is
given to determine efflux and ion exchange kinetics,
from these data dark squares are calculated to show
that the plateau is due to equilibrium between uptake
and efflux. Panel B, calcium uptake in the absence
and presence of calmodulin.

represents the calmodulin-dependence of the active calcium uptake. The
membrane vesicles which accumulate calcium upon hydrolysis of ATP in vitro
are most probably inside-out vesicles derived from the plasma membrane
(Dieter and Marmé, 1980c), suggesting that, in situ, this calcium transport
system extrudes calcium out of the cell.

We could further show that mitochondria take up much more calcium
than do the microsomal vesicles on a mg protein basis (Dieter and Marmé,
1980c). The determination of the kinetic constants K_m and V_{max} of the plant
mitochondrial calcium transport system reveals a V_{max} of 63 nmol x min^{-1} x
mg^{-1} and a K_m of about 250 μM (Dieter and Marmé, 1983). Comparison of the
corresponding values of the microsomal calcium transport system shows that
the V_{max} of mitochondria is about 10-20 times higher, but that mitochondria
have an affinity for calcium which is at most one-tenth that of microsomes.

The second messenger concept for calcium in higher plants implies
that the cytoplasmic free calcium concentration can be modulated by 'primary
signals' like light or hormones. Any change of the transport properties of
calcium transport systems caused by these signals would result in a change
of the free cytoplasmic calcium concentration. Kubowitz et al. (1982)
demonstrated that plant hormones like auxin, zeatin and kinetin alter the
activity of the microsomal calcium transport system isolated from soybean
hypocotyl hooks. Recently, Olah et al. (1983) showed that the affinity
of a calcium-ATPase towards calcium and calmodulin increases when the plant
seedlings were treated with the synthetic cytokinin benzylaminopurine.
We could show that the calmodulin stimulation of the microsomal calcium
uptake disappears when the intact corn seedlings are irradiated with far
red light; the calmodulin-independent calcium uptake, however, was not
significantly changed (Dieter and Marmé, 1981b). The determination of the
kinetic constants V_{max} and K_m reveals that calmodulin does not increase
the V_{max} of the affinity for calcium when the microsomal vesicles are
isolated from far red light-irradiation corn seedlings (Dieter and Marmé,
1983). The mitochondrial calcium uptake is also inhibited by far red light
(Dieter and Marmé, 1981b). Blue and ultra-violet light inhibit both the
calmodulin-dependent and calmodulin-independent microsomal calcium transport
(E. Schnellbächer and D. Marmé, unpublished).

From these data it can be seen that the low calcium concentration in the cytoplasm of a plant cell is maintained by a calmodulin-dependent calcium transport ATPase located most probably in the plasma membrane. By changing calcium fluxes through the membranes surrounding the cytoplasm, extracellular signals like light, hormones, etc., are able to alter cytoplasmic calcium and consequently are able to change calcium-dependent biochemical and physiological reactions.

Recently we were able to demonstrate that a radiolabelled calcium antagonist, $[^3H]$-verapamil, which is known to block calcium channels in mammalian cells, also binds to plant membranes (Andrejauskas et al., 1985). Association of $[^3H]$-verapamil to a membrane fraction was saturable and reversible. The apparent equilibrium dissociation constant was K_D=102 nM and the maximal number of binding sites was B_{max}=60 pmol x mg^{-1}. Sucrose density fractionation of zucchini membrane preparations revealed that $[^3H]$-verapamil binding sites are located primarily at the plasma membrane. The physiological function of these binding sites is not yet clear. However, it could well be – as in the case of mammalian cells – that they indicate the existance of specific calcium channels in plant cells.

CALCIUM AND CALMODULIN-DEPENDENT ENZYMES

The plant enzyme activities which have been reported to be under the control of calcium and calmodulin are listed in table 1. The NAD kinase is of particular interest because it is located in at least three different cellular compartments.

A soluble NAD kinase was found in a cytoplasmic fraction obtained from dark-grown zucchini (Dieter and Marmé, 1980a). The enzyme has been partially purified by calmodulin-Sepharose affinity chromatography and could be stimulated about eight-fold by calmodulin in a calcium dependent manner. The stimulation could be achieved by either bovine brain or zucchini calmodulin at concentrations for half saturation of about 0.1 and 1.0 μM, respectively. At optimal calmodulin concentrations, half saturation occurs at a pCa of about 4 μM. The M_r of the NAD kinase from zucchini has been estimated by Sephadex G-100 chromatography to be about 50 kD (Dieter and Marmé, 1980b).

NAD kinase activity from dark grown corn coleoptiles was shown to be almost totally dependent on calcium and calmodulin. Nearly all of the enzyme activity was found in a particulate fraction (Dieter and Marmé, 1984). Upon differential and density gradient centrifugation the NAD kinase activity co-migrates with the mitochondrial cytochrome c oxidase whereas

Table 1. Calcium, calmodulin-dependent enzymes
          in plants

Enzyme	Location
NAD kinase	Cytoplasm
NAD kinase	Outer mitochondrial membrane
NAD kinase	Chloroplast envelope
$(Ca^{2+} + Mg^{2+})$ATPase	Plasma membrane (?)
Protein kinase(s)	Soluble and membrane-bound

marker activities for nuclei, etioplasts, endoplasmic reticulum, and micro-
bodies could well be separated, indicating that the NAD kinase is associated
with mitochondria. This NAD kinase, associated with intact mitochondria,
can be activated by exogenously added calcium and calmodulin. In order
to investigate the submitochondrial localization of the NAD kinase, the
organelles were ruptured by osmotic treatment and sonication and the submitochondrial fractions were separated by density gradient centrifugation.
The NAD kinase activity exhibits the same density pattern as the antimycin
A-insensitive, NADH-dependent, cytochrome c reductase, a marker of the
outer mitochondrial membrane. This indicates that the calcium, calmodulin-
dependent NAD kinase from coleoptiles of dark grown seedlings is located
at the outer mitochondrial membrane.

In homogenates from light grown pea seedlings more than half of a
calcium, calmodulin-dependent activity and most of a calcium, calmodulin-
independent activity of the homogenate were associated with chloroplasts
(Simon et al., 1984). The calcium, calmodulin-dependent activity could
be detected by adding calcium and calmodulin to the incubation medium
containing intact chloroplasts. This activity could not be separated from
the chloroplasts by successive washes or by phase partition in aqueous
two polymer phase systems. After chloroplast fractionation, the calcium,
calmodulin-dependent NAD kinase activity was localized at the envelope, and
a calcium, calmodulin-independent activity was recovered from the stroma.

In all these cases the NAD kinase is able to sense changes of the
calcium concentration in the cytoplasm, a prerequisite of the calcium
messenger concept. Furthermore, the NAD kinases are of special interest
because of their light dependence.

First evidence for a calmodulin-dependent calcium transport ATPase
came from calcium transport experiments (see above). The calmodulin
dependence of the $(Ca^{2+} + Mg^{2+})$ATPase permits purification of the enzyme
by calmodulin-Sepharose affinity chromatography (Dieter and Marmé, 1981a).
After solubilization and centrifugation of a microsomal fraction from
coleoptiles of dark-grown corn seedlings, the supernatant was loaded on
a affinity column in the presence of calcium. The column was washed with
buffer containing calcium and NaCl to remove contaminating material. The
bound proteins were eluted by chelating calcium with EGTA. The wash fraction
contains most of the ATPase activity but no calmodulin-dependent one. The
eluate fraction has only a very small percentage of the total ATPase
activity, which now can be stimulated by calmodulin in the presence of
calcium by more than 130 per cent. Comparison of the calmodulin-dependent
ATPase activity and the calmodulin-dependent calcium transport activity
suggests that the partially purified ATPase is most probably identical
with the calmodulin-dependent calcium transport ATPase.

Protein phosphorylation is a very important mechanism by which organisms
regulate their cellular activities. Recently several papers have been
published demonstrating effects of calcium and calmodulin on protein phosphorylation in plants (Hetherington and Trewavas, 1982; Polya and Davies,
1982; Salimath and Marmé, 1983). All attempts in our laboratory to
demonstrate either cAMP-dependent or cGMP-dependent protein phosphorylation
have failed. The first identification of a substrate for a calcium,
calmodulin-dependent protein kinase came from work by Ranjeva et al. (1983).
We could demonstrate that the quinate: NAD^+ oxidoreductase is activated
by calcium, calmodulin-dependent phosphorylation. A major focus of research
in the next few years will be the identification of additional substrates
for calcium, calmodulin-dependent protein kinases.

In a recent report on a possible calcium and phospholipid-dependence
of a protein kinase from plants Hetherington and Trewavas (1984) have shown

that phospholipids inhibit a membrane-bound protein kinase activity from peas. However, a calcium and phospholipid (phosphatidylserine)-dependent protein kinase C has been described for animals and seems to be of some physiological importance (Marmé and Matzenauer, 1985).

We have tried to identify a similar activity from a soluble fraction from zucchini (Schäfer et al., 1985). We could show that indeed such a protein kinase is present also in plants. Its activity is increased by calcium and further stimulated by various phospholipids. In addition the phospholipids shift the calcium sensitivity towards lower calcium concentration.

CONCLUSION

It is now evident that the main components of our working hypothesis for calcium being a cellular messenger have been identified. However, enormous efforts have to be initiated to integrate this system into the regulatory mechanisms of plant physiology. Research should be focused on three main areas: (1) Translation of extracellular signals into changes of cytoplasmic free calcium; (2) Identification of calcium, calmodulin-dependent enzymes and processes; (3) Integration of these mechanisms into the physiology of the whole organism.

REFERENCES

Andrejauskas, E., Hertel, R., and Marmé, D., 1985, Specific binding of the calcium antagonist ^{3}H-verapamil to membrane fractions from plants, J. Biol. Chem., 260:5411-5414.
Babu, Y. S., Sack, J. S., Greenhough, T. J., Bugg, C. E., Means, A. R., and Cook, W. J., 1985, Three-dimensional structure of calmodulin, Nature, 315:37-40.
Dieter, P., and Marmé, D., 1980a, Partial purification of plant NAD kinase by calmodulin-Sepharose affinity chromatography, Cell Calcium, 1:279-286.
Dieter, P., and Marmé, D., 1980b, Calmodulin-activated plant microsomal calcium uptake and purification of plant NAD kinase and other proteins by calmodulin-Sepharose affinity chromatography, Ann. N. Y. Acad. Sci., 356:371-373.
Dieter, P., and Marmé, D., 1980c, Calcium transport in mitochondrial and microsomal fractions from higher plants, Planta, 150:1-8.
Dieter, P., and Marmé, D., 1980d, Calmodulin activation of plant microsomal calcium uptake, Proc. Natl. Acad. Sci. USA, 77:7311-7314.
Dieter, P., and Marmé, D., 1981a, A calmodulin-dependent, microcomal ATPase from corn (Zea mays L.), FEBS Letters, 125:245-248.
Dieter, P., and Marmé, D., 1981b, Far-red light irradiation of intact corn seedlings affects mitochondrial and calmodulin-dependent microsomal calcium transport, Biochem. Biophys. Res. Commun., 101:749-755.
Dieter, P., and Marmé, D., 1983, The effect of calmodulin and far-red light on the kinetic properties of the mitochondrial and microsomal calcium ion transport system from corn, Planta, 159:277-281.
Dieter, P., and Marmé, D., 1984, A calcium, calmodulin-dependent NAD kinase from corn is located in the outer mitochondrial membrane. J. Biol. Chem., 259:184-189.
Eldik van, L. J., and Watterson, D. M., 1985, Calmodulin structure and function, in: "Calcium and Cell Physiology", D. Marmé, ed., Springer, Heidelberg.
Gross, J., 1982, Oxalate-enhanced active calcium uptake in membrane fractions from zucchini squash, in: "Plasmalemma and Tonoplast: Their Function in the Plant Cell", D. Marmé, D. Marré, and R. Hertel, eds., Elsevier Biomedical Press, Amsterdam.

Hetherington, A., and Trewavas, A., 1982, Calcium-dependent protein kinase in pea shoot membranes, FEBS Letters, 145:67-71.

Hetherington, A. M., and Trewavas, A., 1984, The regulation of membrane bound protein kinases by phospholipid and calcium, Ann. Proc. Phytochem. Soc. of Eur., 24:181-197.

Hodges, T. K., and Hanson, J. B., 1965, Calcium accumulation by maize mitochondria, Pl. Physiol, 40:101-108.

Kubowitz, B. P., Vanderhoef, L. N., and Hanson, J. B., 1982, ATP-dependent calcium transport in plasmalemma preparation from soybean hypocotyls, Pl. Physiol., 69:187-191.

Lukas, T. J., Iverson, D. B., Schleicher, M., and Watterson, D. M., 1984, Covalent structure of a higher plant calmodulin: Spinacea oleracea, Plant Physiol., 75:788-795.

Marmé, D., and Dieter, P., 1983, The role of calcium and calmodulin in plants, in: "Calcium and Cell Function", Vol. IV, Cheung, W. Y., ed., Academic Press, New York.

Marmé, D., and Matzenauer, S., 1985, Protein kinase C and polyphosphoinositide metabolites: Their role in cellular signal transduction, in: "Calcium and Cell Physiology", D. Marmé, ed., Springer, Heidelberg.

Means, A. R., Lagace, L., Simmen, R. C. M., and Putkey, J. A., 1985, Calmodulin gene structure and expression, in: "Calcium and Cell Physiology", D. Marmé, ed., Springer, Heidelberg.

Olah, Z., Berczi, A., and Erdei, L., 1983, Benzylaminopurine-induced coupling between calmodulin and Ca-ATP in wheat root microsomal membranes, FEBS Letters, 154: 395-399.

Polya, G. M., and Davies, J. R., 1982, Resolution of calcium-calmodulin-activated protein kinase from wheat germ, FEBS Letters, 150:167-171.

Ranjeva, R., Refeno, G., Boudet, A., and Marmé, D., 1983, Plant quinate: NAD^+ oxidoreductase is activated by calcium-calmodulin-dependent phosphorylation, Proc. Nat. Acad. Sci. USA, 80:5222-5224

Salimath, B. P., and Marmé, D., 1983, Protein phosphorylation and its regulation by calcium and calmodulin in membrane fractions from zucchini hypocotyls, Planta, 158:560-568.

Schäfer, A., Bygrave, F., Matzenauer, S., and Marmé, D., 1985, Identification of a calcium and phospholipid-dependent protein kinase in plant tissue, FEBS Letters, in press.

Simon, P., Bonzon, M., Greppin, H., and Marmé, D., 1984, Subchloroplastic localization of NAD kinase activity: evidence for a calcium, calmodulin-dependent activity at the envelope and for a calcium, calmodulin-dependent activity in the stroma of pea chloroplasts, FEBS Letters, 167:332-338.

Tsien, R. Y., Pozzan, T., and Rink, T. J., 1982, Calcium homeostasis in intact lymphocytes: cytoplasmic free calcium monitored with a new intracellularly trapped fluorescent indicator, J. Cell Biol., 94:325-334.

Williamson, R. E., 1981, Free calcium concentration in the cytoplasm: a regulator of plant cell function, What's New Plant Physiol., 12:45-48.

SUB-CELLULAR RESPONSES TO CALCIUM

CALCIUM RECEPTORS AND SUB-CELLULAR CALCIUM REGULATED PROCESSES

MOLECULAR MECHANISMS OF CALMODULIN ACTION

Daniel M. Roberts, Thomas J. Lukas, H. Michael Harrington
and D. Martin Watterson

Department of Pharmacology, Vanderbilt University and
Laboratory of Cellular and Molecular Physiology
Howard Hughes Medical Institute
Nashville, TN 37232, United States of America

Calcium has many biological functions and several groups of macromolecules have the ability to bind calcium with various selectivities and
affinities. A rapidly expanding body of knowledge about the structure,
thermodynamic and kinetic properties, and cell biology of calcium binding
proteins strongly indicates that the targets or receptors for calcium acting
as a signal transducer in eukaryotic cells are a class of calcium binding
proteins referred to as calcium modulated proteins. These proteins are
characterized by their ability to bind calcium in a reversible manner with
dissociation constants in the nanomolar to micromolar range under physiological conditions. Although it is not possible yet to predict with any
degree of certainty what type of calcium binding structure or molecular
mechanism might be associated with a given biological function of calcium, a
trend has begun to emerge from the detailed analyses of calcium modulated
proteins [1].

Calmodulin is a calcium modulated protein that appears to be ubiquitous
among eukaryotes, has multiple calcium binding sites in a single polypeptide
chain and has a number of biochemical activities, including activation of
enzymes, and interaction with other non-enzymatic proteins, peptides and
drugs. Because of these properties, calmodulin has become a standard of
comparison for other calcium modulated proteins and a necessary object of
investigation for anyone interested in the cellular and molecular basis of
calcium action in animal or plant species.

The major interests of our laboratory are the molecular mechanisms of
calcium action in eukaryotic cell processes, with a current emphasis on the
molecular mechanisms of calmodulin action. Our initial entry into this field
was through an attempt in the 1970's to detect, purify and characterize an
intracellular calcium binding protein with properties appropriate for a
potential mediator of calcium effects on the metabolic and mechanochemical
activities of eukaryotic cells. We purified and characterized a non-muscle
troponin C-like protein, showed that it had appropriate calcium binding
properties, elucidated its amino acid sequence, and showed that it interacted
with several proteins in a calcium dependent manner. As part of the latter
studies we showed that the troponin C-like protein had phosphodiesterase
activator activity and was indistinguishable from a phosphodiesterase
activator protein that had been previously described by Cheung, Kakiuchi and
their coworkers [1].

From these initial studies we proceeded to further define the specific and general features of calmodulins from a variety of tissues and species. After screening a large number of biological sources for the presence of calmodulin and calmodulin activity, we selected a defined set of related and divergent species from which to generate a database of knowledge about calmodulin structure and function. Based on the database generated from this study of naturally occurring calmodulins, we synthesized and expressed a calmodulin gene (VU-1) designed for site-specific mutagenesis. Currently, the relationship between calmodulin structure and function is being investigated further by using a combined approach of comparative biochemistry and protein engineering.

Concomitant with the studies of calmodulin, we initiated studies on the next step in the molecular mechanism: calmodulin binding proteins. Several binding proteins have been identified and are being characterized in detail. Most recently, we have mapped calmodulin binding sites on calmodulin binding proteins and have initiated studies on the mechanism of action of selected calmodulin regulated enzymes. We have examined also the regulation of calmodulin and calmodulin binding proteins under various physiological and pathophysiological states. Studies of the biosynthesis of calmodulin in plant and animal cells have provided necessary precedents as well as suggestive data for the possibility of coordinate control of expression of calmodulin and calmodulin binding proteins. We also examined [2] in detail the interaction of the calmodulin antagonists phenothiazines and naphthalene sulfonamide derivatives (e.g., W-7) with calmodulin and other calcium modulated proteins and showed that the interaction with calmodulin is complex. We also showed that effects of these drugs on biological systems cannot be used by itself as a implication of calmodulin involvement in a biological process.

It is from this mechanistic and molecular perspective that we have approached our studies of plant and algal calmodulins, calmodulin binding proteins and calcium regulated enzymes. The intent of this report, therefore, will be to summarize some of our recent and current studies from this perspective. The emphasis will be on studies of the molecular bases of calmodulin actions that are relevant to someone interested in plant physiology and biochemistry.

COMPARATIVE BIOCHEMISTRY AND PROTEIN ENGINEERING

To date, the only amino acid sequences (Fig.1) available for plant and algal calmodulins are those for spinach calmodulin [3], Chlamydomonas calmodulin [4], and a partial amino acid sequence for barley calmodulin [5]. In a series of unpublished studies (M. Schleicher, T. Lukas, D.M. Watterson) we screened a variety of higher plants for calmodulin activity. Several of these calmodulins were purified, characterized and compared to spinach calmodulin. Based on numerous criteria, including partial amino acid sequence, monocotyledon and dicotyledon calmodulins were indistinguishable. Altogether the data suggest that spinach calmodulin may be representative of higher plant calmodulins. In contrast, Chlamydomonas calmodulin does not appear to be representative of eukaryotic algal calmodulins. Current studies on other algal species should provide some insight into calmodulin diversity at this phylogenetic level.

The limited number of conservative amino acid differences among calmodulins (Fig. 1) do not alter calmodulin's activator activity with some enzymes such as phosphodiesterase and myosin light chain kinase but appear to result in differences in NAD kinase activation [5-8] at both maximal and submaximal concentrations of calmodulin (Fig. 2). In terms of maximal activation of NAD kinase, the calmodulins tested appear to fall into two

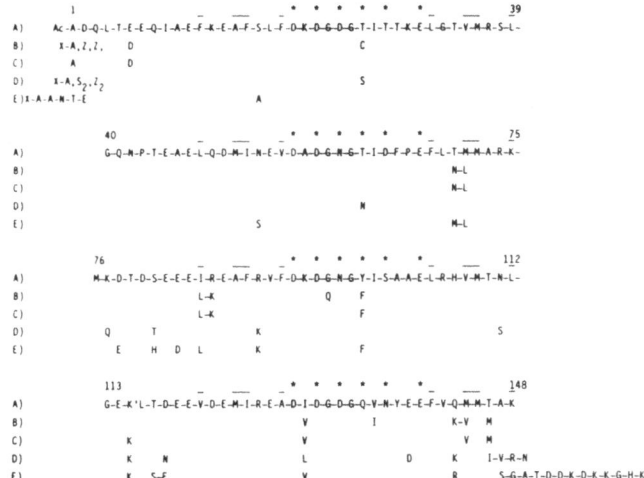

Fig. 1. Comparison of Calmodulin Amino Acid Sequences. A. Bovine brain;
B. Spinach; C. VU-1; D. <u>Dictyostelium discoideum</u>; E. <u>Chlamydomonas
reinhardtii</u>. The single letter amino acid code is shown. K'
indicates trimethyllysine. The complete sequence is shown in line
A and only differences in sequence are shown in lines B-E.

groups: higher maximal activation as exemplified by <u>Chlamydomonas</u>,
<u>Dictyostelium</u>, and VU-1 synthetic calmodulin; or lower maximal activation
in the presence of saturating concentrations of calmodulin as exemplified
by vertebrate, higher plant, <u>Tetrahymena</u>, <u>Paramecium</u>, <u>Tribonema</u>, and
<u>Arbacia</u> calmodulins. The only amino acid sequence identity among
<u>Chlamydomonas</u>, <u>Dictyostelium</u> and VU-1 calmodulins that is not found in the
other calmodulins is the presence of lysine at position 115 instead of
trimethyllysine. To directly test whether the methylation state of
lysine-115 may be contributing to the difference in maximal activation, we
generated calmodulins that apparently differ only in the methylation state
of lysine-115 and tested them for NAD kinase activator activity.
<u>Chlamydomonas</u> and VU-1 calmodulins were incubated with a partially purified
lysine N-methylase in the presence (test) or absence (control) of the
methyl donor S-adenosylmethionine (SAM), the calmodulins purified from each
mixture after termination of the reaction, and the purified calmodulins
characterized in terms of activator activity and lysine-115 methylation
state. The test calmodulins were found to be methylated at lysine-115 and
to have a reduced maximal NAD kinase activator activity, indistinguishable
from the naturally occurring calmodulins that have a trimethyllysine-115
(Fig. 2), while the control calmodulins retained the higher maximal
activity. Although amino acid analyses of the methylated calmodulins
showed the presence of one trimethyllysine per mole of protein and a
decrease in lysine content of one mole per mole, it is possible that other
changes may have occurred that were not detected by the chemical
characterization procedures used. Therefore, we introduced a DNA sequence
change in VU-1 so that it coded for an arginine at position 115. The side
chain of arginine has a charge intermediate to both lysine and trimethyllysine and has hydrogen bonding potential, similar to lysine. Hydrogen
bonding potential is lost in trimethyllysine due to the replacement of
three hydrogens by three methyl groups. This engineered calmodulin, termed
VU-1R115, was expressed in <u>E. coli</u>, purified and characterized. VU-1R115
calmodulin was found to have NAD kinase activator properties similar to
VU-1. Thus, overall, the data indicate that the properties of the amino
acid side chain at position 115 can have a dramtic effect on maximal NAD
kinase activity and raise the question of whether the methylation state of
calmodulin may have regulatory significance in higher plants.

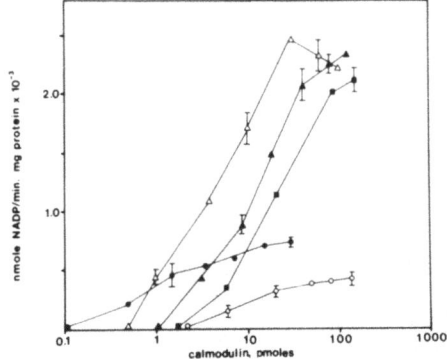

Fig. 2.   Comparative NAD kinase activator activities. Assay mixtures (0.5
          ml) and conditions were as previously described [7],[8]. Calmodulins
          tested were spinach (●), VU-1 (△), Chlamydomonas reinhardtii
          (▲), Dictyostelium discoideum (■), and chicken gizzard (○).
          Each point is the average of two determinations. The error bars
          show the range of values obtained. The absence of error bars
          indicates the range is smaller than the figure symbol.

CALMODULIN BINDING PROTEINS, NAD KINASES AND PROTEIN KINASES

     Similar to our direct, biochemical approach to identifying and char-
acterizing the calcium receptors in higher plants and algae, we have taken a
direct approach to the detection and characterization of the next step in
the molecular mechanism: calmodulin binding proteins and calcium modulated
enzymes. We previously reported [9] that the major calmodulin binding
protein in higher plant chloroplasts appeared to be the major Coomassie blue-
staining protein in envelope fractions. Based on its apparent abundance,
suborganellar localization, and electrophoretic mobility, it was possible
that this protein might be the phosphate translocator protein [10]. In
order to test this possibility, we followed established translocator
purification protocols [10] and analyzed for the presence of the envelope
calmodulin binding protein. As shown in Fig. 3, the calmodulin binding
protein is found in purified preparations of translocator, supporting the
concept that this calmodulin binding protein may be the same protein studied

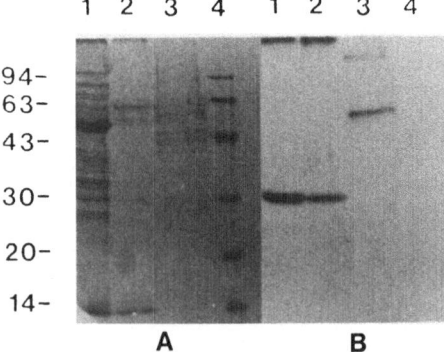

Fig. 3.   Calmodulin Overlay Analysis of Spinach Chloroplast Envelopes.
          Envelope fractions and purified translocator fraction were
          prepared [9],[10] and gel overlay analyses [9] were done as
          previously described. A. Coomassie-Blue stained 12.5% (wt/vol)
          acrylamide SDS gel, B. autoradiogram. Lane 1, envelope fraction;
          Lane 2, translocator fraction [10]; Lane 3, calmodulin binding
          protein standard; Lane 4, molecular weight standards.

by other investigators as the translocator protein. However, it should be noted that we have not been able to demonstrate yet a reproducible modulatory effect of calcium or calmodulin on the translocator activity. Clearly, further biochemical analysis of the translocator and an improved reconstitution system are needed before this effect or possible mechanism can be rigorously pursued.

Although the major calmodulin binding activity was found in the envelope fraction of the chloroplast, we and others have not found significant calmodulin levels associated with the envelope by using standard methodologies. In fact, most of the chloroplast calmodulin appears to be associated with the stromal (or soluble) chloroplast fraction (Fig. 4). Of the total calmodulin found in a soluble extract of leaf tissue, only about 0.02 to 0.1% is recovered in the chloroplast stromal fraction. These results raise several questions and possibilities that need to be addressed in current and future work. First, it is possible that the endogenous protein that interacts with the envelope protein is not calmodulin but a protein that calmodulin substitutes for in vitro. An important precedent in animal biochemistry is the ability of calmodulin to substitute for troponin C in actomyosin reconstitution assays or the ability of troponin I to bind to calmodulin-Sepharose. Second, if calmodulin does associate in vivo with the binding protein, it is possible that under the conditions of fractionation of the chloroplast the calmodulin is extracted away from the binding protein and into the stromal fraction. Again, in the case of animal cells there is the precedent that most of the calmodulin is found in the soluble fraction after subcellular fractionation yet several calmodulin binding proteins, including calmodulin regulated enzymes, are found in membrane fractions. Third, it is possible that much of the calmodulin found in the stromal fraction may be associated with calmodulin binding proteins and cannot be detected under the conditions of analysis. In this regard, calcium sensitive NAD kinase activity has been detected in the stromal fraction but was not clearly stimulatable by exogenous calmodulin. However, based on the level of calmodulin detected in the stromal fraction, an effect of exogenous calmodulin may not be seen unless the fraction is depleted of endogenous calmodulin before the assay. Thus, although we and others find detectable levels of calmodulin associated with the chloroplast fraction, the role that calmodulin may be playing in chloroplast function is still not established.

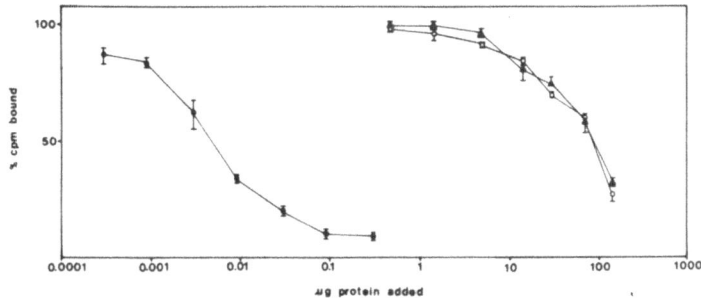

Fig. 4. Analysis of Spinach Chloroplast Stroma by Calmodulin Radioimmuno-assay. Anti-spinach calmodulin serum was prepared and the assay was carried out as described [12]. Various concentrations of spinach calmodulin (●) or stromal protein (▲) unheated or (○) heated (6 min, 80°) were mixed with antiserum (1:7500 final dilution) and incubated overnight at 4°. A fixed amount of Bolton-Hunter ^{125}I-spinach calmodulin (0.5 ng; 60,000 cpm) was added to the mixture and the mixture was processed as described [12]. The degree of competition is expressed as the percentage of cpm in the absence of competing antigen.

As discussed under the Comparative Biochemistry section, we have been examining in some detail the interaction between calmodulins and NAD kinase from pea leaves. As part of these studies we have examined the question of why all NAD kinase preparations from higher plants do not show the same profile of calmodulin activation. Most plant NAD kinase purification protocols that yield a calmodulin sensitive enzyme are based on the pioneering work of Muto and Miyachi [11]. The DEAE-cellulose chromatography step appears to deplete the endogenous calmodulin from the NAD kinase preparation and converts the kinase from a calcium activatable enzyme to a calcium and calmodulin activatable enzyme. However, if the eluate from this step is not processed immediately, the difference in maximal activation of NAD kinase by methylated and unmethylated calmodulins (vide infra) is no longer seen [8]. The molecular basis of this loss of discrimination is not known. However, the time scale of the initial steps in the purification protocol appears to be the important factor, regardless of whether or not PMSF, pepstatin or leupeptin are utilized in the purification protocol. These results emphasize the importance of further purification and characterization of NAD kinase for gaining additional insight into the regulation of NAD kinase activity in higher plants.

In addition to NAD kinases, protein kinases appear to be regulated by calmodulin in both animal and plant cells. As part of our studies on the of molecular mechanisms of calmodulin action, we have examined the interaction of calmodulin with protein kinases, the inhibition of this interaction by various drugs and peptides, and the mechanism of kinase action. As a prototype, the focus has been on myosin light chain kinase (MLCK). The approach has been to estimate the minimal requirements for interaction and to characterize the structural basis and possible modulation of the interaction. For smooth muscle MLCK, a peptide that contains much of the calmodulin binding activity was isolated, its amino acid sequence determined, and a synthetic peptide analog made. Current efforts are directed toward completion of these studies by mapping the interaction of the peptides with calmodulin.

The mechanism of MLCK phosphorylation has been approached by elucidation of the amino acid sequence of the phosphorylation site in selected light chains followed by examination of substrate specificity by analysis of synthetic peptide substrates, analogs and inhibitors. Based on the sequence of the phosphorylation site in smooth muscle light chain [14], a series of peptides were tested for their ability to serve as effective substrates of smooth muscle MLCK (Table I). The synthetic peptides contain a common Lys-Lys-Arg-Pro-Gln-X-Ala-Y-Ser-Asn-Val-Phe-Ala-Met sequence with variants at residues 6 (X) and 8 (Y). Substitution of Thr (residue Y), found in the light chain amino acid sequence, with Ala or Asn does not dramatically alter the ability of the peptide to be phosphorylated by MLCK and results in a peptide with only one possible site for phosphorylation. Substitution of the Arg at position X, found in the smooth muscle light chain amino acid sequence, with a Gly, found at the homologous position in the cardiac light chain amino acid sequence [14], results in a loss of phosphorylation by MLCK. This peptide also does not effectively inhibit light chain phosphorylation, suggesting the Arg at position 6 is required for binding. Placement of an Arg at position 6 (X) and Ala at positions 8 (Y) and 9 resulted in inhibitors of light chain phosphorylation activity (Table II). Current studies are directed toward the long term goal of engineering selective inhibitors of calcium dependent kinases based on the accumulating database from synthetic peptide studies.

In addition to being substrates for MLCK, these peptides are also substrates for a calmodulin dependent protein kinase with more general protein substrate specificity and for a calmodulin independent, calcium dependent protein kinase (kinase C). In animal and some algal cells, cyclic

Table I. Peptide and Protein Substrates of Myosin Light Chain Kinase

Peptide number	Substrate	Km (μM)	Vmax (μmol/min/mg)
	Smooth muscle light chain[13]	8.6 ± 1.5	36.2 ± 1.6
1	K-K-R-P-Q-R-A-T-S-N-V-F-A-M	27.1 ± 2.4	2.2 ± .3
2	K-K-R-P-Q-R-A-A-S-N-V-F-A-M	28 ± 1.0	1.6 ± .02
3	K-K-R-P-Q-R-A-N-S-N-V-F-A-M	17 ± 1.5	2.2 ± .2
4	K-K-R-P-Q-G-A-T-S-N-V-F-A-M	1 420[a]	-
	Bovine cardiac light chain[14]	107 ± 8.4	23.8 ± 1.1

a   An actual Km was not determined. The incorporation of ^{32}P at this concentration was 0.28%.

Table II. Inhibition of Myosin Light Chain Kinase by Synthetic Peptides

Peptide number	Peptide	K_i [a] (μM)
5	K-K-R-P-Q-R-A-A-A-N-V-F-A-M	84 ± 6
6	R-R-R-P-Q-R-A-A-A-N-V-F-A	84 ± 8
7	K-K-K-P-Q-R-A-A-A-N-V-F-A	185 ± 15

a   Apparent inhibition constants were determined using peptide 1 as a substrate, and fitting this data by linear least squares to double reciprocal or single reciprocal relationships. Inhibitors were tested over a concentration of 40 to 160 μM. The peptide substrate concentrations were 16.7, 33.5 and 67.0 μM.

nucleotide dependent protein kinases must also be considered as possible phosphotransferases capable of interacting with these substrates and inhibitors. It has been shown [13] that peptides similar to peptide 1 (Table I) are poor substrates for cyclic AMP-dependent protein kinase. Based on the known preference of this kinase for the amino sequence Arg-X-Ser in which the Ser is phosphorylated, it was not surprising to find that the calmodulin kinase substrates are phosphorylated by cyclic AMP-dependent protein kinase, although less effectively than substrates with the serine penultimate to the arginine. However, the phosphorylation of the substrate was not calcium dependent and the peptide inhibitors of the calcium dependent protein kinases did not inhibit the cyclic nucleotide dependent protein kinases. In a pilot experiment we utilized peptide 1 as a substrate for protein kinase activity in a crude extract of Chlamydomonas and detected calcium dependent phosphorylation activity. Therefore, the use of peptide substrates and inhibitors that were designed based on the specificity of calcium dependent protein kinases in animal tissues may be useful in the assay and purification of calcium dependent protein kinases from higher plants and algae, especially in cases where a physiological substrate is not readily available.

ACKNOWLEDGEMENTS

This work was supported in part by N.I.H. Grant GM 30861 and N.S.F. Grant DMB 8405374. We thank Ms. Becky Lawson for help with the preparation of this manuscript.

REFERENCES

1. L. J. Van Eldik, J.G. Zendegui, D. R. Marshak and D. M. Watterson, Calcium-Binding Proteins and the Molecular Basis of Calcium Action. Int Rev Cytol 77:1-61 (1982).

2. D. R. Marshak, T. J. Lukas and D. M. Watterson, Drug-Protein Interactions. Binding of Chlorpromazine to Calmodulin, Calmodulin Fragments, and Related Calcium Binding Proteins. Biochemistry 24:144-150 (1985).

3. T. J. Lukas, D. B. Iverson, M. Schleicher and D. M. Watterson, Structural Characterization of a Higher Plant Calmodulin: Spinacea oleracea. Plant Physiol 75:788-795 (1984).

4. T. J. Lukas, M. Wiggins and D. M. Watterson, Amino Acid Sequence of a Novel Calmodulin from the Unicellular Alga Chlamydomonas. Plant Physiol 78:477-483 (1985).

5. M. Schleicher, T. J. Lukas and D. M. Watterson, Further Characterization of Calmodulin from the Monocotyledon Barley (Hordeum vulgare). Plant Physiol 73, 666-670 (1983).

6. D. R. Marshak, M. Clarke, D. M. Roberts and D. M. Watterson, Structural and Functional Properties of Calmodulin from the Eukaryotic Microorganism Dictyostelium discoideum. Biochemistry 23:2891-2899 (1984).

7. D. M. Roberts, W. H. Burgess and D. M. Watterson, Comparison of the NAD Kinase and Myosin Light Chain Kinase Activator Properties of Vertebrate, Higher Plant, and Algal Calmodulins. Plant Physiol 75:796-798 (1984).

8. D. M. Roberts, R. Crea, M. Malecha, G. Alvarado-Urbina, R. H. Chiarello and D.M. Watterson, Chemical Synthesis and Expression of a Calmodulin Gene Designed for Site-Specific Mutagenesis. Biochemistry, in press (1985).

9. D. M. Roberts, R. E. Zielinski, M. Schleicher and D. M. Watterson, Analysis of Suborganellar Fractions from Spinach and Pea Chloroplasts for Calmodulin-binding Proteins. J Cell Biol 97:1644-1647 (1983).

10. U. L. Flugge and H. W. Heldt, The Phosphate Translocator of the Chloroplast Envelope. Isolation of the Carrier Protein and Reconstitution of Transport, Biochim Biophys Acta 638:196-204 (1981).

11. S. Muto and S. Miyachi, Properties of a Protein Activator of NAD Kinase from Plants. Plant Physiol 59:55-60 (1977).

12. L. J. Van Eldik and D. M. Watterson, Reproducible Production of Antiserum against Vertebrate Calmodulin and Determination of the Immunoreactive Site. J Biol Chem 256:4205-4210 (1981).

13. B. E. Kemp, R. B. Pearson and C. House, Role of Basic Residues in the Phosphorylation of Synthetic Peptides by Myosin Light Chain Kinase. Proc Natl Acad Sci USA 80:7471-7475 (1983).

14. T. J. Lukas, A. Redelfs, W. H. Burgess and D. M. Watterson, Amino Acid Sequence of the Phosphorylation Site of Bovine Cardiac Myosin Light Chain. Arch Biochem Biophys 238:664-669 (1985).

18

ALUMINUM-INDUCED CHANGES IN CALMODULIN

Alfred Haug and Christopher Weis

Center for Environmental Toxicology and Pesticide Research Center, Michigan State University, East Lansing, MI 48824 USA

INTRODUCTION

Calcium serves as a second messenger in bioregulation via various intracellular calcium trigger proteins such as calmodulin. As a result of a large variety of external stimuli, intracellular calcium transients are generated which can be interpreted as a signal. Within the lifetime of these transients, calcium ions are bound (signal input) to trigger proteins which undergo conformational changes. These changes play a key role in signal amplification and transmission (output) from the trigger protein to respective target enzymes and structural elements [1]. In view of this central reliance of cellular control on calcium ions and a few trigger proteins, severe repercussions on biochemical and physiological processes can be expected when the coupling between signal input and output is interrupted.

In the following we report on the interaction of toxic aluminum ions with calmodulin which can severely antagonize the biochemical trigger activity of the regulatory protein. Aluminum toxicity is a serious global problem for crop productivity [2,3,4] and for human health [5,6]. As a fringe benefit of investigating aluminum-induced, pathological alterations of calmodulin and associated activities, deeper insight is gained into the protein's physiological mode of action as signal transmitter. To rationalize pathophysiological observations in light of calcium's role as intracellular signal and the impact of aluminum ions on calmodulin, the hypothesis has been formulated that calmodulin is a key lesion in the broadly defined syndrome of aluminum toxicity [7,8].

PHYSICO-CHEMICAL PROPERTIES OF ALUMINUM AND CALCIUM

Basic parameters governing metal cation binding to organic ligands (Table I) lie at the heart of understanding the interaction of aluminum ions with calmodulin. Key factors determining the formation of metal complexes with ligands are the ionic charge, ionic size, and the coordination geometry. The aluminum ion has a high ratio of charge/radius, e/r, which profoundly influences the extent of ionic solvation. At acidic pH values, the aluminum ion is hexahydrated forming an octahedral arrangement of water molecules. As a result of the strongly polarized O-H bond, these solvated aluminum ions hydrolyze in a

Table 1. Physico-Chemical Properties of Aluminum and Calcium Ions [9,13,14].

	Ca^{2+}	Mg^{2+}	Al^{3+}
unhydrated radius (nm)	0.099-0.118	0.066	0.051
hydrated radius (nm)	0.412	0.428	0.475
e/r*	20.2	30.3	58.8
coordination number	8 >7 >6 >9	6	6
hydration enthalpy kcal/mol of ion)	-399	-477	-1141
hydrolysis constant	12	11	5
k_{ex} (sec^{-1})#	10^8	10^5	0.13

*e/r represents charge in elementary units per unhydrated radius in nm.
#k_{ex} is the approximate rate constant for water exchange at 25°C.

pH -dependent manner [9]. At physiological pH, and at micromolar aluminum concentration, singly and doubly charged monomeric species, e.g., $[Al(OH)(H_2O)_5]^{2+}$, are present in aqueous solution. Calcium ions, on the other hand, are not hydrolyzed, or only slightly, in the neutral pH region. Ca^{2+} ions can have variable coordination numbers [10,11], and thus the associated, variable ionic radii, afford the calcium cation significant flexibility in selecting a stereochemically fitting binding site, usually oxygen atoms, compared with the fixed coordination geometry of the aluminum and magnesium ion. An additional distinguishing feature for metal affinities, K_d, to ligands, e.g., a protein, is determined by the release rate of water molecules from the primary solvation shell of the respective cation, k_{on}, and by the dissociation rate of the cation from the protein, k_{off}, where $k_{off} = K_d \cdot k_{on}$ [9,12]. On-rates for calcium binding to calcium-modulating proteins like calmodulin are about 10^8 $M^{-1} \cdot sec^{-1}$, yielding k_{off} values ranging from 100 to about 10 sec^{-1}. Calcium-calmodulin is therefore a suitable trigger for biochemical reactions occurring within seconds to milliseconds. As a result of the higher e/r ratio of the aluminum ion, a much slower rate of water release is to be expected. The magnesium ion has a rate k_{on} of about $10^5 M^{-1} sec^{-1}$, a value close to that measured for the rate of water exchange from the hydrated Mg^{2+} cation [9]. The rate of metal dissociation and the accompanying conformational change of the protein are slow.

ALUMINUM IONS INDUCE HELIX-COIL TRANSITIONS IN CALMODULIN

Upon application of stoichiometric quantities of aluminum ions to micromolar concentrations of calmodulin the negative ellipticity of the protein, measured at 222 nm, decreases with increasing aluminum concentrations (Fig. 1). This indicates that aluminum addition to calmodulin, at a molar ratio of 4:1, decreases the helix content by about 20-30 percent [7], whereas calcium addition to the apoprotein promotes helix formation [1]. Major structural changes in the protein are induced upon addition of the first two aluminum ions. The steepness of the response, as characterized by a Hill coefficient of about 1.55, apparently reflects positive cooperativity and is typical of small

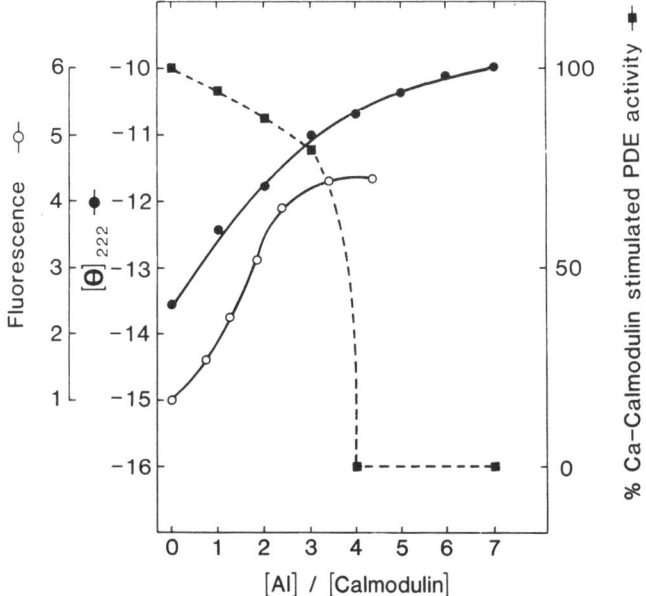

Fig. 1. Mean Residue Ellipticity, $[\theta]_{222}$, ANS
Fluorescence Intensity, and 3':5 -Cyclic
Nucleotide Phosphodiesterase Activity
upon addition of Aluminum Ions to Bovine
Calmodulin. For optical studies, the cal-
modulin concentration was 10 μM, pH 6.5,
23°C, in 10 mM MOPS buffer, 2 μM ANS.
ANS excitation at 360 nm, emission recor-
ded at 490 nm. The enzyme assay was per-
formed with 5 μM calmodulin, 10 mM Tris-
HCl, pH 6.5, 37°C; hydrolysis of cyclic
GMP was assayed.

compact proteins [15].   Aluminum dissociation constants, K_d, have been
obtained varying between 0.1 and 0.4 μM [7,16], similar to those of
"high-affinity" calcium binding [17]. Stoichiometric binding of aluminum
ions to calmodulin takes place irrespective of the presence or absence of
saturating   calcium   concentrations. The aluminum-induced structural
changes of calmodulin do not appear to result simply from a displacement
of calcium ions from their specific sites, as demonstrated by measuring
the protein's calcium content with atomic absorption techniques. Rather,
application of aluminum ions to calmodulin, at a molar ratio of up to
8:1, causes binding of additional calcium ions,   presumably   at
non-specific sites [18]. It is also doubtful whether aluminum ions can
replace calcium ions at all, considering the differences in charge, etc
(Table I).

    The aluminum-induced breakage of hydrogen bonds is thus associated
with a helix-coil transition   as   opposed to the calcium-promoted
coil-helix formation. As opposed to calcium binding [19], thermodynamic
studies demonstrate that binding of the first aluminum ion to the protein
is an enthalpy-driven process, while binding of three additional aluminum
ions seems to be entropy-driven when coordinated water molecules are
released   from   the   coordination   sphere upon formation of the
aluminum-calmodulin complex [16].

Concomitant with the aluminum-induced helix-coil transition, the hydrophobic surface exposure of the protein is enhanced [7], relative to that observed in the presence of stoichiometric quantities of calcium ions [1]. This was demonstrated in studies by measuring the fluorescence emission of a probe, 8-anilino-1-naphthalene sulfonate (ANS), probably adsorbed near region III [20], probably at a molar ratio around 1:1 [7]. These interpretations are consistent with fluorescence studies using internal tyrosyl residues [17], and NMR data [21,22].

INHIBITION OF CALMODULIN-DEPENDENT ENZYMES BY ALUMINUM IONS

Aluminum-induced helix-coil transitions are accompanied by functional changes of calmodulin which is an activator for a variety of enzymes. For example, at a molar ratio of 4:1 for [aluminum]/[calmodulin], the activity of the calcium- and calmodulin-dependent 3':5'-cyclic nucleotide phosphodiesterase was completely blocked as determined from the hydrolysis of cyclic GMP. Control experiments showed that aluminum interfered with calmodulin rather than with the enzymatic protein [7]. Similarly, micromolar concentrations of aluminum ions interfered with calcium- and calmodulin-dependent Ca^{2+}/Mg^{2+}-ATPase activity which plays a role in the maintenance of the transmembrane potential of plasma membrane enriched vesicles isolated from barley roots. The aluminum-induced reduction of membrane potential build-up, in the presence of ATP, was assayed with fluorescent voltage-sensitive probes [23]. Considering the importance of the transmembrane potential and cellular processes, aluminum-triggered imbalances of the potential are thus expected to have severe repercussions on cell growth and root elongation, consistent with pathophysiological observations [2].

DYNAMICS OF CALMODULIN INTERACTION WITH TARGET PROTEINS

Perturbations of this regulatory protein, e.g., by aluminum ions, are expected to lead to changes in the finely tuned association between transmitter protein and its target, thereby eliciting perhaps nonphysiological responses. Since little information is available with respect to molecular mechanisms responsible for calmodulin/target interaction, melittin was selected as target protein for aluminum-altered calmodulin. Melittin is an amphiphilic polypeptide (M=2846) composed of 26 amino acids, it harbours a single tryptophanyl residue, has a calmodulin affinity in the nanomolar range, and competes with target enzymes for calcium-calmodulin. Therefore, this small protein has been proposed as a model system for the study of calmodulin/target interactions [24].

To derive molecular parameters pertinent for the interaction of melittin with calmodulin, circular dichroism and time-resolved fluorescence anisotropy studies were performed. Melittin's single tryptophanyl residue was used as an intrinsic probe monitoring the microenvironment at the melittin/calmodulin interface, in the presence or absence of stoichiometric amounts of aluminum ions. Compared with calcium-calmodulin, the motional freedom of the tryptophanyl residue is somewhat restricted when melittin is associated with aluminum-altered calmodulin (4:1), probably as a result of hydrogen bonding. Furthermore, in the presence of aluminum ions, the increase of melittin's helix content is reduced, presumably as a result of weaker hydrophobic interactions between the two proteins. A more polar environment of the tryptophanyl residue is also indicated by a shorter fluorescence lifetime

Table 2. Fluorescence Lifetime, t, Bimolecular Quenching Constant for Acrylamide, k_q, and Rotational Correlation Time, $\emptyset$, of Melittin's Tryptophanyl Residue upon Association with Calmodulin (10 µM), in the Presence and Absence of Aluminum. Aluminum was present at a molar ratio of 4:1 for [aluminum]/[protein]. All experiments were performed in the presence of calcium at a molar ratio of 8:1 for [calcium]/[protein] at 25°C. The buffer consisted of 45 mM MOPS-KOH, pH 6.5, 0.2 N KCl. The optical studies were carried out on a SLM spectrofluorimeter, model 4800.

Melittin-Calmodulin	t nsec	$\emptyset_A$ nsec	$k_q 10^{-9}$ $M^{-1}sec^{-1}$
Aluminum absent	2.84	8.42	1.20
Aluminum present	2.39	3.54	1.62

(2.39 nsec vs.2.84 nsec when aluminum is absent) and by a smaller blue-shift in the tryptophanyl absorption maximum, when melittin is associated with aluminum-altered calmodulin. Aluminum-induced breakage of hydrogen bonds on calmodulin necessarily leads to a rearrangement of water molecules. Aluminum-altered calmodulin has therefore a solvation structure different from that of the compact calcium-calmodulin. The existence of a more open structure for aluminum-altered calmodulin is consistent with our findings that quenching molecules have better access to melittin's tryptophanyl residue when melittin associates with aluminum-altered calmodulin as compared with calmodulin in the absence of aluminum. Control experiments indicated that application of an eight old excess of aluminum ions over melittin alone did not change the protein's circular dichroism spectrum.

CONCLUSIONS

The results presented in this study demonstrate that application of micromolar concentrations of aluminum ions leads to profound alterations of calmodulin's structure, in the presence or absence of saturating calcium concentrations. Possible binding sites of aluminum ions are carboxyl groups at the surface of the acidic protein. One such binding site may be at, or in close proximity of, region III, because of the enhanced tyrosyl emission, possibly from tyr 99, and because this region has been implicated as a possible binding region of the amphiphilic melittin [24]. The hydrophobic domain of region III is probably also involved in the specific interaction of calcium- calmodulin with certain target enzymes [24].

Upon interaction of aluminum ions with calmodulin, whether in the presence or absence of calcium ions, the conformational states and charge of the regulatory protein are altered. This can be inferred from the aluminum-induced changes in helix content, the reduced electrophoretic mobility of aluminum-altered calmodulin in gels, the enhanced mobility of spin probes attached to the protein [8,16], and the increased hydrophobic surface exposure, compared with corresponding properties of calcium-calmodulin. Aluminum-induced overall structural changes of the protein necessarily impact the local conformation at the specific calcium binding loops. Although calcium ions are rather flexible as to the respective coordination geometry, the ease of departure of the calcium cation, i.e., its rehydration, may be significantly altered. If the off-rate becomes too slow, say slower than about 1 sec^{-1}, calmodulin's trigger capability for signal transmission in biochemical reactions is probably impaired. To illustrate this point, upon addition of trifluoperazine to calmodulin, the off-rate for Ca^{2+} from calmodulin is reduced by about tenfold [25], compared with the off-rate in the absence of the inhibitor.

Since structural fluctuations appear to play a crucial role in protein dynamics and catalysis [26,27], aluminum binding to calmodulin perturbs the dynamic equilibrium existing between the protein's energy states. This type of equilibrium is partially governed by the presence of solvation water, whose amount and type of association with respective amino acid residues are interrelated with the protein's conformation, e.g., that established in the presence of aluminum ions. As a result, certain states of the aluminum-altered solvated protein are no more (or less) accessible which, in turn, prevents the proper fit between trigger molecule and target protein. Moreover, the respective "non-physiological" energy state is virtually frozen, because the off-rate for aluminum ions is extremely small (Table I), qualitatively speaking of the order of $k_{off} < 10^{-7}$ sec^{-1}, assuming a binding constant of about 10^7 [7]. Consequently, the dynamic interaction between such a calmodulin and its target is greatly diminished. Long retention times of aluminum bound to carboxylate ligands have also been found in NMR experiments [28].

As a result of aluminum binding, melittin's association with calmodulin, probably in region III [24], seems to be diminished compared with calcium-calmodulin in the absence of aluminum. As to the size of the interface, it was hypothesized that a minimum structural requirement of melittin binding to calmodulin is a basic, amphiphilic helix of about 1.5 nm length [29]. A weaker association is reflected in less helix formation of the peptide, because the interface site becomes more polar and displayed altered dynamic changes upon binding of aluminum to the calmodulin. It is interesting to point out that region III forms a compact structure only in the presence of Ca^{2+} and shows positive cooperativity with the adjacent region IV , while the domain pairs (I and II) and (III and IV) show negative cooperativity [30]. We lack information to ask whether aluminum binding perturbs the respective cooperative interaction among the four domains of the regulatory protein?

CHELATORS PROTECT CALMODULIN FROM ALUMINUM INJURY

Since multifunctional calmodulin has been highly conserved during evolution, the hypothesis was put forward [31], that protective mechanisms should exist that protect the pivotal protein from aluminum injury. This notion is supported by observations that certain aluminum-tolerant plant species are rich in aluminum-chelating compounds such as organic acids [2]. Indeed, our circular dichroism and ANS fluorescence experiments demonstrate that citric acid, at a molar excess of [citrate]/[aluminum], can protect calmodulin from aluminum injury, if the metal is presented to the protein in stoichiometric amounts. These findings are consistent with results from ^{27}Al-NMR experiments indicating that citrate forms a stable complex with aluminum between pH 5 and 8.0 [28]. Control experiments showed that the presence of a tenfold excess of citrate over calmodulin (10 µM) did not interfere with calcium binding to the protein (4:1), again in accord with known stability constants for citrate/calcium binding.

ACKNOWLEDGEMENTS

This work was supported by a grant, PCM-8314662, from the National Science Foundation, and a gift from Pioneer Hi-Bred International.

REFERENCES

1. C.B. Klee, and T.G. Vanaman, Calmodulin, Adv. Protein Chem. 49:489-515 (1982).
2. C.D. Foy, R.L. Chaney, and M.C. White, The physiology of metal toxicity in plants, Ann. Rev. Plant Physiol. 29:511-566 (1978).
3. R.J. Bennet, C.M. Breen, and M.V. Fey, Aluminium uptake sites in the primary root of Zea mays L., S.Afr. J. Plant Soil 2:1-7 (1985).
4. R.J. Bennet, C.M. Breen, and M.V. Fey, The primary site of aluminium injury in the root of Zea mays L., S. Afr. J. Plant Soil 2:8-17 (1985).
5. S.W. King, J. Savory, and M.R. Wills, The clinical biochemistry of aluminum, CRC Crit. Rev. Clin. Lab. Sci. 14:1-20 (1981).
6. D.P. Perl, D.C. Gajdusek, R.M. Garrato, R.Y. Yanagihara, and C.J. Gibbs, Intraneuronal aluminum accumulation in amyotrophic lateral sclerosis and Parkinsonism-dementia of Guam, Science 217:1053-1055 (1982).
7. N. Siegel, and A. Haug, Aluminum interaction with calmodulin: Evidence for altered structure and function from optical and enzymatic studies, Biochim. Biophys. Acta 744:36-45 (1983).
8. A. Haug, Molecular aspects of aluminum toxicity, CRC Crit. Rev. Plant Sci. 1:345-373 (1984).
9. J. Burgess, "Metal Ions in Solution," Ellis Horwood, Ltd., Chichester, pp. 259-416 (1978).
10. A.K. Campbell, "Intracellular Calcium," J.Wiley & Sons, New York, pp. 85 -134 (1983).
11. H. Einspahr, and C.E. Bugg, Crystal structure studies of calcium complexes and implications for biological systems, in "Metal Ions in Biological Systems," H. Sigel, ed., Marcel Dekker, New York, vol. 17:51-97 (1984).
12. B.A. Levine, and D.C. Dalgarno, The dynamics and function of calcium-binding proteins, Biochim. Biophys. Acta, 726:187-204 (1983).
13. E.R. Nightingale, Phenomenological theory of ion solvation: Effective radii of hydrated ions, J.Phys. Chem. 63:1381-1387 (1959).
14. A. Noller, Catalysis from the standpoint of coordination chemistry, Acta Chim. Scient. Hung. 148: 429-448 (1982).
15. P.L. Privalov, Stability of proteins, Adv. Protein Chem. 35:1-104 (1982).
16. N. Siegel, R.T. Coughlin, and A. Haug, A thermodynamic and EPR study of structural changes in calmodulin induced by aluminum binding, Biochem. Biophys. Res. Comm. 115:512-517 (1983).
17. K.B. Seamon, and R.H. Kretsinger, Calcium-modulated proteins, in "Calcium in Biology," T.G. Spiro, ed., Wiley & Sons, New York, pp.1-51 (1983).
18. N. Siegel, and A. Haug, Aluminum-induced inhibition of calmodulin-regulated phosphodiesterase activity: Enzymatic and optical studies, Inorgan. Chim. Acta 79:230-231 (1983).
19. M. Tanokura, and K. Yamada, A calorimetric study of Ca^{2+} and Mg^{2+}-binding by calmodulin, J. Biochem. Tokyo 94:607-609 (1983).
20. R.F. Steiner, and H. Sternberg, Properties of the complexes formed by

ANS with phosphorylase kinase and calmodulin, Biopolym. 21: 1411-1425 (1982).

21. J. Krebs, A survey of structural studies of calmodulin, Cell Calcium 2:295-311 (1981).

22. M. Ikura, T. Hiraoki, K. Hikichi, T. Mikuni, M. Yazawa, and K. Yagi, NMR studies on calmodulin: Calcium-induced conformational change, Biochem. 22:2573-2579 (1983).

23. N. Siegel, and A. Haug, Calmodulin-dependent formation of membrane potential in barley root plasma membrane vesicles: A biochemical model of aluminum toxicity in plants, Physiol. Plant. 59:285-291 (1983).

24. J.A. Cox, M.Comte, A.Malnoe, D.Burger, and E.A. Stein, Mode of action of the regulatory protein calmodulin, in "Metal Ions in Biological Systems," H.Sigel, Ed., Marcel Dekker, New York, vol. 17:215-273 (1984).

25. S. Forsen, E. Thulin, T. Drakenberg, J. Krebs, and K. Seamon, A ^{113}Cd NMR study of calmodulin and its interaction with calcium, magnesium and trifluoperazine, FEBS Lettr. 117:189-194 (1980).

26. M. Karplus and J.A. McCammon, The internal dynamics of globular proteins, CRC Crit. Rev. Biochem. 9:293-349 (1981).

27. G. Careri, "Order and Disorder in Matter," The Benjamin/Cummings Publ., Menlo Park, pp. 115-137 (1984).

28. S.J. Karlik, E.Tarien, G.A. Elgavish, and G.L. Eichhorn, ^{27}Al NMR study of aluminum (III) interactions with carboxylate ligands, Inorgan. Chem. 22:525-529 (1983).

29. J.A. Cox, M. Comte, J.E. Fitton, and W.F. DeGrado, The interaction of calmodulin with amphiphilic peptides, J. Biol. Chem. 260:2527-2534 (1985).

30. T.N. Tsalkova and P.L. Privalov, Thermodynamic study of domain organization in troponin C and calmodulin, J. Mol. Biol. 181:533-544 (1985).

31. C.G. Suhayda and A. Haug, Organic acids prevent aluminum-induced conformational changes in calmodulin, Biochem. Biophys. Res. Comm. 119:376-381 (1984).

Ca²⁺ BINDING TO CALMODULIN AND INTERACTIONS WITH ENZYMES

Jacques Haiech, Nicole Vidal, Jean Sallantin and
Jean-Claude Cavadore

CREM, LP 8402 CNRS and U.249 INSERM - CRIM -
ERA CNRS 1045, BP 5051 - 34033 Montpellier Cedex, France

INTRODUCTION

Multicellular organisms possess the ability to communicate in order to behave harmoniously. Generally, cells send messages by means of chemical messengers called the first messenger. These molecules interact with target cells on receptors. This interaction modifies the cellular properties by increasing or inducing the synthesis of new molecules called second messengers.

By definition, a second messenger has the following properties:

(1) It is an intracellular metabolite. Its concentration increases under cellular stimulation by a first messenger.
(2) This increase is transitory and triggers a cellular response.
(3) An artificial intracellular increase of the second messenger would trigger completely or partially the same cellular event as the first messenger.

Calcium is among the most important second messengers.

Cytoplasmic calcium concentration is around 10^{-7}M in resting cells and reaches 10^{-5}M upon stimulation. Calcium concentration is 10^{-3}M in the extracellular medium and 10^{-2}M in the endoplasmic reticulum.

CALCIUM FLUX IN THE CELL

Calcium gradients are created and maintained with a consumption of energy (from ATP or from the osmotic energy of the Na^+ gradient). Ca^{2+}-ATPases, located in the plasma membrane or reticulum membrane hydrolyze one molecule of ATP to translocate one or two calcium ions from the cytoplasm to either the extracellular medium or the endoplasmic reticulum. The two pumps exhibit different physical and chemical properties. Moreover, their regulation is different. Three Na^+ ions are exchanged against one calcium ion by the plasma membrane antiport which is indirectly coupled to the Na^+/K^+ ATPase (Pitts, 1979; Le Peuch, 1979).

Several first messengers are able to trigger an increase of cytoplasmic calcium. Calcium coming from the external medium or intracellular vesicles enters the cytoplasm either through specific calcium channels or could be carried by ionophores.

Three types of calcium channels can be described:

- Channels opening upon membrane depolarization (POC).
- Channels opening upon direct occupancy of a membrane receptor (ROC) (Bolton, 1979).
- Channels opening upon indirect occupancy of a membrane receptor. For instance, activation of a receptor triggers the synthesis of a metabolite which, in turn, opens the calcium channel of the endoplasmic reticulum (IROC) (Berridge et al., 1984).

This double system allows:

(1) to obtain steep calcium gradients between cytoplasm and intracellular calcium stores on the one hand and cytoplasm and external medium on the other hand;
(2) calcium to flow briefly down the gradient into the cytoplasm upon stimulation, and possesses the property to generate a transitory calcium wave.

CALCIUM WAVE MANAGEMENT

This calcium wave codes information. Several parameters play an important role to maintain the specificity of the information coming from the first messenger and could therefore explain the specificity of the cellular response to a given first messenger:

- the cytoplasmic localization of the calcium wave,
- the maximal concentration reached by the wave,
- the kinetic parameters of the calcium wave (speed of calcium concentration increase, decrease and duration of the wave).

This transient calcium wave is managed (and relayed) by proteins which exhibit the following properties:

- they bind calcium in the 10^{-7} to 10^{-5}M range in a physiological medium (pH = 7, KCl around 150 mM and Mg^{2+} around 1 mM),
- the kinetic parameters of calcium binding are compatible with those of the calcium wave,
- they are able to trigger a cascade of events leading to the cellular response.

Due to the characteristics of the calcium wave, the management of this signal by a calcium binding protein is dependent upon:

- the same localization of calcium signal and calciproteins (topological regulation),
- the mechanism of calcium binding (kinetic regulation),
- the cellular content of a calciprotein at a given place and at a given time (genetical regulation).

The binding of calcium ions to the calciprotein induces a conformational change which allows the calciproteins either to interact with a target protein or to activate its own enzymatic activity. It is the starting point of a cascade of molecular events which leads to the cellular response.

Three important cellular functions are regulated by calcium:

- cellular motility,
- energy supply: it is necessary to furnish energy to the motile cell,
- regulation of second messengers (feedback control of mechanism leading to the cellular response).

Moreover, some responses are coordinated by calcium. Upon triggering of serotonin granules secretion (cellular motility), the synthesis of serotonin is also initiated. The cell appears to have created a set of coordinated regulations coupling the excitation-secretion phenomena and the regeneration of vesicle content (Yamaguchi et al., 1981).

MODULATION OF Ca^{2+} SIGNAL

Inside the animal cell, calcium is not the only second messenger. cAMP also plays an important role. The cAMP signal is triggered by hormones which are coupled with adenylate cyclase. The cAMP in the cell is metabolized by phosphodiesterases. These two systems are able to produce a transient cAMP increase which is managed by a cAMP dependent protein kinase. This tetrameric enzyme (R_2C_2) is composed of two regulatory and two catalytic subunits. Upon cAMP binding, the enzyme dissociates into its two active catalytic subunits (2 C) and homodimers of the cAMP bound regulatory subunit R_2cAMP_4. Calcium channels, calcium pumps and target proteins of calciproteins are some of the substrates for the catalytic subunit. Upon phosphorylation, the activity of these proteins is modified. Therefore, the kinetics of the calcium wave can be modified, modulating the interaction between calcium-binding proteins and target proteins. Modulation of Ca^{2+} signal by other second messengers is called concerted regulation.

EVOLUTION OF THE INTRACELLULAR CALCIUM BINDING PROTEINS

The search for intracellular calciproteins began more than twenty years ago and has led to the purification and characterization of more than thirty proteins. These proteins present similar functional and structural features:

- they are low molecular weight proteins,
- they are acidic (pH from 4 to 5.5),
- they bind from 1 to 4 calcium ions.

Analysis of their primary structure shows that they are homologous and probably evolved from a common ancestor, a 36 residual peptide. This peptide has been duplicated twice to give finally a protein with four domains. This protein would have evolved to generate the contemporary subfamilies (Moews et al., 1975).

The secondary and tertiary structure of the ancestral domain has been described by Kretsinger. It is composed of three parts:

- an alpha helix of 12 residues,
- a loop of 12 residues which ligands the calcium ion by means of 6 amino-acids at positions 1, 2, 5, 7, 9, 12.

The tertiary structure of this domain has been called "EF hand". The tertiary structure of four calciproteins has been published and seems to show a strong interactivity between individual domains (Reid et al., 1980).

Moreover, the primary structure of some proteins which do not bind calcium (the light chain of myosin, for instance) shows a high degree of homology with the other calcium binding proteins. Therefore, in the calciprotein family, contemporary proteins have evolved from the common ancestor either keeping their calcium binding properties or losing them and acquiring a new function. Proteins of the calcium family seem to have specialized for a specific function in a specific cell type.

However, calmodulin, the ubiquitous Ca^{2+}-binding protein (CaBP) of eukaryotic cells, is in terms of evolution the closest to the four domain ancestor protein. In addition, it is the only known CaBP involved in many different metabolic pathways and devoid of tissue specificity. Therefore calmodulin can be considered as the only non-specialized protein of the CaBP family. Calmodulin has been extensively studied in order to understand the interaction between CaBPs and target proteins (see reviews Kilhoffer et al., 1983; Klee et al., 1980; Cheung, 1980).

TRANSFORMATION OF QUANTITATIVE Ca^{2+} WAVE INTO QUALITATIVE CELLULAR RESPONSE: INTERACTION BETWEEN CALMODULIN AND TARGET PROTEINS

Calmodulin is a 16700 dalton molecular weight protein which binds four calcium ions and participates in numerous metabolic pathways by interacting with target proteins. After binding of 3 to 4 calcium ions, calmodulin modifies its conformation and then forms a ternary complex with the target proteins. The formation of this complex leads to the stimulation of an enzymatic activity. Briefly, three theories have been elaborated to describe the fine coupling between the calcium pulse and this stimulation:

- The first one is based on the following concepts:

  Calmodulin has four independent and equivalent sites.

  The conformational changes are associated with the number of sites occupied by calcium whatever the specific sites liganding the calcium. Conformational changes appear as a two-step process.

  At the end of this process, calmodulin exhibits one or two hydrophobic zones which allow the target proteins to bind to calmodulin through hydrophobic interaction. All the calmodulin binding proteins interact with the same sites of calmodulin. Therefore, using drugs which bind to the hydrophobic zones it is possible to inhibit the interaction of calmodulin with its target proteins, but the design of drugs able to inhibit one specific calmodulin-dependent activity is not possible (Levin et al., 1977; Cox et al., 1981).

- The second theory established by biophysicists is based on the following concepts:

  Calmodulin presents two sets of sites: there are two sites with high affinity which are associated with a major conformational change – these sites would be sites III and IV as numbered from NH_2 terminus to COOH terminus – and two low affinity sites associated with a small conformational change but important enough to be essential for interaction with target proteins.

  Calmodulin is therefore similar to troponin C, the CaBP involved in the regulation of muscle contraction. The high affinity sites would be structural sites (Levine et al., 1983).

- Finally, the third theory tries to explain how calmodulin can translate a quantitative signal (calcium pulse) into a qualitative one (enzymatic activity) and is based on the following concepts:

Calmodulin binds calcium in a sequential and ordered manner. It means that calmodulin in the resting cell presents one site for calcium binding, upon occupancy of this site, a second calcium binding site is shaped, a second calcium binds and so on ...

From indirect evidence, the sequence of binding could be II then I then III and at last IV.

Therefore, several calcium-calmodulin complexes are sequentially present and exhibit different conformations. In a given confirmation, calmodulin interacts with a given enzyme or set of enzymes. This complex can bind more calcium ions and becomes active when four ions are bound to the complex (Haiech et al., 1981).

In this case, it becomes possible to design specific drugs able to inhibit one given enzymatic activity.

Strong support for the last theory has been brought by C. Klee and colleagues during the past few years. They cut calmodulin into peptides and purified them. They then showed that stimulation by calmodulin of specific enzymatic activities was inhibited by some peptides and that the same peptides were able to inhibit the activities of others (Newton, et al., 1984).

CONCLUSION

Finally, for a given cell, the response to a stimulus which activates the calcium signal, is dependent on several factors:

- the localization of proteins involved in the calcium signal (topological regulation),
- the quantitative ratio between these different proteins (genetical regulation),
- the mechanism of interaction of these proteins (kinetic regulation),
- the covalent modification of these proteins by enzymes activated through other second messenger pathways (concerted regulation).

To integrate these multiple regulations, we assumed in 1980 that supramolecular complexes which order a cellular function exist in the cytoplasm and we called them calcisomes (Haiech et al., 1981).

REFERENCES

Berridge, M.J. and Irvine, R.P., 1984, Inositol triphosphate, a novel second messenger in cellular transduction, Nature (London), 312:315-321.
Bolton, T.B., 1979, Mode of action of transmitters and other substances on smooth muscle, Physiol. Rev., 59:606-718.
Cheung, W.Y., 1980, Calmodulin plays a pivotal role in cellular regulation, Science, 207:19-27.
Cox, J.A., Malnoë, A. and Stein, E.A., 1981, Regulation of brain cyclic nucleotide phosphodiesterase by calmodulin, J. Biol. Chem., 256: 3218-3222.
Haiech, J. and Demaille, J.G., 1980, Supra molecular organisation of regulatory proteins into calcisomes: A model of the concerted regulation by calcium ions and cyclic adenosine 3:5'-monophosphate in eucaryotic cells. Metabolic interconversion of enzymes. Proc. in Life Sciences, 303-313, Springer Verlag, Berlin.
Haiech, J., Klee, C.B. and Demaille, J.G., 1981, Effects of cations on affinity of calmodulin for calcium: ordered binding of calcium ions

allow the specific activation of calmodulin-stimulated enzymes, Biochemistry, 20:3890-3897.

Kilhoffer, M.C., Haiech, J. and Demaille, J.G., 1983, Ion binding to calmodulin. A comparison with other intracellular calcium-binding protein, Mol. Cell. Biochem., 51:33-54.

Klee, C.B., Crouch, T.H. and Richman, P.G., 1980, Calmodulin, Ann. Rev. Biochem., 49:489-515.

Le Peuch, C.J., Haiech, J. and Demaille, J.G., 1979, Concerted regulation of cardiac sarcoplasmic reticulum. Calcium transport by cyclic adenosine monophosphate-dependent and calcium-calmodulin-dependent phosphorylations, Biochemistry, 18:5150-5157.

Levine, B.A. and Dalgarno, D.C., 1983, The dynamic and function of calcium binding proteins, Biochim. Biophys. Acta, 726:187-204.

Moews, P.C. and Kretsinger, R.H., 1975, Refinement of the structure of carp calcium binding parvalbumin by model building and Fourier difference analysis, J. Mol. Biol., 91:201-208.

Newton, D.L., Olderwurtel, M.D., Krinks, M.H., Siloach, J. and Klee, C.B., 1984, Agonist antagonist properties of calmodulin fragments, J. Biol. Chem., 259:4419-4426.

Pitts, B.J.R., 1979, Stoichiometry of sodium-calcium exchange in cardiac sarcolemmal vesicles coupling to the sodium pump, J. Biol. Chem., 254:6232-6235.

Rasmussen, H., 1970, Cell communication, calcium ion and cyclic adenosine monophosphate, Science, 170:405-412.

Rasmussen, H. and Goodman, D.B.P., 1977, Relationships between calcium and nucleotides in cell activation, Physiol. Rev., 57:421-509.

Reid, R.E. and Hodges, R.S., 1980, Cooperativity and calcium/magnesium binding to troponin C and muscle calcium binding parvalbumin, J. Theor. Biol., 84:401-404.

Yamaguchi, T. and Fujisawa, H., 1981, A calmodulin-dependent protein kinase that is involved in the activation of tryptophan 5-monooxygenase is specifically distributed in brain, FEBS Lett., 129:117-119.

CALCIUM AND CALCIUM-BINDING PROTEINS IN PHLOEM

Dinkar D. Sabnis and Alan R. McEuen

Department of Plant Science
University of Aberdeen
Aberdeen, Scotland

INTRODUCTION

The sieve elements of most dicotyledenous and many monocotyledenous plants are characterized by a cytoskeletal framework of P-protein tubules and filaments that can be both extensive and elaborate (Parthasarathy, 1975). The maintenance and turnover of this macromolecular complex, in the face of the translocation stream, must require specialized regulatory mechanisms and the expenditure of metabolic energy. In addition to transported metabolites, phloem sap contains cAMP, high levels of ATP, and proteins such as lectins, phosphodiesterase, and an ATPase, whose activity in other systems is often modulated by divalent cations (see Eschrich & Heyser, 1975; Sabnis and Hart, 1982). Calcium is an obvious candidate for a regulatory role in protein conformation and enzyme activity. The available evidence arguably indicates that calcium is not mobile in the phloem (see McEuen, 1979); nevertheless, it is present in the sieve tube sap (Hall, Baker and Milburn, 1971). In an earlier paper (McEuen, Hart and Sabnis, 1981) we reported on a calcium-binding protein in sieve tube sap from Cucurbita maxima. In this paper we extend our observations to other species, we report on the presence of significant levels of calmodulin in sieve tube sap and plant extracts, and we present data on calcium levels in the sieve tube.

MATERIAL AND METHODS

Plants: The species used for analysis of calcium-binding or calmodulin levels were Cucurbita maxima Duch. cvs. Golden Delicious and Gelber Zentner, Cucurbita pepo L. cv. Long Green Trailing, Cucumis sativus L. cv. Telegraph, Cucumis melo L. cv. Blenheim Orange and Zea mays L. cv. Golden Bantam.

Exudate: Phloem exudate from cucurbit species was collected either by slicing and rinsing the stem and immersing the cut end into collection buffer, or by using microcapillaries and a graduated microsyringe and expelling the exudate into buffer. Gelling of exudate proteins was prevented by the addition of 0.14 M 2-mercaptoethanol or 25 mM iodoacetamide. The collection/binding buffer contained 60 mM KCl and 2 mM $MgCl_2$ to reduce non-specific calcium-binding activity (Chevalier & Butow, 1971) and to approximate the in vivo ionic content of sieve tube sap (Ziegler, 1975), 0.14 M mercaptoethanol and 30 mM Tris-HCl, pH 7.5. The exudate samples were usually centrifuged at 20,000 g for 20 min at 4°C, the pellets discarded and the supernatant fractions desalted by centrifugation through Sephadex G-25.

Calcium-binding Assay: 1.0 ml of protein sample was incubated with 10 ul of 10 mM ^{45}CaCl$_2$ (37 kBq; Radiochemical Centre, Amersham, U.K.) at 30°C for 30 min. Calcium binding to high MW components was assayed by either gel chromatography or by ultrafiltration. For the former, the total incubation mixture was eluted through a 15x1.6 cm column of Sephadex G-100 at 0-4°C and a flow rate of 30 ml/h. 1.0 ml fractions were divided into 0.4 ml aliquots for scintillation counting, 0.2 ml for protein assay, and 0.2 ml for sodium dodecyl polyacrylamide gel electrophoresis (SDS-PAGE).

Ultrafiltration of incubated samples was through Diaflow membranes in an Amicon model 12 cell and driven by compressed nitrogen at 60 psi for exudate and 10 psi for control samples to give similar flow rates. After 12 rinses of 1.0 ml with buffer, the membranes were counted in scintillation fluid.

SDS-PAGE of exudate or column eluant samples was either in 7.5% gel cylinders containing 4 M urea, or on 10% gel slabs following the discontinous system of Laemmli (1970).

Protein was assayed by the Lowry method as modified by Ross & Schatz (1973) for the presence of thiols, or using the Biorad dye reagent.

Calmodulin was assayed using a radioimmunoassay (RIA) kit obtained from Amersham International plc, following the procedure of Chafouleas, et al. (1979). The assay is based on the competition between unlabelled calmodulin (CAM) and a fixed quantity of ^{125}I-labelled CAM for affinity purified antibody raised against CAM. Separation of the antibody-bound CAM from unbound CAM is achieved by precipitation of the antibody using formalin-fixed Staphylococcus aureus followed by centrifugation. The antibody-bound radiolabelled CAM was measured in a gamma counter and the concentration of CAM in each sample determined from a standard curve.

Plant Extracts for Calmodulin Assay: Maize seeds were grown for 5 days at 18°C and the radicles were harvested and frozen in liquid N$_2$. 10 gm of frozen roots were ground and extracted in 13.0 ml of 125 mM borate buffer containing 1 mM EGTA and 75 mM NaCl, pH 8.4. The homogenate was clarified at 40,000 g for 30 min, providing a supernatant fraction of 10 ml. Cucurbita maxima stems were sliced into liquid N$_2$ and similarly processed. The homogenates were heated to 90°C in a water bath for 5 min and then rapidly cooled in a methanol-dry ice bath. Precipitated proteins were removed by centrifugation at 20,000 g for 30 min at 4°C and the supernatant fractions assayed for CAM using the RIA. Exudate from C. maxima cv. Gelber Zentner was collected into (a) degassed buffer, (b) buffer containing 0.14 M mercaptoethanol, and (c) buffer containing 25 mM iodoacetamide, until some gelling or precipitation was evident. The samples were then centrifuged for 5 min at 1,500 g and subsequently heat-treated and clarified for the CAM assays.

Assays for Total and Free Calcium: Total calcium was assayed by atomic absorption spectrometry and total potassium by flame em spectrometry. Measured volumes of exudate were expelled from microcaps into 0.14 M mercaptoethanol in water. To a 0.5 ml aliquot of exudate was added 0.2 ml of 5% EDTA, 0.4 ml of 1% lanthanum chloride and 0.9 ml of water for the final sample to be measured. In addition, some samples for the calcium assay were subjected to the mixed acid digestion procedure of Allen, et al. (1974) as a check on the previous procedure; no significant differences were found in the levels recorded.

Free calcium was estimated using Chelex-100 ion exchange resin (100-200 mesh; Biorad). The resin, obtained as the fully hydrated sodium form, was converted to the potassium form by elution with 1 M HCl and 1 M KOH, equilibrated with binding buffer, and 50 mg of damp, drained resin was added to a centrifuge tube. The sample for assay was added, the tube vortimixed for 30 sec, centrifuged and aliquots of the supernatant fluid counted in a Packard Tri-Carb 2420 liquid scintillation spectrometer.

Murexide (ammonium purpurate), used for the assay of free calcium, has the advantage that it binds only weakly to divalent cations and therefore does not effectively alter the concentration of free ion (Caswell, 1979). However, it is bleached very rapidly by mercaptoethanol. It was used, therefore, to estimate free calcium in exudate samples collected into degassed buffer lacking thiol reagents. The procedure was a modification of that described by Ohnishi & Ebashi (1963) and is described in detail by McEuen (1979).

RESULTS AND DISCUSSION

Calcium-binding Protein: Exudate, assayed for calcium-binding activity (CBA) using a column of Sephadex G-100, showed a radioactive peak in the void volume indicating a calcium-binding component with a MW in excess of 100,000. The CBA is unstable; longer elution times (on Sephadex G-150 or G-200) resulted in the disappearance of the high MW radiolabelled peak.

Since the protein composition of phloem exudate varies widely, even among closely related cucurbit species (Sabnis & Hart, 1982), CBA was assayed in two species each of Cucurbita and Cucumis. All four showed a similar radiolabelled peak (Fig. 1). The principal differences were quantitative: Cucumis sativus consistently provided lower yields; Cucumis melo was highly variable. Cut cells and cell walls do not contribute to CBA (McEuen, et al., 1981).

Additional evidence for the sieve tube origin of the calcium-binding macromolecule came from studies using exudate collected in glass capillaries. This also contained a calcium-binding macromolecule with a yield comparable to exudate collected by stem immersion - 2.1 nmol/ml of macromolecule-bound calcium, and a v(peak) value of 4.5 nmol Ca/ mg protein.

The absence of reducing agents in the collecting buffer resulted in gelling and precipitation of exudate proteins, together with loss of CBA. The major proteins of cucurbit phloem are rich in half-cystine residues. Concentrations of 2-mercaptoethanol (0.14 M) that reduce accessible disulphide linkages at phloem protein concentrations of ca. 10 mg/ml (Read & Northcote, 1983) provided the highest levels of CBA. Lower mercaptoethanol concentrations (0.014 M) resulted in a 6-fold decrease in calcium-binding activity.

The protein nature of the calcium-binding component was suggested by its high molecular weight, its response to thiol reagents, and its lability over time. CBA was sensitive also to digestion with subtilopeptidase A; in 30 min CBA was reduced by more than 60 per cent. The calcium-binding macromolecule is not identical with any major exudate protein. In 11 out of 16 assays of C. maxima exudate on Sephadex G-100, the [45]Ca peak clearly preceded the first protein peak, while in 3 more, the rise of the [45]Ca peak preceded the rise of the protein peak, even though the two maxima

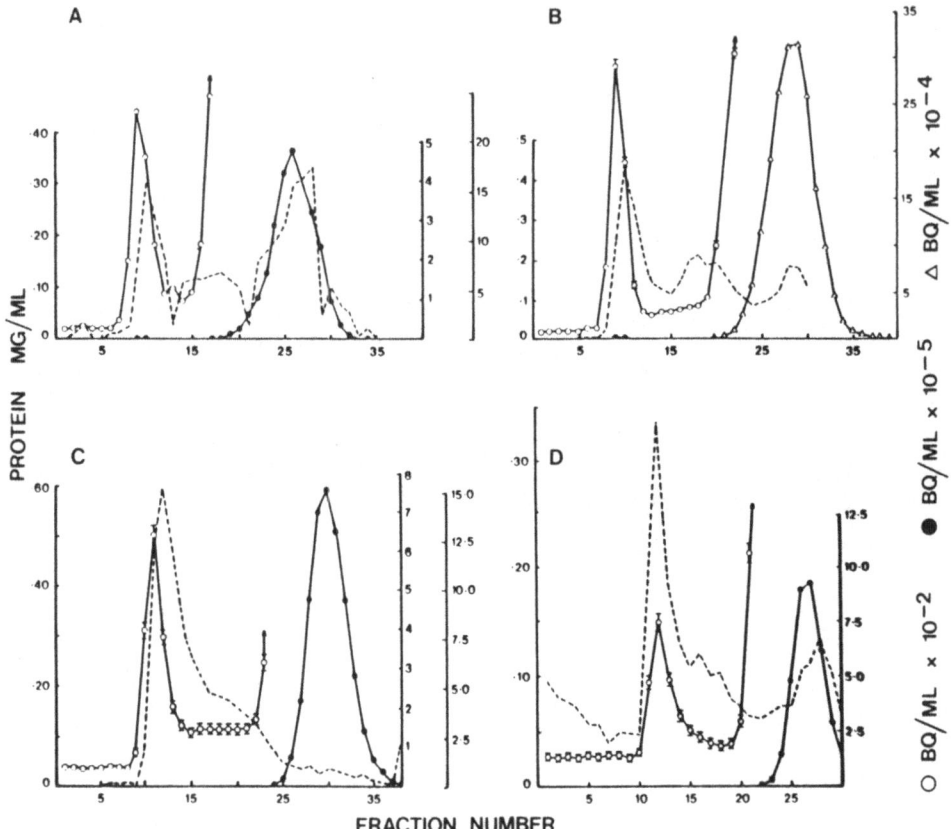

Fig. 1. Calcium-binding activity in sieve tube exudate from A) Cucumis
melo, B) Cucurbita maxima, C) Cucumis sativus and D) Cucurbita
pepo fractionated on Sephadex G-100. Broken curves are protein
concentrations and solid curves are ^{45}Ca levels.

coincided. So, if the Ca-binding molecule is a protein, it is present
in relatively minor concentrations. SDS-PAGE of samples representing
congruent ^{45}Ca and protein peaks revealed only faint bands located at
MW zones over 100,000.

The binding constant, K_d, was estimated from a Scatchard plot.
Aliquots of desalted exudate were incubated in 10 calcium concentrations
(0.3 mM - 1.02 mM) and assayed by ultrafiltration (Fig. 2). Although
the scatter of points is large, due to the high background error inherent
in the ultrafiltration assay, it is possible to set limits on K_d. If the
data are interpreted as a curve (Fig. 2A) which is the sum of 2 classes
of binding site, then the K_d for the high affinity site is the upper
limit for K_d. If the data are interpreted as a straight line due to a
single class of binding site (Fig. 2B), then the value for K_d is
represented by the lower limit. The resultant range of values for K_d is
$4 \times 10^3 M^{-1}$ - $6 \times 10^4 M^{-1}$. Since the Scatchard plot indicates that
saturation was not achieved at any concentration tested, and since most
of the v(peak) values for C. maxima exudate were in the 6-16 nmol Ca/
mg protein range, at least 2 binding sites are indicated on a protein of
MW >100,000.

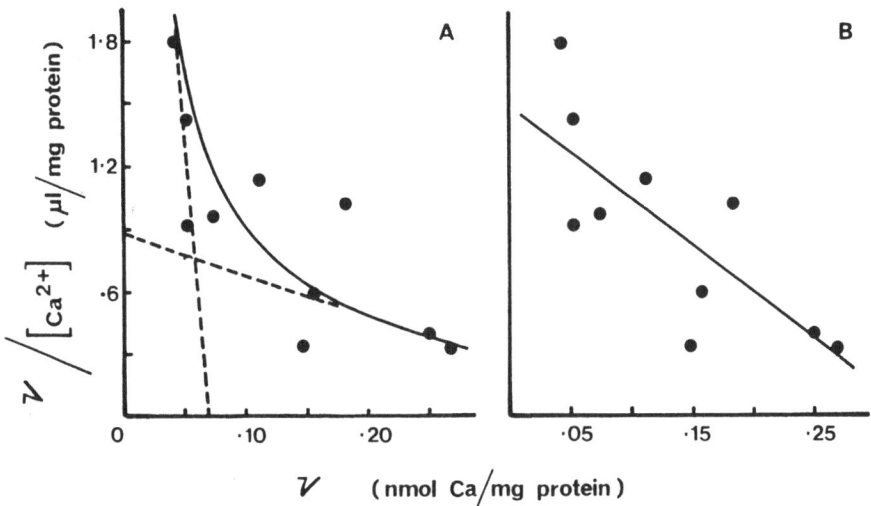

Fig. 2. Scatchard analysis of calcium-binding activity of exudate from
Cucurbita maxima assayed by ultrafiltration. A. Two-site
interpretation. High affinity site computed by linear regression
of the 3 data points which gave the highest possible affinity.
B. One-site interpretation. K_d= 4 x 10^3M^{-1} - 6 x 10^4M^{-1}

Calmodulin: The presence of a calcium-binding protein prompted us
to assay for calmodulin using a radioimmunoassay. Since the effect of
thiol reagents on the assay was uncertain, exudate was collected into
three different media: borate buffer thoroughly degassed just before
use; buffer containing 25 mM iodoacetamide; and buffer containing
mercaptoethanol. For comparison, the exudate samples were assayed
together with extracts of maize root and C. maxima stem. The results
are presented in Table 1 and demonstrate that the concentrations of CAM
in exudate, expressed as a function of total protein content, are equal
to or greater than cytoplasmic levels. Furthermore, the levels of CAM
in the sieve element are comparable to those in other plant tissues
(see Cormier, et al., 1982). In addition to antibody specificity, the
heat-insensitivity of CAM in all 5 preparations is a further diagnostic
feature.

Table 1. Radioimmunoassay of calmodulin in phloem exudate
and plant extracts

Sample	CAM conc μg/ml	Protein conc[a] mg/ml	CAM/Protein μg/mg
Exudate (H$_2$0)[b]	1.9	0.6	3.2
Exudate (mercapt)	11.3	2.7	4.2
Exudate (iodoacetam)	5.1	3.4	1.5
Maize root	22.8	15.0	1.5
Cucurbita stem	14.3	16.0	0.9

[a]Total protein measured using the Biorad dye reagent
[b]Exudate collected into buffer plus/minus thiol reagents

Attempts to locate exudate CAM on immunoblots in order to determine subunit MW are at a preliminary stage. The obvious questions concern the relationship between CBA associated with the high MW component of exudate and the CAM detected by radioimmunoassay. The MW of CAM from both plant and animal sources is ~17,000 (Cormier, et al., 1980). Plant CAM differs from its animal counterpart in possessing a single cysteine residue. There is also much evidence to indicate that CAM may associate with other high MW proteins such as caldesmon, spectrin and tau factor, all proteins associated with the cytoskeleton.

In sieve elements, a primary requirement of structural or regulatory components must be stabilization against the rapid flow of solution through the sieve tube conduits. Cross-linking of component proteins by disulphide linkages provides a mechanism that can be controlled by intracellular redox conditions. Immunocytochemistry may reveal whether CAM constitutes part of the cytoskeletal P-protein network of the sieve element.

Calcium: Total calcium in sieve tube exudate collected into glass capillaries was in the region of 1 mM. This value compares well with earlier reports (e.g. Hall, Baker & Milburn, 1971) on exudate from other species. We were concerned that the relatively high levels of calcium may have resulted from release of turgor pressure during cutting, and a consequent surge of water and ions into the sieve tubes from surrounding appoplastic and symplastic compartments. To test this possibility, two plants were subjected to repeated cuts and exudate analysed for total calcium and potassium levels. Potassium levels in the sieve tubes are very high and would be expected to drop if the flux of water on cutting

Table 2. Total calcium and potassium in phloem exudate from C. maxima plants subjected to repeated cuts

Plant	Cut	Ca^{2+} mM	K^+ mM
1[a]	1	0.88	34.0
	2	1.21	32.9
	3	1.42	41.2
	4	1.14	40.2
	5	1.03	37.0
	6	1.39	44.7
	7	0.77	45.9
	8	0.87	42.8
	9	1.04	42.3
	10	1.08	43.4
2[b]	1	1.12	57.5
	2	1.09	57.5
	3	1.43	56.4
	4	0.71	67.0
	5	0.75	53.4

[a] Cuts made at intervals of 2 mm; exudate collected every 30 sec and volume measured
[b] Cuts made at intervals of 5 mm; exudate collected after 1 min.

was to make any significant difference to in vivo concentrations. In the first, larger plant, cuts were made at 2 mm intervals and exudate collected 30 sec after each cut. In the second, cuts were made at intervals of 1 cm and exudate collected 1.0 min after cutting. The results are presented in Table 2. Calcium levels remained constant over 10 cuts involving 2 cm of stem or 5 cuts removing 5 cm of stem. Both plants also showed very similar calcium concentrations in sieve tube exudate, levels that matched individual assays on other plants. However, while potassium levels in exudate from a single plant were very similar, there were significant differences in the potassium levels of different plants - a feature recorded also in the literature. Potassium levels in C. maxima were 40-60 times higher than calcium.

Vacuolar contents of damaged cells and the cell walls at the cut surface may contribute to the calcium levels recorded for exudate, but if so, this must be a very constant fraction, even in plants of different sizes.

Measurements of free calcium using Chelex-100 or Murexide indicated that one to two thirds of total calcium ions were complexed with low MW ligands (McEuen, 1979). The contribution to the amount of calcium bound in the presence of Chelex-100 by the calcium-binding protein was negligible. After allowance for Ca-binding by low MW ligands normally found in exudate, calculations predict that the concentrations of free calcium ions within the sieve tube (0.4 - 0.8 mM) are a thousand-fold greater than normal cytoplasmic levels (McEuen, 1979). Such levels, if indeed present in vivo, would seem to be inconsistent with any regulatory role of calcium mediated through CAM or the high MW Ca-binding protein. Calcium compartmentation within elements of the endoplasmic reticulum in the sieve tubes may be a partial answer to this dilemma. The intracellular localization of free calcium and its protein receptors must be the next step in attempts to resolve this problem.

Acknowledgements: We thank Dr B.T. Watson for advice, Dr B. Heaton for the loan of a gamma counter, and Lynne Lamont for the figures.

REFERENCES

Allen, S.E., Grimshaw, H.M., Parkinson, J.A. and Quarmby, C., 1974, "Chemical Analysis of Ecological Materials," Blackwell Scientific Publications, Oxford.

Caswell, A.H., 1979, Methods of measuring intracellular calcium, Int. Rev. Cytol., 56: 145.

Chafouleas, J.G., Dedman, J.R., Munjaal, R.P. and Means, A.R., 1979, Calmodulin: development and application of a sensitive radio-immunoassay, J. Biol. Chem., 254: 10262.

Chevallier, J. and Butow, R.A., 1971, Calcium binding to the sarcoplasmic reticulum of rabbit skeletal muscle, Biochemistry 10: 2733.

Cormier, M.J., Anderson, J.M., Charbonneau, H., Jones, H.P. and McCann, R.O., 1980, Plant and fungal calmodulin and the regulation of plant NAD kinase. In: "Calcium and Cell Function," W.Y. Cheung, ed., Academic Press, New York.

Cormier, M.J., Jarrett, H.W. and Charbonneau, H., 1982, Role of Ca[++]-calmodulin in metabolic regulation in plants, In: "Calmodulin and Intracellular Ca[++] Receptors," S. Kakiuchi, H. Hidaka and A.R. Means, eds., Plenum Press, New York.

Eschrich, W. and Heyser, W., 1975, Biochemistry of phloem constituents, In: "Transport in Plants. I. Phloem Transport," M.H. Zimmermann and J.A. Milburn, eds., Encyclopedia of Plant Physiology, 1: 101, Springer-Verlag, Berlin.

Hall, S.M., Baker, D.A. and Milburn, J.A., 1971, Phloem transport of ^{14}C-labelled assimilates in Ricinus, Planta (Berl.), 100: 200.

Laemmli, U.K., 1970, Cleavage of structural proteins during the assembly of the head of bacteriophage T4, Nature, 227: 680.

McEuen, A.R., 1979, Studies on calcium and a calcium-binding protein in the sieve tube exudate of Cucurbita maxima and related species, Ph.D. thesis, University of Aberdeen.

McEuen, A.R., Hart, J.W. and Sabnis, D.D., 1981, Calcium-binding protein in sieve tube exudate, Planta (Berl.), 151: 531.

Ohnishi, T. and Ebashi, S., 1963, Spectrophotometrical measurements of instantaneous calcium-binding of the relaxing factor of muscle, J. Biochem. (Tokyo), 54: 506.

Parthasarathy, M.V., 1975, Sieve element structure, In: "Transport in Plants. I. Phloem transport," M.H. Zimmermann and J.A. Milburn, eds., Encyclopedia of Plant Physiology, 1: 1, Springer-Verlag, Berlin.

Read, S.M. and Northcote, D.H., 1983, Subunit structure and interactions of the phloem proteins of Cucurbita maxima (pumpkin), Eur. J. Biochem., 134: 561.

Ross, E. and Schatz, G., 1973, Assay of protein in the presence of high concentrations of sulfhydryl compounds, Ann. Biochem., 54: 304.

Sabnis, D.D. and Hart, J.W., 1982, Microtubule proteins and P-proteins, In: "Nucleic Acids and Proteins in Plants. I.," D. Boulter and B. Parthier, eds., Encyclopedia of Plant Physiology, 14A: 401, Springer-Verlag, Berlin.

Ziegler, H., 1975, Nature of transported substances, In: "Transport in Plants. I. Phloem Transport," M.H. Zimmermann and J.A. Milburn, eds., Encyclopedia of Plant Physiology, 1: 59, Springer-Verlag, Berlin.

A NOVEL PLANT CALCIPROTEIN AS TRANSIENT SUBUNIT OF ENZYMES

Raoul Ranjeva, Annick Graziana, Marietta Dillenschneider
Martine Charpenteau and Alain M. Boudet

Centre de Physiologie Végétale
Université Paul Sabatier, U.A. CNRS n° 241
118 Route de Narbonne, 31062 Toulouse Cedex

INTRODUCTION

It is generally accepted that the conversion of a stimulus into bio-logical/biochemical responses involves a multistep process that includes the variation in the amounts of second messenger(s) and the transduction of the chemical message through sensing molecules (Cohen, 1982). Such a scheme, inferred from animal cell physiology has been shown to be essenti-ally valid for plants. Thus, several lines of evidence have established that calcium controls different biochemical/biological processes which are being discussed during this workshop; especially calcium is implicitly recog-nized as the main second messenger in plants.

In this way, plants contain different calcium-binding molecules such as phospholipids and proteins that may transduce the calcium effect (Hetherington and Trewavas, 1983). Among them, calmodulin (CaM) has received the most attention in the literature (Marmé and Dieter, 1983) but the occur-rence of other calcium-binding proteins in animals and plants is largely extablished (Waisman et al., 1983; Staun et al., 1984). However, the actual role of such molecules is difficult to assess because they have no identi-fied biological activities.

The present paper deals with a particular species of calcium-binding protein that behaves as a reversible subunit of a typical plant enzyme, namely, quinate: NAD^+ oxidoreductase (E.C. : 1.1.1.24) hereafter referred to as QORase.

QORase catalyzes the reversible oxidation of quinic acid into dehydro-quinic acid, an intermediate of the shikimate pathway that leads to aromatic amino acids and phenolic compounds.

The enzyme from carrot cell-suspension cultures appears to be sensitive or insensitive to Ca^{2+} depending upon the light environment of the cells. From the molecular point of view, it turns out that in light-grown cell, QORase is a monomeric enzyme that is insensitive to Ca^{2+}. In dark-grown cell, the enzyme becomes oligomeric, Ca^{2+}-dependent and contains a specific calcium-binding moiety (regulatory subunit).

The comparative properties of the two forms of QORase are described here. In addition, putative mechanism involved in the reversible association of the catalytic and the regulatory subunit is discussed.

MATERIALS AND METHODS

Carrot Cell Cultures

Carrot cell-suspension cultures were grown either under continuous white light (Refeno et al., 1982) or in the dark (Graziana et al., 1984a). QORase extracted from light-grown cells (L-cells) will be referred to as L-QORase, that from dark-grown cells (D-cells) as D-QORase.

Enzyme Extraction and Assays

The harvested cells (6-8 day-old) were frozen in liquid nitrogen; the proteins were extracted as described by Refeno et al. (1981), QORase activity was measured by spectrophotometry. The assays contained 6 mM dehydroquinate, 0.2 mM NADH, 0.25 M Tris-Cl buffer pH 8.5 and protein extract in a final volume of 1 ml. The control was done without dehydroquinate. The reactions were run at 30°C and the decrease in absorbance at 340 nm monitored with a Kontron model Uvikon spectrophotometer.

Purification of the Enzyme

Partial purification was done by ammonium sulphate precipitation 80% saturation followed by gel filtration through a sephadex G_{25} column in 0.05 M Tris-Cl pH 8. The enzyme was purified to near to homogeneity as described by Graziana et al. (1984a, b). Briefly, the partially purified preparation was chromotofocused, gel filtered through a Sephacryl S200 column and submitted to affinity chromatography through a blue-ultrogel column. The purity of the enzyme was checked by SDS-PAGE.

Electrophoretic Transfer and Autoradiography

After SDS-PAGE, the proteins were transferred on a nitrocellulose membrane (pore size 0.1 μm) according to Graziana et al. (1984b) and labelled with [^{45}Ca] following Maruyama et al. (1984). The autoradiographies were done as described previously (Graziana et al., 1984a, b).

RESULTS AND DISCUSSION

QORase from Dark-Grown Cells is Sensitive to Calcium

The properties of QORase extracted from carrot cell-suspension cultures change as a function of the light environment of the culture (Fig. 1). Thus, the enzyme from light-grown cells was insensitive to Ca^{2+} and Ca^{2+} trapping (EGTA) or sensing (calmodulin) molecules. In contrast, D-QORase was activatable by Ca^{2+} and inhibited by excess EGTA. The manipulation of the EGTA over Ca^{2+} ratios resulted in the inhibition/activation of D-QORase. CaM and CaM antagonists (R 24571 or fluphenazine) had little or no effect on the process and therefore CaM is not the calcium sensor involved. The Ca^{2+} dependence was not induced on acellular extracts from light-grown cells suggesting that the mechanism is associated, by some manner, with an in vivo dependent process.

The Ca^{2+} Dependence is Reversible in vivo

Since L-QORase differs from D-QORase by its sensitivity to calcium, such a parameter may be used to check the reversibility of the process.

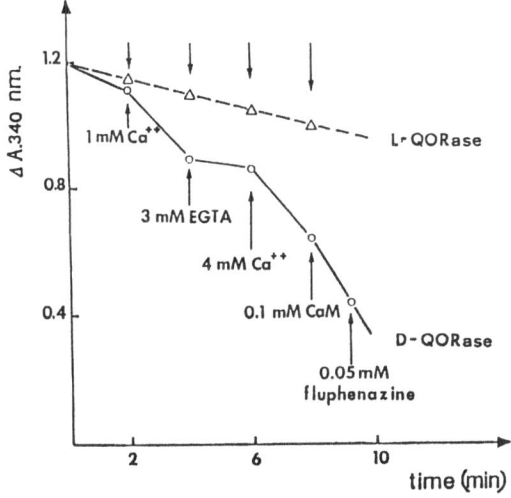

Fig. 1. Effects of Ca²⁺ and Ca²⁺ -sensing
molecules on D- and L-QORase.

In this way, dark-grown or light-grown cells were transferred respectively
to light or dark. Then, proteins were extracted at indicated times and the
Ca^{2+}-dependence of QORase measured (Table 1).

From the data obtained it clearly appears that, on transfer from L to
D conditions, L-QORase that was initially insensitive became progressively
dependent on Ca^{2+} whereas a transfer from D to L led to the reverse situation, namely D-QORase became insensitized.

The shift in properties was slightly appreciable within less than one
hour and was virtually complete after a 6-hour transfer in both cases.

Table 1. Ca^{2+} dependence of QORase activity as a function of the light
environment of carrot cells. The carrot cells were transferred
from dark (D) to light (L) or inversely for indicated times. The
activity of the enzyme without Ca^{2+} was set as 1 and compared
with the same assay supplemented with 1 mM Ca^{2+}.

Time (h)	L ⟶ D activation by Ca^{2+} (fold)	D ⟶ L activation by Ca^{2+} (fold)
0	1.2	4
1	1.8	3.2
2	2.1	2.8
3	2.8	2.1
4	3	1.9
5	3.5	1.4
6	3.7	1.3

The data obtained are consistent with a gradual change in the proportion of calcium-sensitive over insensitive enzyme molecules in response to
the light environment.

The behaviour of QORase on analytical gel filtration through a Sephacryl S200 column depends strikingly upon the enzyme source. Thus, light treated cells contained L-QORase that exhibited 42 kDa (Refeno et al., 1982) whereas D-QORase eluted as a larger molecule and the activity was associated with a 110 kDa protein. Interestingly, the Ca²⁺-dependence was retained and even increased when D-QORase was purified to homogeneity and freed from CaM by affinity chromatography (not shown).

Moreover, the enzyme content of L or D cells that were transferred to respectively D or L for 3 hours (dependence on Ca²⁺, as shown in Table 1) appeared to be heterogeneous (Fig. 2). Thus, in both cases, the pattern was identical and constituted a composite population, namely a 110 kDa protein (D-QORase) that was sensitive to Ca²⁺ and a 42 kDa molecule (L-QORase) that was insensitive.

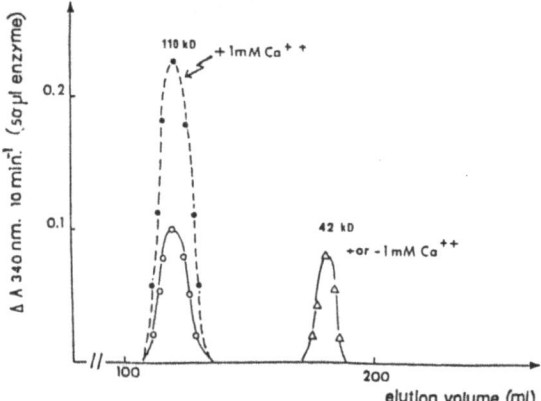

Fig. 2. Enzyme pattern of D cells transferred to L for 3 hours (partially
purified extracts).

In contrast to the enzyme from L-cells which is a monomeric protein, pure D-QORase, analysed by SDS-PAGE contained two distinct subunits (Fig. 3). One of them behaved as L-QORase (42 kDa) the other as a 63 kDa peptide.

When supplemented with [³H] leucine in combination with cycloheximide and transferred subsequently, to dark for 6-h, the protein synthesis was almost stopped in comparison with the control done without cycloheximide. Meanwhile, the enzyme was virtually unlabelled but its Mr shifted to 110 kDa and QORase became sensitive to Ca²⁺.

Taken together, these data show that the light environment induces the correlative changes in QORase properties and Mr. Since protein synthesis is not involved, the overall process occurs post-translationally.

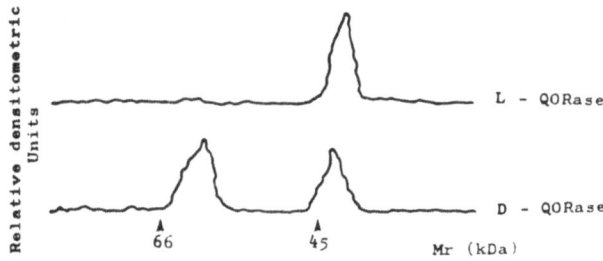

Fig. 3. SDS-PAGE analysis of L and D-QORase.

The 63 kDa Peptide is the Calcium-Binding Moiety

When sieved through a Sephacryl S200 column that has been pre-equilibriated with [^{45}Ca] Cl$_2$, pure D-QORase coeluted with a peak of radioactivity as a 110 kDa molecule (Graziana et al., 1984b). The calcium-binding capacity has been analysed further following the procedure of Maruyama et al. (1984). Briefly, the homogenous enzyme was first dissociated by SDS-mercaptoethanol treatment, then, the subunits were separated by SDS-PAGE and the peptides electrophoretically transferred onto a nitrocellulose membrane. The membrane was soaked in [^{45}Ca] Cl$_2$, extensively rinsed and autoradiographied (Maruyama et al., 1984).

The densitometric trace of the resulting autoradiography demonstrates that only the large subunit was able to bind Ca^{2+} (Fig. 4).

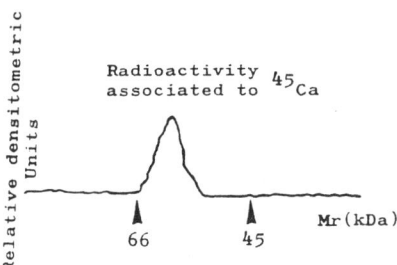

Fig. 4. Binding of [^{45}Ca] by the regulatory subunit.

Therefore, it clearly appears that the calcium dependence is specifically due to the 63 kDa peptide that seems capable to act as a calcium sensor.

The Effects of Ca^{2+} are Complex

The dependence of D-QORase on increasing concentrations of free calcium in the medium appeared to be biphasic (Fig. 5a). The enzyme was insensitive up to 0.1 mM Ca^{2+} but beyond that concentration, two plateaus were observed. The first one occurred at 0.4 mM Ca^{2+} and the other one at 0.7 mM.

Therefore, D-QORase may contain two types of binding sites with different affinities for Ca^{2+}. In contrast to CaM-dependent enzymes that are activatable by micromolar concentrations of Ca^{2+}, D-QORase was sensitive to higher concentrations. In this respect, it resembles the m-calcium-activatable neutral protease that requires near millimolar concentration of Ca^{2+} for enzyme activation (Kishimoto et al., 1981).

Further analysis of the Ca^{2+} effects shows that D-QORase was more activatable at low protein concentrations (Fig. 5b). For example, less than a two-fold activation was observed with 10 µg protein but more than a 5-fold stimulation was obtained with 1 µg protein.

Regardless of the presence of Ca^{2+} in the medium, the activity was no more proportional with the protein amounts beyond 5 µg protein while the enzyme was saturated by the substrates.

Such data suggest that some steric problems may occur when the amounts of proteins are large enough. Therefore, the effects of dissociation or disaggregating agents were studied.

First, the incubation of D-QORase with trypsin used at different concentrations and pH conditions led to the loss in activatability and activity

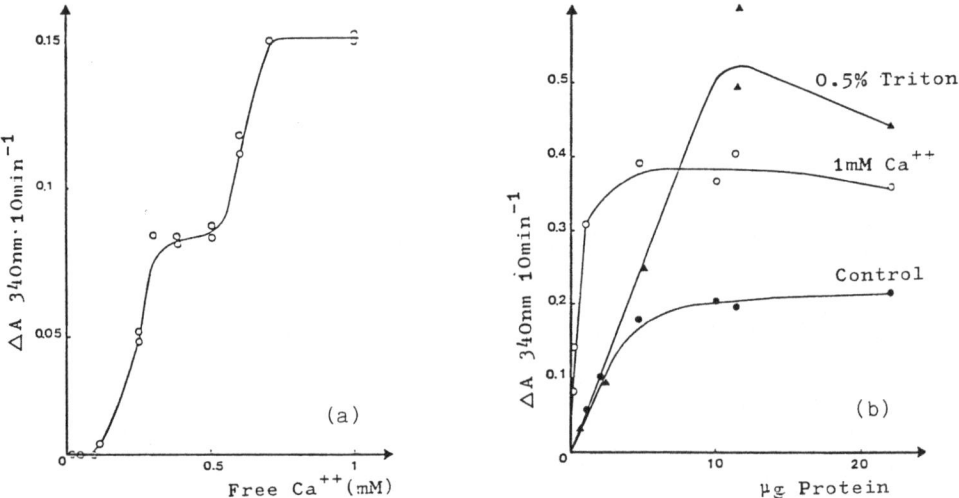

Fig. 5. Effects of calcium on D-QORase activity
    (a) increasing Ca^{2+} concentrations
    (b) increasing protein amounts.

of the enzyme (Graziana et al., 1984).

    In contrast, in the presence of 0.25-1% triton X 100, D-QORase became independent upon Ca^{2+} at high concentrations of proteins, but not at lower ones (Fig. 5b). Other detergents either non-ionic or zwitterionic had no significant effects.

    The sequential additions of Ca^{2+} and triton did not modify the extent of Ca^{2+} sensitivity (not shown); therefore, the detergent did not potentiate the action of calcium but may only act as a chaotropic agent.

    In this way, the influence of antichaotropic agents on D-QORase was studied and representative results obtained by varying the concentrations of NaCl are reported in Figure 6.

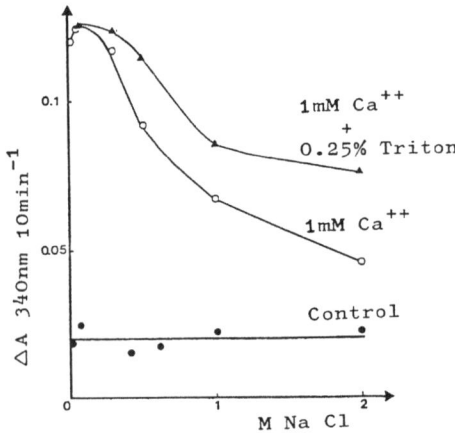

Fig. 6. Effects of chaotropic and antichaotropic agents.

It is clear that up to 2 M NaCl has no effect on the enzyme activity; however, the activatability of D-QORase was dramatically influenced.

Thus, the Ca^{2+} effects decreased when NaCl concentrations increased; as an example in the absence of NaCl, Ca^{2+} stimulated the enzyme by more than four-fold. When 2 M NaCl was present the activation was only two-fold.

If only 0.25% triton was subsequently added to the assays, the stimulation by Ca^{2+} was partially recovered, even in the presence of 2 M NaCl. Using NH_4Cl the antichaotropic effects of which are higher than for NaCl resulted in a lower recovery of the activatability (not shown).

From these data, it is suggested that the sensitivity of D-QORase to Ca^{2+} depends on the organization degree of the medium. The more the organization degree is high, the less the binding sites of Ca^{2+} are accessible. Accordingly, it is assumed that high protein concentrations lead, by some manner, to a "condensation" of the enzyme preventing the accessibility to substrates and Ca^{2+}. Therefore the chaotropic agent may not mimic the calcium effects as for other enzyme systems (Kauss et al., 1984) but only limit the intermolecular association. Such an interpretation is consistent with the lack of triton effects with low enzyme concentrations and is now under examination by using physico-chemical approach.

CONCLUSIONS AND PROSPECTS

The data reported in this paper demonstrate that a calcium-binding protein acts as a reversible subunit of D-QORase. The characteristics of the system are as follows:

(1) the subunit is a relatively large peptide,
(2) the sensitivity to Ca^{2+} is in the range of millimolar concentration;
(3) the CaBP is bound to the catalytic moiety regardless of the presence of Ca^{2+}, and
(4) results obtained in our laboratory have shown that the regulatory subunit protects the catalytic moiety from dephosphorylation (Graziana et et., 1984).

Therefore the system differs sharply from CaM-dependent process by at least the three first characteristics.

While they are not analysed in detail from the molecular point of view, at least two other enzymes appear to be Ca^{2+} regulated independently of CaM. Thus glucan synthase from soybean cells (Kauss et al., 1983) and mitochondrial glutamate dehydrogenase from corn shoots (Yamaya et al., 1984) are both stimulated by millimolar concentrations of Ca^{2+}. However, the binding moiety has not been identified so far and their homology degree with QORase has to be assessed.

An interesting point concerning the QORase system lies in the photoreversibility of the binding process. Such a characteristic may be exploited to screen for other molecules that may have an affinity for the CaBP, and to check for its eventual role as calcium transducer in plant systems.

REFERENCES

Cohen, P., 1982, The role of protein phosphorylation in neural and hormonal control of cellular activity, Nature, 296:613.
Graziana, A., Ranjeva, R., Salimath, B. and Boudet, A.M., 1984a, The reversible association of quinate: NAD+ oxidoreductase from carrot cells

with a putative regulatory subunit depends on light conditions, F.E.B.S. Lett., 163:306.

Graziana, A., Dillenschneider, M. and Ranjeva, R., 1984b, A calcium-binding protein is a regulatory subunit of quinate: NAD^+ oxidoreductase from dark-grown carrot cells, Biochem. Biophys. Res. Comm., 125:170.

Hetherington, A.M. and Trewavas, A.J., 1984, The regulation of membrane-bound protein kinase by phospholipids and calcium, in "Annual Proceedings of the Phytochemical Society of Europe", 24, A.M. Boudet, G. Alibert, G. Marigo and P.J. Lea, eds, Clarendon Press, Oxford.

Kauss, H., Köhle, H. and Jeblick, W., 1984, Proteolytic activation and stimulation by Ca^{2+} of glucan synthase from soybean cells, F.E.B.S. Lett., 158:84.

Kishimoto, A., Kajikawa, N., Tabuchi, H. and Nishizuka, Y., 1981, Neutral proteases activatable by calcium, J. Biochem., 90:889.

Marmé, D., and Dieter, P., 1983, Calcium and calmodulin in plants, in "Calcium and Cell Functions", IV, W.Y. Cheung, ed., Academic Press, New York.

Maruyama, K., Mikawa, T. and Ebashi, S., 1984, Detection of calcium-binding proteins by ^{45}Ca autoradiography on mitrocellulose membrane after SDS-PAGE, J. Biochem., 95:511.

Refeno, G., Ranjeva, R. and Boudet, A.M., 1982, Modulation of quinate: NAD^+ oxidoreductase activity through reversible phosphorylation in carrot cell suspensions, Planta, 154:193.

Staun, M., Norén, O. and Sjostrom, H., 1984, Ca^{2+}-binding protein from human kidney. Purification and properties, Biochem. J. 217:229.

Waisman, D.M., Muranyi, J. and Ahmed, M., 1983, Identification of a novel calcium-binding protein from bovine brain, F.E.B.S. Lett., 164:80.

Yamaya, T., Oaks, A. and Matsumoto, H., 1984, Characteristics of glutamate dehydrogenase in mitochondria prepared from corn roots, Plant Physiol., 76:1009.

ROLE OF CA^{2+} IN THE REGULATION OF α-AMYLASE SYNTHESIS AND SECRETION IN BARLEY ALEURONE

Russell L. Jones, Jill Deikman, and Diane Melroy

Department of Botany
University of California
Berkeley, CA 94720, USA

The requirement for Ca^{2+} in the response of the isolated barley aleurone layer to gibberellic acid (GA) was first reported by Chrispeels and Varner (1967). These workers showed that when Ca^{2+} was omitted from the incubation medium the synthesis of α-amylase in response to GA was reduced by 50% in imbibed half-grains and by 80% in isolated aleurone layers. Jacobsen et al. (1970) showed that two isoenzymes of α-amylase were produced when aleurone layers were incubated in GA and another two when Ca^{2+} was added. The α-amylase isoenzymes synthesized in response to GA have been referred to as group A: they belong to the same antigenic group, have isoelectric points (pIs) around 5 and are sensitive to Hg salts and N-ethylmaleimide. α-Amylases produced on addition of Ca^{2+} are referred to as group-B isoenzymes: they belong to a different antigenic group, have pIs around 6.0, and are stable to Hg and N-ethylmaleimide (Jacobsen et al., 1970; Jacobsen and Higgins, 1982).

All animal and plant α-amylases examined in detail have been shown to be Ca^{2+}-containing metalloenzymes that require Ca^{2+} for their activity. Barley α-amylase isoenzymes have different affinities for Ca^{2+}; group-B isoenzymes are inactivated by 5 mM EDTA, whereas group-A isoenzymes are not (Jacobsen, 1970). Because α-amylases of the B group bind Ca^{2+} less tenaciously, Jacobsen et al. (1970) reasoned that their apparent absence from media of layers incubated without Ca^{2+} was due to their inactivity. In an examination of the effects of Ca^{2+} on α-amylase production using immunochemical and radiolabelling methods, Jones and Jacobsen (1983) concluded that withdrawing Ca^{2+} from GA-treated aleurone layers affected the secretion of B-group α-amylases, not their activity, since neither labelled polypeptides nor antigens corresponding to B-group enzymes accumulated in either medium or tissue. Moll and Jones (1980) in a kinetic study of enzyme secretion from single aleurone layers in a flow-through cell, also concluded that Ca^{2+} withdrawal primarily affected the process of enzyme secretion.

In addition to affecting the secretion of B-group α-amylases in barley aleurone, Ca^{2+} also affects their synthesis. When aleurone layers are incubated in GA alone, only group-A isoenzymes accumulate in the tissue. Mitsui et al. (1984) noted a similar effect of Ca^{2+} on the synthesis of α-amylase in rice scutellum: α-amylase synthesis, which was Ca^{2+}-dependent, was saturated at 0.5 mM external Ca^{2+}, whereas secretion was not saturated until the external Ca^{2+} concentration reached 10 mM.

Work in my laboratory has focused on the roles of GA and Ca^{2+} in α-amylase production by barley aleurone layers. In this paper we describe experiments aimed at an understanding of how GA and Ca^{2+} affect the synthesis and secretion of α-amylase isoenzymes. Using cDNA clones to α-amylases of groups A and B we have shown that GA, not Ca^{2+}, regulates the accumulation of mRNAs for α-amylase isoenzymes. We also present evidence that the effects of Ca^{2+} on the accumulation of specific isoenzymes of α-amylase and other secreted hydrolases are on the post-translational processing or transport of these proteins.

MATERIALS AND METHODS

Barley (Hordeum vulgare L. cv. Himalaya, 1979 harvest) grains were de-embryonated and imbibed for 4 d (Jones and Jacobsen, 1982). Aleurone layers were removed and incubated (10 layers/2 ml) in a medium containing 30 μM chloramphenicol and 5 μM GA, 7 μM monensin, and $CaCl_2$ as specified.

α-Amylase was purified by affinity chromatography on cyclohepta-amylose(CHA)-sepharose-6B, dialyzed against 20 mM Tris-HCl (pH 6.5) containing 20 mM $CaCl_2$, and further purified by ion-exchange chromatography on DEAE-cellulose (Jones and Jacobsen, 1982). α-Amylase isoenzymes were separated by agar-gel electrophoresis (Jacobsen et al., 1970) or isoelectric focusing (IEF, Jacobsen and Higgins, 1982) and localized on gels with the starch-IKI procedure.

Proteins were separated by sodium dodecylsulfate polyacrylamide gel electrophoresis (SDS-PAGE, Spencer et al., 1980). For fluorography gels were soaked in 1 M sodium salicylate, briefly rinsed in H_2O, dried and exposed to Kodak XAR film (Deikman and Jones, 1985).

RNA was purified, cDNA clones were constructed, and RNA blotting and hybridization were carried out as described by Deikman and Jones (1985). Clone 1-28, isolated in our laboratory, contains α-amylase cDNA sequences complementary to B-group isoenzymes (Deikman and Jones, 1985), clone E, a gift of J. Rogers, contains sequences complementary to A-group α-amylase (Rogers and Milliman, 1983), and clone pTA71, a gift of J. Bedbrook, contains a complete rDNA repeat from wheat (Gerlach and Bedbrook, 1979).

Organelles were isolated from aleurone layers as described by Jones (1980). Aleurone layers incubated in GA (5 μM) and Ca^{2+} (20 mM) with or without monensin were homogenized in low Mg^{2+} buffer containing 0.56 M sucrose. After centrifugation at 2,000 g to remove cell debris, the supernatant was subjected to molecular sieve chromatography on Sepharose 4B. The turbid fractions excluded by the column were loaded onto linear sucrose gradients and centrifuged to equilibrium. Gradient fractions were assayed for α-amylase (Jones and Varner, 1967), acid phosphatase (Ray et al., 1969), cytochrome c reductase (Lord et al., 1973), and inosine diphosphatase (IDPase, Shore and MacLachlan, 1975).

RESULTS AND DISCUSSION

Calcium has a profound effect on the amount of α-amylase and other enzymes produced by GA-treated aleurone layers of barley (Table 1). Whether layers are incubated in GA or GA plus Ca^{2+}, or are preincubated in GA plus Ca^{2+} then transferred to either GA or GA plus Ca^{2+}, they produce much more α-amylase in the presence of Ca^{2+} (Table 1).

TABLE 1. The effect of Ca^{2+} on α-amylase activity in GA-treated aleurone layers

| Incubation conditions | α-Amylase units/aleurone layer/h | | |
	Medium	Tissue extract	Total
Incubated for 24 h in:			
GA (5 μM)	0.03	0.05	0.08
GA + Ca^{2+} (20 mM)	0.48	0.04	0.52
Incubated 12 h in GA + Ca^{2+}, then 4 h in:			
GA (5 μM)	0.06	0.11	0.17
GA + Ca^{2+} (20 mM)	0.44	0.13	0.57

In addition to a quantitative effect on α-amylase production, the composition of the incubation medium has a qualitative effect on the α-amylase isoenzymes produced by aleurone layers (Fig. 1). Aleurone layers incubated in H_2O synthesize and secrete only one of the low-pI α-amylase isoenzymes, whereas layers incubated in GA secrete both low-pI isoenzymes (Fig. 1a-f). In GA plus Ca^{2+}, both low- and high-pI isoenzymes accumulate in media (Fig. 1a-f) and tissue extracts (Deikman and Jones, 1985). When layers preincubated in GA plus Ca^{2+} are transferred to new media containing GA or GA plus Ca^{2+} the isoenzyme spectra of the incubation media are similar to those for layers not preincubated: only in the presence of GA and Ca^{2+} are both low- and high-pI isoenzymes released (Fig. 1g-j). Because tissue extracts of aleurone layers pretreated with GA plus Ca^{2+} contain both low- and high-pI α-amylases we have concluded that Ca^{2+} promotes the secretion of the high-pI isoenzymes (Fig. 1g-j; Jones and Jacobsen, 1983; Jones and Carbonell, 1984). This conclusion supports that drawn by Moll and Jones (1980) that Ca^{2+} plays a role in regulating α-amylase secretion.

Varner and Mense (1972) also argued that Ca^{2+} influenced the appearance of α-amylase in the incubation medium, but they proposed that GA controlled the secretion of α-amylase across the plasma membrane and Ca^{2+} controlled the release of enzyme through the cell wall. We have used protoplasts isolated from barley aleurone layers to determine whether Ca^{2+} affects enzyme secretion or release. Our data (Bush et al., 1985) show that Ca^{2+} affects the process of enzyme secretion across the plasma membrane, since the

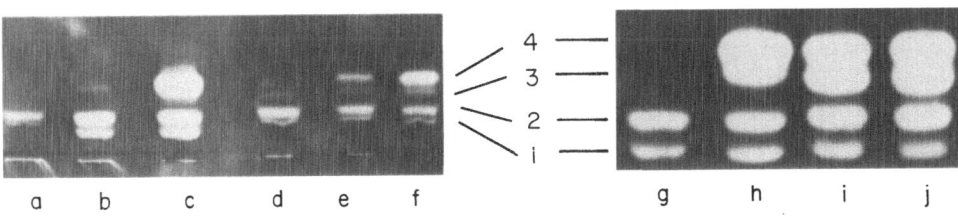

Fig. 1. Agar gel electrophoresis of α-amylase isoenzymes in incubation media (a-c, g, h) and tissue extracts (d-f, i, j) of barley aleurone layers incubated for 12 h in H_2O (a, d), GA (b, c), or GA plus Ca^{2+} (e, f) or preincubated for 22 h in GA plus Ca^{2+} then incubated for 4 h in GA (g, i) or GA plus Ca^{2+} (h, j). a-f From Deikman and Jones, 1985; g-j from Jones and Jacobsen, 1983.

appearance of high-pI isoenzymes in incubation media surrounding proto-plasts depends on the presence of 10-20 mM Ca^{2+}. Our data do not rule out an effect of Ca^{2+} on the movement of these polypeptides through the cell wall, however.

Our experiments with both aleurone layers (Fig. 1) and protoplasts (Bush et al., 1985) show that in addition to affecting enzyme secretion, Ca^{2+} affects the synthesis of high-pI isoenzymes of α-amylase. To inves-tigate whether Ca^{2+} affects the synthesis of α-amylase isoenzymes at the transcriptional or translational level, we isolated an α-amylase cDNA clone to study the accumulation of α-amylase mRNA (Deikman and Jones, 1985). Clone 1-28 contains α-amylase cDNA sequences cloned into the Pst1 site of pBR322. Sequence data show that 1-28 is similar to clone pHV19 of Chandler et al. (1984) and pM/C of Rogers (in press), and comparison of these nucleotide sequences with the amino acid sequences of barley α-amylases show these cDNAs are complementary to mRNA for group-B α-amylases.

Poly (A)$_2$ RNA was isolated from aleurone layers incubated in H_2O, GA, or GA plus Ca^{2+} for 12 h, electrophoresed on agarose, and transferred to nitrocellulose. The hybridization of ^{32}P-labelled clone 1-28 to this RNA is shown in Figure 2. No mRNA accumulated in aleurone layers incubated in H_2O, confirming the observations of others (Chandler et al., 1984); however, there was no detectable difference between GA and GA plus Ca^{2+} treatments (Fig. 2A). We have eliminated the possibility that RNA loading can explain these results. Based on the observation of Jacobsen and Zwar (1974) that rRNA comprises 98% of total aleurone RNA and that this RNA is unaffected by GA treatment, we used an rDNA clone to quantitate the RNA transferred to nitrocellulose (Fig. 2B). There are no appreciable differences in RNA loading between the three treatments in this experiment.

The absence of differences in α-amylase mRNA accumulation between GA and GA plus Ca^{2+} treatments could be explained by arguing that the mRNA that accumulates in GA-treated tissue is A-group α-amylase RNA whereas in GA plus Ca^{2+}-treated tissue, the mRNA that accumulates is either a mixture

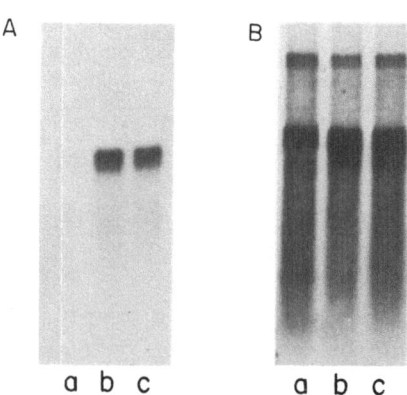

Fig. 2. RNA blot probed with clones 1-28 (A) and pTA71 (B). Poly (A) RNA was isolated from aleurone layers incubated for 12 h in H_2O (a), GA (b), and GA plus 5 mM Ca^{2+} (c). From Deikman and Jones, 1985.

of A- and B-group mRNAs or is primarily B-group. We have tested this possibility using a cDNA clone for A-group α-amylase (clone E). Clones 1-28 and E do not significantly hybridize indicating that these cDNA clones can distinguish between mRNAs for A- and B-group α-amylases.

That these clones detect different species of aleurone mRNA is shown by a detailed time course of mRNA accumulation in aleurone incubated in H_2O, GA, and GA plus Ca^{2+} for up to 24 h (Fig. 3). While clone E detects appreciable levels of mRNA in H_2O-treated layers, probing the same blot with clone 1-28 shows little mRNA at time zero and only a small increase during incubation in H_2O. Both clones E and 1-28 show mRNA accumulation in layers incubated in GA and GA plus Ca^{2+}-treated layers. These two treatments are qualitatively and quantitatively similar, establishing GA as the principal regulator of α-amylase mRNA accumulation in barley aleurone.

We have investigated the possibility that Ca^{2+} influences some aspect of the translation of α-amylase mRNAs or the post-translational processing of these isoenzymes. If Ca^{2+} does affect translation or post-translational processing, its effects must be selective: Ca^{2+} withdrawal affects aspects of both the synthesis and the secretion of B-group α-amylase isoenzymes.

Experiments with the sodium ionophore monensin show that α-amylases A and B are segregated in different compartments of the aleurone cell and that this segregation affects the transport of α-amylase to the exterior of the cell. Monensin inhibits the secretion of α-amylase and acid phosphatase from barley aleurone layers. The effect of the ionophore on α-amylase secretion is selective, since it inhibits primarily the release of group-B isoenzymes and causes them to accumulate within the cell (Melroy and Jones, in press). When organelles are isolated from aleurone layers incubated in GA plus Ca^{2+} with or without monensin, qualitative and quantitative effects of monensin on α-amylase and acid phosphatase distribution are seen. In the absence of monensin, α-amylase and acid phosphatase accumulate in the ER-region of the density gradient; in the presence of 7.5 μM monensin, both amylase and phosphatase activities accumulate in the region of the gradient having Golgi marker enzyme (IDPase) activity (Fig. 4). When the α-amylase isoenzymes of ER and Golgi regions of sucrose density gradients are examined by IEF it becomes apparent that monensin treatment causes B-group isoenzymes to accumulate in the Golgi region of the density gradient (Fig. 5). We have other evidence that accumulation of secreted

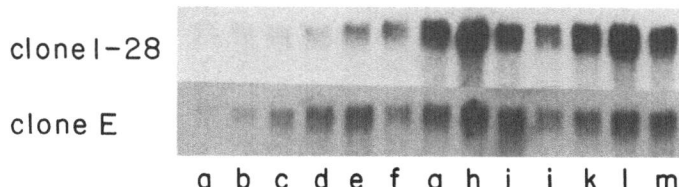

Fig. 3. RNA blots of total RNA probed with clones 1-28 and E. a, RNA extracted from layers without any incubation. b-e, RNA from layers incubated in H_2O for 4 h (b), 8 H (c), 12 h (d), or 24 h (e); f-i, RNA from layers incubated in GA for 4 h (f), 8 h (g), 12 h (h), or 24 h (i); j-m, RNA from layers incubated in GA plus Ca^{2+} for 4 h (j), 8 h (k), 12 h (l), or 24 h (m). From Deikman and Jones, submitted.

enzymes in this region of the density gradient represents accumulation in Golgi. Monensin treatment results in distension of Golgi cisternae and vesicles, and cytochemistry shows that Golgi and its vesicles in aleurone contain prominent acid phosphatase.

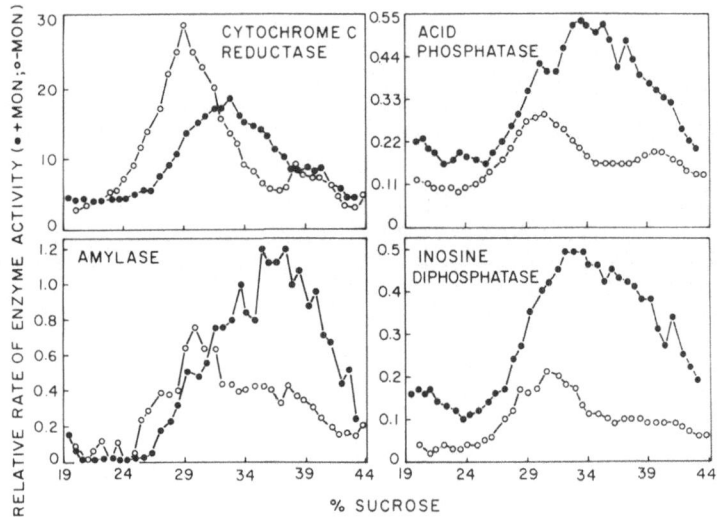

Fig. 4. Enzyme activities of sucrose density gradient fractions of organelles of barley aleurone layers incubated with (●) and without (○) monensin. From Melroy and Jones, in press.

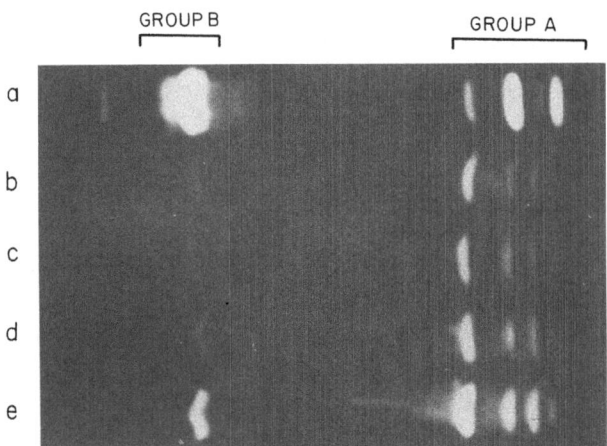

Fig. 5. IEF gel of incubation medium from layers incubated in GA plus Ca^{2+} (a) and fractions from sucrose density gradient fractions (b-e). b, ER and c, Golgi fractions from GA plus Ca^{2+}-treated layers. d, ER and e, Golgi fractions from GA plus Ca^{2+}-treated layer incubated in monensin.

We interpret these data to indicate the existence of two routes for the secretion of α-amylase from the aleurone cell. Both routes involve ER and Golgi. We propose that isoenzymes A and B are segregated from each other in the Golgi based on their glycosylation. Group-A α-amylases are not glycosylated while group-B isoenzymes contain at least one glycosylation site (J. Rogers, personal commun.). Since monensin inhibits glycosylation of plant glycoproteins (Chrispeels, 1983), we suggest that it affects the secretion of only B-group α-amylases via their glycosylation. Although these experiments indicate A- and B-group α-amylases become segregated from each other in the Golgi, we have not established that Ca^{2+} affects the production of these isoenzymes by interfering with intracellular transport as monensin does. To account for the effects of Ca^{2+} on both the synthesis and secretion of barley α-amylase, we hypothesize that because of the tight coupling between enzyme synthesis and secretion in the barley aleurone, any effect of Ca^{2+} on enzyme secretion will manifest itself on enzyme synthesis. According to this hypothesis, the principal effect of Ca^{2+} in the barley aleurone would be on enzyme secretion. A similar scheme for the regulation of enzyme secretion by Ca^{2+} from rice scutellum has been proposed by Akazawa and Hara-Nishimura (1985).

REFERENCES

Akazawa, T. and Hara-Nishimura, I. 1985. Topographic aspects of bio-synthesis, extracellular secretion and intracellular storage of proteins in plant cells. Ann. Rev. Plant Physiol. 36: 441-472.
Bush, D.S., Cornejo, M.-J., Huang, C. and Jones. R.L. 1985. Ca-stimulated α-amylase secretion from barley aleurone protoplasts. NATO Advanced Studies Workshop, Molecular and Cellular Aspects of Calcium in Plant Development, Edinburgh, 1985.
Chandler, P.M., Zwar, J.A., Jacobsen, J.V., Higgins, T.J.V., and Inglis, A.S. 1984. The effects of gibberellic acid and abscisic acid on α-amylase mRNA levels in barley aleurone layers: studies using an α-amylase cDNA clone. Plant Mol. Biol. 3: 407-418.
Chrispeels, M.J. 1983. The Golgi apparatus mediates the transport of phytohemagglutinin to the protein bodies in bean cotyledons. Planta 158: 140-151.
Chrispeels, M.J. and Varner, J.E. 1967. Gibberellic acid-enhanced synthesis and release of α-amylase and ribonuclease by isolated barley aleurone layers. Plant Physiol. 42: 398-406.
Deikman, J. and Jones, R.L. 1985. Control of α-amylase mRNA accumulation by gibberellic acid and calcium in barley aleurone layers. Plant Physiol. 78: 192-198.
Gerlach, W.L. and Bedbrook, J.R. 1979. Cloning and characterization of ribosomal RNA genes from wheat and barley. Nucleic Acid Res. 7: 1869-1885.
Jacobsen, J.V. and Higgins, T.J.V. 1982. Characterization of the α-amylase synthesized by aleurone layers of Himalaya barley in response to GA_3. Plant Physiol. 70: 1647-1653.
Jacobsen, J.V., Scandalios, J.G., and Varner, J.E. 1970. Multiple forms of amylase induced by gibberellic acid in isolated barley aleurone layers. Plant Physiol. 42: 367-371.
Jacobsen, J.V. and Zwar, J.A. 1974. Gibberellic acid and RNA synthesis in barley aleurone layers: metabolism of rRNA and tRNA and of RNA containing polyadenylic acid sequences. Aust. J. Plant Physiol. 1: 343-356.
Jones, R.L. 1980. The isolation of endoplasmic reticulum from barley aleurone layers. Planta 150:58-69.

Jones, R.L. and Carbonell, J. 1984. Regulation of the synthesis of barley aleurone α-amylase by gibberellic acid and calcium ions. Plant Physiol. 76: 213-218.

Jones, R.L. and Jacobsen, J.V. 1982. The role of the endoplasmic reticulum in the synthesis and transport of α-amylase in barley aleurone layers. Planta 156: 421-432.

Jones, R.L. and Jacobsen, J.V. 1983. Calcium regulation of the secretion of α-amylase isoenzymes and other proteins from barley aleurone layers. Planta 158: 1-9.

Jones, R.L. and Varner, J.E. 1967. The bioassay of gibberellins. Planta 72: 155-161.

Lord, J.M., Kagawa, T., Moore, T.S., and Beevers, H. 1973. Endoplasmic reticulum as the site of lecithin formation in castor bean endosperm. J. Cell Biol. 57: 659-667.

Melroy, D. and Jones, R.L. (1985) The effect of monensin on intracellular transport and secretion of α-amylase isoenzymes in barley aleurone. Planta, in press.

Mitsui, T., Christeller, J.T., Hara-Nishimura, I., and Akazawa, T. 1984. Possible roles of calcium and calmodulin in the biosynthesis and secretion of α-amylase in rice seed scutellar epithelium. Plant Physiol. 75: 21-25.

Moll, B.A. and Jones, R.L. 1980. α-Amylase secretion by single aleurone layers. Plant Physiol. 70: 1149-1155.

Ray, P.M., Shinninger, T.L., and Ray, M.M. 1969. Isolation of β-glucan synthetase particles from plant cells and the identification with Golgi membranes. Proc. Natl. Acad. Sci. USA 64: 605-612.

Rogers, J.C. and Millman, C. 1983. Isolation and sequence analysis of a barley α-amylase cDNA clone. J. Biol. Chem. 258: 8169-8174.

Shore, G. and MacLachlan, G. 1975. The site of cellulose synthesis: hormone treatment alters the intracellular location of alkali-insoluble β, 1-4 glucan (cellulose) synthetase activities. J. Cell Biol. 64: 557-571.

Spencer, D., Higgins, T.J.V., Button, S.C., and Davey, R.A. 1980. Pulse labelling studies on protein synthesis in developing pea seeds and evidence of a precursor form of legume small subunit. Plant Physiol. 66: 510-515.

Varner, J.E. and Mense, R.M. 1972. Characteristics of the process of enzyme release from secretory plant cells. Plant Physiol. 49: 187-189.

CELLULOSE MICROFIBRIL SYNTHESIS AND ORIENTATION IN

OOCYSTIS SOLITARIA: EVIDENCE FOR THE INVOLVEMENT OF CALCIUM

Hartmut Quader

Zellenlehre, Universität Heidelberg
Im Neuenheimer Feld 230
D-6900 Heidelberg 1, FRG

INTRODUCTION

Cell walls, the extracellular skeleton of plant cells, have a vital function in plant morphogenesis. Although numerous investigations have been conducted through the past our understanding of their formation and differentiation still is rather fragmentary and limited. The assembly and deposition of cellulose, the major constituent of plant cell walls is, in particular, of interest. Cellulose assembly involves various subsequent metabolic steps: synthesis of the cellulose synthetase, guided transport of the enzymes to their destination, the plasma membrane, incorporation into and orientation in the plasma membrane, and the movement of the enzymes through the plane of the membrane.

The cellulose microfibrils are deposited in many organisms in a highly oriented pattern during secondary cell wall formation. Since their discovery microtubules have been proposed to play a fundamental role in the orientation of the microfibrils (Heath, 1974, Gunning and Hardham, 1982, Robinson and Quader, 1982). With respect to many organisms evidences in favour of this proposal have accumulated (for references see Robinson and Quader, 1982) but there are some examples, e.g. root hairs (Emons, 1982), in which cellulose orientation is thought to occur without the involvement of microtubules.

Calcium obviously functions as regulatory agent in the assembly/disassembly mechanism of the microtubules (Schliwa et al., 1981). Calcium, therefore, may also play an important role in the microtubule-dependent regular deposition of cellulose though its role in plant microtubule dynamics has, yet, not been demonstrated.

A unique system to study the microtubule-mediated deposition of cellulose is the unicellular green alga Oocystis solitaria (Quader and Robinson, 1981). The cell wall of O.solitaria is synthesized in about 35 hours. The fully-developed cell wall consists of about 30-35 layers of microfibrils indicating that one layer is synthesized, roughly, within 1 hour.

Studies in vitro have, yet, been unsuccessful with any plant material for two reasons: cellulose synthesis has not been achieved and the handling of plant microtubules still is a very critical task. With O.solitaria it is, however, possible to manipulate the regular assembly/disassembly of the cortical microtubules, the synthesis of cellulose per se, and the crystallization of the ß-1,4-glucan chains to cellulose through the treatment with selective inhibitors (Quader and Robinson, 1981, Quader, 1984).

This approach, the application of selective antagonists was also chosen to investigate the posible involvement of calcium in metabolic events concerning the regular deposition of cellulose in O.solitaria. The proposed modes as well as sites of action of the antagonists used are summarized in Table 1.

MATERIAL AND METHODS

The methods employed and the origin of the chemicals used have already been given elsewhere (Quader et al.,1978, Quader, 1982).

Table 1: Substances known to interfere with the function of calcium and which have been used in this study.

Substance	Mode of Action	Site of Action	Reference
EGTA	chelator	extracellular	Caswell (1979)
Cryptate 21 211 221	chelator and ionophore ?	plasma membrane	Lehn et al. (1973)
CTC	chelator/ionophore	membranes	Caswell (1979)
A-23187	ionophore	membranes	Pressman (1976)
APM	enhancement of calcium efflux	mitochondria	Hertel et al. (1981)
Nupercaine	inhibition of calcium efflux	mitochondria	Dawson et al.(1979)
Ruthenium red	inhibition of calcium uptake	plasma membrane mitochondria	Moore (1971) Hinds et al.(1981)
Nifidepine	calcium channel blocker	plasma membrane	Fleckenstein (1977)
Phenothiazines	inhibition of the function of the calcium-calmodulin complex	sites of calmodulin localisation	Klee et al.(1980)

RESULTS

The growth of the cell wall of O.solitaria can be measured either chemically by the procedure introduced by Updegraff (1969), or ultrastructurally by counting the deposited microfibril layers after marking the state of cell wall formation through a short colchicine treatment (Quader et al., 1978). Colchicine, 10 mM, causes a characteristic alteration of the alternating orientation of the layers (see Fig.1-3). The influence of the various substances on cellulose synthesis or microfibril orientation is, then, investigated during the recovery period of the colchicine treatment. The obtained effects of chlorotetracycline (CTC), nupercaine, and the herbicide amiprophosmethyl (APM) on cellulose synthesis are

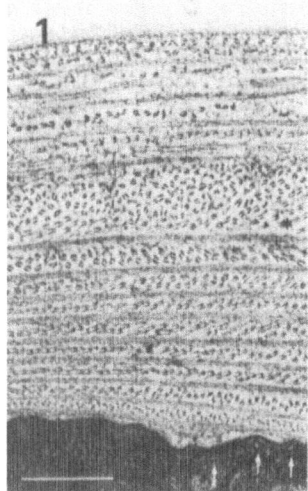 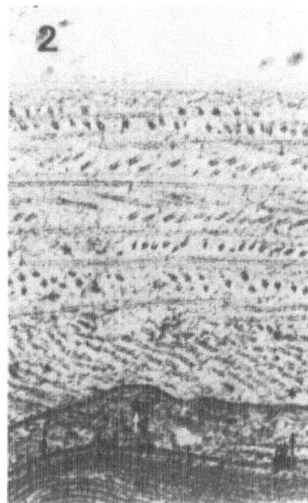 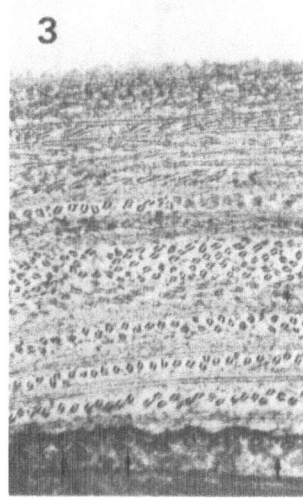

Fig.1 to 3: Small segments of the cell wall and the cortical cytoplasm of Oocystis solitaria after a
6 hs colchicine treatment, 10 mM, followed by
different post-colchicine treatments. The end
of the colchicine treatment is marked by a star.
Bar 0,2 µm

Fig.1: Cell wall development during the colchicine recovery
period in culture medium for 16 hs. Microtubules and
the reestablished alternating microfibril pattern
are visible.

Fig.2: Cell wall development in the presence of chlorotetracycline, 100 µM, after the removal of colchicine.
Microtubules are re-assembled (arrows) but no microfibrils are deposited.

Fig.3: Cell wall development in the presence of nupercaine,
100 µM, for 8 hs followed by a recovery period of
again 8 hs. After the removal of colchicine about 8
microfibril layers have been deposited within 16 hs
due to the inhibitory influence of nupercaine during
the first 8 hs of the post-colchicine time. Microtubules are clearly visible (arrows). Note that the
microtubules are oriented perpendicular to the last
microfibril layer indicating that the microtubules
change their direction distinctly before the new
layer is deposited.

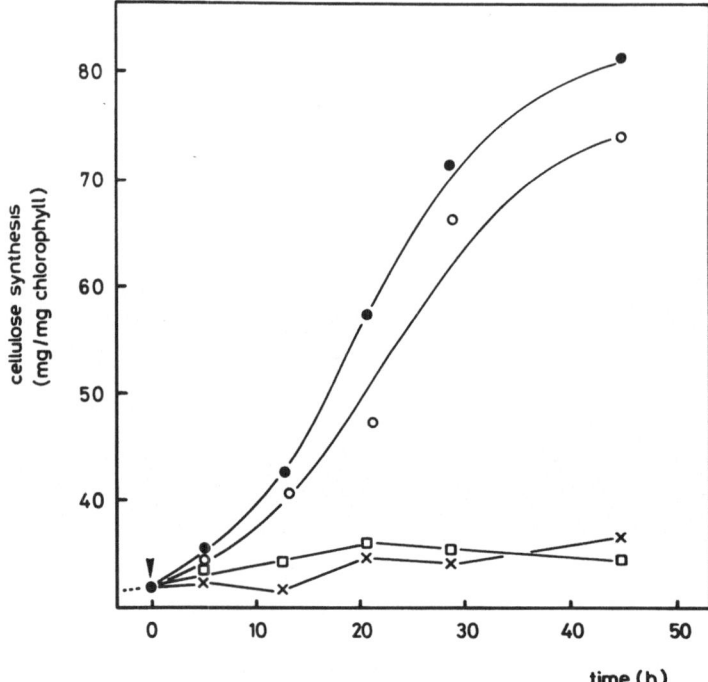

Fig.4: Cell wall development in Oocystis solitaria in the presence of different calcium antagonists. The increase in cellulose was chemically determined.

● —— ●   untreated cultures
○ —— ○   amiprophosmethyl, APM, 10 μM
× —— ×   chlorotetracycline, CTC, 100 μM
□ —— □   nupercaine, 100 μM

shown in Fig.1-3 by the ultrastructural or in Fig.4 by the chemical procedure.

In Table 2 the results are summarized obtained with all the antagonists tested. The substances believed either to interfere with the regular flux of calcium across the particular membranes, or to chelate calcium intracellularly (see Table 1) prevent the synthesis of cellulose but do not affect the recovery of the cortical microtubules after the release from the colchicine treatment. There are, however, two exceptions, EGTA and APM. EGTA apparently is without any effect whereas APM shows a colchicine-like influence. Microtubules remain disassembled and the alternating deposition of the microfibril layers is not restored though microfibrils are synthesized (Quader et al., 1978). Since the action of APM was thought to be due to an enhancement of the free cytoplasmic calcium concentration it was applied in combination with substances which supposingly antagonize its action (Table 2). In all cases the effect of APM is subdued by the other antagonists regarding the influence on cellulose synthesis and microtubule re-assembly.

Many calcium-regulated processes are mediated by calcium binding proteins, e.g. calmodulin (Klee et al.,1980). Fluphenazine and trifluoperazine which interfere with the action of calmodulin inhibit the synthesis of cellulose but seem to

Table 2: Cell wall formation and microtuhule assembly in the presence of substances known to interfere with the regular calcium function administered during the recovery period of a 6 hs colchicine treatment (1o mM) through which the state of cellulose synthesis is marked in single autospores.

Treatment	Microfibril layers deposited within 6 h	microtubules re-assembled
none	6 - 7	+
EGTA (10 mM)	5 - 7	+
CTC (100 µM)	1 - 2	+
Cryptates (1 mM)	0 - 1	+
APM (10 µM)	5 - 7	-
Nupercaine (100 µM)	0 - 2	+
Ruthenium red (10 µM)	0 - 2	+
A-23187 (10 µM)	0 - 1	+
Nifidepine (10 µM)	1 - 2	+
APM + EGTA [*]	5 - 7	-
APM + CTC	1 - 2	+
APM + Nupercaine	1 - 2	+
APM + Ruthenium red	3 - 4	+/-
APM + A-23187	1 - 2	+

[*] The first mentioned substance was applied immediately after the removal of colchicine, the second followed after 1 hour.

promote the re-assembly of the cortical microtubules (Table 3) The latter is evident from studies in which fluphenazine has been applied to APM-treated cells (Table 3). The number of reformed microtubules equals that of untreated cells (not shown). Interestingly, the more lipophilic derivative deca-noate-fluphenazine does not affect cellulose deposition (Table 3).

DISCUSSION

The failure of Oocystis solitaria to form a regular cell wall in the presence of antagonists of calcium functions indicates that calcium may play an important role in cellulose deposition. The antagonists studied are each thought to affect distinct activities at different cellular sites. The maintenance of the cytoplasmic free calcium concentration is a result of a very fine regulated balance between the Ca-channels, Ca-pumps, and Ca-binding proteins that regulate the flux of calcium in and out as well as within the cell (Akerman and Nicholls, 1983).
There are two major problems in evaluating the obtained results: firstly, how specific/selective is the believed function of each substance, and secondly the cells, of course, try to counteract any interference, e.g. an induced enhancement of calcium efflux frommitochondria may be balanced

Table 3: Cellulose synthesis and microtubule re-assembly in
the presence of amiprophosmethyl or/and phenothiazines administered during the recovery period of a
6 hs colchicine treatment (10 mM).

Treatment	microfibril layers deposited within 6 h	microtubules re-assembled
none	5 - 7	+
APM (10 µM)	5 - 7	-
Fluphenazine (25 µM)	1 - 2	+
Trifluoperazine (25 µM)	0 - 1	+
D-Fluphenazine (25 µM)	5 - 7	+
Fluphenazine + APM[*)]	0 - 1	+
APM + Fluphenazine	1 - 2	+

*) The first mentioned substance was applied immediately
after the removal of colchicine, the second followed after
1 hour.

through an increased uptake by the ER or/and by pumping more
calcium to the exterior. Naturally, this will modulate other
simultaneously occuring metabolic events, in particular, if
they are activated by calcium and calmodulin.
Following observations are in favour of a, at least, decent
selectivity of the antagonists: The autospores recover from
the treatments in a very short time (Quader and Robinson,
1981) and substances structurally almost identical with those
employed show no effect, e.g. the cryptate 222 which his highly specific for potassium (Quader and Robinson, 1981).
    It is,therefore, strongly suggested that with respect
to the assembly/disassembly cycle of the cortical microtubules in O.solitaria calcium as well as calmodulin functions as
regulating agent. Treatments believed to cause a decrease or
to interfere with the necessary change of the cytoplasmic
calcium concentration either destabilize, e.g. APM, or stabilize the microtubules, e.g. nupercaine, CTC, nifidepine
(Table 2). This is most evident from studies in which APM
is applied in combination with an other antagonist (Table 2).
In the presence of fluphenazine the same number of microtubules was observed, they are even a little higher, as in untreated cells (H.Quader, in preparation).
    What is the role of calcium in cellulose synthesis? The
chemicals may interfere with the integrity of the membrane.
In the presence of A-23187, CTC, and fluphenazine the terminal complexes, the putative cellulose synthesizing enzyme
complexes, loose their characteristic morphology when applied
shortly before a new microfibril layer is produced (Robinson
and Quader, 1981, Quader, in preparation). On the other hand,
it is well known that the influx of calcium is a necessity
for the fusion of secretory vesicles with the plasma membrane
(Akerman and Nicholls, 1983). The inhibition of microfibril
formation may be the result of an interference with the regular calcium influx which is not difficult to imagine with

respect to nifidepine, CTC, or ruthenium red known to affect calcium entry (see references given in Table 1), thereby, preventing the fusion of secretory vesicles which carry the enzymes for the subsequent microfibril layer. Terminal complexes linked to microfibril imprints are morphologically less affected. Unfortunately, at the moment, it is not possible to verify if cellulose synthesis becomes immediately blocked after the application of the antagonists or only when cell wall formation ought to commence by starting a new microfibril layer implying that the layer in synthesis will be terminated. The procedure to measure cellulose chemically is to coarse to determine the small increase in cellulose in the case of the imagined microfibril layer termination. Since the cortical microtubules appear to be stabilized through the treatment with antagonists which inhibit cellulose synthesis secretory vesicles may also spatially be restrained to fuse with the plasma membrane.

Acknowledgement: Miss C. Angerstein is thanked for technical assistance and the Deutsche Forschungsgemeinschaft for financial support.

REFERENCES

Akerman, K.E.O., Nicholls, D.G., 1983, Ca^{2+}-transport and the regulation of transmitter release in isolated nerve endings, TIBS, 8:63-64.

Caswell, A.H., 1979, Methods of measuring intracellular calcium, Int.Rev.Cytol., 56:145-181.

Dawson, A.P., Selwyn, M.J., Fulton, D.V., 1979, Inhibition of Ca-efflux from mitochondria by nupercaine and tetracaine, Nature, 277:484.

Emons, A.M.C., 1982, Microtubules do not control microfibril orientation in a helicoidal cell wall, Protoplasma, 113: 85-87.

Fleckenstein, A.A., 1977, Specific pharmacology of calcium in myocardium, cardiac pacemakers, and vascular smooth muscle, Ann.Rev.Pharm.Tox., 17:149-166.

Gunning, B.E.S., Hardham, A.R., 1982, Microtubules, Ann.Rev. Plant Physiol., 33:651-698.

Heath, I.B., 1974, A unified hypothesis for the role of membrane bound enzyme complexes and microtubules in plant cell wall synthesis, J.theor.Biol., 48:445-449.

Hertel, C., Quader, H., Robinson, D.G., Marmé, D., 1980, Antimicrotubular herbicides and fungicides affect Ca-transport in plant mitochondria, Planta, 149:336-340.

Hinds, T.R., Raess, U.B., Vincenzi, F.F., 1981, Plasma membrane Ca-transport: antagonism by several potential inhibitors, J.Membr.Biol., 58:57-65.

Klee, C.B., Crouch, T.H., Richman, P.G., 1980, Calmodulin, Ann.Rev.Biochem., 49:489-515.

Lehn, J.M., Dietrich, B., Sauvage, J.P., 1973, Cryptates - XI complexes macrobicycliques, formation, structure, propriétés, Tetrahedron, 29:1647-1658.

Moore, C.L., 1971, Specific inhibition of mitochondrial Ca-transport by ruthenium red, Biochem.Biophys.Res.Commun., 42:298-301.

Pressman, B.C., 1976, Biological applications of ionophores, Ann.Rev.Biochem., 45:501-530.

Quader, H., The microtubule-microfibril-syndrome in Oocystis solitaria: regulatory aspects, in: "Microtubules in Microorganisms", P.Cappucinelli, H.R.Morris, eds., Marcel Dekker, New York, 313-324.

Quader, H., 1983, Morphology and movement of cellulose synthesizing (terminal) complexes in Oocystis solitaria: evidence that microfibril assembly is the motive force, Eur.J.Cell Biol., 32:174-177.

Quader, H., Robinson, D.G., 1981, Oocystis solitaria: a model organism for understanding the organization of cellulose synthesis, Ber.Deutsch.Bot.Ges., 94:75-84.

Quader, H., Wagenbreth, I., Robinson, D.G., 1978, Structure, synthesis, and orientation of microfibrils: V. On the recovery of Oocystis solitaria from microtubule inhibitor treatment, Cytobiologie, 18:39-51.

Robinson, D.G., Quader, H., 1982, The microtubule-microfibril-syndrome, in: "The Cytoskeleton in Plant Growth and Development", C.W.Lloyd, ed., Acad. Press, London, 109-126.

Schliwa, M., Euteneuer, U., Bulinski, J.C., Izant, J.G., 1981, Calcium lability of cytoplasmic microtubules and its modulation by microtubule-associated proteins, Proc. Natl.Acad.Sci.USA, 78:1037-1041.

Updegraff, D.M., 1969, Semimicro determination of cellulose in biological materials, Anal.Biochem., 32:420-424.

EFFECT OF CALCIUM ON LIPID METABOLISM DURING THE GROWTH OF A CALCIFUGE

PLANT (Lupinus luteus L.) OR A CALCICOLOUS ONE (Vicia faba)

Annette Oursel and Brigitte Citharel

Laboratoire de Physiologie Cellulaire (ERA 323)

Université P. & M. Curie, Paris, France

INTRODUCTION

It has been generally assumed that the major, if not the only role of lipids in biological membranes is to provide a semipermeable barrier between intracellular and extracellular environments. Moreover, numerous data established that phospholipids generally have ionophoric capability.

Using plant species typically sensitive to an excess of calcium (calcifuge) or not sensitive to it (calcicolous), we attempted to analyse this difference by studying : i) the action of calcium on lipid metabolism and ii) the absorption of this cation on phospholipid bilayers.

MATERIALS AND METHODS

Horse bean (Vicia faba L. var minor) and Lupin (Lupinus luteus L.) seedlings were used for our own experiments (OURSEL, 1979) and for those of ROSSIGNOL (1984). Seedlings were grown on a nutritive medium (OURSEL et al., 1973) and the calcium concentrations used were 0.1 mM in the normal medium and 50 mM in the calcium enriched medium. In vivo and in vitro incubations, lipid analyses and radioactivity counting were performed as previously described (OURSEL, 1979 ; CITHAREL et al., 1983). Pure phospholipid preparations were obtained as described by ROSSIGNOL (1982, 1984).

RESULTS

Inhibition by calcium of phospholipid metabolism in Horse bean roots

When the amount of calcium increased in the growing medium of yellow lupin plants, the lipid composition of their membranes was not modified. On the contrary, the phospholipid content of the membranes of Horse bean roots decreased as calcium increased in the medium (fig. 1).

To explain this fact we have investigated the action of calcium ions on the biosynthesis of two classes of phospholipids : phosphatidylcholine (PC) and phosphatidylethanolamine (PE). We have studied the effect of calcium on the activities of two enzymes involved in the biosynthesis of these membrane phospholipids : CDP-choline-diacylglycerol-choline-phosphotransferase (EC-2.7.8.2.) and CDP-ethanolamine-diacylglycerol-ethanolamine-phosphotransferase (EC-2.7.8.1.).

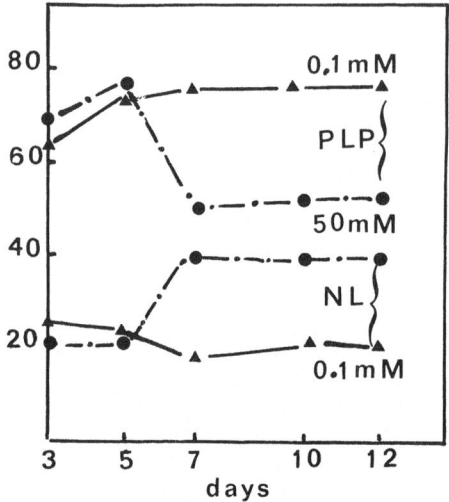

Fig. 1. Variation of lipid composition with calcium content of the growth medium (0.1 to 50 mM calcium chloride). Some plants, seven days old, are again put of on a normal medium (0.1 mM calcium chloride).

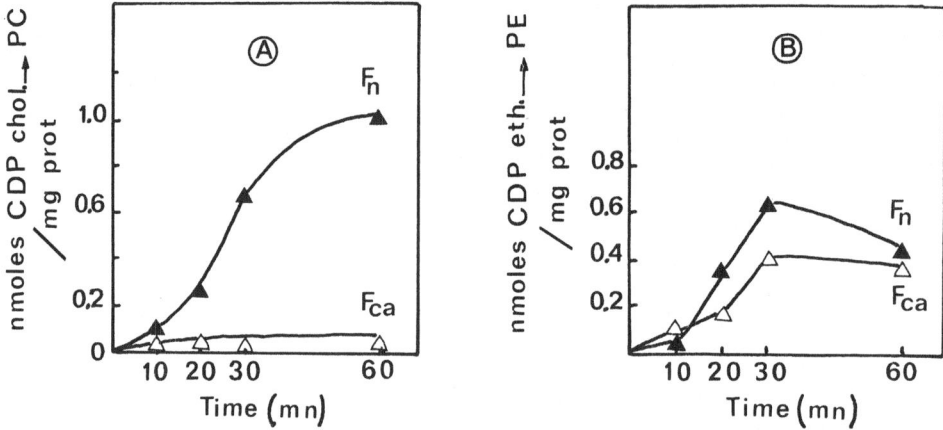

Fig. 2. Incorporation of $\left[^{14}C\right]$-choline (A) or $\left[^{14}C\right]$-ethanolamine (B) into the phospholipids (PC or PE) of Horse bean root microsomes. F_N = microsomes from roots grown on a normal medium (0.1 mM $CaCl_2$) ; F_{Ca} = microsomes from roots grown on a 50 mM $CaCl_2$ medium.

The phosphotransferase activities of microsomal preparations from Horse bean grown on a medium poor in calcium (F_N, 0.1 mM $CaCl_2$) were more important than those from plants grown on a medium enriched in calcium (F_{Ca}, 50 mM $CaCl_2$) (fig. 2). This was particularly obvious for the CDP-choline phosphotransferase which was 100 times more active in plantlets grown on a normal medium. In these experiments, the incubation media did not contain any calcium ; we can thus suppose that it is the calcium fixed _in vivo_ on the microsomal membranes of plants which inhibited phospholipid synthesis. When microsomes of plantlets grown on a normal medium were incubated in the presence of increasing amounts of $CaCl_2$ (0 to 1.50 mM),

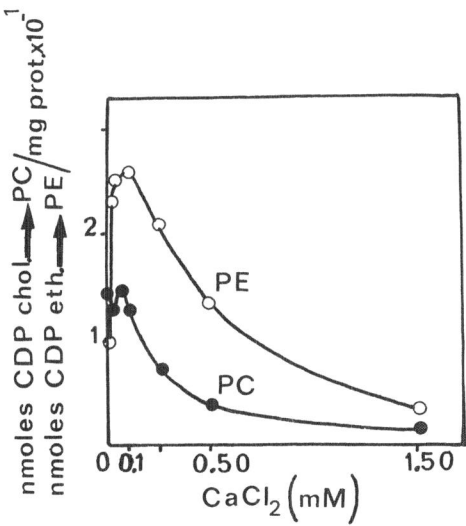

Fig. 3. Influence of calcium ions added to the incubation medium on the activities of the CDP-choline (●——● PC) and the CDP-ethanolamine-transferases (○——○ PE) of Horse bean root microsomes.

inhibition of the enzymatic activities could be observed for $CaCl_2$ concentrations greater than 0.1 mM. The inhibition was virtually complete with 1.50 mM (fig. 3).

Effect of calcium on oleate and linoleate desaturase activities into yellow Lupin roots

Fatty acid desaturation varies according to stages of root growth

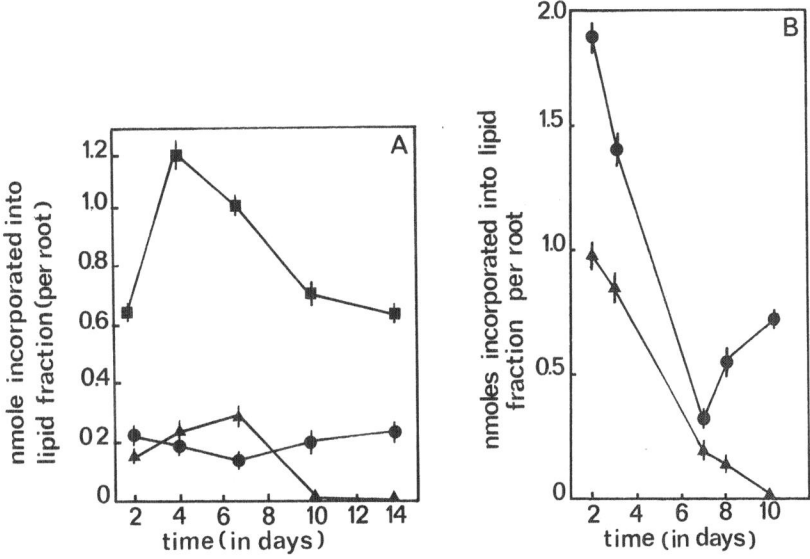

Fig. 4. Variation of oleate (A) or linoleate (B) desaturase activities with root age. (■——■) $[^{14}C]$oleoyl ; (●——●) $[^{14}C]$linoleoyl ; (▲——▲) $[^{14}C]$linolenoyl residues.

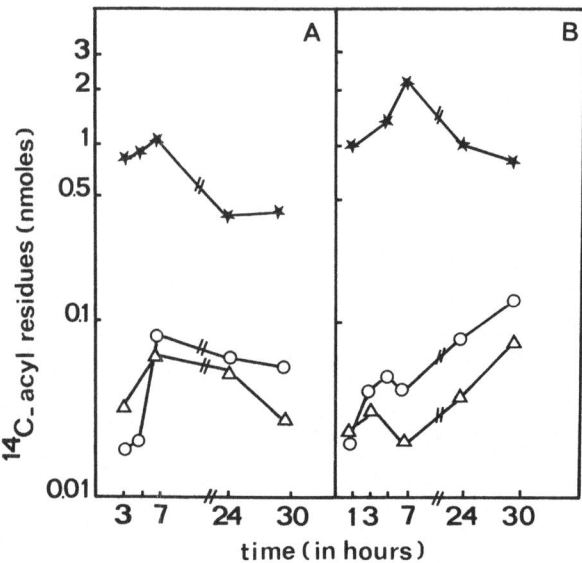

Fig. 5. Acylation of $[^{14}C]$oleate (A) or ^{14}C linoleate (B) into PC and PE and incorporation into neutral lipids. (✶——✶) $[^{14}C]$ free fatty acid or $[^{14}C]$ acyl residues linked to neutral lipids ; (△——△) $[^{14}C]$ acyl residues linked to PC ; (○——○) $[^{14}C]$ acyl residues linked to PE.

particularly with $[^{14}C]$ linoleate as a precursor (fig. 4). About 10 % of the $[^{14}C]$ oleate furnished was incorporated into phospholipids mainly PC and PE. PC was more heavily labelled at the beginning of incubation but after 7 hrs, PE was always more intensively labelled than PC (fig. 5 A). $[^{14}C]$ linoleate was also esterified into PC and PE. The radioactivity of both phospholipids increased almost continuously with time (fig. 5 B).

When esterified in phospholipid molecules, the labelled acyl residues were desaturated. We have observed that $[^{14}C]$ oleate incorporation into root lipids, acylation into phospholipids and desaturation into $[^{14}C]$ linoleate increased when 10 day old seedlings were grown on a medium containing 50 mM calcium chloride instead of 0.1 mM as usual (fig. 6).

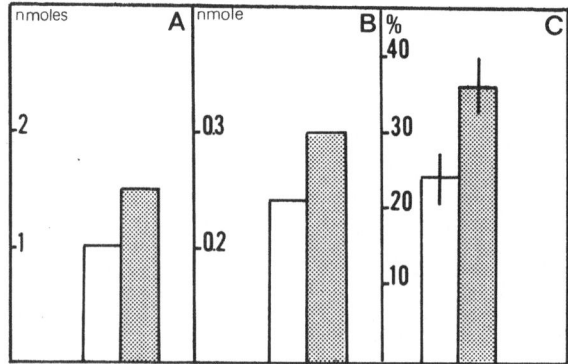

Fig. 6. Incorporation into roots (A), acylation into phospholipids (B) and desaturation (C) of $[^{14}C]$ oleate in yellow lupin roots (10 day old) grown on either a normal (0.1 mM, ☐) or calcium enriched (50 mM, ▨) medium.

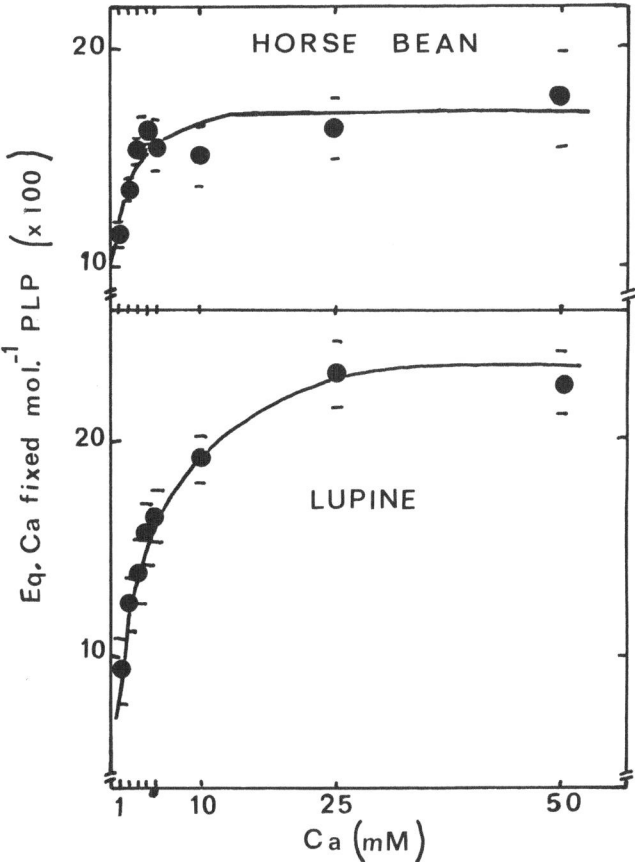

Fig. 7. Influence of the calcium concentration on its binding by root
phospholipids. pH of the medium is kept constant at 7.4 (ROSSIGNOL, 1984).

Phospholipids as ionophores

According to GREEN et al. (1980) phospholipids are ionophoric species
in biological membranes. They interact with cations to form complexes which
are soluble in the membrane and which can then cross it. TYSON et al.
(1976) established that cardiolipid (DPG) and phosphatidic acid (PA) are
highly efficient ionophores for divalent and monovalent cations. CULLIS
et al. (1980) proposed a potential carrier mechanism which proceeds via
an intermediary intrabilayer inverted micellar Ca^{2+}-phospholipid complex.
More recently, ROSSIGNOL (1984) has shown that root phospholipids can
interact with calcium. Binding of this cation is dependent on the negative
charges. Lupin root phospholipids exhibit more negative charges than Horse
bean root phospholipids. The former binds twice as much calcium as the
latter (fig. 7). This binding corresponds to interaction between calcium
ions and polar heads. Three acidic phospholipids have a particular role
on this calcium binding in Lupin roots: phosphatidic acid (PA), diphos-
phatidylglycerol (DPG) and phosphatidylserine (PS).

CONCLUSION

To explain the reactions of plants to many environmental stresses an
important role has been proposed for membranous lipids (LEVITT, 1980).

All classes of lipids would be involved : sterols, sulfolipids, phospholipids, glycolipids and neutral lipids (KUIPER, 1980). However, phospholipids and unsaturated fatty acids would play a special role in plant reactions (LAMANT and HELLER, 1975 ; MULLER and SANTARIUS, 1978 ; STUIVER et al., 1982).

On the other hand, many results have shown a direct effect of the composition of the growing medium of plants on lipid metabolism (GHARSALLI, 1978 ; BETTAIEB et al., 1980). Lipid modifications are observed when resistant species are put in stress conditions (LAMANT and HELLER, 1975 ; ELLOUZE, 1977 ; WILLEMOT, 1979).

Similarly effects of calcium on lipid metabolism have been observed (OURSEL, 1979 ; CITHAREL et al., 1982). These effects are specific in the case of the Lupin-Horse bean pair and they show how more sensitive is the calcifuge plant in a calcium enriched medium. On the same medium, lipid modifications could allow a better control of membrane permeability in the calcicolous plant (SALSAC and LAMANT, 1973 ; ROSSIGNOL, 1984).

REFERENCES

Bettaieb, L., Gharsalli, M., and Cherif, A., 1980, Effect of sodium chloride and calcium sulfate on the lipid composition of sunflower leaves (Helianthus annuus L.), in: Mazliak P., Benveniste, P., Costes, C. and Douce, R., ed., Elsevier Biomedical Press, Amsterdam, p. 243.

Citharel, B., Oursel, A., and Mazliak, P., 1982, Effect of calcium on the biosynthesis of linoleic and linolenic acids during the growth of a calcifuge plant (Lupinus luteus L.), in: "Biochemistry and metabolism of plant lipids", Wintermans, J.F.G.M. and Kuiper, P.J.C., ed., Elsevier Biomedical Press, Amsterdam, p. 39.

Citharel, B., Oursel, A., and Mazliak, P., 1983, Desaturation of oleoyl and linoleoyl residues linked to phospholipids in growing roots of yellow lupin. FEBS Lett., 161:251.

Cullis, P.R., De Kruijff, B., Hope, M.J., Nayar, R., and Shmid, S.L., 1980, Phospholipids and membrane transport. Can. J. Biochem., 58:1091.

Ellouze, M., 1977, Contribution à l'étude de l'action du chlorure de sodium sur la composition lipidique des feuilles de quatre espèces de Citrus. D.E.A., Tunis, pp. 42.

Gharsalli, M., 1978, Contribution à l'étude de l'action du chlorure de sodium sur les lipides des feuilles et des racines de Tournesol (Helianthus annuus L.). D.E.A., Tunis, pp. 35.

Green, D.E., Fry, M., and Blondin, G.A., 1980, Phospholipids as the molecular instruments of ion and solute transport in biological membranes. Proc. Natl. Acad. Sci. U.S.A., 77:257.

Kuiper, P.J.C., 1980, Lipid metabolism as a factor of environmental adaptation, in: "Biogenesis and function of plant lipids. Mazliak, P., Benveniste, P., Costes, C., and Douce, R., ed., Elsevier Biomedical, Amsterdam, p. 169.

Lamant, A., and Heller, R., 1975, Intervention des systèmes membranaires dans l'absorption du calcium par les racines de Féverole (calcicole) et de Lupin (calcifuge). Physiol. Vég., 13:685.

Levitt, J., 1980, Responses of plants to environmental stresses. Academic Press, New York, II, p. 533.

Muller, M., and Santarius, K.A., 1978, Changes in chloroplast membrane lipids during adaptation of barley to extreme salinity. Plant Physiol., 62:326.

Oursel, A., Lamant, A., Salsac, L., and Mazliak, P., 1973, Etude comparée des lipides et de la fixation passive du calcium sur les racines et les fractions subcellulaires du Lupinus luteus et de la Vicia faba. Phytochemistry, 12:1865.

Oursel, A., 1979, Effets du calcium sur le métabolisme des lipides dans les racines de Féverole (plante calcicole) ou de Lupin (plante calcifuge). Thèse Doct. Etat, Paris, pp. 100.

Rossignol, M., Grignon, N., and Grignon, C., 1982, Effect of temperature and ions on the microviscosity of bilayers from natural phospholipids mixtures. Biochimie, 4:263.

Rossignol, M., 1984, Relations entre la structure et la perméabilité des membranes phospholipidiques : effets du calcium et des protons sur les phospholipides de racines. Thèse Doct. Etat, Montpellier, pp. 270.

Salsac, L., and Lamant, A., 1973, Etude des échanges protons-calcium dans les racines d'une plante calcicole (Féverole) et d'une plante calcifuge (Lupin jaune). Oecol. Plant, 8:263.

Stuiver, C.E.E., De Kok, L.J., Hendriks, A.E., and Kuiper, P.J.C., 1982, The effect of salinity on phospholipid content and composition of two Plantago species, differing in salt tolerance, in: Biochemistry and metabolism of plant lipids, Wintermans, J.F.G.M., and Kuiper, P.J.C., ed., Elsevier Biomedical, Amsterdam, pp. 455.

Tyson, C.A., Van de Zande, H., and Green, D.E., 1976, Phospholipids as ionophores. J. Biol. Chem., 251:1326.

Willemot, C., 1979, Chemical modification of lipids during frost hardering of herbaceous species, in: Low temperature stress in crop plants, Lyons, J.M., Graham, D. and Raison, J.K., ed., Academic Press, New York, p. 411.

SUB-CELLULAR RESPONSES TO CALCIUM

CALCIUM-REGULATED ENZYMES

PLANT LEAF CALCIUM-DEPENDENT PROTEIN KINASES

Gideon M. Polya, Vito Micucci, Steven Basiliadis,
Trevor Lithgow and Anna Lucantoni

Department of Biochemistry
La Trobe University
Bundoora, Victoria, 3083, Australia

INTRODUCTION

Protein phosphorylation catalyzed by cyclic nucleotide- or Ca^{2+}-dependent protein kinases represents a major means by which external stimuli change animal cellular processes (Krebs and Beavo, 1979; Cohen, 1982; Reichardt and Kelly, 1983; Nishizuka, 1984). Cyclic nucleotide-dependent protein kinase has not been unequivocally resolved from higher plants (Brown and Newton, 1981; Kato et al., 1983) although many possible elements of a cyclic nucleotide-regulatory system have been found in plants (Brown and Newton, 1981; Polya and Bowman, 1981; Francko, 1983; Newton et al., 1984). Ca^{2+} may act as a second messenger in plant as in animal cells (Dieter, 1984). A variety of agents have been demonstrated or inferred to change cytosolic free Ca^{2+} concentration in plant cells with consequent changes in plant cellular processes (Dieter, 1984). Plants contain calmodulin (Cormier et al., 1981) and various Ca^{2+}-calmodulin- or Ca^{2+}-activated enzymes, including Ca^{2+}-dependent protein kinases (Dieter, 1984).

Soluble Ca^{2+}-dependent and Ca^{2+}-calmodulin-activated protein kinases have been partially-purified from wheat embryo (Polya and Davies, 1982; Polya et al., 1983; Polya and Micucci, 1984) and from silver beet leaves (Polya et al., 1985b). Ca^{2+}-dependent phosphorylation of plant membrane proteins by endogenous protein kinases has been demonstrated (Hetherington and Trewavas, 1982, 1984; Salimath and Marmé, 1983; Polya et al., 1984; Veluthambi and Poovaiah, 1984a,b,c; Polya et al., 1985a). Soluble proteins in crude plant extracts are phosphorylated in Ca^{2+}-dependent reactions (Veluthambi and Poovaiah, 1984a,b,c; Polya et al., 1985a). Histone phosphorylation catalyzed by soluble and particulate fractions from germinated Nicotiana alata pollen is largely Ca^{2+}-dependent as is the phosphorylation of a number of soluble and membrane-bound pollen proteins (Polya et al., 1985a). Plant quinate:NAD^+ oxidoreductase from carrot cells is clearly activated by phosphorylation (Refeno et al., 1982a,b) and appears to be activated by phosphorylation catalysed by Ca^{2+}-calmodulin-activated protein kinase (Ranjeva et al., 1983; Graziana et al., 1983a). Quinate:NAD^+ oxidoreductase from dark-grown carrot cells is directly Ca^{2+}-activated and contains an additional Ca^{2+}-binding subunit that protects the phosphoenzyme from dephosphorylation (Graziana et al., 1983b, 1984); relatively high free Ca^{2+} concentration (0.7 mM) is required for maximal activation of the enzyme from dark-grown cells (Graziana et al., 1984).

The soluble Ca^{2+}- and Ca^{2+}-calmodulin activated protein kinases (termed protein kinases I and II) from wheat embryo are fully activated at free Ca^{2+} concentrations of about 10^{-5} M (Polya et al., 1983; Polya and Micucci, 1984) i.e. by the free Ca^{2+} concentrations observed in the cytosol of excited (as opposed to resting) plant cells (Williamson and Ashley, 1982). A similar requirement for low free Ca^{2+} concentrations of about 10^{-6} M to 10^{-5} M has been found for Ca^{2+}-dependent phosphorylation associated with plant membranes (Hetherington and Trewavas, 1982; Salimath and Marmé, 1983; Polya et al., 1984) and Ca^{2+}-dependent casein peptide phosphorylation catalyzed by soluble and particulate fractions from germinated pollen (Polya et al., 1985a). These enzymes are therefore likely to be involved in plant responses to external stimuli that cause the transient elevation of intracellular free Ca^{2+} concentrations.

Photosynthetic as well as non-photosynthetic plant cells respond to particular environmental stimuli in processes determined or surmised to involve changes in intracellular Ca^{2+} concentrations (Dieter, 1984). We have therefore compared the properties of a soluble Ca^{2+}-dependent protein kinase and a solubilized particulate Ca^{2+}-dependent protein kinase from silver beet leaves with those of the soluble wheat embryo Ca^{2+}-dependent protein kinases.

MATERIALS AND METHODS

Protein Kinase Assays and Analysis of Protein Phosphorylation

Protein phosphorylation was determined radiochemically at $30^{\circ}C$; $[^{32}P]$-labelled phosphorylated peptides were precipitated on paper disks in 15% trichloroacetic acid (Polya et al., 1983). Phosphorylated basic oligopeptides (Kemptide and S6-1) were bound to phosphocellulose paper (Whatman P81) with subsequent successive washes in 75 mM H_3PO_4. The standard protein kinase assay conditions and the phosphoamino acid analysis and protein determination procedures were as described previously (Polya et al., 1983).

Materials

Mature leaves of silver beet (Beta vulgaris) were purchased locally. Calmodulin was purified from rat brains by Ca^{2+}-dependent chromatography on phenyl-Sepharose CL-4B as described previously (Polya et al., 1983). $[\gamma^{-32}P]$-ATP (3 Ci/mmol) was obtained from the Radiochemical Centre, Amersham, U.K. Fluphenazine was obtained from Squibb (Australia), heparin from the Commonwealth Serum Laboratories, Melbourne and calf thymus histones H1 and H2B from Boehringer, West Germany. The synthetic peptide S6-1 (ArgArgLeuSer SerLeuArgAla) which has the sequence of the major phosphorylated region of rat liver ribosomal protein S6 (Wettenhall and Morgan, 1984) was kindly supplied by Dr. R.E.H. Wettenhall, La Trobe University. Other substrates (including Kemptide and mixed calf thymus histones designated as type IIAS), nucleotides and calmodulin antagonists were obtained from the Sigma Chemical Co., St. Louis, USA.

Partial Purification of Embryo and Leaf Ca^{2+}-Dependent Protein Kinases

All operations were conducted at $0-4^{\circ}C$. About 100 g de-veined silver beet leaf tissue was homogenized in 500 ml Buffer A (50 mM Tris (Cl^{-}, pH 8.0) - 10 mM 2-mercaptoethanol) containing 0.5 mM phenylmethylsulphonylfluoride and 0.25% (v/v) ethanol and the homogenate filtered through muslin and Miracloth. The filtered homogenate was centrifuged at 40,000 g/30 min and the resulting supernatant added to 50 g DEAE-Sephacel. After washing with 2 l buffer A, the protein kinase was eluted in 0.2 M NaCl-buffer A. This fraction was concentrated by ultrafiltration, applied to an Ultrogel AcA 44 column (7 cm^2 x 55 cm) and the protein kinase eluted in 0.2 M NaCl-buffer A. The fractions containing Ca^{2+}-dependent histone kinase were pooled, diluted 5-fold in H$_2$O and applied to a DEAE-Sephacel column (7 cm^2 x 5 cm). The protein kinase was eluted in a gradient of increasing NaCl concentration in buffer A or by stepwise elution (the enzyme was eluted by 0.1 M to 0.15 M NaCl in buffer A).

The particulate fraction from centrifugation of the initial homogenate at 15,000 g/30 min was resuspended in 0.9%-Nonidet P-40 in buffer B (50 mM Tris (Cl$^-$, pH 7.4) - 10 mM 2-mercaptoethanol) and re-centrifuged at 15,000 g/30 min. The supernatant was applied to DEAE-Sephacel (20 g wet weight) equilibrated with buffer B and after washing with 300 ml buffer B, the protein kinase was eluted in 0.075 M NaCl-buffer B. The eluate was concentrated by ultrafiltration, applied to an Ultrogel AcA 44 column (7 cm^2 x 55 cm) and the protein kinase eluted in 0.2 M NaCl-buffer A as a single peak of Ca^{2+}-dependent kinase activity.

Protein kinase I was partially purified from a wheat embryo chromatin-containing fraction by a variation of the procedure previously described (Polya and Davies, 1982). The chromatin extract in 0.5 M NaCl-buffer A was diluted 5-fold and passed through 30 g DEAE-cellulose (Whatman DE52). The eluate was then diluted a further 10-fold and added to a further 30 g DE52. The protein kinase was eluted from DE52 in 0.2 M NaCl-buffer A, concentrated by precipitation at 90% (NH$_4$)$_2$SO$_4$ saturation and then chromatographed on a Sephacryl S-200 column in buffer A - 1 mM EGTA to resolve Ca^{2+}-dependent protein kinase I from Ca^{2+}-independent protein kinase.

RESULTS

Partial Purification of the Soluble Leaf Ca^{2+}-dependent Protein Kinase

The soluble Ca^{2+}-dependent protein kinase was partially-purified from silver beet leaves by a protocol involving step-wise elution from DEAE-Sephacel, gel filtration and a second elution from DEAE-Sephacel. The gel filtration step yields a single peak of Ca^{2+}-dependent histone kinase activity. While the histone kinase activity is always largely Ca^{2+}-dependent at this stage, the degree of Ca^{2+}-dependence of casein phosphorylation varied between enzyme preparations. Thus in some preparations the peak of Ca^{2+}-dependent histone kinase exactly co-purified with a peak of Ca^{2+}-dependent casein kinase but in most preparations the co-purifying peak of casein kinase activity was largely Ca^{2+}-independent and exhibited Ca^{2+}-dependence only on the advancing edge. This latter phenomenon evidently derives from the presence of an inhibitor of net Ca^{2+}-dependent casein phosphorylation since Ca^{2+}-dependence of casein phosphorylation catalyzed by such preparations can be elicited by appropriate dilution of the enzyme preparations. A subsequent second chromatography on DEAE-Sephacel also elicits Ca^{2+}-dependence of casein phosphorylation and this purification step was therefore routinely employed in this study. The overall purification scheme yields partially-purified Ca^{2+}-dependent protein kinase with a specific activity of about 0.2 nmol min^{-1} mg protein^{-1} with casein as substrate as compared to specific activities of the order of 1 μmol min^{-1} mg^{-1} for homogeneous preparations of Ca^{2+}-independent plant protein kinases (Davies and Polya, 1983; Yan and Tao, 1982). The Ca^{2+}-dependent casein- and histone-phosphorylating activities exactly copurify on gradient elution from DEAE-Sephacel and on gel filtration and we conclude that these reactions are catalyzed by the same enzyme. The molecular weight of the Ca^{2+}-dependent protein kinase (as determined from gel filtration of the enzyme in 0.2 M NaCl-buffer A on an Ultrogel AcA 44 column) is 49,000 (mean from 8 determinations).

Table 1. Inhibition of the Soluble Leaf Protein Kinase

Addition[a]	Protein kinase[b] (% control)	Addition[a]	Protein kinase[b] (% control)
None	100	None	100
$CeCl_3$	14	ITP	5
La(acetate)$_3$	18	GTP	8
$(NH_4)_2SO_4$ (0.1 M)	29	CTP	3
$(NH_4)_2SO_4$ (0.25 M)	9	UTP	1
NaCl (0.25 M)	43	Trifluoperazine	25
Spermidine (25 mM)	52	Fluphenazine	19
Spermine (20 mM)	44	Chlorpromazine	28
ADP	3	Calmidazolium	25

[a]Protein kinase was assayed in triplicate in the standard reaction medium containing 1 mg/ml casein and 1.25 mM $CaCl_2$ in the presence or absence of the indicated additions (all at 1 mM final except where indicated otherwise)

[b]Protein kinase activity is expressed as % control activity (no additions)

Activation and Inhibition of the Soluble Leaf Protein Kinase

With casein as substrate the protein kinase exhibits a broad pH-activity profile over the pH range 6-9 with pH optima at pH 6.5 and pH 8.5. As found with the soluble wheat germ Ca^{2+}-dependent protein kinases (Polya et al., 1983; Polya and Micucci, 1984), casein phosphorylation catalyzed by the leaf enzyme is markedly activated by free Ca^{2+} concentrations above about 10^{-7} M, half-maximal and maximal activation being obtained at 1 μM and 10 μM respectively. As found with the wheat embryo enzymes, much higher free Ca^{2+} concentrations are required to activate histone phosphorylation, namely 0.2 mM and 1.0 mM for half-maximal and maximal activation, respectively. Maximal activation requires millimolar Mg^{2+} in addition to Ca^{2+}: rates of casein phosphorylation (in the presence of 1 mM Ca^{2+}) with 0 and 1 mM $MgCl_2$ are 2% and 33%, respectively of the rate with 10 mM $MgCl_2$ added. Mn^{2+} can largely replace the (Ca^{2+} + Mg^{2+}) requirement and with 1 mM $MnCl_2$ present the protein kinase activity is Ca^{2+}-independent and is ca. 50% of the activity with 1 mM Ca^{2+} plus 10 mM $MgCl_2$.

The protein kinase is inhibited by 10^{-3} M Ce^{3+} and La^{3+} and by high ionic strength (Table 1). The phenothiazine-derived calmodulin antagonists trifluoperazine, fluphenazine and chlorpromazine inhibit the enzyme (Table 1), the concentrations for 50% inhibition (IC_{50} values) of these compounds being ca. 0.1 mM. The calmodulin antagonists calmidazolium and W7 (N-(6-aminohexyl)-5-chloro-1-naphthalene-sulfonamide) also inhibit the enzyme (IC_{50} values 0.4 and 0.1 mM, respectively).

3',5'-cyclic AMP and 3',5'-cyclic GMP neither activate nor substantially inhibit the enzyme at 1 mM. 1 mM ADP, ITP, GTP, CTP and UTP largely inhibit phosphoryl transfer from ATP catalyzed by the enzyme (Table 1). A suspension of L-α-phosphatidyl-L-serine and diolein at concentrations (125 μg/ml and 12.5 μg/ml, respectively) sufficient to activate animal protein kinase C (Davis and Clark, 1983) does not substantially activate the leaf Ca^{2+}-dependent protein kinase (casein and histone phosphorylation activated 9% and 33%, respectively). Calmodulin (6 μM) and heparin (164 μg/ml) stimulate histone phosphorylation 2-fold and 6-fold, respectively, but do not affect casein phosphorylation catalyzed by the soluble leaf enzyme.

Substrate Specificity of the Soluble Leaf Ca^{2+}-dependent Protein Kinase

The soluble leaf protein kinase preparations have an absolute dependence on added protein substrate (Table 2). Histone H1 is the best substrate found

Table 2. Substrate Specificities of the Soluble Leaf Ca^{2+}-dependent
Protein Kinase and Wheat Embryo Protein Kinase I

| Protein Substrate Added | Protein Kinase (% control)[a] | |
	Leaf	Embryo
None	0	4
Histone H1 (0.5 mg/ml)	100	100
Histone H2B (0.5 mg/ml)	14	47
Casein (1 mg/ml)	17	10
Casein (0.5 mg/ml)	9	–
Mixed histones (1 mg/ml)	2	–
Mixed histones (0.5 mg/ml)	4	13
Phosvitin (1 mg/ml)	3	–

[a]Protein kinase was determined in the standard assay including 0.5 mM
free Ca^{2+} and rates are expressed relative to the rate with histone H1 as
substrate (100%).

for both the silver beet leaf enzyme and for wheat embryo protein kinase I
(Table 2). No phosphorylation of the synthetic peptide Kemptide was
detected but rates of phosphorylation with synthetic peptide S6-1 (38 µM)
were 7% and 12.5% of the rates with 0.5 mg/ml histone H1 with the leaf and
wheat embryo protein kinases, respectively. The phosphorylation of all
protein substrates catalyzed by the soluble leaf protein kinase was largely
Ca^{2+}-dependent. The K_m of the leaf enzyme for the dephosphorylated casein
preparation used is 2 mg/ml at pH 8.0 in the standard assay conditions.
The acid hydrolysates of ^{32}P-labelled casein from reactions catalyzed by
the protein kinase were chromatographed on Dowex 50W-X8 as described
previously (Polya et al., 1983); 65% of applied radioactivity was associated
with inorganic phosphate, 33% with phosphoserine and only 0.8% with phosphothreonine. The K_m for ATP in the standard assay conditions is 19 µM.

Properties of a Solubilized Membrane-Associated Ca^{2+}-Dependent Protein Kinase from Silver Beet Leaves

The pellet fraction from centrifugation of the initial silver beet
leaf homogenate contains protein kinase activity. Assayed with 1 mg/ml
histone as added substrate, about 30% of the pellet protein kinase activity
and 75% of the supernatant protein kinase is Ca^{2+}-dependent (as determined
by conducting assays with 0.2 mM EGTA ±1.25 mM $CaCl_2$). We have found a
similar marked Ca^{2+}-dependence of histone phosphorylation catalyzed by the
supernatant fraction (97% Ca^{2+}-dependent) and pellet fraction (50% Ca^{2+}-
dependent) derived from homogenates of germinated pollen of Nicotiana alata
(Polya et al., 1985a). The particulate Ca^{2+}-dependent silver beet protein
kinase is not solubilized by extraction with 1 mM EGTA in buffer B nor with
0.5 M NaCl in buffer B. However histone kinase activity that is largely
Ca^{2+}-dependent can be extracted from the particulate fraction in 0.9%
Nonidet P-40-buffer B. The partially-purified, solubilized particulate Ca^{2+}-
dependent protein kinase (see Materials and Methods) is similar in many
properties to the soluble enzyme. The solubilized particulate enzyme has a
molecular weight of ca. 50,000 (as determined from gel filtration), is
inhibited by calmodulin antagonists (IC_{50} for trifluoperazine, 0.05 mM) and
is completely Ca^{2+}-dependent, casein phosphorylation being maximal at ca.
10^{-6} M free Ca^{2+} and histone phosphorylation maximal at ca. 0.3 mM free Ca^{2+}.
Rates of phosphorylation with 38 µM Kemptide, 38 µM S6-1 or 1 mg/ml mixed
histones are 0%, 74% and 10%, respectively of the rate with 1 mg/ml
casein as substrate. Rates of phosphorylation with histone H1 (0.5 mg/ml)
and histone H2B (0.5 mg/ml) are approximately 5-fold and 2-fold greater than
the phosphorylation rate with 0.5 mg/ml casein (cf. Table 2). However while
1 mM CTP and ITP inhibit the solubilized particulate enzyme by only ca.

20%, the soluble leaf Ca^{2+}-dependent protein kinase is largely inhibited by these nucleotides at 1 mM (Table 1).

DISCUSSION

The present report is the first description of the partial purific- ation and characterization of a soluble Ca^{2+}-dependent protein kinase from leaf tissue. The leaf enzyme has a much lower molecular weight (49,000) than the wheat embryo Ca^{2+}-dependent protein kinases I (90,000) and II (86,000), but is very similar to these enzymes in the Ca^{2+}-dependence of histone and casein phosphorylation, preferential serine residue phos- phorylation, the $(Ca^{2+} + Mg^{2+})$ requirement, the activation by calmodulin and in the inhibition by phenothiazine-derived calmodulin antagonists, by certain metal ions and by high ionic strength (cf. Polya et al., 1983; Polya and Micucci, 1984). The soluble (but not the solubilized, part- iculate) leaf enzyme resembles wheat embryo protein kinase I in that phosphoryl transfer to protein from ATP catalyzed by both enzymes is substantially inhibited by 1 mM CTP and ITP; protein kinase II is not inhibited by these nucleoside 5'-triphosphates. The particulate leaf Ca^{2+}- dependent protein kinase is nevertheless very similar to the soluble enzyme and could be a modified form of the soluble protein kinase. The histone-specific activation of the soluble protein kinases by calmodulin and the histone-specific activation of the leaf enzyme by heparin could both be due to interactions of these polyanions with the histone substrate although direct interaction with the protein kinases remains possible.

Low and high Ca^{2+} concentrations, respectively, are required for casein and histone phosphorylation catalyzed by the leaf and wheat embryo Ca^{2+}-dependent protein kinases. The basis for this difference is not known. However in the case of animal Ca^{2+}- and phospholipid-activated protein kinase C, protamine phosphorylation catalyzed by this enzyme is Ca^{2+}-independent while phosphorylation of other substrates (e.g. histones) is Ca^{2+}-dependent (Wise et al., 1982). The low free Ca^{2+} concentrations required for activation of casein phosphorylation by the soluble (and solubilized) plant Ca^{2+}-dependent protein kinases are commensurate with the cytosolic free Ca^{2+} in stimulated plant cells (Williamson and Ashley, 1982) and we therefore propose that these enzymes function in Ca^{2+}-mediated stimulus-response coupling in plant cells. The nature of the in vivo protein substrates of these enzymes is unknown although we have demonstrated phosphorylation of histones H1 and H2B and of a synthetic peptide analogue of ribosomal protein S6. The precise consequences of the in vivo activ- ation of these enzymes remain to be determined.

REFERENCES

Brown, E.G., and Newton, R.P., 1981, Cyclic AMP and higher plants, Phytochem., 20:2453.

Cohen, P., 1982, The role of protein phosphorylation in neural and hormonal control of cellular activity, Nature, 296:613.

Cormier, M.J., Charbonneau, A. and Jarrett, H.W., 1981, Plant and fungal calmodulin:Ca^{2+}-dependent regulation of plant NAD kinase, Cell Calcium, 2:313.

Davies, J.R., and Polya, G.M., 1983, Purification and properties of a high specific activity protein kinase from wheat germ, Plant Physiol., 71:489.

Davis, J.S., and Clark, M.R., 1983, Activation of protein kinase in the bovine corpus luteum by phospholipid and Ca^{2+}, Biochem. J., 214:569.

Dieter, P., 1984, Calmodulin and calmodulin-mediated processes in plants, Plant Cell Environment, 7:371.

Francko, D.A., 1983, Cyclic AMP in photosynthetic organisms: recent developments, Adv. Cyclic Nucl. Res., 15:97.

Graziana, A., Ranjeva, R., and Boudet, A.M., 1983a, Provoked changes in cellular calcium controlled protein phosphorylation and activity of quinate:NAD$^+$ oxidoreductase in carrot cells, FEBS Letters, 156:325.

Graziana, A., Ranjeva, R., Salimath, B.P. and Boudet, A.M., 1983b, The reversible association of quinate NAD$^+$ oxidoreductase from carrot cells with a putative regulatory subunit depends on light conditions, FEBS Letters, 163:306.

Graziana, A., Dillenschneider, M., and Ranjeva, R., 1984, A calcium-binding protein is a regulatory subunit of quinate:NAD$^+$ oxidoreductase from dark-grown carrot cells, Biochem. Biophys. Res. Commun., 125:774.

Hetherington, A.M., and Trewavas, A., 1982, Calcium-dependent protein kinase in pea shoot membranes, FEBS Letters, 145:67.

Hetherington, A.M., and Trewavas, A., 1984, Activation of a pea membrane protein kinase by calcium ions, Planta, 161:409.

Kato, R., Uno, I., Ishikawa, T., and Fujii, T., 1983, Effects of cyclic AMP on the activity of soluble protein kinases in Lemna paucicostata, Plant Cell Physiol., 24:841.

Krebs, E.G. and Beavo, J.A., 1979, Phosphorylation-dephosphorylation of enzymes, Ann. Rev. Biochem., 48:929.

Newton, R.P., Kingston, E.E., Evans, D.E., Younis, L.M., and Brown, E.G., 1984, Occurrence of guanosine 3',5'-cyclic monophosphate (cyclic GMP) and associated enzyme systems in Phaseolus vulgaris, Phytochem., 23:1367.

Nishizuka, Y., 1984, The role of protein kinase C in cell surface signal transduction and tumour promotion, Nature, 308:693.

Polya, G.M., and Bowman, J.A., 1981, Resolution and properties of two high-affinity cyclic adenosine 3',5'-monophosphate-binding proteins from wheat germ, Plant Physiol., 68:577.

Polya, G.M., and Davies, J.R., 1982, Resolution of Ca^{2+}-calmodulin-activated protein kinase from wheat germ, FEBS Letters, 150:167.

Polya, G.M., Davies, J.R., and Micucci, V., 1983, Properties of a calmodulin-activated Ca^{2+}-dependent protein kinase from wheat germ, Biochim. Biophys. Acta, 761:1.

Polya, G.M., and Micucci, V., 1984, Partial purification and characterization of a second calmodulin-activated Ca^{2+}-dependent protein kinase from wheat germ, Biochim. Biophys. Acta, 785:68.

Polya, G.M., Schibeci, A., and Micucci, V., 1984, Phosphorylation of membrane proteins from cultured Lolium multiflorum (ryegrass) endosperm cells, Plant Sci. Lett., 36:51.

Polya, G.M., Micucci, V., Rae, A.L., Harris, P.J., and Clarke, A.E., 1985a, Calcium-dependent protein phosphorylation in germinated pollen of Nicotiana alata, Proc. Aust. Biochem. Soc., 17:74.

Polya, G.M., Micucci, V., Gantinas, A., and Basiliadis, S., 1985b, Plant leaf calcium-dependent protein kinase, cyclic GMP-promoted protein phosphorylation and cyclic GMP-inhibited phosphohydrolase, Proc. Aust. Biochem. Soc., 17:88.

Ranjeva, R., Refeno, G., Boudet, M., and Marmé, D., 1983, Activation of plant quinate:NAD$^+$ 3-oxidoreductase by Ca^{2+} and calmodulin, Proc. Natl. Acad. Sci. USA, 80:5222.

Refeno, G., Ranjeva, R., and Boudet, A.M., 1982a, Modulation of quinate: NAD$^+$ oxidoreductase activity through reversible phosphorylation in carrot cell suspensions, Planta, 154:193.

Refeno, G., Ranjeva, R., Fontaine-Delvare, S., and Boudet, A.M., 1982b, Functional properties of protein kinase(s) and phosphatase(s) converting quinate:NAD$^+$ oxidoreductase into active and deactivated forms in carrot cell suspension cultures, Plant Cell Physiol., 23:1137.

Reichardt, L.F., and Kelly, R.B., 1983, A molecular description of nerve terminal function, Ann. Rev. Biochem., 52:871

Salimath, B.P., and Marmé, D., 1983, Protein phosphorylation and its regulation by calcium and calmodulin in membrane fractions from zucchini hypocotyls, Planta, 158:560.

Veluthambi, K., and Poovaiah, B.W., 1984a, Polyamine-stimulated phosphorylation of proteins from corn (Zea mays L.) coleoptiles, Biochem. Biophys. Res. Commun., 122:1374.

Veluthambi, K., and Poovaiah, B.W., 1984b, Calcium-promoted protein phosphorylation in plants, Science, 223:167.

Veluthambi, K., and Poovaiah, B.W., 1984c, Calcium- and calmodulin-regulated phosphorylation of soluble and membrane proteins from corn coleoptiles, Plant Physiol., 76:359.

Wettenhall, R.E.H., and Morgan, F.J., 1984, Phosphorylation of hepatic ribosomal protein S6 on 80 and 40S ribosomes. Primary structure of S6 in the region of the major phosphorylation sites for cAMP-dependent protein kinases, J. Biol. Chem., 259:2084.

Williamson, R.E., and Ashley, C.C., 1982, Free Ca^{2+} and cytoplasmic streaming in the alga Chara, Nature, 296:647.

Wise, B.C., Glass, D.B., Chou, C.-H.J., Raynor, R.L., Katoh, N., Schatzman, R.C., Turner, R.S., Kibler, R.F., and Chou, J.F., 1982, Phospholipid-sensitive Ca^{2+}-dependent protein kinase from heart. II. Substrate specificity and inhibition by various agents, J. Biol. Chem., 257:8489.

Yan, F.F.J., and Tao, M., 1982, Purification and characterization of a wheat germ protein kinase, J. Biol. Chem., 257:7037.

THE ROLE OF CALCIUM AND CALMODULIN IN HORMONE ACTION IN PLANTS:

IMPORTANCE OF PROTEIN PHOSPHORYLATION

B. W. Poovaiah and K. Veluthambi

Department of Horticulture and Landscape Architecture
Washington State University
Pullman, Washington 99164-6414, USA

INTRODUCTION

The transduction of extracellular signals such as hormones and neuro-transmitters into intracellular events in mammalian cells is mediated by cellular messengers such as cAMP (Rasmussen, 1981), calcium (Cheung, 1980), inositol-3-phosphate and diacylglycerol (Nishizuka, 1984). Although cAMP has been reported to be present in plants, it does not activate protein kinases suggesting that its role in plants could be different from that in animals (Marme and Dieter, 1983). Since the discovery of calmodulin and calmodulin-dependent enzymes in plants (Anderson and Cormier, 1978; Marme and Dieter, 1983; Poovaiah, 1985) there is increasing interest to study the role of calcium as a cellular messenger in plants. The response of plants to light (Serlin and Roux, 1984), and hormones (Elliott et al., 1983; Raghothama et al., 1983, 1985) has been shown to be mediated by calcium and calmodulin. In our laboratory, the role of calcium in auxin-induced coleoptile elongation, cytokinin-mediated delay of leaf senescence and in the tuberization response of potato plants has been investigated. Protein phosphorylation is studied both under in vitro and in vivo conditions to evaluate the role of calcium in mediating auxin response in coleoptiles.

METHODS

Corn (Zea mays L. var. Patriot) and oat (Avena sativa L. var. Cayuse) seeds were germinated in the dark for 5 to 6 days. Eight mm long coleoptile segments were cut 3 mm from the tip and the effects of auxin and calmodulin antagonists on the elongation was studied (Raghothama et al., 1985). For studies on leaf senescence, corn seedings were grown in a growth chamber, leaf discs (8 mm diameter) were excised, and the effects of cytokinin, EGTA and calcium were studied in the dark (Poovaiah and Leopold, 1973). Single-node leaf-cuttings of potato (Solanum tuberosum L. var. Russet Burbank) plants were grown in liquid media under 12-h day photoperiod (Ewing and Wareing, 1978). Oat coleoptiles devoid of leaves were excised from dark-grown seedlings and used for in vitro protein phosphorylation studies (Veluthambi and Poovaiah, 1984, a,b). To study in vivo protein phosphorylation, the coleoptile segments were incubated in the respective media for 1 h and subsequently fed with ^{32}Pi for 30 min. Protein was extracted in 50 mM Mes-NaOH (pH 7), 10 mM KH_2PO_4,

1 mM EDTA, 10 mM NaF, 0.5 mM PMSF, 1 mM DTT, 5 μg/ml RNAase and analyzed by two-dimensional gel electrophoresis (O'Farrell, 1975). Glucan synthase activity in the microsomal preparation of corn coleoptiles was assayed by measuring the incorporation of [^{14}C]UDP-glucose into the alcohol-insoluble fraction (Paliyath and Poovaiah, unpublished).

RESULTS AND DISCUSSION

Auxin-induced Coleoptile Elongation

Free calcium concentration in mammalian cells increases from less than 0.1 μM to 1 μM in response to neural and hormonal stimuli. At this concentration calcium binds to calmodulin and activates it (Cheung, 1980). The activated calmodulin in turn activates a variety of enzymes. Calmodulin antagonists such as phenothiazine drugs and naphthalene-sulfonamides (W compounds) bind to activated calmodulin and block calmodulin-mediated functions (Hidaka and Tanaka, 1982; Paliyath and Poovaiah, 1984 1985a,b). The role of calmodulin in auxin-induced elonga-tion was studied using chlorpromazine (CP), fluphenazine (FPZ) and naphthalenesulfonamide compounds (W$_7$ and W$_5$). At 15 to 20 μM CP and FPZ caused 50% inhibition of NAA-induced elongation (Fig. 1A). Since pheno-thiazine drugs are known to cause nonspecific effects due to their hydrophobic nature, the effects of naphthalensulfonamide compounds were also studied. Both W$_7$ and W$_5$ have similar hydrophobicity but only W$_7$ inhibits calmodulin activity at low concentrations. NAA-induced elonga-tion of oat coleoptiles was inhibited by W$_7$ at 20 to 100 μM (Fig. 1B). At this concentration W$_5$ had no inhibitory effect. These findings, together with those of Elliott et al. (1983), suggest a role for calcium and cal-modulin in mediating auxin-induced coleoptile elongation.

Cytokinin-mediated Delay of Leaf Senescence

Calcium acts synergistically with cytokinin in delaying leaf senescence and abscission (Poovaiah and Leopold, 1973a, b). The possible role of calcium in mediating the cytokinin effect was studied by first depleting calcium from the corn leaf disks by preincubating with 5 mM EGTA for 5 h and then transferring to a medium containing benzyladenine (BA) with or without calcium (Rhee and Poovaiah, unpublished). After the

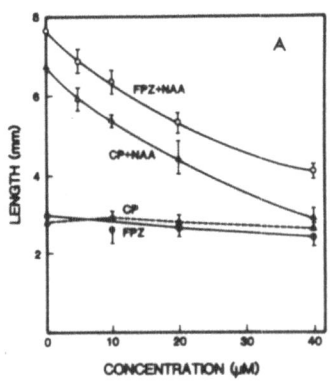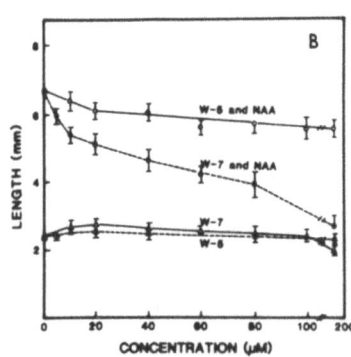

Fig. 1. The effect of calmodulin antagonists on NAA-induced oat coleop-tile elongation. Eight mm long coleoptile segments were incubated in the dark for 18 h at room temperature. The incubating medium consisted of 10 mM KH$_2$PO$_4$(pH 6.3), 1.5% (w/v) sucrose, 10 mM sodium citrate and 0.1% (v/v) DMSO. All the treatments except the control had 10 μM NAA. The calmodulin antagonists were added to the media at the beginning of incu-bation (from Raghothama et al., 1985).

EGTA-pretreatment BA was no longer effective in delaying the loss of chlorophyll and protein (Fig. 2). The cytokinin effect was however restored by the addition of calcium which suggested that the response to cytokinin was mediated by calcium. Previously, Saunders and Hepler (1982) have shown that the calcium ionophore A 23187 could mimic the effect of cytokinin in the bud formation of <u>Funaria</u>. The results discussed here add further evidence to the role of calcium in mediating cytokinin response.

Tuberization in Potato Plants

Tuberization in potato plants is regulated by cytokinin, photo-period and night temperature (Melis and van Staden, 1984). The nature of the primary stimulus controlling tuberization response is not yet clearly established. We used the single-node leaf-cuttings to study whether the tuberization stimulus is mediated by calcium (Balamani et al., unpublished). The axillary buds of the leaf cuttings were pretreated for 12 h in 5 mM EGTA to deplete extracellular calcium or in 5 mM EGTA + 50 μM calcium ionophore A 23187 to deplete extracellular + intracellular calcium. Subsequently, the leaf cuttings were grown in media containing or lacking calcium. The H_2O-pretreated and EGTA-pretreated leaf cuttings formed tubers (Fig. 3). The EGTA + A 23187-pretreated leaf cuttings did not form tubers, but subsequent addition of calcium resulted in the recovery of the tuberization response. These results suggested that intracellular calcium is critical in the mediation of tuberization response. Chlorpromazine inhibited tuberization at 5 to 10 μM suggesting a role for calmodulin in this process.

Effect of Calcium on Glucan Synthase Activity

The effects of calcium and calmodulin on β-glucan synthase activity was studied because of its key role in plant growth and development (Table 1). Calmodulin alone did not promote the activity significantly, however, addition of calcium along with ATP and calmodulin enhanced the activity substantially suggesting that stimulation of β-glucan synthase could involve calcium and calmodulin-dependent protein phosphorylation. Sodium fluoride, a phosphoprotein phosphatase inhibitor, further increased the calcium- and ATP-stimulated activity (data not shown), indicating that phosphorylation and dephosphorylation are involved in the regulation of the enzyme activity.

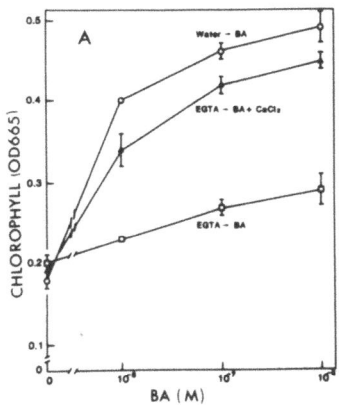 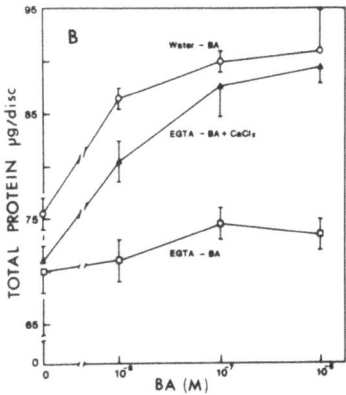

Fig. 2. Effect of pretreatment of 1 mM EGTA on the chlorophyll (A) and protein (B) content of corn leaf discs. After pretreatment for 5 h, leaf discs were transferred to 10^{-8} to 10^{-6} M BA with or without 1 mM $CaCl_2$ and incubated in the dark for 4 days. Initial value of chlorophyll was 0.72 ± 0.05 and total protein 120.2 ± 6.2 (Joung and Poovaiah, unpublished).

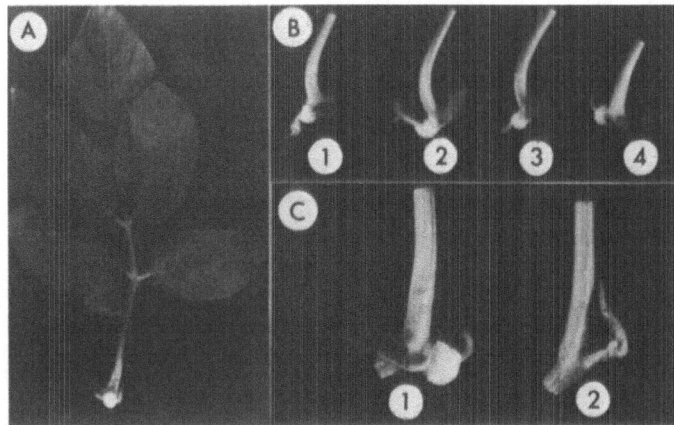

Fig. 3. Tuberization of axillary buds of single-node leaf cuttings. (A)
A leaf cutting with the tuber formed after growth for 7 d in water. (B)
Effects of pretreatments with water (1, 2) or 5 mM EGTA (3, 4).
Subsequent to pretreatment for 15 h, the leaf cuttings were grown for 7 d
in water (1, 3) or in 1 mM $CaCl_2$ + 1mM $MgCl_2$ (2, 4). (C) Effect of
pretreatment with 5 mM EGTA + 50 µM calcium ionophore A 23187. After a
15-h pretreatment, the leaf cuttings were grown for 7 d in 1 mM $CaCl_2$ +
1mM $MgCl_2$ (1) or in water (2) (Balamani, Veluthambi and Poovaiah,
unpublished).

Table 1. Effects of Calcium (1 mM), Calmodulin (2.5 µg) and ATP (500 µM)
on β-glucan synthase activity in corn microsomal membranes.

Treatment	Specific Activity + S.D. pmol/mg protein/min
-Ca	13.70 + 0.65
+Ca	20.10 + 1.25
-Ca + ATP	13.40 + 0.65
+Ca + ATP	23.50 + 1.50
-Ca + Calmodulin	13.60 + 0.85
+Ca + Calmodulin	24.75 + 1.35
-Ca + Calmodulin + ATP	22.45 + 2.70
+Ca + Calmodulin + ATP	37.40 + 2.25

Protein Phosphorylation

     A wide variety of regulatory agents produce diverse types of biologi-
cal responses by regulating the state of phosphorylation of specific proteins (Cheung, 1980; Cohen, 1982). The diverse effects of cAMP in animal
cells is accompanied by the activation of cAMP-dependent protein kinase
(Rasmussen, 1981). During the activation of surface receptors of several
hormones and neurotransmitters, phosphatidylinositol-4,5-biphosphate is
hydrolyzed leading to the release of diacylglycerol and inositol-3-phosphate
(Nishizuka, 1984). Diacylglycerol activates protein kinase C and inositol-
-3-phosphate releases calcium from the intracellular calcium stores into
the cytoplasm. At micromolar concentrations, calcium activates calmodulin which in turn activates a variety of protein kinases (Cohen, 1982).
Calcium and calmodulin-promoted protein phosphorylation has been reported
from several plant systems (Hetherington and Trewavas, 1982; Polya and
Davies, 1982; Salimath and Marme, 1983; Veluthambi and Poovaiah, 1984 a,b;

Graziana et al., 1984). Other cellular regulators of plants such as polyamines also regulate protein phosphorylation (Veluthambi and Poovaiah, 1984 c). All these studies have been performed under in vitro conditions but there is very little information on in vivo protein phosphorylation. In vivo studies would be helpful to understand the effects of hormones and other stimuli in altering calcium levels and thereby affecting protein phosphorylation.

If intracellular calcium concentration increases in response to an external stimulus, then calmodulin and calmodulin-dependent protein kinases will be activated leading to the phosphorylation of several proteins. Since phosphorylation of proteins is a covalent modification, those changes which occur in vivo in response to a stimulus could be evaluated by analyzing the phosphorylated proteins under suitable conditions. In this direction, we selected the auxin-responsive oat coleoptile segments and first studied the effects of calcium and calmodulin antagonists on in vitro protein phosphorylation. Subsequently, the effect of auxin on in vivo protein phosphorylation was studied by feeding ^{32}Pi to coleoptile segments.

The processes regulated in vitro at micromolar calcium concentrations are known to be of physiological significance in stimulus response coupling. In the extracts from oat coleoptiles, only a few proteins were phosphorylated in the absence of calcium, or at concentrations of free calcium below 1 μM (Fig. 4A). Promotion of phosphorylation of polypeptides of 89,000, 59,000, 55,000, 30,000, 17,000 and 15,000 M_r was very distinct at 1 μM free calcium. Phosphorylation levels increased up to 15 μM free calcium above which no further promotion occurred. Polypeptides of M_r 65,000 and 23,000, which showed significant calcium-independent phosphorylation, exhibited only slight promotion in the presence of calcium. The calmodulin antagonists trifluoperazine and W7 were used to evaluate the role of calmodulin in protein phosphorylation. Calcium-promoted protein phosphorylation was markedly reduced at 50 μM trifluoperazine and at 100 μM W7 (Fig. 4B).

In vivo protein phosphorylation in oat coleoptile segments was studied to understand whether any of the polypeptides which exhibited increased phosphorylation in vitro in the presence of calcium also showed higher levels of phosphorylation in vivo in response to auxin. A comparison of patterns of in vivo and in vitro phosphorylations is shown in Fig. 5A. The 65,000 and 55,000 M_r polypeptides were phosphorylated both under in vivo and in vitro conditions. But the in vivo and in vitro phosphorylation patterns of several other proteins were strikingly different. Auxin treatment did not result in any major qualitative (Fig. 5A) or quantitative changes in protein phosphorylation. Trifluoperazine reduced protein phosphorylation by 30% both in the control and auxin-treated coleoptiles (data not shown) without affecting the pattern of protein phosphorylation.

In order to resolve small changes affected by auxin, the in vivo phosphorylated proteins were analyzed by two-dimensional gel electrophoresis. As seen in Fig. 5B, phosphorylation of polypeptide A was higher in auxin-treated coleoptiles, whereas the phosphorylation of polypeptides B,C and D were reduced in auxin-treated tissue. Promotion of protein phosphorylation at micromolar free calcium concentration and the inhibition of calcium-promoted phosphorylation at low concentrations of calmodulin-antagonists clearly suggest that calcium could play a role as a cellular messenger in plants. Further studies on the auxin-induced changes on in vivo phosphorylation of proteins should help us in understanding whether auxin affected those changes by altering intracellular calcium levels.

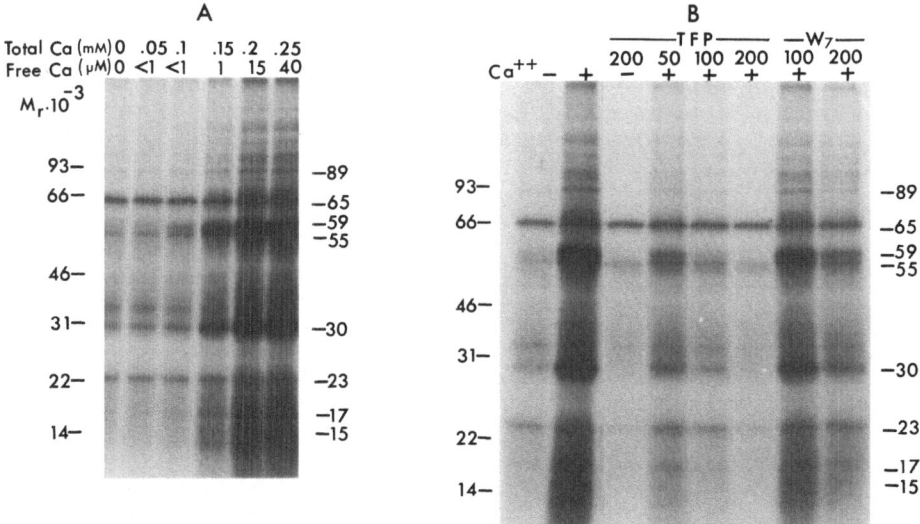

Fig. 4A. Effect of micromolar concentrations of free Ca^{2+} on in vitro protein phosphorylation. All reaction mixtures contained 0.2 mM EGTA and total [Ca^{2+}] was varied from 0 to 0.25 mM. Free Ca^{2+} concentrations as determined using a Ca^{2+} sensitive electrode are also indicated. Molecular weight standards and M_r values of some representative polypeptides are indicated on the margin. Fig. 4B. In vitro effects of various calmodulin antagonists on Ca^{2+} regulated protein phosphorylation. The reaction mixtures with 0.2 mM EGTA alone are designated as –Ca and those with 0.2 mM EGTA + 0.2 mM $CaCl_2$ are designated as +Ca (15 μM free Ca^{2+}). Calmodulin antagonists were added to the reaction mixtures at indicated concentrations (μM).

Protein phosphorylation studies in animal systems have provided valuable information on the role of calcium as a cellular messenger (for review see Nishizuka, 1984). For instance, under in vitro conditions a 20,000 M_r myosin light chain was phosphorylated by a calcium, calmodulin-dependent protein kinase, whereas a 40,000 M_r polypeptide was phosphorylated by a diacylglycerol-dependent protein kinase C. In vivo studies of platelets showed that the calcium ionophore A 23187, which increases intracellular calcium, increased phosphorylation of only the 20,000 M_r polypeptide, whereas the tumor promoting phorbol ester (an analog of diacylglycerol) caused the phosphorylation of only the 40,000 M_r polypeptide. Thrombin, a natural signal for platelet activation, promoted the phosphorylation of both polypeptides in vivo suggesting that both calcium and diacylglycerol served as cellular messengers in the response of platelets to thrombin.

Auxin (Murray and Key, 1978), cytokinin (Ralph and Wojcik, 1981) and gibberellin (Wielgat and Kleczkowski, 1981) have been shown to regulate protein phosphorylation under in vitro conditions. The approach described here should be useful to study the effects of these hormones on in vivo protein phosphorylation and to study the role of calcium and calmodulin in modulating hormonal responses in plants.

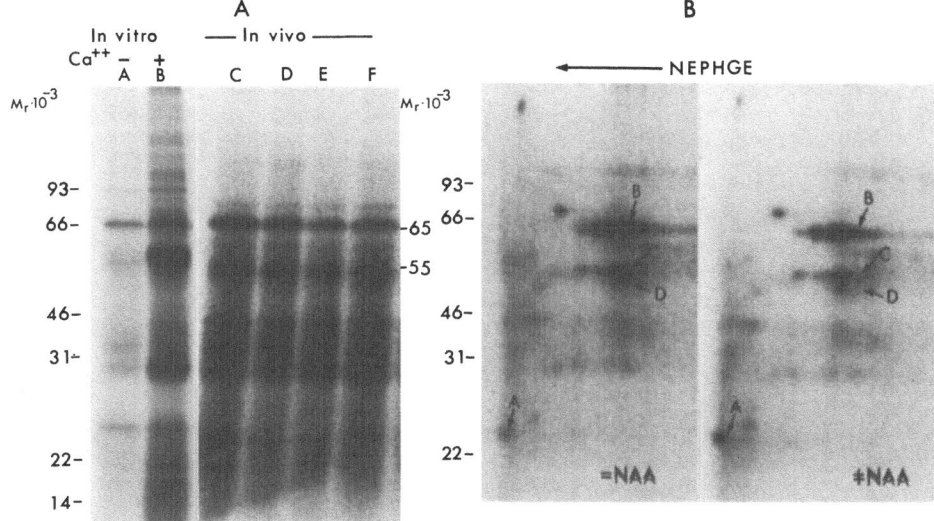

Fig. 5A. Comparison of in vivo and in vitro protein phosphorylation patterns. In vitro phosphorylation was performed in $-Ca^{2+}$ (A) and $+Ca^{2+}$ (15 μM free Ca^{2+}, B) conditions. In vivo phosphorylations were performed in the basal medium (C), medium + 10 μM NAA (D), medium + 50 μM trifluoperazine (E) and medium + 10 μM NAA + 50 μM trifluoperazine (F). Fig. 5B. Effect of auxin on in vivo phosphorylation of proteins. Oat coleoptile segments were incubated in basal medium (-NAA) or medium containing 10 μM NAA (+NAA) for 1 h. The coleoptiles were transferred to the same media but containing 1 mCi carrier-free ^{32}Pi and incubated further for 30 min. Phosphorylated proteins were analysed by two-dimensional electrophoresis.

REFERENCES

Anderson, J. M. and Cormier, M. J., 1978, Calcium-dependent regulation of NAD kinase in higher plants, Biochem. Biophys. Res. Commun., 84:595.

Cheung, W. Y., 1980, Calmodulin plays a pivotal role in cellular regulation, Science, 207:19.

Cohen, P., 1982, The role of protein phosphorylation in neural and hormonal control of cellular activity, Nature, 296:613.

Elliott, D. C., Batchelor, S. M., Cassar, R. A., and Marinos, N. G., 1983, Calmodulin-binding drugs affect responses to cytokinin, auxin and gibberellic acid, Plant Physiol., 72:219.

Ewing, E. E., and Wareing, P. F., 1978, Shoot, stolon, and tuber formation on potato (Solanum tuberosum L.) cuttings in response to photoperiod, Plant Physiol., 61:348.

Graziana, A., Dillenschneider, M., and Ranjeva, R., 1984, A calcium-binding protein is a regulatory subunit of quinate: NAD+ oxidoreductase from dark-grown carrot cells, Biochem. Biophys. Res. Commun., 125:774.

Hetherington, A., and Trewavas, A., 1982, Calcium-dependent protein kinase in pea shoot membranes, FEBS lett., 145:67.

Hidaka, H., and Tanaka, T., 1982, Biopharmacological assessment of calmodulin function: utility of calmodulin antagonists, in: "Calmodulin and Intracellular Calcium Receptors", S. Kakiuchi, H. Hidaka and A. R. Means eds., Plenum, NY.

Marme, D., and Dieter, P., 1983, Role of Calcium and calmodulin in plants, in: "Calcium and Cell Function", W. Y. Cheung, ed., Academic Press, NY.

Melis, R. J. M., and van Staden, J., 1984, Tuberization and hormones, Z. Pflanzenphysiol., 113:271.

Murray, M. G., and Key, J. L., 1978, 2,4-Dichlorophenoxyacetic acid-enhanced phosphorylation of soybean nuclear proteins, Plant Physiol., 61:190.

Nishizuka, Y., 1984, The role of protein kinase C in cell surface signal transduction and tumor promotion, Nature, 308:693.

O'Farrell, P. H., 1975, High resolution two-dimensional electrophoresis of proteins, J. Biol.Chem., 250:4007.

Paliyath, G., and Poovaiah, B. W., 1984, Calmodulin inhibitor in senescing apples and its physiological and pharmacological significance, Proc. Natl. Acad. Sci. USA, 81:2065.

Paliyath, G., and Poovaiah, B. W., 1985a, Identification of naturally occurring calmodulin inhibitors in plants and their effects on calcium and calmodulin-promoted protein phosphorylation. Plant Cell Physiol. 26:201.

Paliyath, G., and Poovaiah, B. W., 1985b, Calcium and calmodulin-promoted phosphorylation of membrane proteins during senescence in apples. Plant Cell Physiol. (In press).

Polya, G. M., and Davies, J. R., 1982, Resolution of Ca^{2+}, calmodulin-activated protein kinase from wheat germ, FEBS Lett., 150:167.

Poovaiah, B. W., and Leopold, A. C., 1973a, Deferral of leaf senescence with calcium, Plant Physiol., 52:236.

Poovaiah, B. W., and Leopold, A. C., 1973b, Inhibition of abscission by calcium. Plant Physiol. 51:848.

Poovaiah, B. W., 1985, The role of calcium and calmodulin in plant growth and development. HortScience 20:347.

Raghothama, K. G., Mizrahi, Y., and Poovaiah, B. W., 1983, Effects of calmodulin inhibitors on auxin-induced elongation. Plant Physiol. 72:144 (suppl.).

Raghothama, K. G., Mizrahi, Y., and Poovaiah, B. W., 1985, The effect of calmodulin antagonists on auxin-induced elongation, Plant Physiol., (In press).

Ralph, R. K., and Wojcik, S. J., 1981, Plant protein kinases and cytokinins, Plant Sci. Lett., 22:127.

Rasmussen, H., 1981, "Calcium and cAMP as Synarchic Messengers", Wiley, NY.

Salimath, B. P., and Marme, D., 1983, Protein phosphorylation and its regulation by calcium and calmodulin in membrane fractions from zucchini hypocotyls, Planta, 158:560.

Saunders, M. J., and Hepler, P. K., 1982, Calcium ionophore A23187 stimulates cytokinin-like mitosis in Funaria. Science, 217:943.

Serlin, B. S., and Roux, S. J., 1984, Modulation of chloroplast movement in the green alga Mougeotia by the Ca^{2+} ionophore A23187 and by calmodulin antagonists, Proc. Natl. Acad. Sci. USA, 81:6368.

Veluthambi, K., and Poovaiah, B. W., 1984a, Calcium-promoted protein phosphorylation in plants, Science, 223:167.

Veluthambi, K., and Poovaiah, B. W., 1984b, calcium and calmodulin-regulated phosphorylation of soluble and membrane proteins from corn coleoptiles, Plant Physiol., 76:359.

Veluthambi, K., and Poovaiah, B. W., 1984c, Polyamine-stimulated phosphorylation of proteins from corn (Zea mays L.) coleoptiles, Biochem. Biophys. Res. Commun., 122:1374.

Wielgat, B., and Kleczkowski, K., 1981, Gibberellic acid-enchanced phosphorylation of pea chromatin proteins, Plant Sci. Lett., 21:381.

ACKNOWLEDGEMENT: The work reported here was supported by a research grant from the National Science Foundation.

PLANT NAD KINASE: REGULATION BY CALCIUM AND CALMODULIN

Peter Dieter

Biochemical Institute, University of Freiburg
Hermann Herder Strasse 7, D-7800 Freiburg, FRG

INTRODUCTION

It has been shown that illumination of plant seedlings or plant tissues causes an increase of NADP and a decrease of NAD (Muto et al., 1981). So far, NAD kinase (AMP:NAD 2'-phosphotransferase; EC 2.7.1.23) has been the only enzyme which is known to catalyzes the phosphorylation of NAD to NADP. Therefore a light-dependent activation of the NAD kinase has been postulated. In 1972 Tezuka and Yamamoto reported that NAD kinase activity in cell free extracts from peas is photoregulated via phytochrome. However, these results could not be verified later (Hopkins and Briggs, 1973). Since hitherto no evidence for a photoregulation of the enzyme could be demonstrated another mechanism of regulation has to be postulated.

In 1977 Muto and Miyachi presented the first evidence for a protein activator which is involved in the activation of the plant NAD kinase. In 1978 Anderson and Cormier demonstrated that the activation occurs only in the presence of calcium and identified the protein activator as calmodulin. In the following years calcium, calmodulin-dependent NAD kinase has been found in zucchini squash (Dieter and Marmé, 1980), corn (Dieter and Marmé, 1984) and spinach (Simon et al., 1984). They are all located either in the cytoplasm or when associated to an organelle membrane faced with the regulatory subunit to the cytoplasm and consequently are able to sense changes in the cytoplasmic calcium concentration. Based on these data a new hypothesis for the light-dependent activation of the NAD kinase has been suggested (Marmé and Dieter, 1983; Dieter, 1984,1985): Light increases primarily the cellular free calcium concentration which then leads to the formation of the complex calcium/calmodulin and consequently to an activation of the calcium, calmodulin-dependent NAD kinase. This hypothesis is based on two main assumptions:
- The cytoplasmic free calcium concentration in plant cells has to be maintained at a low level (below the affinity of calcium to calmodulin) in an "unstimulated cell".
- Stimuli such as light must be able to change (increase) this low level of cytoplasmic calcium.

These two assumptions have been carefully verified in the recent years. It has been shown that the free cytoplasmic calcium concentration can be maintained at a μM level by extrusion of calcium out of the cell and by accumulation of calcium into intracellular organelles like mitochondria, endoplasmic reticulum and vacuoles (Marmé and Dieter, 1983; Dieter, 1984). It could further be demonstrated that light decreases the calcium-transport activities of the calcium pumps located in the plasmamembrane and the mitochondria (Dieter and Marmé, 1981). Both light effects consequently lead in the cell to an increase of the cytoplasmic calcium concentration. Thus the two assumptions of the proposed hypothesis seem to be fulfilled.

In the following chapter a detailled analysis of the particular components in the light-dependent regulation of the NAD kinase - light-calcium-calmodulin-NAD kinase-NAD/NADP - will be presented.

MATERIALS AND METHODS

All experiments has been performed with corn seedlings (Zea mays L., Inracorn Hybrid, Categorie 5A, 3070, from Hambrecht, Freiburg, FRG). Corn was grown on vermiculite at $25°$C for 5.5 days in total darkness or far red light (Dieter and Marmé, 1981). Coleoptiles were removed and homogenized as described previously (Dieter and Marmé, 1984). Purification of the mitochondrial NAD kinase and measurement of the NAD kinase activity has been carried out as described by Dieter and Marmé (1984). Solubilization and purification of the NAD kinase using a calmodulin-Sepharose affinity column will be described otherwise (manuscript in preparation).

The extraction and determination of the nicotinamide nucleotides $NAD(H_2)$ and $NADP(H_2)$ was performed as follows: Corn coleoptiles were frozen to liquid nitrogen temperatures in a precooled teflon cell. The frozen material was reduced to a powder together with a 9 mm tungsten carbide ball (pre cooled) for 20 sec in a Micro Dismembrator (Braun-Melsungen AG, Melsungen, West Germany). Extraction was accomplished by adding a distinct volume of 0.5 N $HClO_4$ ($50°$C) for oxidized nucleotides or a distinct volume of 0.1 N NaOH ($50°$C) for reduced nucleotides. Both probes were immediately reshaken for 10 sec. Then 200 μmoles triethanolamine hydrochloride were added to the $HClO_4$ treated samples. The liquified extracts were quantitatively transferred to centrifuge tubes and the tubes containing the reduced nucleotides kept at $85°$C for 1 min. All probes were then subsequently cooled to $4°$C and adjusted to pH 7.0 with 0.1 M KH_2PO_4 and 0.5 N KOH respectively. After centrifugation at 30,000xg for 30 min the clear supernatant was decanted and used in the assay. $NAD(H_2)$ and $NADP(H_2)$ were determined as described by Matsumara (1980). Recovery of nucleotides in the corresponding fractions were 89%, 75%, 72% and 76% for NAD, NADP, $NADH_2$ and $NADPH_2$ respectively whereas in the opposite fractions NAD, NADP, $NAD(H_2)$ and $NADP(H_2)$ were destructed up to 90%, 83%, 95% and 91%.

Calmodulin was purified as described previously (Dieter and Marmé, 1985).

cAMP phosphodiesterase activity was determined as described earlier (Dieter and Marmé, 1980).

RESULTS AND DISCUSSION

We demonstrated recently that the NAD kinase activity from coleoptiles of dark grown corn seedlings is totally dependent on calcium and calmodulin (Dieter and Marmé, 1984). We could further show that almost of the NAD kinase is membrane-bound and associated with the outer mitochondrial membrane, but nevertheless is able to sense changes in cytoplasmic calcium concentration via cytoplasmic calmodulin (Dieter and Marmé, 1984). The calcium-dependency of the NAD kinase activity in intact mitochondria show that the enzyme is almost totally inactive below 100 nM calcium (Fig.1). In the absence of calmodulin the enzyme activity increases slightly with higher calcium concentrations. We have some evidence that this increase may be due to endogenous calmodulin, present in the mitochondrial fraction (data not shown). In the presence of exogenous calmodulin the NAD kinase activity increases almost linearly from 10^{-7}M to 10^{-3}M (Fig.1). This activation profile is not typical for a calmodulin-dependent enzyme as e.g. the soluble NAD kinase from zucchini squash (Fig.2; Dieter, 1984) which becomes maximal activated between 1 μM and 10 μ . However if the mitochondrial-associated NAD kinase from corn is solubilized, it shows a very similar calcium-dependency as the soluble NAD kinase from zucchini squash (manuscript in preparation). We suggest therefore that in the membrane bound form some interactions of membrane lipids with the enzyme or the calcium metabolism of the mitochondria are responsible for this discrepancy and thus reflect some physiological significance.

Fig. 2 shows the calcium-dependency of the NAD kinase activity and the calcium-dependency of the binding to calmodulin. It can be seen that the binding of calcium to calmodulin occurs at about 1 μM whereas the activation of the calmodulin-dependent NAD kinase starts at higher calcium concentrations. This let suggest that the regulation of the calmodulin-dependent NAD kinase takes place between 0.1 μM and 10 μM calcium. As shown previously (Dieter and Marmé, 1983; Dieter, 1985b) the plasmamembrane-located calcium pump operates at about the same calcium concentrations and therefore is able to regulate the cytoplasmic free calcium in this range. All these data support the hypothesis that in the living cell an increase of the cellular calcium concentration leads to the formation

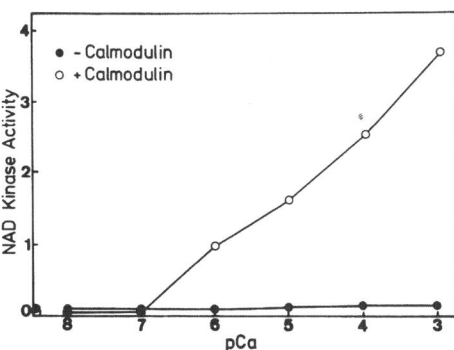

Fig. 1. Calcium-dependence of the mitochondrial NAD kinase from corn coleoptiles

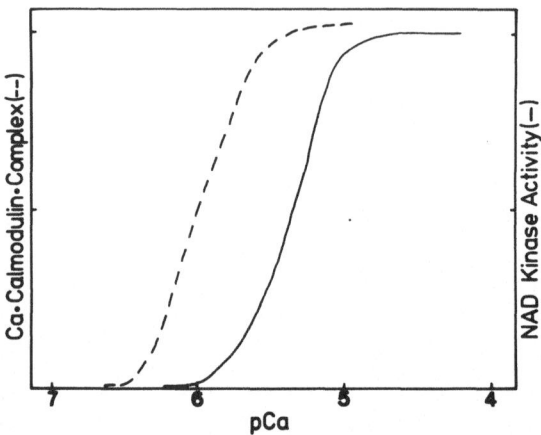

Fig. 2. Calcium-dependence of the calcium-binding to
        calmodulin (data from Anderson et al., 1980)
        and of the activation of the NAD kinase from
        zucchini squash (data from Dieter, 1984).

of the "active" calcium-calmodulin complex which then binds
to the calmodulin-dependent enzyme as the NAD kinase (Fig.3).

Effect of light on the properties of the NAD kinase

        In order to investigate whether far red light irradiation
of corn seedlings (in vivo) causes a change in the properties
of the NAD kinase which may be responsible for the increase
of NADP (see Table 2) we isolated NAD kinase from dark and far
red light grown corn tissue (manuscript in preparation).
We didn't find a significant change in the total NAD kinase
activity, in the absence or in the presence of both calcium
and calmodulin. We could further demonstrate that the cellular
distribution of the NAD kinase didn't change after the light
irradiation; also the dependency of calcium and calmodulin
was about the same.
        We investigated further if the NAD kinase can be activated
by light irradiation in vitro. For these experiments mitochondria from dark grown corn coleoptiles were isolated and
incubated at 4°C for one hour in darkness or in red or far red
light. No significant change in the NAD kinase activity could
be detected.
        Thus the light irradiation do not alter the properties of
the NAD kinase with respect to total activity, calcium- and
calmodulin-dependency or localisation.

Table 1.   Calmodulin Content of Coleoptiles from
           Dark and Far Red Light Grown Corn Seedlings

	Calmodulin Content (mg/kg fresh weight)
dark	3.3 ± 0.7
far red	2.6 ± 0.5

Effect of light on the calmodulin content of corn coleoptiles

In order to investigate whether far red light irradiation of corn seedlings (in vivo) causes a change in the calmodulin content of coleoptiles we determined the amount of calmodulin in both tissues using a calmodulin-deficient cAMP phospho-diesterase and calmodulin from bovine brain as a standard (Dieter, 1980). As it can be seen from Table 1 no significant change in the calmodulin content could be detected after the light irradiation.

Effect of light on the ratio of the nicotinamide nucleotides

The ratio of $NAD(H_2)/NADP(H_2)$ was determined in coleoptiles from dark and far red light grown corn seedlings (in vivo)and in segments from dark grown corn coleoptiles which were irradiated at $4^{\circ}C$ for one hour with red or with fared light or were kept in darkness (in vitro). It can be seen in Table 2 that if the light irradiation has been carried out in vivo a significant decrease of the ratio could be measured whereas an irradiation in vitro causes no significant change in the ratio of the nucleotides. The decrease in the ratio was primarily due to a decrease of NAD and a corresponding increase of NADP (manuscript in preparation).

In order to investigate whether the light induced increase of NADP may be due to an increase of the cellular calcium concentration we incubated segments of coleoptiles from dark grown corn seedlings in the presence of the calcium ionophore A 23187 and the absence or presence of high calcium (with 1 mM EGTA or with 1 mM $CaCl_2$). Table 2 show that the presence of high calcium in the incubation medium causes a decrease of the ratio of $NAD(H_2)/NADP(H_2)$. Again this decrease was due to a decrease of NAD and an increase of NADP (manuscript in preparation). Thus an experimental increase of the cellular calcium concentration in corn coleoptiles leads to about the same effect as a light irradiation in vivo.

Table 2. Effect of Light and the Calcium Ionophore A 23187 on the Ratio of the Nicotinamide Nucleotides in Corn Coleoptiles

		$NAD(H_2)/NADP(H_2)$
Light Irradiation		
in vivo	-	1.6
	+	0.8
in vitro	-	1.9
	+	2.0
A 23187	+ EGTA	4.4
	+ $CaCl_2$	2.1

CONCLUSION

The experiments shown in this paper demonstrate that a light irradiation of intact corn seedlings leads to an increase of NADP but does not change the properties of the NAD kinase, the only enzyme known which phosphorylates NAD to NADP. The properties investigated include total activity of the NAD kinase in the tissue, the dependency on calcium and calmodulin and the cellular distribution. We further demonstrated that the cellular content of calmodulin is the same in dark and light grown corn coleoptiles. Thus neither the enzyme nor the activator protein of the enzyme, calmodulin, seems to be altered by the light irradiation.

Since a light irradiation of the mitochondrial associated NAD kinase in vitro has also no effect on the enzyme activity we suggest that the light dependent increase of NADP in vivo can not be due to a photoregulation on the enzyme level.

Based on these data we propose the following hypothesis for the light-dependent regulation of the NAD kinase (Fig. 3).

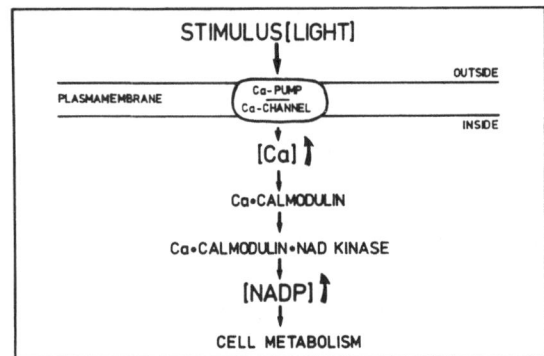

Fig. 3. Stimulus response coupling during the light-dependent regulation of the NAD kinase in plant cells.

Light increases the cytoplasmic free calcium concentration via an alteration of the activity of the calcium pump or via change in the properties of calcium channels. The increase of the free calcium concentration in the cytoplasm leads consequently to the formation of the "active" calcium-calmodulin complex (see Fig. 2). The calcium-calmodulin complex binds to the NAD kinase in its inactive form, this binding leads to the activation of the enzyme and consequently to an increase of NADP. Since NAD and NADP are cofactors in many metabolic key processes in plant cells and enzymes have clear preferences for NAD or for NADP, an alteration in the concentrations of NAD and NADP is of physiological importance. The activation of the NAD kinase can be reversed by extrusion of calcium out of the cytoplasmic space.

This hypothesis is supported by the folloeing experimental evidence:
- Light is able to decrease the calcium transport activity of the calcium pump in the plasmamembrane and stimulates the calcium uptake into maize protoplasts (Das and Sopri, 1985).

- The NAD kinase activity is almost totally dependent on calcium and calmodulin (Fig. 1 and Fig. 2).
- Light does not alter either the properties of the NAD kinase nor the content of the cellular calmodulin.
- An increase of the cellular calcium concentration is able to imitate the light effect (Table 2).

The direct measurement of the cellular free calcium concentration in plant cells or in plant tissues using e.g. the fluorescent techniques such as the quin 2 method (Tsien, 1980) will show if calcium fluxes exist during stimulation of plant cells and if calcium acts in plant cells as in animal cells as a second messenger.

REFERENCES

Anderson, J.M., and Cormier, M.J., 1978 , Calcium dependent regulator of NAD kinase in higher plants, Biochem. Biophys. Res. Commun., 84, 595-602

Anderson, J.M., Charbonneau, H., Jones, H.P., McCann, R.O., and Cormier, M.J., 1980, Characterization of the plant nicotin-amide adenine dinucleotide kinase activator protein and its identification as calmodulin, Biochemistry, 19, 3113-3120

Das, R., and Sopori, S.S., 1985, Evidence of regulation of calcium uptake by phytochrome in maize protoplasts, Biochem. Biophys. Res. Commun. 128, 1455-1460

Dieter, P., 1980, PhD Thesis, University of Freiburg, Freiburg, FRG.

Dieter, P., and Marmé, D., 1980, Partial purification of plant NAD kinase by calmodulin-Sepharose affinity chromatography, Cell Calcium, 1, 279-286

Dieter, P., and Marmé, 1981, Far red light irradiation of intact corn seedlings affects mitochondrial and microsomal calcium transport, Biochem. Biophys. Res. Commun. 101, 749-755

Dieter, P., 1984, Calmodulin and calmodulin-mediated processes in plants, Plant, Cell, Env., 7, 371-380

Dieter, P., and Marmé, 1984, A calcium, calmodulin dependent NAD kinase from corn is located in the outer mitochondrial membrane, J. Biol. Chem., 259, 184-189

Dieter, P., 1985a, Calcium und seine Bedeutung in Regulations-mechanismen der Pflanzen, in: " Physiologische Schlüssel-prozesse in Pflanze und Insekt", P. Böger, ed., Universitäts-verlag Konstanz, Konstanz, FRG.

Dieter, P., 1985b, Calcium and calmodulin, in: " Models in Plant Physiology/Biochemistry/Technology ", D.Newman, K.Wilson, eds., CRC Press Inc., in press

Dieter, P., and Marmé, D., 1985, Calcium-binding and its effect on circular dichroism of plant calmodulin, Planta, in press

Hopkins, D.W., and Briggs, W.R., 1973, Phytochrome and NAD kinase: a reexamination, Plant Phys. Suppl., 51, 52 (284)

Marmé, D., and Dieter, P., 1983, The role of calcium and cal-modulin in plants, in: " Calcium and Cell Function ", Vol. 4, pp. 263-311, W.Y.Cheung, ed., Academic Press, New York, London.

Matsumara, H., and Miyachi, S., 1980, Cycling assay for nicotin-amide adenine dinucleotides, Methods Enzymol., 69, 465-470

Muto, S., and Miyachi, S., 1977, Properties of a protein activator of NAD kinase from plants, Plant Physiol., 59, 55-60

Muto, S., Miyachi, S., Usuda, H., Edwards, G.E., and Bassham, J.A., 1981, Light-induced conversion of nicotinamide adenine dinucleotide to nicotinamide adenine dinucleotide phosphate in higher plant leaves, Plant Physiol., 68, 324-328

Simon, P., Bonzon, M., Greppin, H., and Marmé, D., 1984, Subchloroplastic localization of NAD kinase activity at the envelope and for a calcium-calmodulin-independent activity in the stroma of pea chloroplasts, FEBS Letters, 159, 332-338

Tezuka, T., and Yamamoto, Y., 1972, Photoregulation of nicotinamide adenine dinucleotide kinase activity in cell free extracts, Plant Physiol., 50, 458-462

Tsien, R.Y., 1980, New calcium indicators and buffers with high selectivity against magnesium and protons: design, synthesis and properties of prototype structures, Biochemistry, 19, 2396-2404.

CALCIUM-DEPENDENT PROTEIN PHOSPHORYLATION IN SUSPENSION-CULTURED SOYBEAN CELLS[1]

Cindy L. Putnam-Evans, Alice C. Harmon, and Milton J. Cormier

Department of Biochemistry
University of Georgia
Athens, GA   30602

The discovery of the calcium-binding protein calmodulin in plants (Anderson and Cormier, 1978; Anderson et al., 1980) provided the basis for suggesting that Ca^{2+} serves a second messenger role in plants, and that Ca^{2+}-dependent metabolic regulation in plant cells may be mediated by such Ca^{2+}-binding proteins (Anderson et al., 1980). Support for these hypotheses has come from the demonstration that enzymes such as pea NAD kinase (Anderson and Cormier, 1978; Anderson et al., 1980) and Ca^{2+}-transport ATPases of zucchini (Dieter and Marme, 1980) and corn (Dieter and Marme, 1981) are activated by calcium and calmodulin. Recently, several investigators have observed calcium-dependent and possibly calmodulin-dependent phosphorylation of endogenous proteins in plant extracts (Hetherington and Trewavas, 1982; Salimath and Marme, 1983; Veluthambi and Poovaiah, 1984a; 1984b; Putnam-Evans and Cormier, 1984). Also, Ca^{2+}-dependent protein kinases have been partially purified from wheat germ (Polya and Davies, 1982; Polya et al., 1983; Polya and Micucci, 1984) and soybean cells (Putnam-Evans and Cormier, 1984). The regulation of protein phosphorylation by calcium may be a mechanism of metabolic and physiological control in plants, as it is in animals.

To better understand the role of Ca^{2+}-dependent protein phosphorylation in plant cell function, we are purifying and characterizing a Ca^{2+}-dependent protein kinase from suspension-cultured soybean cells. We have previously shown that crude extracts of the soybean cells contain soluble protein kinase(s) that preferentially phosphorylates at least eight endogenous proteins in vitro in the presence of 0.5 mM $CaCl_2$ (Putnam-Evans and Cormier, 1984). To facilitate purification of a Ca^{2+}-dependent kinase, we developed an assay based on phosphorylation of the artificial substrate histone Hl. Phosphorylation of histone Hl by kinase activity in crude extracts was both greater and more dependent on calcium that was phosphorylation of casein, phosvitin, or other histones (Putnam-Evans and Cormier, 1984). We have purified the Ca^{2+}-dependent kinase approximately 100-fold, and we examine the effect of calcium and calmodulin on the enzyme's activity in this report.

[1]This work was supported by grants from the Department of Energy (DE-AS09-83ER13107) and the National Science Foundation (PCM-8213177).

MATERIALS AND METHODS

TAPP-Sepharose[2] was synthesized according to Hart et al. (1983b). W-7 and W-5 were synthesized according to Hidaka, et al. (1978) and Hart et al. (1983a). Histone H1 (Type III-S) was obtained from Sigma, St. Louis, MO.

The hypocotyl-derived soybean cell cultures (Glycine max L. Wayne) were obtained from Dr. Joe L. Key, University of Georgia. The cultures were grown at 30°C in the dark with shaking (175 rpm) in Murishige and Skoog (1962) medium supplemented with 20 g/l sucrose, 1 mg/l napthalene acetic acid, and 0.5 mg/l kinetin. The cells were subcultured weekly.

Soybean cells were harvested in log-phase growth, washed and resuspended in cold homogenization buffer (0.4 sorbitol, 10 mM $MgCl_2$, 20 mM Tris pH 8.5) and then ruptured in a chilled French press. All subsequent operations were performed at 4°C. The homogenate was centrifuged for 30 min. at 19,700 x g and 0.5 mM PMSF and 5 mM EDTA were added to the supernatant. The supernatant was loaded on a column of DEAE-Cellulose equilibrated in 20 mM Tris, 5 mM EDTA, pH 7.5. After the column had been washed with 2 column volumes of equilibration buffer, the protein kinase was eluted with 0.4 M NaCl in equilibration buffer. Fractions containing protein kinase activity were pooled and dialyzed against 20 mM HEPES, pH 7.2. The dialysate was made 0.5 mM in $CaCl_2$ and was loaded on a column of TAPP-Sepharose which had been equilibrated in 20 mM HEPES, 0.5 mM $CaCl_2$, pH 7.2. The column was washed with two column volumes 0.2 M NaCl, 20 mM HEPES, 0.5 mM $CaCl_2$, and then the protein kinase and calmodulin were eluted with 20 mM Tris, 5 mM EGTA, pH 8.0. Fractions containing protein kinase activity were pooled and concentrated to a volume of 2 ml in an Amicon ultrafiltration device. The protein kinase was chromatographed on Sephadex G-100 (1.7 x 117 cm) equilibrated in 20 mM Tris, pH 8.0, and 5 mM EDTA. Fractions of 2 ml volume were collected. The void volume (V_o) was determined with blue dextran, MW = 2,000,000 and the total volume (V_t) was determined with $NaNO_2$. The M_r of the protein kinase was determined from a plot of K_D vs log M_r for the following marker proteins: phosphorylase b, M_r = 97,400; bovine serum albumin, M_r = 66,000; ovalbumin, M_r = 45,000; carbonic anhydrase, M_r = 29,000; and cytochrome C, M_r = 12,400. The protein kinase that eluted from this column was frozen at -80°C.

Protein kinase activity was determined by measuring incorporation of counts from $[\gamma-^{32}P]ATP$ into the artificial substrate histone H1. The standard assay mixture contained, in a volume of 0.15 ml, 1 mg/ml histone H1, 30 µM $[\gamma-^{32}P]ATP$ (37 cpm/pmol), 50 mM HEPES, pH 7.2, 15 µl protein kinase, and $Ca^{2+}/Mg^{2+}/EDTA$ buffer. The $Ca^{2+}/Mg^{2+}/EDTA$ buffers were composed of 5.00 mM EDTA, 10.00 mM $MgCl_2$ (BDH, Analar) and various concentrations of $CaCl_2$. The concentration of free calcium was calculated by a computer program based on the method of Perrin and Sayce (1967). The stability constants reported by Sillen and Martel (1971) for EDTA and ATP complexes were used in the calculations. The free magnesium concentration ranged from 5 mM, when no Ca^{2+} was added, to ⌣9 mM for free Ca^{2+} > 200 µM. Assay mixtures were incubated at 30°C for 6 min, and the reactions were terminated by the addition of 0.5 ml of cold 20% trichloroacetic acid and 0.2% sodium pyrophosphate. The precipitate was collected on Whatman GFA filter paper and washed with 20% TCA followed by ethanol/ether (1:1). Radioactivity was determined by liquid scintillation counting.

---

[2]abbreviations: TAPP, 2-trifluoromethyl 10H-10(3'-aminopropyl) phenothiazine; PMSF, phenylmethylsulfonylfluoride; HEPES, 4-(2-hydroxyethyl)-1-piperazineethanesulfonic acid; CaM, calmodulin.

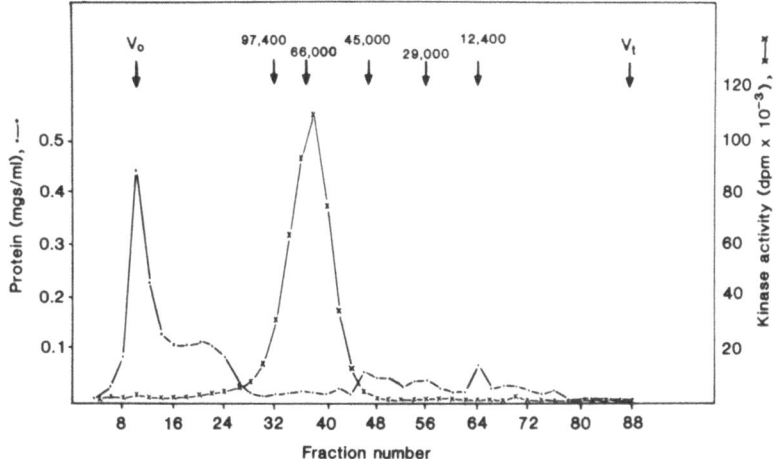

Fig. 1.   Chromatography of Ca^{2+}-dependent protein kinase on
Sephadex G-100.   Aliquots of fractions were assayed
for protein kinase activity in the standard assay mix
with 1 μM calmodulin.   The elution positions of
molecular weight markers are indicated by arrows.

Calmodulin was detected by the assay based on NAD kinase activation
described by Harmon et al. (1984).   NAD kinase used in this assay was
purified through the Cm-cellulose step.   Bovine brain calmodulin was used
as the standard.

Protein was determined by the method of Bradford (1976) with bovine
serum albumin as standard.

RESULTS

The Ca^{2+}-dependent protein kinase co-eluted with calmodulin from both
DEAE-cellulose and TAPP-Sepharose (data not shown).   When the Ca^{2+}-dependent
protein kinase was chromatographed on Sephadex G-100 in the presence of
EDTA, it eluted in a single peak of activity at a position which corresponded
to M_r = 65,000 (Fig. 1).

The time course of histone H1 phosphorylation by G-100-purified kinase
is shown in Fig. 2.   Activity increased linearly for at least 15 min which
indicates that protein kinase at this stage of purification is free from
interfering protein phosphatase activity.   This preparation of enzyme was
60-fold activated by calcium.

The activation by calcium varied from 15- to 60-fold for different
enzyme preparations (compare Fig. 2, Fig. 3, and Table 1).   The specific
activity measured in the presence of calcium varied little, but the
specific activity measured in the absence of calcium varied 3- to 4-fold
(data not shown).   The effect of calcium and calmodulin on the activity of
the G-100 purified enzyme is shown in Table 1.   Phosphorylation of histone
H1 by this enzyme preparation was activated 25-fold by the addition of
calcium, and 35-fold by the addition of both calcium and calmodulin.   The
activity of protein kinase from the two previous purification steps was
also stimulated ∽30% by calmodulin in addition to calcium (data not shown).

Table 1. The Effect of Calcium and Calmodulin on the Activity
of Soybean Calcium-Dependent Protein Kinase

$[Ca^{2+}]_{free}$	[CaM]	Activity pmol/min/mg	Relative Activity
0	0	1,856	0%
86 μM	0	49,494	100%
86 μM	1 μM	64,995	133%

To determine whether the relatively small activation of the protein
kinase by calcium and calmodulin (33% compared to the activation by calcium
alone, Table 1) could have been due to the presence of calmodulin in the
enzyme preparation itself, aliquots from two different preparations of G-100-
purified kinase were tested for calmodulin activity. Both boiled and non-
boiled aliquots of the protein kinase preparations were able to fully
activate the calmodulin-dependent enzyme NAD kinase. These results demonstrated that the protein kinase preparations (total protein = 64 μg/ml or
211 μg/ml) both contained > 1 μg/ml calmodulin.

The effect of free calcium concentration on the activity of G-100-
purified protein kinase is shown in Fig. 3. The concentration of free
calcium required for 50% activation was 2.8 μM. A similar $K_{0.5}$ was
observed for a kinase preparation that was activated 25-fold (data not
shown). Activity was maximal between 10 and 50 μM free calcium and concentrations above 50 μM were inhibitory. The $K_{0.5}$ for crude enzyme was 16 μM,

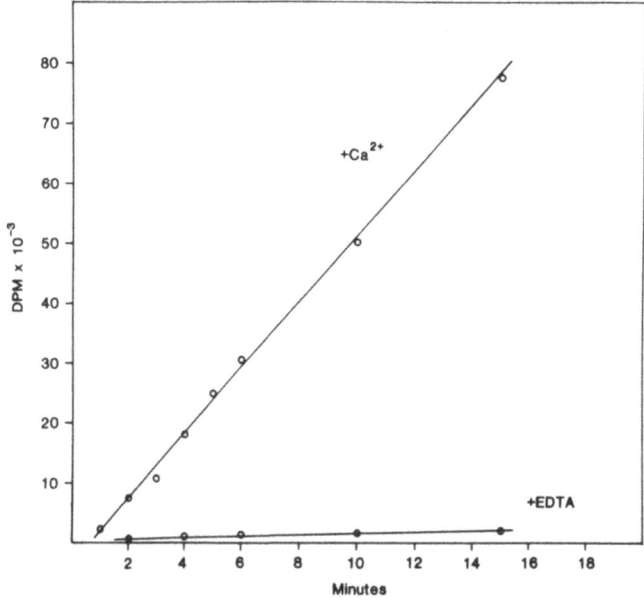

Fig. 2. Time course of histone H1 phosphorylation
catalyzed by soybean Ca^{2+}-dependent protein
kinase. Protein kinase activity was determined in the standard assay mix with either
86 μM free calcium (+Ca^{2+}) or with no added
calcium (+EDTA).

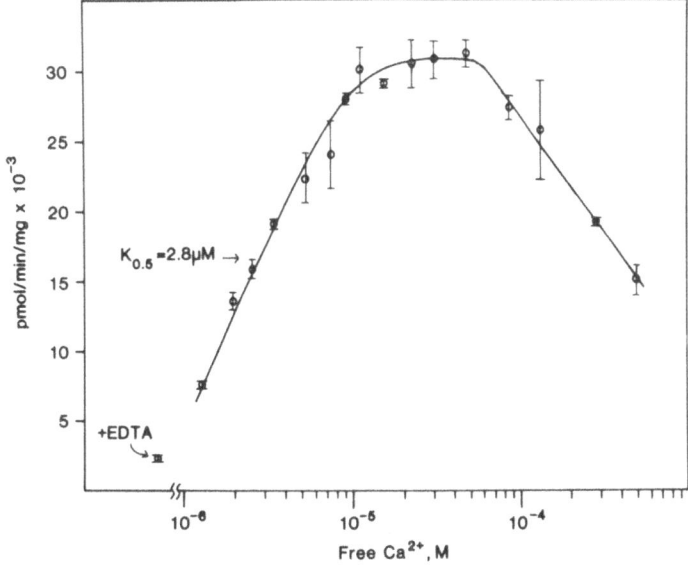

Fig. 3. Calcium concentration dependence of protein
kinase activity. The activity of Sephadex
G-100-purified protein kinase with histone H1
was measured at various free calcium concentrations which were controlled with $Ca^{2+}/Mg^{2+}/EDTA$
buffers as described in Materials and Methods.
The point marked "+EDTA" indicates the activity
with no added calcium.

and the calcium concentration range required for maximal activity of the
crude enzyme was 30 to 200 μM. These observations indicate that competing
Ca^{2+}-binding proteins were removed by the purification procedure.

The effect of the napthalene sulfonamides W-7 and W-5 on the activity
of G-100-purified enzyme are shown in Fig. 4. The IC_{50} for W-7 was 400 μM.
W-5 is insoluble at concentrations in the milimolar range, therefore
concentrations high enough to completely inhibit activity could not be
obtained.

DISCUSSION

We have demonstrated that soybean cells contain a protein kinase that
is highly activated by micromolar concentrations of free calcium. However,
we have not been able to unequivocally show that calmodulin mediates the
enzyme's Ca^{2+}-sensitivity, since we were unable to remove calmodulin from
the enzyme preparations. All calmodulin-dependent enzymes except phosphorylase kinase dissociate from calmodulin when Ca^{2+} is removed by EDTA.
Chromatography of the Ca^{2+}-dependent protein kinase on a gel filtration
column in the presence of EDTA failed to remove calmodulin (MW = 16,700)
from the enzyme (M_r = 65,000). Plant calmodulin elutes from gel filtration
columns at a position corresponding to M_r = 23,300, (Anderson et al., 1980).
Calmodulin may not have been completely resolved from the enzyme although
proteins of these M_rs are well resolved by this column. The presence of
> 1 μg/ml calmodulin in the enzyme preparation could account for the large
stimulation by calcium alone and the relatively small stimulatory effect of
added calmodulin. The $K_{0.5}$s for activation of calmodulin-dependent enzymes

103

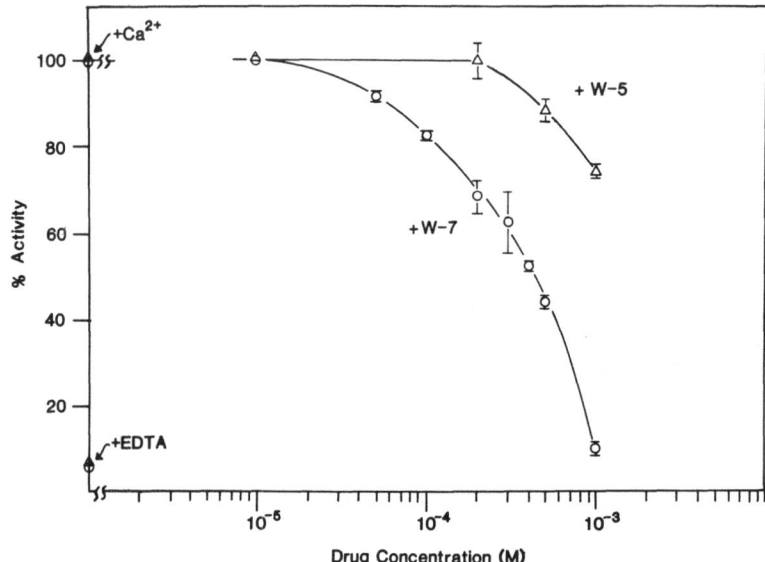

Fig. 4. The effect of W-7 and W-5 on the activity of soybean Ca^{2+}-dependent protein kinase. The activity of Sephadex G-100-purified protein kinase in the standard assay mixture with 86 μM free calcium and various concentrations of either W-7 or W-5 was determined. Activity in the presence of 86 μM free calcium $(+Ca^{2+})$ was defined as 100% activity. Activity with no added calcium is labeled "+EDTA".

other than phosphorylase kinase range from 0.1 to 50 ng/ml.

The effect of the calmodulin-binding compound W-7 on the activity of the protein kinase contradicts the hypothesis that the protein kinase is a conventional calmodulin-stimulated enzyme. The IC_{50} determined for this compound is approximately ten-fold higher than it is for most calmodulin-dependent enzymes (reviewed by Hidaka and Tanaka, 1983).

Another possibility that is consistent with all our data is that calmodulin is a tightly bound subunit of the protein kinase, as is the case for phosphorylase kinase. Calmodulin does not dissociate from phosphorylase kinase when calcium is removed. The $K_{0.5}$ for activation of the phosphorylated form of this enzyme by Ca^{2+} is 1-2 μM (Cohen, 1980). Binding of a second molecule of calmodulin by phosphorylase kinase enhances its activity an additional 2-7 fold (Shenoliker et al., 1979). Pure preparations of phosphorylase kinase are able to activate the calmodulin-dependent enzymes cyclic nucleotide phosphodiesterase and myosin light chain kinase (Cohen, et al., 1978). The calmodulin-binding compound trifluoperazine at a concentration of 500 μM inhibits the activity of phosphorylase kinase by only 50% (Chan & Graves, 1982). This concentration is about 50-fold higher than the IC_{50} for most calmodulin-dependent enzymes (Weiss, 1983). All of these properties of phosphorylase kinase are qualitatively similar to the properties of the soybean Ca^{2+}-dependent protein kinase.

A third possibility could also explain our data. The protein kinase could be activated by calcium directly or by a calcium-binding protein other than calmodulin, with calmodulin simply being a contaminant. Determination of which of these possibilities is correct awaits purification of the enzyme to homogeneity.

REFERENCES

Anderson, J.M., and Cormier, M.J., 1978, Calcium-dependent regulator of NAD-kinase, Biochem. Biophys. Res. Commun., 84: 595.

Anderson, J.M., Charbonneau, H., Jones, H.P., McCann, R.O. and Cormier, M.J., 1980, Characterization of the plant nicotinamide adenine dinucleotide kinase activator protein and its identification as calmodulin, Biochemistry, 19: 3113.

Bradford, M.M., 1976, A rapid and sensitive method for the quantitation of microgram quantities of protein utilizing the principles of protein-dye binding, Anal. Biochem., 72: 248.

Chan, K.-F.J., and Graves, D.J., 1982, Rabbit skeletal muscle phosphorylase kinase, interactions between subunits and influences of calmodulin on different complexes, J. Biol. Chem., 257: 5956.

Cohen, P., 1980, The role of calcium ions, calmodulin, and troponin in the regulation of phosphorylase kinase from rabbit skeletal muscle, Eur. J. Biochem., 111: 563.

Cohen, P., Burchell, A., Foulkes, J.G., Cohen, P.T.W., Vanaman, T.C., and Nairn, A.C., 1978, Identification of the Ca^{2+}-dependent modulator protein as the fourth subunit of rabbit skeletal muscle phosphorylase kinase, FEBS Lett., 92: 287.

Dieter, P. and Marme, D., 1980, Calmodulin activation of plant microsomal Ca^{2+} uptake, Proc. Natl. Acad. Sci. USA, 77: 7311.

Dieter, P., and Marme, D., 1981, A calmodulin-dependent microsomal ATPase from corn (Zea mays L.), FEBS Lett., 125: 245.

Harmon, A.C., Jarrett, H.W., and Cormier, M.J., 1984, An enzymatic assay for calmodulins based on plant NAD kinase activity, Anal. Biochem., 141: 168.

Hart, R.C., Bates, M.D., Cormier, M.J., Rosen, G.M., and Conn, P.M., 1983a, Synthesis and characterization of calmodulin antagonistic drugs, Meth. Enzymol., 102: 195.

Hart, R.C., Hice, R.E., Charbonneau, H., Putnam-Evans, C., and Cormier, M.J., 1983b, Preparation and properties of calcium-dependent resins with increased selectivity for calmodulin, Anal. Biochem., 135: 208.

Hetherington, A., and Trewavas, A., 1982, Calcium-dependent protein kinase in pea shoot membranes, FEBS Lett., 145: 67.

Hidaka, H., Asano, M., Iwadare, I., Totsuka, T., and Aoki, N., 1978, A novel vascular relaxing agent, N-(6-aminohexyl)-5-chloro-1-naphthalene-sulfonamide which affects vascular smooth muscle actomyosin, J. Pharmacol. Exp. Ther., 207: 8.

Hidaka, H., and Tanaka, T., 1983, Napthalene-sulfonamides as calmodulin antagonists, Meth. Enzymol., 102: 185.

Murishige, T., and Skoog, F., 1962, A revised medium for rapid growth and bioassays with tobacco tissue cultures, Physiol. Plant., 15: 473.

Perrin, D.D., and Sayce, I.G., 1967, Computer calculation of equilibrium concentrations in mixtures of metal ions and complexing species, Talanta, 14: 833.

Polya, G.M., and Davies, J.R., 1982, Resolution of Ca^{2+}-calmodulin-activated protein kinase from wheat germ, FEBS Lett., 150: 167.

Polya, G.M., Davies, J.R., and Micucci, V., 1983, Properties of a calmodulin-activated Ca^{2+}-dependent protein kinase from wheat germ, Biochim. Biophys. Acta, 761: 1.

Polya, G.M., and Micucci, V., 1984, Partial purification and characterization of a second calmodulin-activated Ca^{2+}-dependent protein kinase from wheat germ, Biochim. Biophys. Acta, 785: 68.

Putnam-Evans, C., and Cormier, M.J., 1984, Calcium-dependent protein kinase activity in a plant cell line, Fed. Proc. Abst., 43: 296.

Salimath, B.P., and Marme, D., 1983, Protein phosphorylation and its regulation by calcium and calmodulin in membrane fractions from zucchini hypocotyls, Planta, 158: 560.

Shenolikar, S., Cohen, P.T.W., Cohen, P., Nairn, A.C., and Perry, S.V., 1979, The role of calmodulin in the structure and regulation of phosphorylase kinase from rabbit skeletal muscle, Eur. J. Biochem., 100: 329.

Sillen, L.G., and Martell, A.E., 1971, Stability Constants of Metal-ion Complexes, Second Edition, Burlington House, London.

Veluthambi, K., and Poovaiah, B.W., 1984, Calcium-promoted protein phosphorylation in plants, Science, 223: 167.

Veluthambi, K., and Poovaiah, B.W., 1984, Calcium- and calmodulin-regulated phosphorylation of soluble and membrane proteins from corn coleptiles, Plant Physiol., 76: 359.

Weiss, B., 1983, Techniques for measuring the interaction of drugs with calmodulin, Meth. Enzymol., 102: 171.

ROLES OF CALMODULIN DEPENDENT AND INDEPENDENT NAD KINASES IN REGULATION

OF NICOTINAMIDE COENZYME LEVELS OF GREEN PLANT CELLS

Shoshi Muto and Shigetoh Miyachi

Institute of Applied Microbiology
University of Tokyo
Bunkyo-ku, Tokyo 113, Japan

INTRODUCTION

Light-induced conversion of NAD to NADP occurs in algal cells (Oh-hama and Miyachi, 1959) and higher plant leaves (Ogren and Krogman, 1965). The conversion was highly associated with photosynthetic electron transport and photophosphorylation (Matsumura-Kadota et al., 1982). Studies with proto-plasts and intact chloroplasts isolated from higher plant leaves showed that the conversion reaction occurred in the chloroplasts and that NAD kinase by which this reaction was catalyzed, was mostly localized in the chloroplasts (Muto et al., 1981). During purification from pea seedlings, NAD kinase was totally inactivated because of the removal of a protein activator (Muto and Miyachi, 1977), which was subsequently identified as calmodulin (CaM) (Anderson and Cormier, 1978; Anderson et al., 1980). Recently, we found that CaM concentration in the chloroplast was sufficent to saturate NAD kinase located therein (Muto, 1982), and that the intact chloroplasts from wheat and spinach leaves actively took up Ca^{2+} when illuminated (Muto et al., 1982). Based on these findings, we proposed that the light-induced conversion of NAD to NADP was catalyzed by the CaM dependent NAD kinase, and was regulated by Ca^{2+} flux into the stroma in the light. However, it was reported that NAD kinase was located in cytoplasm and chloroplasts of spinach leaves, and that the cytoplasmic enzyme was CaM dependent while the chloroplastic enzyme was CaM independent and located exclusively in the stroma (Simon et al., 1982). Recently, the CaM dependent and the indepen-dent NAD kinases were reported to be localized in the envelope and the stroma of pea chloroplasts, respectively (Simon et al, 1984). The subcel-lular and the subchloroplastic distribution of these enzymes is very impor-tant to learn the roles of them and of Ca^{2+}/CaM in regulation of cellular and subcellular levels of nicotinamide coenzymes. However, information on the distribution of these enzymes is contradictory as mentioned above.

In this article, we report the localization of the CaM dependent and the independent NAD kinases in several plants and discuss their roles in the regulation of nicotinamide conezyme levels.

MATERIALS AND METHODS

Plant Materials and Protoplast Preparation
Pea (Pisum sativum L.) and wheat (Triticum aestivum L.) were grown in a greenhouse (Muto and Miyachi, 1977). Spinach (Spinacia oleracea L.) was

purchased from a local market. Pea and wheat protoplasts were prepared as previously described (Muto et al., 1981). Spinach protoplasts were prepared essentially according to Takebe et al. (1968).

Subcellular and Subchloroplastic Fractionations

Protoplasts were fractionated into chloroplastic, mitochondrial and cytolasmic fractions as described (Muto, 1982). Chloroplasts were also isolated according to Jensen and Bassham (1966) and purified by precipitating through 40% Percoll (Mills and Joy, 1980). Intact chloroplasts were osmotically broken and fractionated into stroma and thylakoid (Muto, 1982). Chloroplast envelope membranes were isolated and subfractionated by the method of Cline et al. (1981).

Enzyme Assay

The assay mixture for total NAD kinase (calmodulin dependent plus independent) activity contained in 0.5 ml, 3 mM ATP, 3 mM NAD, 5 mM $MgCl_2$, 0.1 mM $CaCl_2$, 60 nM spinach calmodulin and 100 mM Tricine-KOH buffer (pH 8.0). The calmodulin independent activity was assayed in the presence of 1 mM EGTA in the same assay mixture. The calmodulin dependent activity was expressed by total minus calmodulin independent activity. Since some subcellular fractions, especially cytoplasmic fraction, contained NADP reducing activity, the product of NAD kinase reaction was reduced during the enzyme assay. This caused underestimations of the enzyme activity when the reaction was terminated by adding HCl, because of instability of reduced NADP at acidic pH. Thus two assays, one terminated by HCl and another by NaOH, were run and two activities were added up. After neutralization, NADP and NADPH were assayed by an enzymatic cycling method (Muto et al., 1981)

Permeability of Chloroplast Envelope Membrane for Calmodulin

Spinach calmodulin was iodinated as described previously (Muto and Miyachi, 1984). Intact pea chloroplasts were incubated with ^{125}I-labeled CaM for 10 min at room temperature in the media with various osmolarity. After centrifugation through a silicon oil layer (Muto et al., 1982), radioactivities in the chloroplast pellet and the supernatant reaction medium were determined.

Trypsin Treatment of Chloroplasts

Intact pea chloroplasts (0.45 mg chl/ml) were incubated with trypsin (10 μg/ml) at 25°C and the reaction was terminated by adding soybean trypsin inhibitor (40 μg/ml). Activities of NAD kinases were assayed after osmotic shock.

Light-induced Conversion of NAD to NADP

This was done as previously described (Muto et al., 1981) in the presence and absence of trifluoperazine. Illumination was carried out for 15 min at 25°C.

Other Assays

Chlorophyll and protein were assayed as described (Muto et al., 1981).

RESULTS

Subcellular Distribution of NAD kinases

Protoplast homogenates were fractionated into the chloroplastic. mitochondrial and cytoplasmic fractions, and the CaM dependent and the independent NAD kinases were assayed (Table 1). Subcellular distribution of two type of enzymes were different from plant to plant, but the highest total activity was located in the chloroplasts of all plants examined. The CaM dependent enzyme of the chloroplasts of pea and wheat were ca. 40 and 10% of total cellular activity, respectively. With most of spinach (Spinach, A in Table 1), NAD kinase in the chloroplasts was CaM independent. However, the chloroplasts from some spinach contained the CaM dependent enzyme (Spinach, B in Table 1).

Subchloroplatic Distribution of NAD kinases

When intact chloroplasts were suspended in isotonic or hypertonic reaction medium, both the CaM dependent and the independent NAD kinases

Table 1. Subcellular distribution of calmodulin dependent and independent
NAD kinases in leaf mesophyll cells

Plant	Subcellular fraction	Enzyme activity (nmol NADP/mg chl* ·h) (%)		
		Total	CaM independent	CaM dependent
Pea	Chloroplastic	243.1 (85.2)	131.0 (46.1)	111.1 (39.1)
	Mitochondrial	8.9 (3.1)	2.7 (1.0)	6.2 (2.2)
	Cytoplasmic	33.0 (11.6)	15.2 (5.4)	17.8 (6.3)
Wheat	Chloroplastic	63.8 (63.7)	53.0 (52.9)	10.8 (10.8)
	Mitochondrial	1.3 (1.3)	0.6 (0.6)	0.7 (0.7)
	Cytoplasmic	35.0 (35.0)	21.6 (21.6)	13.4 (13.4)
Spinach,A	Chloroplastic	390.9 (85.9)	368.6 (81.0)	22.3 (4.9)
	Mitochondrial	3.1 (0.7)	1.5 (0.3)	1.6 (0.4)
	Cytoplasmic	61.1 (13.4)	61.4 (13.5)	-0.3 (—)
Spinach,B	Chloroplastic	130.5 (96.9)	8.1 (6.0)	122.4 (90.9)
	Mitochondrial	0.5 (0.4)	0.5 (0.4)	0 (0)
	Cytoplasmic	3.7 (2.7)	3.7 (2.7)	0 (0)

* Chlorophyll was assayed before fractionation of the protoplasts.

could be assayed. Upon osmotic shock, the activities were increased. The
increase of the CaM independent activity was higher than that of the CaM
dependent one, suggesting that the later enzyme was localized in the
chloroplast envelope while the former in the stroma. This was confirmed by
trypsin treatment. When intact pea chloroplasts were treated with trypsin,
the CaM dependent activity was completely inactivated, while the CaM
independnet activity was not changed (Fig. 1). When the intact chloroplasts
were treated with insoluble trypsin (attached to polyacrymide, Sigma), a
part of the CaM independent NAD kinase was inactivated retaining full
activity of the CaM independent enzyme. This indicates that at least a part
of the CaM dependent enzyme is located outer surface of the outer envelope

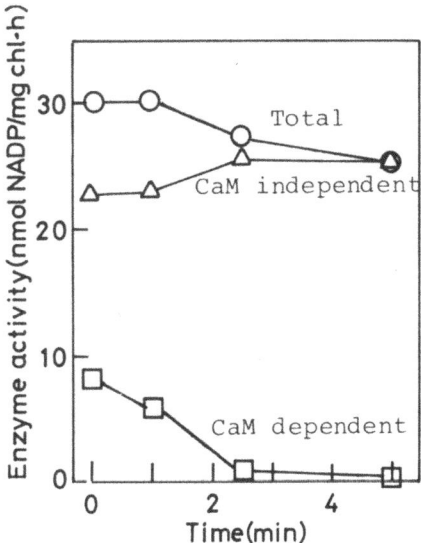

Fig. 1. Effect of trypsin on the
activities of NAD kinases
of intact pea chloroplasts.

Table 2. The NAD kinase activities in the inner and the outer envelope membranes of pea chloroplast

Membrane	Addition	Total activity (nmol NADP/h)	Specific activity (nmol NADP/mg protein·h)
Inner	1 mM EGTA	0	——
	0.1 mM CaCl$_2$	0	——
	0.1 mM CaCl$_2$+60 nM calmodulin	41.75	334.0
	0.1 mM CaCl$_2$+60 nM calmodulin +10 uM calmidazolium	0	——
Outer	1 mM EGTA	0	——
	0.1 mM CaCl$_2$	2.06	2.88
	0.1 mM CaCl$_2$+60 nM calmodulin	177.3	248.0
	0.1 mM CaCl$_2$+60 nM calmodulin +10 uM calmidazolium	2.06	2.88

Table 3. Permeability of pea chloroplasts for ^{125}I-labeled calmodulin

Sorbitol concentratioin	Fraction	cpm	%
0.132 M	Supernatant	169,563	99.92
	Pellet	141	0.08
0.33 M	Supernatant	146,213	99.91
	Pellet	130	0.09
0.5 M	Supernatant	169,088	99.30
	Pellet	1,198	0.70

and that soluble trypsin can penetrate the outer envelope.

In order to further examine the subchloroplastic distribution of NAD kinases, intact pea chloroplasts were fractionated into stroma, thylakoid and envelope. The activity of NAD kinases rapidly decreased during washing of thylakoids with the chloroplast suspension medium. After washing for 3 times, only 2.6% of the total chloroplastic activity remained in the thylakoid. This indicates that the thylakoid does not contain NAD kinase. The stromal and the envelope fractions were obtained by centrifuging the post-thylakoid supernatant at 150,000 x$\underline{g}$ for 1 h, and the activities of the CaM dependent and the independent NAD kinases were assayed. About 90% of the stromal activity was CaM independent, while as much as 90% of the envelope activity was CaM dependent. The envelope membrane was further fractionated into the inner and the outer membranes. As shown in Table 2, NAD kinase was localized in both membranes, and essentially all of the activities were CaM dependent. The outer membrane contained 4 times higher activity than the inner membrane. However, the specific activity was rather higher in the innner membrane. The CaM dependent NAD kinase in the inner membrane should be located at the cytoplasmic side of the chloloplast, since trypsin treatment of the intact chloroplasts completely inactivated the CaM dependent enzyme (Fig. 1). Similar results on subchloroplastic distribution of the enzymes were obtained with wheat chloroplasts. However, most of spinach chloroplasts (spinach, A in Table 1) contained the CaM independent NAD kinase exclusively in the stroma.

Table 4.   Effect of illumination on the calmodulin dependent and the independent NAD kinases of pea chloroplast

| Plant | Treatment | Enzyme activity (nmol NADP/mg chl·h)   (%) | |
		CaM independent	CaM dependent
Pea	Dark * Light*	79.8 (100) 111.4 (140)	22.4 (100) 23.5 (105)
Wheat	Dark * Light*	25.0 (100) 35.2 (141)	227.5 (100) 239.5 (105)

*Illuminated at 20,000 erg/cm^2·s with an incandescent lamp for 5 min. Temperature was at 25°C.

The facts described above indicate that the chloroplasts containes both the CaM dependent and the independent NAD kinases and the former is exclusively located in the envelope and the later in the stroma, and the thylakoid contains no NAD kinase.

Calmodulin Permeability to Membranes of Chloroplast Envelope

NAD kinase in the envelope membranes was inactive without externally added CaM indicating that the absence of CaM in these membranes. A radio-immunoassay has shown that most of the cellular CaM is localized in cyto-plasm and only 1-2% was in the chloroplasts of wheat leaves (Muto, 1982). If CaM could permeate the outer envelope membranes and distribute between the outer and the inner membranes, the intact chloroplasts would contain the above mentioned small amount of CaM, and the CaM dependent NAD kinase in the inner membranes would be activated. The permeability of the outer envelope of pea chloroplasts for CaM was examined by silicon layer filtering centrifugation. Incorporation of ^{125}I-labeled CaM into the intact chloroplasts was increased when the chloroplasts were suspended in a hypertonic medium where the inter membrane space between the outer and the inner envelope was increased (Table 3). This indicates that CaM permeates the outer envelope membrane and could activate the CaM dependent NAD kinase in the inner membrane and inner side of the outer membrane.

Photoactivation and Kinetic Nature of the CaM independent NAD Kinase

We have reported the photoactivation of NAD kinase in pea and wheat leaf cells (Muto et al., 1981). This was reexamined to know whether chloro-plast NAD kinases are activated or not, and whether both the CaM dependent and the independent NAD kinases are activated or only one of them is activated. Table 4 shows that the CaM independent NAD kinase was activated by 40% both in pea and wheat chloroplasts, while the CaM dependent enzyme was essentially unaffected. This suggests that the CaM independent enzyme is important in the light-induced conversion of NAD to NADP in the chloro-plasts, but the CaM dependent enzyme is not involved in this reaction. As the activation of CaM independent NAD kinase was rather small when compared with other photoactivatable enzymes (Anderson et al., 1982), one may assume that this activation is insufficient to support the light-induced conversion of NAD to NADP. The activation was assayed in the reaction mixture with saturated concentrations of substrates for the partially purified NAD kinase from pea (Muto and Miyachi, 1983), while the stromal concentration of ATP (Kobayashi et al., 1979) and NAD (Muto et al., 1982) in the dark were reported as about 0.5 and 0.7 mM, respectively. To know whether these con-cenrations were sufficiently high for saturating the CaM independent NAD kinase in the chloroplasts, the kinetic nature of the CaM independent enzyme was determined using chloroplast stroma of pea as enzyme source. The enzyme was inhibited by free ATP, and exhibited the highest activity at

Table 5. Photoactivation of the calmodulin independent NAD kinase of pea
chloroplast at physiological substrates concentrations

Treatment	Enzyme activity (nmol NADP/mg chl·h)
Dark	3.32
Light	8.26
Dark minus internal NADP[*]	1.32
Light plus internal NADP[*]	4.96
(Activation)	(376%)

[*] Internal NADP: Level of NADP plus NADPH in chloroplasts was increased
from 2.0 to 3.3 nmol/mg chl during illumination, thus
these values were substracted from the respective enzyme
activity.

ATP to Mg^{2+} ratio of 1. Thus the substrate was $MgATP^{2-}$ as reported in the
CaM dependent enzyme (Muto and Miyachi, 1983). The enzyme showed sigmoidal
responces to both substrates and $S_{0.5}$ values were 0.4 and 0.45 mM for NAD
and $MgATP^{2-}$, respectively. These values were comparable for the stromal
concentrations of ATP and NAD. The sigmoidal saturation with $MgATP^{2-}$ is
especially important in the regulation of this enzyme, since the stromal
concentrations of ATP (Kobayashi et al., 1979) and Mg^{2+} (Portis and Heldt,
1976) are increased by illumination. When the enzyme was assayed with the
stromal levels of substrates, it was markedly activated by illumination
(Table 5). This suggests that the photoactivation of the CaM independent
NAD kinase, in addition to the sigmoidal responce in the substrate
saturation, is important to proceed the light-induced conversion of NAD to
NADP in the chloroplasts. The optimum pH of the enzyme is also preferable
to operate this reaction, since the stromal pH is increased to around 8 upon
illumination (Heldt et al., 1973).

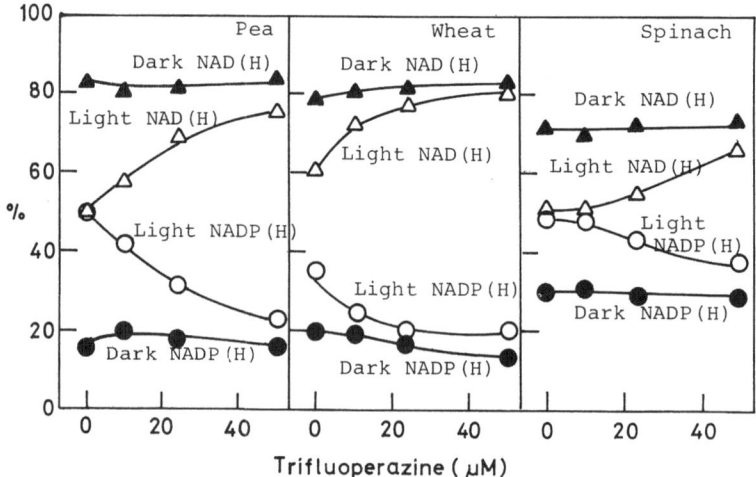

Fig. 2. Effect of trifluoperazine on the light-induced conversion of NAD
to NADP in chloroplasts.

Light-induced Conversion of NAD to NADP in the Chloroplasts

The light-induced conversion of NAD to NADP has been reported in wheat, pea and spinach (Muto et al., 1981), and algal cells such as Chlorella (Oh-hama and Miyachi, 1959; Matsumura-Kadota et al., 1982) and Chlamydomonas (Ogren and Krogman, 1965). Among these plants, the CaM dependent NAD kinase was present in pea, wheat and some of spinach, however, the presence of this enzyme has not been tested with algal cells, thus the survey was extended to several species of algae and their mutant strains including Anacystis nidulans, Chlamydomonas reinhardi wt., C. reinhardi cw-15 (wall-less), Chlorella vulgaris 11h., C. vulgaris 211-11h/20 (chlorophyll-less), C. vulgalis #125 (chlorophyll- and carotenoid-less) and Senedesmus obliquus. The NAD kinase activity in these cells was inhibited by neither EGTA (1 mM) nor a CaM antagonist trifluoperazine (60 µM) indicating the absence of the CaM dependent NAD kinase in algal cells. Jarrett et al. (1982) reported that trifluoperazine inhibited the light-induced conversion of NAD to NADP in pea chloroplasts via the inhibition of the CaM dependent NAD kinase. Our results presented above, however, are not in accord with their conclusion. We reexamined the effect of trifluoperazine on the conversion reaction in the chloroplasts from pea and wheat in which the CaM dependent NAD kinase is present, and the spinach in which this type of enzyme is absent. The converion occured in all of these chloroplasts and was inhibited by trifluoperazine (Fig. 2). This suggests that the effect of this compound is not mediated by CaM, but caused by the inhibition of electron transport through photosystem II as reported by Barr et al. (1982).

DISCUSSION

Though a great variation in the subcellular distribution of the CaM dependent and the independent NAD kinases was obvious among plant species, in general the chloroplast was the main compartment of the enzyme localization. Photosynthesis is the central physiological process in green plant cells and requires NADP for its operation. This may be the reason for the most abundant occurence of the NAD kinases in the chloroplast. The sub-chloroplastic fractionation showed that the CaM dependent NAD kinase was localized in the envelope, while the independent enzyme in the stroma exclusively. The CaM independent NAD kinase had a sigmoidal responce, while CaM independent one showed hyperbolic response to $MgATP^{2-}$. Both activities were optimum at pH 8. In the illuminated chloroplasts, ATP is produced by photophosphorylation, protons are transported into the thylakoid with antiport of Mg^{2+} resulting in increases of $MgATP^{2-}$ concentration and pH of the strome (to ca. 8). In addition to photoactivation, the storomal environment of illuminated chloroplasts is favourable to the CaM independent NAD kinase for catalyzing the phosphorylation of NAD. The light-induced conversion of NAD to NADP in the chloroplast is, therefore, caltalyzed by the photoactivated CaM independent NAD kinase using ATP produced by photophosphorylation.

The CaM dependent NAD kinase was localized in both the inner and the outer envelope membranes. The susceptibility to trypsin indicates that the enzyme faces to the cytoplasmic side of chloroplast. Thus the enzyme should function in the intermembrane space and the surface of the chloroplast. Because of permeability of the outer envelope for CaM, the enzyme could be activated when Ca^{2+} is available. The light-induced uptake of Ca^{2+} by intact wheat and spinach chloroplasts has been reported (Muto et al,, 1982). This was confirmed by Kreimer et al.(1985) in the spinach chloroplast. The influx of Ca^{2+} may bring about increase of stromal Ca^{2+} concentration, but this Ca^{2+} seems not likely to activate the NAD kinase. If the light-induced uptake of Ca^{2+} by the intact chloroplast would bring about a temporal increase of Ca^{2+} in the interspace of the envelope membranes or the surface of the chloroplast, the CaM dependent NAD kinase would be activated. Regulation of the CaM dependent NAD kinase in the envelope by Ca^{2+} flux is unclear at present. However, the enzyme may have an important role in the operation of the indirect transport of reducing power produced by photosynthesis via dihydroxyacetone phosphate/phosphoglycerate shuttle system (Heldt, 1976).

There were two types of spinach with respect to the distribution of NAD kinase. One is spinach, A in Table 1 in which chloroplasts contained only the CaM independent NAD kinase in the stroma. Another is spinach, B in which chloroplasts contained only the CaM dependent enzyme in the envelope. The later spinach may not carry out the light-induced conversion of NAD to NADP in the chloroplast. Such chloroplast might have special permeability for NADP to supply it for photosynthesis. As spinach plants were obtained from a local market their cultivars were not always the same and the growth conditions were of course different. Use of a defined species grown under the same conditions is nessesary for further study on the roles of NAD kinases in spinach.

Acknowledgements

This work was supported in part by a Grant-in-Aid for Scientific Research from the Japanese Ministry of Educatioin, Science and Culture, 58540421. We are grateful to Yamada Science Foundation for supporting travel expence to this workshop.

REFERENCES

Anderson,J.M., Charbonneau, H., Joanes,H.P., McCann,R.O., and Cormier,M.J., 1980, Biochemistry, 19:3113.

Anderson,J.M., and Cormier,M.J., 1978, Biochem. Biophys. Res. Commun., 84:595.

Anderson,L.E., Ashton,A.R., Mohamed,A.H., and Sheibe,R., 1982, BioSci., 32:103.

Barr,R., Troxel,K.S., and Crane,F.L., 1982, Biochem. Biophys. Res. Commun., 104:1182

Cline,K., Andrews,J., Mersey,B., Newcomb,E.H., and Keegstra,K., 1981, Proc. Natl. Acad. Sci. USA, 78:2595.

Heldt,H.W., 1976, in "The Intact Chloroplast", J.Barber, ed., Elsevier/ North-Holland Biomedical Press, Amsterdam.

Heldt,H.W., Werdan,K., Milovancev,M., and Geller,G., 1973, Biochim. Biophys. Acta, 314:224.

Jarrett,H.W., Brown,C.J., Black,C.C., and Cormier,M.J., 1982, J. Biol. Chem. 257:13795.

Jensen,R.G., and Bassham,J.A., 1966, Proc. Natl. Acad. Sci. USA, 56:1095.

Kreimer,G., Melkoian,M., and Latzko,E., 1985, FEBS Lett., 180:253.

Kobayashi,Y., Inoue,Y., Furuya,F., Shibata,K., and Heber,U., 1979, Planta, 147:69.

Matsumura-Kadota,H., Muto,S., and Miyachi,S., 1982, Biochim. Biophys. Acta, 679:300.

Mills,W.R., and Joy,K.W., 1980, Planta, 148:75.

Muto,S., 1982, FEBS Lett., 147:161.

Muto,S., 1983, Z. Pflanzenphysiol., 109:385.

Muto,S., and Miyachi,S., 1977, Plant Physiol., 59:55.

Muto,S., and Miyachi,S., 1984, Z. Pflanzenphysiol., 114:421.

Muto,S., Izawa,S., and Miyachi,S., 1982, FEBS Lett., 139:250.

Muto,S., Miyachi,S., Usuda,H., Edwards,G.E., and Bassham,J.A., 1981, Plant Physiol., 68:324.

Ogren,W.L., and Krogman,D.W., 1965, J. Biol. Chem., 240:4603.

Oh-hama,T., and Miyachi,S., 1959, Biochim. Biophys. Acta, 30:202.

Portis,Jr.A.R., and Heldt,H.W., 1976, Biochim. Biophys. Acta, 449:434.

Simon,P., Bonzon,M., Greppin,H., and Marme,D., 1982, Plant Cell Reports, 1:119.

Simon,P., Bonzon,M., Greppin,H., and Marme,D., 1984, FEBS. Lett., 167:332.

Takebe,I., Otsuki,Y., and Aoki,S., 1968, Plant Cell Physiol., 9:115.

MODULATION OF ENZYME ACTIVITIES IN ISOLATED

PEA NUCLEI BY PHYTOCHROME, CALCIUM AND CALMODULIN

Neeraj Datta, Yuh-Ru Chen, and Stanley J. Roux

Department of Botany
The University of Texas at Austin
Austin, Texas 78713 U.S.A.

INTRODUCTION

The photoactivation of phytochrome by red light initiates many important developmental changes in plants (Hendricks and Van Der Woude, 1983). Some phytochrome-controlled responses are saturated at fluences of light as low as 3×10^{-12}moles/cm^2 (Mandoli and Briggs, 1981). For such minute amounts of light to have major effects on the way a plant grows and develops, this initial stimulus must be amplified by some transduction mechanism, some linked sequence of biochemical changes that ultimately affects many facets of a plant's metabolism.

Based mainly on the abundant evidence that Ca^{2+} played a central role in transducing low-energy sensory stimuli into physiological responses in animal cells, Haupt and Weisenseel (1976) proposed a decade ago that a similar, Ca^{2+}-based amplification system could be mediating phytochrome responses. Recent evidence suggests that Haupt and Weisenseel's hypothesis merits serious consideration (Roux, 1984). Phytochrome induces calcium fluxes in cells of Mougeotia (Dreyer and Weisenseel, 1979), Avena (Hale and Roux, 1980), Onoclea (Wayne and Hepler, 1985), and Zea (Das and Sopory, 1985) and in isolated mitochondria (Roux et al., 1981). Phytochrome-induced membrane potential changes in Nitella appear to involve the uptake of Ca^{2+} into these cells (Weisenseel and Ruppert, 1977). Recent data indicate that there is a causal link between photoinduced changes in calcium fluxes and photoinduced changes in fern spore germination in Onoclea (Wayne and Hepler, 1984) and in chloroplast rotation in Mougeotia (Serlin and Roux, 1984). These calcium-mediated Pfr responses appear to involve also the activation of the regulator protein, calmodulin, because calmodulin inhibitors strongly antagonize them (Wayne and Hepler, 1984; Serlin and Roux, 1984).

Phytochrome photoreversibly regulates gene expression in many plants, as evidenced by its induction of altered levels of translatable messages for specific genes (Tobin and Silverthorne, 1985). It is not known whether this phytochrome response is mediated by calcium and calmodulin, or by some other mechanism(s). Recent data showing that the calcium ionophore, A23187, induces the expression of glucose-regulated genes in Chinese hamster cells reveal that calcium ions may play an important role in the regulation of transcription of certain genes (Resendez et al., 1985). Our approach to testing whether Ca^{2+} may help to mediate Pfr-induced gene expression in plants has been to investigate whether there are any nuclear enzymes whose activity is modulated by light and Ca^{2+}. There appear to be several non-nuclear enzymes that exhibit this dual regulation (Roux, 1984). Our initial results indicate that there are several enzyme activities in isolated nuclei which are stimulated by both red light and Ca^{2+} and which are inhibited by calmodulin antagonists. Here we will review the current status of our research on this topic.

EXPERIMENTAL SYSTEM AND RESULTS

To facilitate our manipulation of the chemical environment of nuclear enzymes, we chose to use isolated nuclei for these studies. Hagen and Guilfoyle (1985) have described an efficient procedure for isolating large quantities of highly purified, intact nuclei from pea plumules. Phytochrome regulates aspects of the growth and development of this tissue (De Greef and Fredericq, 1983), and there are reports of the regulation of an ATPase activity in pea nuclei by both Pfr (Wagle and Jaffe, 1980) and calmodulin (Matsumoto et al., 1984), so this experimental system seemed well suited for our investigation.

We used the method of Hagen and Guilfoyle (1985) to purify nuclei from pea nuclei, and we examined the purity of the preparation by electron microscopy (Datta et al., 1985). All of the recognizable organelles in the preparation were nuclei; i.e., there were no contaminating plastids, mitochondria, Golgi, or vacuoles. Nor were there any contaminating bacterial cells. Membrane particulates of (presumably) multiple origin were present, and these varied in amount in each preparation. Evaluating the relative intactness of the nuclei in the preparation was less straight forward. By light microscopy most of the nuclei appeared to be intact. R. Wayne in our laboratory has found that most of them were also osmotically active. That is, they would swell and shrink reversibly as the osmotic strength of the medium was changed. Wagle and Biro in M. Jaffe's laboratory observed a red-light induced swelling of isolated pea nuclei (R. Biro, personal communication). Wayne is currently investigating this phenomenon in our laboratory.

Red-Light Stimulated NTPase Activity

Wagle and Jaffe (1980) reported a photoreversible ATPase activity in pea nuclei. The actinic irradiation for these experiments was given to intact etiolated pea seedlings. We replicated these results once, then one of us (Y.-R. Chen) began to examine whether the irradiation of isolated nuclei from pea seedlings would induce a similar photoreversible response. The results of these experiments have been presented at a recent meeting by Y.-R. Chen, and are being reviewed for journal publication. To summarize them, red light induces a more than two-fold stimulation of a nuclear NTPase in isolated pea nuclei, and far-red light reverses this response. Thus far this response has been observed only when ATP was used as the substrate, but the enzyme itself can also use CTP, UTP, and GTP as substrates, all with a pH optimum of 7.5. Preliminary data indicate that in the presence of Mg^{2+}, the activity of this enzyme can be stimulated by micromolar Ca^{2+}. This stimulation is inhibited by low concentrations of compound 48/80 and chlorpromazine, two calmodulin antagonists. We have not tested whether the nuclear NTPase regulated by phytochrome in vitro is the same as the nuclear ATPase found by Wagle and Jaffe (1980) to be regulated by phytochrome in vivo.

In SV40-3T3 cells, the activity of nuclear NTPases has been correlated with the transport of poly-A containing RNA out of nuclei (Agutter et al., 1979). If the light-regulated NTPase in nuclei also serves this function, its stimulation by red light could account for some of the Pfr-promoted increase in the level of translatable messages for some genes. We are currently testing this hypothesis.

Phytochrome and Calcium Stimulation of Nuclear Protein Phosphorylation

In rat liver cells, the activity of a nuclear NTPase is correlated with its level of phosphorylation (Purrello et al., 1983). This observation was just one of several that directed our attention to phosphorylation as a potentially important mechanism for regulating the activity of nuclear enzymes. In fact, there were many previous reports that had suggested there was a causal link between the phosphorylation of nuclear proteins and the regulation of gene expression (see reviews of Cohen (1982) and Ranjeva et al.(1984)). Moreover, the phosphorylation of many different proteins in both plants and animals was known to be regulated by Ca^{2+} and calmodulin (Cohen, 1982; Veluthambi and Poovaiah, 1984). As a result of the combined influence of these previous studies, we decided to test the involvement of phytochrome, Ca^{2+} and calmodulin in regulating the phosphorylation of nuclear proteins. Our initial results, obtained primarily by N. Datta, have been described in a recent journal article (Datta et al., 1985). Here we will summarize the published findings and discuss the current status of the project.

For the phosphorylation studies, we used the same experimental system as for the NTPase studies: nuclei isolated from etiolated pea plumules. Phosphorylation was initiated by the addition of gamma-^{32}P ATP to the nuclear preparation. For irradiated samples, this occurred immediately after the light treatments. Both in irradiated and unirradiated samples the net amount of ^{32}P incorporated into protein increased linearly with time during the first 60 sec after ^{32}P-ATP addition and then increased only slowly or remained the same for at least the next 5 min. Phosphorylation was terminated by the addition of electrophoresis sample buffer to the sample and immediately boiling it. Treatment of the phosphorylated macromolecules in the nuclear preparation with trypsin and chymotrypsin significantly hydrolyzed them, thus confirming their identity as proteins. Analysis of a labeled and fixed nuclear preparation by autoradiography (Fig. 1) showed that the ^{32}P label was in the nuclei of the preparation.

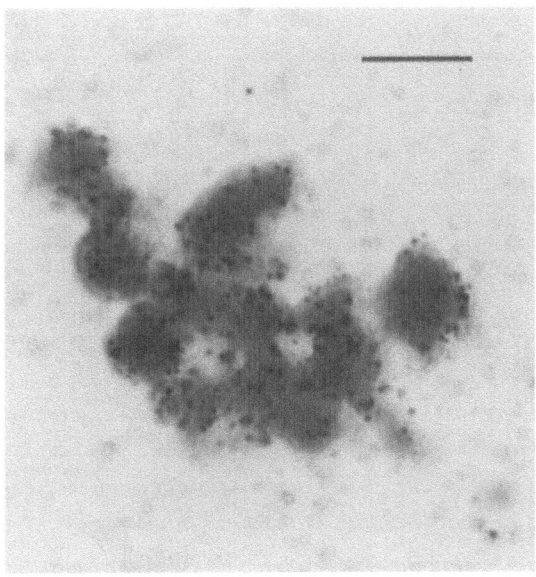

Fig. 1. Autoradiograph of ^{32}P-labeled pea nuclei that were fixed, washed, attached to slides and processed for autoradiography (8-day exposure), using Kodak AR-10 film. Bar = 10 um.

Phosphorylated proteins were separated by sodium dodecyl sulfate gel electrophoresis. Qualitative and quantitative analysis of the phosphorylated proteins was done by autoradiography and densitometric scanning. The equality of protein content on each lane was tested by densitometric analysis of the bands stained by Coomassie blue. The variation in stain intensity between equivalent bands in any two lanes was usually less than 5%; nonetheless, quantitative comparisons of phosphorylation levels among labeled bands in the autoradiogram were normalized to the same protein content, as estimated densitometrically from the Coomassie blue stain intensity of the major bands.

Figure 2 synopsizes our key findings. It shows that the phosphorylation level of several proteins in pea nuclei is promoted by red light and by Ca^{2+} concentrations in the low micromolar range. The stimulation induced by red light is canceled by far-red light treatment, which strongly implicates phytochrome as the photoreceptor for this response. The enhancement of phosphorylation by both light and Ca^{2+} is blocked by calcium chelation (EGTA) and by the calmodulin antagonists, chlorpromazine and compound 48/80.

Because these effects result from label addition to previously unlabeled nuclei, they could be due to an enhancement of any enzyme activity that would increase the turnover rate of phosphorylation: i.e., either higher protein kinase activity or higher phosphoprotein phosphatase activity, or both. We have begun to investigate this question, using the methods of Purrello et al. (1983). Our initial results indicate that red light stimu-

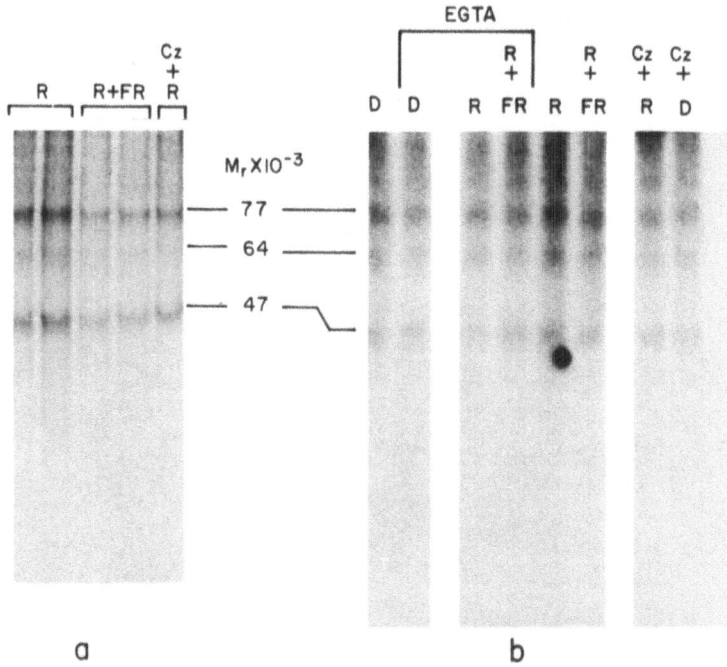

Fig. 2. Autoradiograph of gel containing ^{32}P-labeled proteins from nuclei treated with red (R), far-red (FR) light, EGTA, and/or chlorpromazine (Cz). Nuclei were incubated with (a) 10 uM ^{32}P-ATP, 5 min., or (b) 5 uM ^{32}P-ATP, 1 min. Samples on "D" lanes were unirradiated.

lates both protein kinase and phosphatase activity. We are currently testing the validity of these results using alternative methods. We will also soon test the far-red reversibility of these two red-light stimulated enzyme activities.

Assays for Calmodulin, Calmodulin-Binding Proteins, and Phytochrome in Pea Nuclei

The effects of calmodulin inhibitors on the phosphorylation response indicate that calmodulin must be a protein component of the isolated nuclei. We have confirmed this implication by both radioimmunoassay (Biro et al., 1983) and by immunocytochemical analysis, using rhodamine-labeled second antibody against sheep anticalmodulin immunoglobulin (Fig. 3, courtesy of M. Dauwalder). Matsumoto's laboratory has reported the presence of a calmodulin-like protein in the chromatin of pea nuclei (Matsumoto et al., 1983) and a calmodulin-regulated ATPase there (Matsumoto et al., 1984). Calmodulin is also known to regulate several protein kinases and phospho-protein phosphatases in other cells (Cohen, 1982).

To further assess the role of calmodulin in the regulation of the NTPase and phosphorylation activities of isolated nuclei, we have assayed our preparation of purified nuclei for the presence of calmodulin-binding proteins. Using calmodulin gel-overlay procedures described by Van Eldik and Burgess (1983), Sung-ha Kim in our laboratory has identified at least four protein bands in our nuclei preparations that bind calmodulin in a Ca^{2+}-dependent fashion. There appear to be two other bands that bind labeled calmodulin in a calcium-independent fashion. We are now testing whether any of these bands represent a calmodulin-modulated protein kinase or phosphoprotein phosphatase or subunits of them.

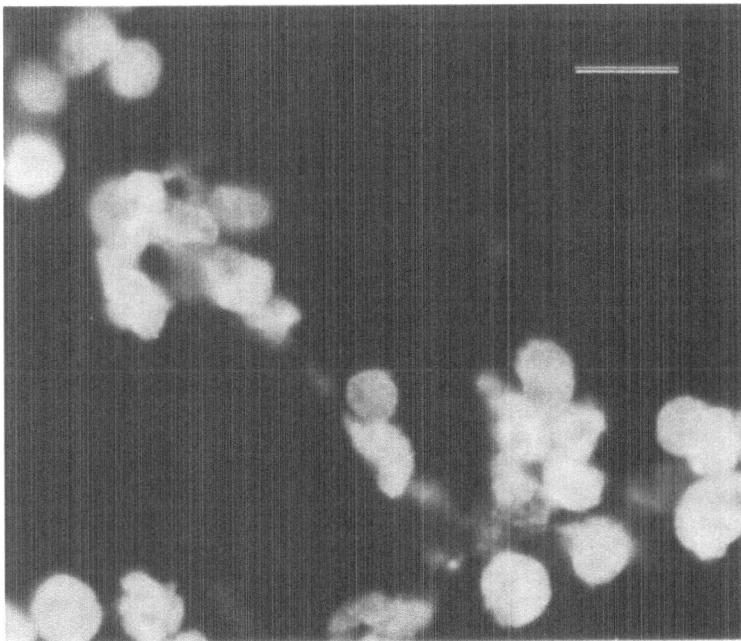

Fig. 3. Immunocytochemical localization of calmodulin in isolated pea nuclei. Formaldehyde-fixed nuclei were incubated with antibody to bovine calmodulin, washed, and stained with a rhodamine-conjugated second antibody. Bar = 10 um.

We are also investigating the question of phytochrome presence in pea nuclei. MacKenzie et al. (1978) found that they could immunocytochemically localize phytochrome in the nuclei of various monocot cells 90 minutes after a red-light irradiation. Our nuclei are extracted from pea tissue which received the equivalent of about a 10 second irradiation of bright red light between 0.5 to 1.5 h before the extraction of the tissue began. To date, our initial attempts to localize phytochrome in pea nuclei have included both immunocytochemical analysis of pea plumule tissue and immunoprecipitation of the detergent-solubilized proteins in the nuclear preparation that bind to anti-oat phytochrome monoclonal antibodies. These monoclonal antibodies recognize epitopes on pea phytochrome as judged both by ELISA and Western blot analysis (Sung-ha Kim, unpublished observations). Our preliminary data indicate that there is immunologically detectable phytochrome in pea nuclei, but the results are not yet definitive.

SPECULATIONS AND FUTURE DIRECTIONS

Our research on Pfr-induced, Ca^{2+}-modulated nuclear enzyme activities is clearly at an early stage. To date we have spent most of our effort establishing the basic experimental conditions that will reproducibly support the enzyme responses described. Over a dozen different nuclear preparations have been assayed under our standardized conditions and over 80% of these exhibited either the NTPase or the phosphorylation responses described here, or both. Nuclear preparations obtained from seedlings that showed significantly impaired growth also showed significantly reduced responsiveness to light. The exact biochemical/physiological basis of this correlation has not been determined.

To speculate on the significance of these findings, we propose that nuclear protein phosphorylation may be an important intermediate step linking the photoactivation of phytochrome to its modulation of gene expression. Our working hypothesis is that Pfr increases the equilibrium concentration of free Ca^{2+} in the nucleus, and that this activates, in turn, calmodulin and calmodulin-dependent protein kinases and/or phosphoprotein phosphatases in the nucleus. We propose that the activity of these enzymes alters the phosphorylation level of some protein(s) and/or enzyme(s) that critically regulates the initiation or rate of transcription of certain genes.

Testing this hypothesis will require a battery of new experiments. Currently, investigators in our laboratory are experimentally addressing the following specific questions relevant to the hypothesis: Can calmodulin-modulated protein kinases or phosphoprotein phosphatases be isolated from nuclei? What are the proteins whose phosphorylation level is regulated by light and Ca^{2+}, and what, if any, is their functional relationship to gene expression? Where in the nucleus are calmodulin and phytochrome localized? Does Pfr stimulate calmodulin-controlled enzymes by regulating Ca^{2+} fluxes into or out of the nucleus? Is the stimulation of a nuclear NTPase by Ca^{2+}-activated calmodulin due to calmodulin binding to it, or to some less direct action, such as calmodulin control of its phosphorylation/dephosphorylation?

Answers to the above questions will undoubtedly clarify the biochemical mechanisms underlying the nuclear responses we have been observing. These _in vitro_ data, in turn, will generate specific predictions about the regulation of nuclear events _in vivo_. The accuracy of these predictions will be the ultimate test of the significance of this work.

Acknowledgements. The authors' research was supported in part by grants from the National Science Foundation (PCM 8402526) and from the National Aeronautics and Space Administration (NSG 7480).

REFERENCES

Agutter, P.S., McCaldin, B., and McArdle, H. J., 1979, Importance of mammalian nuclear-envelope nucleoside triphosphatase in nucleo-cytoplasmic
transport of ribonucleoproteins, Biochem. J., 182:811.

Biro, R. L., Daye, S., Serlin, B. S., Terry, M. E., Datta, N., Sopory, S. K.,
and Roux, S. J., 1984, Characterization of oat calmodulin and radioimmunoassay of its subcellular distribution, Plant Physiol., 75:382.

Cohen, P., 1982, The role of protein phosphorylation in neural and hormonal
control of cellular activity, Nature, 296:613.

Das, R. and Sopory, S. K., 1985, Evidence of regulation of calcium uptake
by phytochrome in maize protoplasts, Biochem. Biophys. Res. Commun.,
128:1455.

Datta, N., Chen, Y.-R. and Roux, S. J., 1985, Phytochrome and calcium
stimulation of protein phosphorylation in isolated pea nuclei, Biochem.
Biophys. Res. Commun., 128:1403.

De Greef, J. A. and Fredericq, H., 1983, Photomorphogenesis and hormones,
in: "Photomorphogenesis", W. Shropshire, Jr. and H. Mohr, eds.,
Springer-Verlag, Berlin.

Dreyer, E. M. and Weisenseel, M. H., 1979, phytochrome-mediated uptake of
calcium in Mougeotia cells, Planta, 146:31.

Hagen, G. and Guilfoyle, T. J., 1985, Rapid induction of selective transcription by auxins, Mol. Cell. Biol., 5:1197.

Hale, C. C. II and Roux, S. J., 1980, Photoreversible calcium fluxes induced
by phytochrome in oat coleoptile cells. Plant Physiol., 65:658.

Haupt, W. and Weisenseel, M. H., 1976, Physiological evidence and some
thoughts on localised responses, intracellular localisation and action
of phytochrome, in: "Light and Plant Development", H. Smith, ed.,
Butterworths, London.

Hendricks, S. B. and VanDerWoude, W. J., 1983, How phytochrome acts-
Perspectives on the continuing quest, in: "Photomorphogenesis", W.
Shropshire, Jr. and H. Mohr, eds., Springer-Verlag, Berlin.

Mackenzie, J. M., Briggs, W. R., and Pratt, L. H., 1978, Intracellular
phytochrome distribution as a function of its molecular form and of
its destruction, Amer. J. Bot., 65:671.

Mandoli, D. F. and Briggs, W. R., 1981, Phytochrome control of two low-
irradiance responses in etiolated oat seedlings, Plant Physiol.,
67:733.

Matsumoto, H., Tanigawa, M., and Yamaya, T., 1983, Calmodulin-like activity
associated with chromatin from pea buds, Plant & Cell Physiol. 24:593.

Matsumoto, H. Yamaya, T., and Tanigawa, M., 1984, Activation of ATPase
activity in the chromatin fraction of pea nuclei by calcium and
calmodulin, Plant & Cell Physiol., 25:191.

Purrello, F., Burnham, D. B., and Goldfine, I. D., 1983, Insulin regulation
of protein phosphorylation in isolated rat liver nuclear evelopes:

potential relationship to mRNA metabolism, Proc. Natl. Acad. Sci. U.S.A., 80:1189.

Ranjeva, R., Graziana, A., Ranty, B., Cavalie, G., and Boudet, A. M., 1984, Phosphorylation of proteins in plants: A step in the integration of extra and intracellular stimuli?, Physiol. Veg., 22:365.

Resendez, Jr., E., Attenello, J. W., Grafsky, A., Chang, C. S., and Lee, A. S., 1985, Calcium ionophore A23187 induces expression of glucose-regulated genes and their heterologous fusion genes, Mol. Cell. Biol., 5:1212.

Roux, S. J., 1984, Ca^{2+} and phytochrome action in plants, BioScience, 34:25.

Roux, S. J., McEntire, K., Slocum, R. D., Cedel, T. E., and Hale II, C. C., 1981, Phytochrome induces photoreversible calcium fluxes in a purified mitochondrial fraction from oats, Proc. Natl. Acad. Sci. U.S.A., 78:283.

Serlin, B. S. and Roux, S. J., 1984, Modulation of chloroplast movement in the green alga Mougeotia by the Ca^{2+} ionophore A23187 and by calmodulin antagonists, Proc. Natl. Acad. Sci., U.S.A., 81:6368.

Tobin, E. M. and Silverthorne, J., 1985, Light regulation of gene expression in higher plants, Ann. Rev. Plant Physiol., 36:569.

Van Eldik, L. J. and Burgess, W. H., 1983, Analytical subcellular distribution of calmodulin and calmodulin-binding proteins in normal and virus-transformed fibroblasts, J. Biol. Chem., 258:4539.

Veluthambi, K. and Poovaiah, B. W., 1984, Calcium-promoted protein phosphorylation in plants, Science, 223:167.

Wagle, J. and Jaffe, M. J., 1980, The association and function of phytochrome in pea nuclei, Plant Physiol., 65 (Suppl.):3.

Wayne, R. and Hepler, P. K., 1984, The role of calcium ions in phytochrome-mediated germination of spores of Onoclea sensibilis, L., Planta, 160:12.

Wayner, R. and Hepler, P. K., 1985, Red light stimulates an increase in intracellular calcium in the spores of Onoclea sensibilis, Plant Physiol., 77:8.

Weisenseel, M. H. and Ruppert, H. K., 1977, Phytochrome and calcium ions are involved in light-induced depolarization in Nitella, Planta, 137:225.

CALCIUM/CALMODULIN DEPENDENT MEMBRANE BOUND PROTEIN KINASE

A.M. Hetherington*, D. Blowers and A. Trewavas

Department of Botany, University of Edinburgh
Midlothian, Scotland
*Department of Biological Sciences, University of Lancaster

Introduction

In recent years it has become apparent that the calcium ion plays a key regulatory role in plant cell metabolism. The details of the molecular events which contribute to this function are beginning to emerge. In animals it has been convincingly demonstrated that a common response to a variety of cell surface stimuli is a transient increase in cytoplasmic calcium concentration (Rasmussen & Barrett 1984). These alterations in cytoplasmic calcium concentration are interpreted by a number of calcium binding proteins. Of these the best characterized is calmodulin which has also been shown to participate in a variety of physiological processes in higher plants (Dieter, 1984). Other calcium binding proteins include enzymes whose activity is modulated by physiological concentrations of calcium. Among these enzymes, the protein kinases because of their known regulatory function are potentially of great importance in plant cells.

Since the first reports of membrane bound and soluble calcium activated protein kinases in plants (Hetherington & Trewavas 1982) (Polya & Davies 1982), these enzymes have been found in a number of species, tissues and are widely distributed within the cell. (Polya,Davies & Micucci 1983) (Salimath & Marme 1983) (Hetherington & Trewavas 1984) (Polya & Micucci 1984) (Blowers, Hetherington & Trewavas 1985). Calcium dependent protein kinases have been found in both soluble and particulate fractions, however with one exception the in vivo substrates for these enzymes are not known. The enzyme quinate : NAD$^+$ oxidoreductase has been shown to be regulated by phosphorylation/dephosphorylation catalysed by a calcium dependent protein kinase (Ranjeva et al 1983).

In this paper we will be restricting ourselves to discussion of membrane bound calcium dependent protein kinases. The role of membrane bound protein kinases in plants is not well understood. However an important role for these enzymes may well be anticipated from two recent sets of findings. Firstly it has been reported that auxins, cytokinins and gibberellins (Elliot et al 1983) and abscisic acid (de Silva, Hetherington and Mansfield 1985) require calcium for their activity and secondly that exogenous application of auxin influences the pattern of protein phosphorylation in soybean membranes (Morre, Morre and Varnold 1984).

The remainder of this paper will consider the properties of a protein kinase from pea shoot membranes.

Materials and Methods

Peas (Pisum sativum L.cv. Feltham First) were grown in the dark at $20-22^{\circ}C$ for 12 days. The experimental material was the bud which contains the unexpanded 4th and 5th leaves.

Membrane preparation and phase partitioning : membranes were prepared as previously described (Hetherington, Trewavas 1984). An acetone solubilized membrane fraction was prepared from these membranes as described in (Blowers, Hetherington & Trewavas 1985). A plasma membrane enriched fraction was prepared by the procedure of Yoshida et al (1983). Total membrane was partitioned in 4g phases containing 5.6% (w/w) each of polyethylene glycol (Av. Mr.3,350) and dextran (Av.Mr. 472,000). Phases were thoroughly mixed and left to settle. The diluted upper and lower phases were centrifuged at 150000 x g for 1h.30 min and the resulting resuspended upper phase pellet referred to as the plasma membrane enriched fraction and the lower phase as the residual membrane fraction.

Electrophoresis, electroelution and blotting of proteins: PAGE of labelled and unlabelled proteins in denaturing and non-denaturing conditions and electroelution of proteins was carried out as described in Hetherington & Trewavas (1982) and Blowers, Hetherington & Trewavas (1985). Proteins separated using non-denaturing PAGE were transferred to nitrocellulose membranes (pore size 0.2 μm) using a Bio-Rad Transblot cell overnight (30V, 0.1A @ $4^{\circ}C$) with a $25mmol.dm^{-3}$ Tris, pH 8.3, $192mmol\ dm^{-3}$ glycine buffer system.

Detection of protein kinase activity on nitrocellulose membranes : After gel blotting nitrocellulose membranes were incubated for 30 mins in 0.5% (w/v) BSA, 0.9% (w/v) NaCl, $10mmol\ dm^{-3}$ Tris pH 7.5. The saturated membranes were washed 3 times in the above solutions without BSA. The membranes were then incubated in plastic bags containing appropriate buffer and approx 370 KBq γ -[^{32}P] ATP under the conditions indicated. On completion the membranes were washed overnight in 10% TCA, $20mmol\ dm^{-3}$ sodium pyrophosphate, $10mmol\ dm^{-3}$ EDTA. After boiling for 15 mins in fresh TCA mix and 30 min cooling the membranes were washed and filtered 2 times with TCA mix and air dried. Labelled proteins were detected by autoradiography.

Protein kinase assays, phosphoamino acid analysis and protein determinations: For membrane preparations protein kinase was assayed using the cellulose disc assay previously described (Hetherington and Trewavas 1982) using γ- [^{32}P]-ATP. For the purified enzyme, assays were performed exactly as described by Roskoski (1983) on Whatman P81 phosphocellulose strips (1x2cm). Incorporated radioactive phosphate was determined by Cerenkov counting of each disc or strip in 5 ml of distilled water. Phosphoamino acid analysis and protein determinations were carried out as detailed in Blowers, Hetherington & Trewavas (1985).

Results and discussion

We originally detected a membrane bound protein kinase in pea buds which was activated by micro molar concentrations of free calcium. Our subsequent studies revealed that the kinase activity was present throughout the 12 day old pea plant. However since the greatest calcium activation was associated with bud tissue (up to a maximum of 20 fold but typically some 6-7 fold) we chose this tissue as our experimental material.

We considered that the characterization and purification of the calcium activated membrane bound protein kinase (Ca^{2+}PK) would be significant steps towards understanding its physiological function. Specifically we attempted to answer two questions, first where at the subcellular level was the Ca^{2+}PK located and secondly was it regulatd by calmodulin? We attempted to localize the Ca^{2+}PK activity using a number of techniques. Isopynic centrifugation revealed the presence of at least 2 components, one was a shoulder and had a density of $1.16g\ cm^{-3}$ the other major component had a peak density of 1.136g cm^{-3}. This latter density is characteristic of pea epicotyl plasma membrane. We confirmed these results using the technique of Rasi-Caldogno et al (1982) devised for pea tissue. These results indicated that protein kinase activity was associated with a number of subcellular fractions, however best calcium activations were observed in plasma membrane enriched fractions (Hetherington & Trewavas 1984).

Recently, to extend our localization studies we have made use of the technique of phase partitioning. Figure 1 shows that the upper plasma membrane enriched phase contains much higher (c 6 fold) CaPK activity. Importantly the upper phase contains only one tenth of the total protein subjected to phase partitioning.

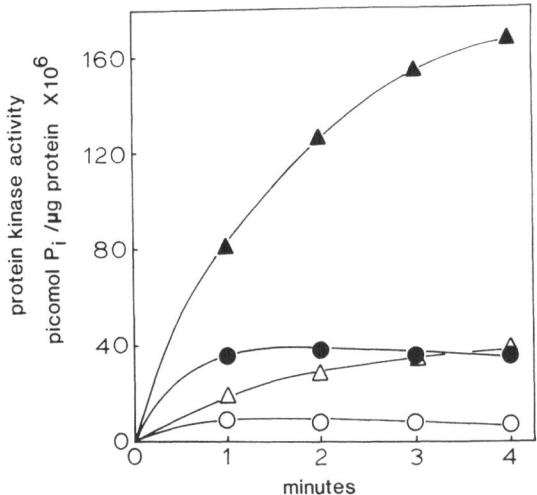

Fig.1. Phosphorylation of proteins in membrane fractions prepared by phase partioning. A membrane fraction was prepared from pea buds and separated by phase partitioning into an upper phase (plasma-membrane enriched) and a lower phase residual membrane fraction. Closed symbols incubation contains 100μmol dm^{-3} CaCl$_2$, open symbols no CaCl$_2$ present, triangles show upper plasma membrane enriched phase, circles show lower residual membrane phase.

We have also investigated possible calmodulin (CaM) dependence of the Ca^{2+} PK. Initially we were only able to demonstrate inhibition of Ca^{2+} PK activity by high concentrations of Trifluoperazine (TFP) and stimulation by large amounts of CaM (Hetherington & Trewavas 1982). We interpreted this difficulty as being due to contamination of the membrane preparations by endogenous CaM, although we subsequently managed to improve the TFP inhibition by preincubation of the membrane preparations in 2mmol dm^{-3} EGTA.

In order to further characterize the enzyme we attempted solubilization using a variety of agents (Na-deoxycholate, NP40, CHAPS). However the best method for both solubilizing the Ca^{2+} PK while retaining its calcium dependence was the acetone procedure devised by Venis (1977) as illustrated in Fig.2.

Interestingly acetone solubilized material was more sensitive to TFP but showed no response to added calmodulin.

We had previously obtained evidence that pea bud membranes contained more than one kinase. Did our solubilized preparation contain more than

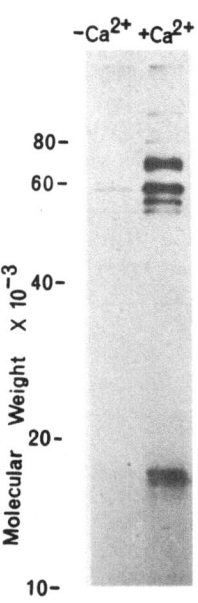

Fig.2. Autoradiograph of proteins phosphorylated by acetone solubilized protein kinase incubated in the absence and presence of calcium ions separated by SDS PAGE.

one kinase? We utilized the procedures of non-denaturing PAGE coupled with Western blotting to answer this question. Fig.3 shows the results of a non denatured sample of labelled protein separated by non denaturing PAGE.

To detect protein kinase activity we took advantage of the observations that protein kinases are known to autophosphorylate and also they frequently co-purify with their endogenous substrates. After Western transfer of non-denatured proteins to nitrocellulose, enzyme activity was retained. A single phosphorylated band appears on the filter indicating the presence of protein kinase. We were subsequently able to demonstrate by phospho-amino acid analysis that the labelled band resulted from kinase activity and not from non specific binding of ^{32}P to protein. Approximately 90% of the phosphorylation was on serine residues with the remainder on either threonine or tyrosine (Blowers, Hetherington & Trewavas 1985).

By running non-denaturing gels of the acetone solubilized Ca^{2+}PK and including a single labelled lane we were able to locate the protein kinase band in other non labelled lanes. These bands were removed and the protein kinase electroeluted. The isolated protein was tested for Ca^{2+}/calmodulin dependence in the absence of exogenous substrate. Fig 4 indicates that the enzyme shows a 3 fold activation in the presence of calcium and calmodulin. Importantly little activity is seen in the presence of calcium alone.

126

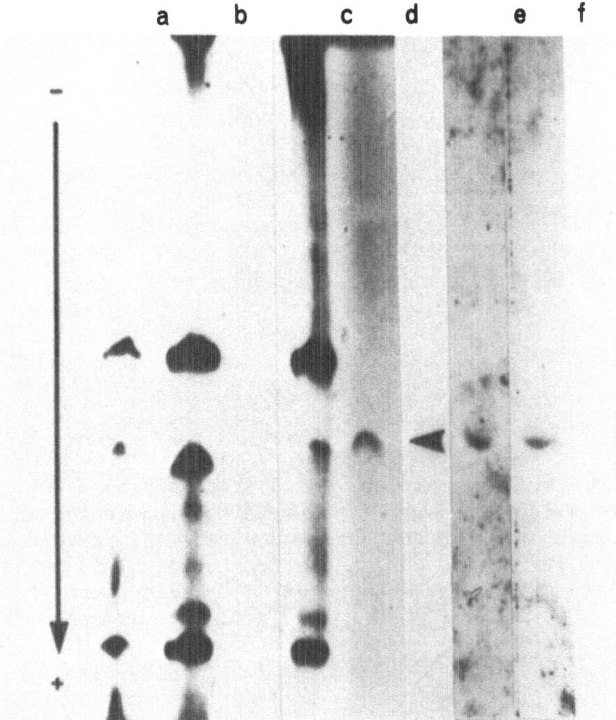

Fig.3. Separation of labelled peptides in acetone solubilized protein
kinase preparations by non denaturing gel electrophoresis and Western
blotting of protein kinase activity.

Tracks a and b: acetone solubilized membrane proteins were
phosphorylated in the absence (track a) or presence (track b) of 100
umol.dm^{-3} CaCl$_2$ and were loaded directly on a non-denaturing gel separated
for 16h at 4°C and subsequently autoradiographed.   Track c : proteins
labelled as above and blotted to nitrocellulose.   Tracks d-f solubilized
proteins were separated as above unlabelled.   The gel containing tracks
c-f was blotted onto nitrocellulose and active protein kinase detected by
incubating the nitrocellulose in γ -[^{32}P]-ATP.   In tracks d and e the
nitrocellulose was additionally incubated in 100 umol dm^{-3} CaCl$_2$ and track
f in the absence of CaCl$_2$.

When the kinase containing band was electroeluted from non-denaturing
gels, incubated with γ -[^{32}P]-ATP and separated on SDS gels a number of
polypeptide bands were visible after coomassie blue or silver staining.
However, autoradiography revealed the presence of only one phosphorylated
band, and this did not correspond with any of the protein stainable bands.
We do not know whether these components represent subunits of the native
protein kinase or polypeptides which co-purify with the enzyme. Thus the
phosphorylated component may be the result of autophosphorylation or may
represent the co-purified endogenous substrate.   However, it is clear that
the latter component is present in low amounts (Fig.5).

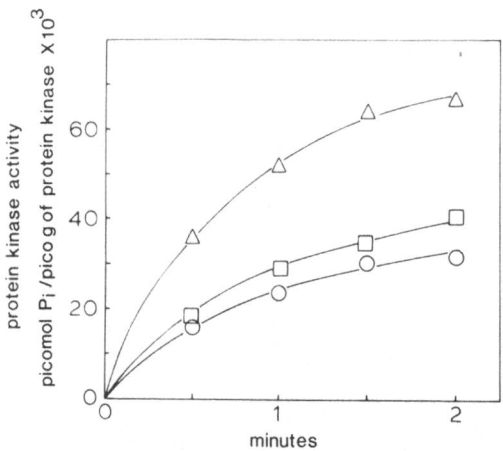

Fig.4. Activation of purified membrane bound protein kinase activity by
calmodulin: acetone solubilized membrane proteins were separated by non
denaturing gel electrophoresis, the band containing the protein kinase was
located, excised and the active enzyme electroeluted. This was then
incubated in γ-[^{32}P]- ATP in the absence of Ca^{2+} (o) or with 100 mol dm^{-3}
$CaCl_2$ ($\square$) or with 100 mol dm^{-3} $CaCl_2$ 100 g ml^{-1} bovine calmodulin ($\triangle$).

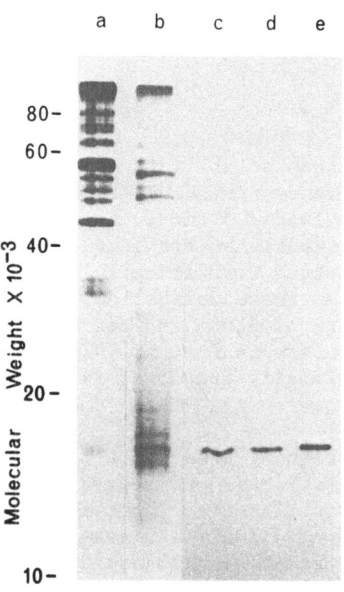

Fig.5. SDS PAGE of phosphorylated polypeptides from various protein kinase
preparations: Pea bud membranes were prepared and separated by phase
partitioning into a lower residual membrane preparation and an upper
membrane enriched fraction. Phosphorylation with γ-[^{32}P] ATP was carried
out in the presence of 100 μmol dm^{-3} free Ca^{2+} and the labelled proteins
separated by SDS gel electrophoresis. Track a, autoradiograph of the
residual membrane protein phosphorylation and track b the plasma membrane
enriched fraction.

Acetone solubilized protein kinase was separated by non-denaturing gel electrophoresis, the active bands excised electroeluted and allowed to autophosphorylate with γ -[^{32}P] ATP. Track c no CaCl$_2$, track d 100μmol dm^{-3} free Ca^{2+}, track e 100μmol dm^{-3} free Ca^{2+}, 50μg ml^{-1} bovine calmodulin. Notice the single band in c-e which is present only in the plasma membrane enriched fraction.

To summarise our results we have extensively purified a calcium calmodulin dependent protein kinase from pea bud membranes. All our evidence indicates that the enzyme is located in a plasma membrane enriched fraction. We feel that our previous failures to satisfactorily demonstrate calmodulin dependence were due to difficulty in depleting endogenous calmodulin from our membrane preparations.

Our future research will be concerned with investigating the function of the enzyme. Towards this goal we are currently raising antibodies to the purified enzyme.

We acknowledge support by the Science and Engineering Research Council.

REFERENCES

Blowers, D.P., Hetherington, A.M., Trewavas, A.J. (1985) Isolation of a plasma membrane bound calcium/calmodulin regulated protein kinase from pea using Western blotting. Planta (in press).

de Silva, D.L.R., Hetherington, A.M. and Mansfield, T.A. (1985) Synergism between calcium ions and abscisic acid in preventing stomatal opening. New Phytologist (in press).

Dieter, P. (1984) Calmodulin and calmodulin mediated processes in plants. Plant Cell and Environment 7, 371-380.

Elliot, D.C., Batchelor, S.M., Cassar, R.A. and Marinos, N.G. (1983) Calmodulin binding drugs affect responses to cytokinin, auxin and gibberellic acid. Plant Physiology, 72 219-244.

Hetherington, A.M. & Trewavas, A. (1984). Activation of a pea membrane bound protein kinase by calcium ions. Planta, 161, 409-417.

Hetherington, A.M. & Trewavas, A. (1982). Calcium-dependent protein kinase in pea shoot membranes. FEBS Letters, 145, 67-71.

Morre, D.S., Morre, J.T. and Varnold R.L. (1984) Phosphorylation of membrane located proteins of soybean in vitro and response to auxin. Plant Physiology 75 (1) 265-268.

Polya, G.M. & Micucci, V. (1984). Partial purification and characterization of a second calmodulin-activated Ca^{2+} -dependent protein kinase from wheat germ. Biochemica et Biophysica Acta, 785, 68-74.

Polya, G.M., Davies, J.R. & Micucci, V. (1983). Properties of a calmodulin-activated Ca^{2+} -dependent protein kinase from wheat germ. Biochemica et Biophysica Acta, 761, 1-12.

Polya, G.M. & Davies, J.R. (1982). Resolution of Ca^{2+}-calmodulin-activated protein kinase from wheat germ. Biochemica et Biophysica Acta, 761 1-12.

Ranjeva, R., Refeno, G., Boudet, A.M. Marme, D. (1983) Activation of plant quinate NAD^{+} oxidoreductase by Ca^{2+} and calmodulin. Proceedings of the National Academy of Sciences (USA) 80. 5222-4.

Rasi-Caldogno, F., De-Michelis, M., & Pugiarello, M.C. (1982) Active transport of Ca^{2+} in membrane vesicles from pea. Biochimica et Biophysica Acta. 693 287-295.

Rasmussen, H. & Barrett, P.Q. (1984) Calcium messenger system : An Integrated view. Physiological Reviews, 64, 938-984.

Roskoski, R. (1983) Assays of protein kinase In: Methods in Enzymology eds. S.P.Calowick and N.O.Kaplan Vol. 99 3-6.

Salimath, B.P. & Marme, D. (1983) Protein Phosphorylation and its regulation by calcium and calmodulin in membrane fractions from zucchini hypocotyls. Planta 158, 560-568.

Venis, M.A. (1977) Solubilization and partial purification of auxin binding sites on corn membranes. Nature 266, 268-269.

Yoshida, S., Vemura, M., Niki, T., Sakai, A., Gusta, L.V. (1983) Partition of membrane particles in aqueous 2 polymer phase system and its practical use for purification of plasma membrane from plants. Physiologia Plantarum 72, 105-114.

Ca^{2+}-DEPENDENCE OF CALLOSE SYNTHESIS AND THE ROLE OF POLYAMINES IN THE ACTIVATION OF 1,3-ß-GLUCAN SYNTHASE BY Ca^{2+}

Heinrich Kauss

Dept. of Biology, University of Kaiserslautern
Postfach 3049, D-6750 Kaiserslautern
German Federal Republic

INTRODUCTION: Ca^{2+}-DEPENDENCE OF CHITOSAN-INDUCED CALLOSE SYNTHESIS IN SOYBEAN CELLS

During recent investigations we tried to establish a convenient system in which several reactions could be simultaneously induced which are regarded to be of importance in the resistance of plants to pathogens. We used suspension-cultured soybean cells with the aim to affect as many cells as possible at the same time in order to have a low background of undisturbed cells, and to allow accurate time studies for biochemical work. As an "elicitor" we selected the natural polycation chitosan (= deacetylated chitin) as experimental results with basic polyamino acids (e.g. Poly-L-Lys[1]) suggested that the primary interaction of such molecules with the cell or its membrane might be relatively nonspecific. This allowed to avoid pragmatically the consideration whether or not group-specific receptors are involved in plant/pathogen interactions.

It was found that addition of chitosan results in a rapid increase in the leakage[2] of electrolytes, nucleotides and peptides and a concomitant displacement of Ca^{2+} from the wall and membrane surface[3]. Working under sterile conditions and in nutrient solution we subsequently found[4] that addition of chitosan to soybean cells also induced a variety of metabolic changes. At chitosan concentrations where growth of the cultures was more or less arrested we observed a browning which was most likely due to changes in the alkali-soluble wall phenolics, as well as an accumulation of the phytoalexin glyceollin. The regulation of the increase in internal pool size and the export of the latter substance is regarded to be mainly due to *de novo* synthesis of respective enzymes[5]. Phytoalexin production thus is a developmental process which requires the realization of genetic information. The stimulus set by the elicitors, presumably at the level of the plasma membrane, has to be translated into biochemical information capable of interacting with the activation of genes in the nucleus. Necessarily, this process requires many hours or even days to be properly expressed at the level of the final products and, therefore, can be regarded as a rather slow defense mechanism.

In chitosan-treated cells we have also observed a striking change in the composition of cell wall carbohydrates which was due to the formation of callose, a polysaccharide containing a high proportion of 1,3-ß-linked glucose[4,6]. We developed a rapid and quantitative fluorometric assay for callose which is based on the aniline blue staining technique and could show[6] that callose synthesis is far more rapid than phytoalexin synthesis. First deposition was evident about 10 min after addition of chitosan and proceeded with constant rates after about 30 min[6]. This was rapid enough to allow that all further *in vivo* experiments were performed without sterile handling and that the complex nutrient solution could be replaced by a buffer containing 2 % sucrose, conditions under which we could measure and control external Ca^{2+}. Using these techniques we could show[6,7] that induction of callose biosynthesis by chitosan required about 10 μM external Ca^{2+}. Callose formation was also immediately stopped by EGTA under conditions where it proceeded at constant rates and was partially restored again when the concentration of free Ca^{2+} was titrated back to the initial value. It has been suggested, therefore, that enforced Ca^{2+}-influx into the cytoplasm is an important parameter for callose synthesis, Ca^{2+} possibly causing its effect by a direct activation of the 1,3-ß-glucan synthase which appears to be an enzyme vectorially arranged in the plasma membrane[6,7]. This working hypothesis includes also the suggestion that callose formation in any plant tissue, which can be easily monitored by fluorescence microscopy after staining with decolorized aniline blue, can be regarded to indicate that cytoplasmic $[Ca^{2+}]$ has been increased. This aspect has been discussed in more detail in connection with the chitosan-induced formation of the phytoalexin glyceollin in soybean cells[6,7].

The present contribution deals mainly with recent results on the regulative properties of the plasma-membrane located 1,3-ß-glucan synthase which might help to understand the regulation of callose synthesis by Ca^{2+}. It will be shown that this enzyme can be rendered remarkably more sensitive to the activation by Ca^{2+} in the physiologically important low concentration range by the polyamines spermine and spermidine. The activity of the 1,3-ß-glucan synthase is also affected by free unsaturated fatty acids and some other amphipathic compounds. This suggests a phospholipid dependence of the enzyme and might also provide another regulative parameter superimposed on Ca^{2+}/spermine regulation of callose synthesis.

MATERIAL AND METHODS

All details have been described in recent publications[8-11] and are, therefore, mentioned only briefly in the legends.

RESULTS AND DISCUSSION

A principle enzyme involved in callose synthesis appears to be the 1,3-ß-glucan synthase (EC 2.4.1.34) located in the plasma-membrane. We could demonstrate before that the activity of this enzyme in washed microsomes from suspension-cultured soybean cells strongly depends on $[Ca^{2+}]$. This was first found[8] in assays containing 5 mM Mg^{2+}; under these conditions a 10- to 15-fold activation was caused at saturating concentrations of Ca^{2+} (50 to 100 μM). In the absence of Ca^{2+} ($<10^{-8}$M), however, 5 mM Mg^{2+} also slightly activates the enzyme. If Mg^{2+} is omitted, than the overall effect of Ca^{2+} becomes 30- to 40-fold[7] or even higher (Table 1). The activating effect of Ca^{2+} was found[8] to be readily reversed on addition of EGTA, and the activation as well as inactivation process in the washed microsomes did not require addition of any nucleotide. It was concluded, therefore, that the effects were not resulting from phosphorylation/dephosphorylation but from a direct allosteric type of interaction of Ca^{2+} with the enzyme. Although Ca^{2+} was found to be essential for the 1,3-ß-glucan synthase activity we still observed, even at saturating concentrations of this ion, the sigmoidal substrate-concentration curve (Fig. 1), which has also been reported by other authors[12] working with microsomes from various plants. The unfavourably low activity of the enzyme at UDP-glucose concentrations in the lower μM range is regarded to be a characteristic feature of the so-called glucan synthase II, reported to be a plasma-membrane marker[13]. We could recently show[9] that this enzyme in the absence of Ca^{2+} can also be activated by basic polyamino acids (poly-L-Lys, poly-L-Orn, poly-L-Arg) and by Ruthenium Red, substances which have in common the presence of many positively charged groups in the same molecule. As these substances otherwise exhibit very different chemical features it appears likely that they all may act by binding to negatively charged sites. The activation by Ca^{2+} is greatly inhibited by La^{3+}. In contrast, when poly-L-Orn is used as an activator there is no inhibition at all by La^{3+} whereas with Ruthenium Red only a relatively slight inhibition is observed at elevated La^{3+} concentrations. These results suggest[9] that the newly described activators and

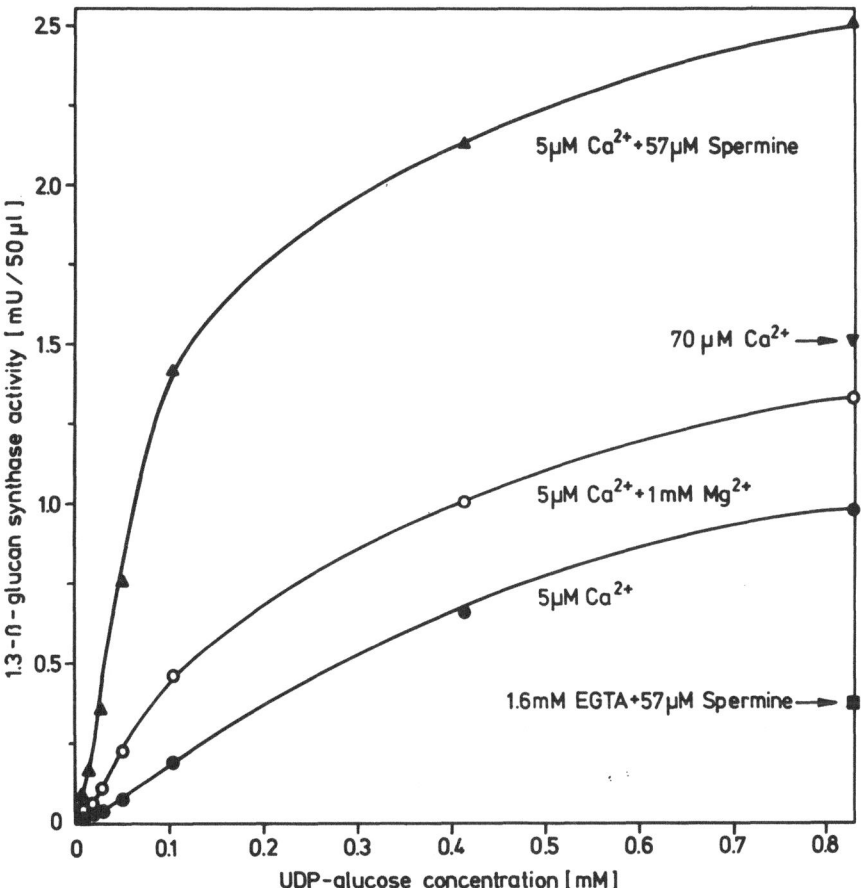

Fig. 1. Influence of Ca^{2+}, Mg^{2+} and spermine on 1,3-ß-glucan synthase activity in microsomes from suspension-cultured soybean cells at various UDP-glucose concentrations. The 5 μM Ca^{2+} were present endogenously in the assay mixture. In the presence of 1.6 mM EGTA to establish a $[Ca^{2+}] < 10^{-8}$M essentially no activity of the enzyme was observed without Mg^{2+} or spermine (see Table 1). For experimental details see ref. 6 and Table 1.

Table 1. Increases in the Effectiveness of Ca^{2+} for the Activation of 1,3-ß-Glucan Synthase by Spermine

| Free Ca^{2+} [μM][b] | 1,3-ß-Glucan Synthase Activity[a] [mU/50 μl Microsomes] at the Spermine Concentration of | | |
	0 μM	20 μM	200 μM
$<10^{-2}$	0.00	0.01	0.05
0.2	0.01	0.07	0.16
0.7	0.04	0.15	0.62
2.0	0.08	0.33	0.65
60	0.23	0.40	0.78

[a] measured at 36 μM UDP-[^{14}C]-glucose in the presence of 0.016 % (w/v) digitonin, 8 mM cellobiose and 6.4 % glycerol, as described in ref. 6. Reaction time was 1 to 10 min, depending on activity, and values from zero time controls were subtracted.
[b] a range of EGTA/Ca^{2+}-buffers without Mg^{2+} was used and the free [Ca^{2+}] determined[11] in a scaled-up assay mixture using either Quin 2 (<1 μM) or a Radiometer-Selectrode (>1 μM).

Ca^{2+} interact at different sites of the 1,3-ß-glucan synthase. The most striking observation was that on activation by poly-L-Orn or Ruthenium Red the apparent affinity of the enzyme towards UDP-glucose was far better than on activation by saturating concentrations of Ca^{2+}.

The polyamines spermine and spermidine at a 57 μM concentration were found to cause relatively little activation as long as EGTA was added to complex the Ca^{2+} endogenously contained in the assay[9]. If, however, the 5 μM Ca^{2+} in the assay was allowed to be present in addition to the poly-amines than the activity was greatly enhanced at any substrate concentration and the sigmoidal shape of the curve was less pronounced (Fig. 1). $MgCl_2$ at a concentration 17-fold higher than spermine shows only a slight but nevertheless similar effect. The activation caused by spermine in co-operation with non-saturating concentrations of Ca^{2+} was found to be of special interest at UDP-glucose concentrations in the low μM-range. It was observed[9] that spermine under these conditions caused almost no activation (not shown in Fig. 1) but greatly increased the effect of Ca^{2+}. At 2 μM Ca^{2+} and 36 μM UDP-glucose significant effects were brought about even by 4 μM spermine and increased up to about 200 μM spermine[11]. Most remarkably, the effects caused by the two activators were far more than additive. This cooperative action is evident again from the data in Table 1 for any [Ca^{2+}] in combination with any of the two spermine concentrations when the values in the first line are added to the values in the first vertical column and compared with those obtained when both activators are present simultane-ously. This observation strongly suggests that both types of activation may occur at the same enzyme, namely the plasma-membrane located 1,3-ß-glucan synthase II. This is confirmed at present by sucrose density gradient studies (J. Fink and H. Kauss, unpublished) which show that both the acti-vation by Ca^{2+}, poly-L-Orn alone or Ca^{2+} plus spermine coincide with the same peak-fractions containing plasma membrane-derived vesicles. Spermine not only increases the apparent substrate affinity and maximal activity of the enzyme (Fig. 1) but also appears to render it more sensitive to Ca^{2+}-activation. This becomes evident when the enzyme activities at low $[Ca^{2+}]$ are considered (Table 1). At 0.7 μM Ca^{2+}, for instance, it attains without spermine 17 % of the activity found at 60 μM Ca^{2+}. If, however, 200 μM spermine are present in addition, this increases to 83 % of activity attained at saturating concentrations of Ca^{2+}.

In the context of the present workshop the data reported here appear to be of double significance. At one hand they suggest that polyamines might represent an internal parameter by which plant cells could regulate their callose synthesis, in addition to Ca^{2+} which mainly may be of exogenous origin. Polyamines occur in considerable amounts in plant cells. Their amounts present and the activity of the respective enzymes are under developmental control and appear to be regulated[14], for instance, by phytochrome-perceived light, hormones and stress of various nature. The suggestion that polyamines may represent "intracellular messengers" is of special interest in regard to the present observation that they cooperate with Ca^{2+} to regulate the activity of the 1,3-ß-glucan synthase. One should bear in mind that classical second messengers such as cAMP or cGMP often function at the level of target enzymes in cooperation with Ca^{2+}.

At the other hand the results might indicate that Ca^{2+}/polyamine regulation of enzymes might be a more general phenomenon which may become evident only when the concentrations of the two types of activators as well as of the substrates are considered simultaneously. This might apply to other enzymes known to be stimulated by polyamines (e.g. protein kinases[15,16]) as well as to those reported to require unphysiologically high Ca^{2+} concentrations for stimulation (e.g. the quinate-oxidoreductase in dark-grown carrot cells[17]).

In addition to polyamines still other parameters might be superimposed on Ca^{2+}-regulation of the 1,3-ß-glucan synthase. On trypsinization of the microsomes the enzyme becomes active[8] without Ca^{2+}, a property which is even more pronounced[9] in the presence of $MgCl_2$. Assayed under the latter condition the enzyme was nevertheless inhibited by Calmidazolium and Trifluoperazine[8]. As these drugs bind to calmodulin only in the presence of Ca^{2+} and relatively high concentrations were required, it was concluded that the effect of Ca^{2+} on the 1,3-ß-glucan synthase is not mediated by calmodulin. The inhibition caused by these two drugs and by Polymyxin B was taken as an indication for a phospholipid-requirement of the membrane-bound enzyme[8]. This assumption was sustained by the more recent observation that various other substances known to interfere with phospholipids can also influence the 1,3-ß-glucan synthase activity[10]. Certain free unsaturated fatty acids, lysophosphatidylcholine, acylcarnitine, Echinocandin B and platelet-activating factor (PAF) showed in principle similar effects, although certain quantitative differences occurred. They all stimulated at certain low concentrations and at more elevated ones strongly inhibited. Inhibition was similarly observed when the enzyme was activated by Ca^{2+} or by trypsinization, suggesting that the effect is not due to an interference of the substances with the Ca^{2+} necessary for activation. The effect of the amphipathic compounds is exemplarily shown for palmitoyl-*DL*-carnitine and PAF in Table 2. In most of our experiments the 1,3-ß-glucan synthase was assayed with the addition of digitonin which is assumed to render membrane vesicles permeable to substrates and activators. This might, however, not be the only explanation for its action as in a certain concentration range the amphipathic compounds still can increase the activity over the effect caused by saturating concentrations of digitonin (Table 2). The stimulating as well as the inhibitory effect at higher concentrations might indicate that they act by replacement of endogenously phospholipids necessary for optimal activity of the enzyme. However, it remains to be established in future as to what extent this suggested effect, or a merily

Table 2. Influence of Amphipathic Compounds on the
Activity of 1,3-ß-Glucan Synthase as an
Indication of Phospholipid-Dependence

Assay Conditions (nmole/50 µl)[2]	1,3-ß-Glucan Synthase[1] [mU/50 µl Microsomes]
- Digitonin:	
control	0.44
+ Palmitoyl-DL-carnitine[3] (10)	1.75
+ " (50)	0.0
+ PAF-acether[4] (10)	1.22
+ " (100)	0.35
+ Digitonin[5]:	
control	2.33
+ Palmitoyl-DL-carnitine (2.5)	3.12
+ " (50)	0.0
+ PAF-acether (2.5)	3.2
+ " (100)	0.3

[1]Assayed for 10 min at 0.83 mM UDP-[^{14}C]-glucose and
1.6 mM EDTA + 1.2 mM $CaCl_2$ + 4 mM $MgCl_2$ to establish
a [Ca^{2+}] of 130 to 190 µM.
[2]Additions are given in nmole/50 µl microsomes to indicate that the amount of amphipathic compound per
amount of membrane is the essential feature (10 nmole/
assay ≈ 80 µM nominal conc.).
[3]Added in water
[4]Platelet-activating factor or 1-0-octadecyl-2-0-
acetyl-sn-glycerol-3-phosphorylcholine, added in ethanol
to give a 1 % solution of the solvent, controls with
1 % ethanol showed about 5 % lower activity.
[5]0.016 % (w/v) final conc.

opening of vesicles contribute to the stimulation observed in the absence
of digitonin (Table 2). It appears, however, well possible that both a
stimulatory and inhibitory action of unsaturated fatty acids, lysophosphatidylcholine or similar unknown endogenous compounds is superimposed
on the Ca^{2+}/spermine regulation of callose synthesis in vivo. This process
is associated with cell and tissue repair and pathogen defense mechanisms
which in plants are all associated with membrane perturbation and are,
therefore, likely to be accompanied by phospholipid deacylation. The
amphipathic substances used in vitro have in addition to a direct interaction with the 1,3-ß-glucan synthase various other effects on membrane-
associated processes. They can, for instance, also directly induce callose
synthesis[6,10]. Clarification of causal relationships, therefore, requires
far more efforts.

ACKNOWLEDGEMENT

The experiments reported have been enabled by grants from the Deutsche
Forschungsgemeinschaft and the skillful technical assistance of W. Jeblick.

REFERENCES

1. H. R. Lerner and M. Reuveni, Induction of pore formation selectively in the plasmalemma of plant cells by poly-L-lysine treatment: a method for the direct measurement of cytosol solutes in plant cells, in: "Plasmalemma and Tonoplast: Their Function in the Plant Cell," D. Marmé, E. Marré and R. Hertel, eds., Elsevier Biomedical Press, Amsterdam, New York, Oxford (1981).
2. D. H. Young, H. Köhle and H. Kauss, Effects of chitosan on membrane permeability of suspension-cultured *Glycine max* and *Phaseolus vulgaris* cells, Plant Physiol. 70:1449 (1982).
3. D. H. Young and H. Kauss, Release of calcium from suspension-cultured *Glycine max* cells by chitosan, other polycations, and polyamines in relation to effects on membrane permeability, Plant Physiol. 73: 698 (1983).
4. H. Köhle, D. H. Young and H. Kauss, Physiological changes in suspension-cultured soybean cells elicited by treatments with chitosan. Plant Sci. Lett. 33:221 (1983).
5. J. Chappell and K. Hahlbrock, Transcription of plant defense genes in response to UV light or fungal elicitors, Nature 311:76 (1984).
6. H. Köhle, W. Jeblick, F. Poten, W. Blaschek and H. Kauss, Chitosan-elicited callose synthesis in soybean cells, Plant Physiol. 77:544 (1985).
7. H. Kauss, Callose biosynthesis as a Ca^{2+}-regulated process and possible relations to the induction of other metabolic changes, J. Cell Sci., in press (1985).
8. H. Kauss, H. Köhle and W. Jeblick, Proteolytic activation and stimulation by Ca^{2+} of glucan synthase from soybean cells, FEBS-Lett. 158:84 (1983).
9. H. Kauss and W. Jeblick, Activation by polyamines, polycations and Ruthenium Red of the Ca^{2+}-dependent glucan synthase from soybean cells, FEBS-Lett. 185:226 (1985).
10. H. Kauss and W. Jeblick, Influence of free fatty acids, lysophosphatidylcholine, platelet-activating factor, acylcarnitine and Echinocandin B on 1,3-ß-D-glucan synthase and callose synthesis, Plant Physiol., submitted (1986).
11. H. Kauss and W. Jeblick, Polyamines render the 1,3-ß-D-glucan synthase more sensitive to activation by Ca^{2+}, submitted (1985).
12. D. P. Delmer, U. Heiniger and C. Kulow, UDP-glucose: glucan synthetase in developing cotton fibers, Plant Physiol. 59:713 (1977).
13. P. M. Ray, Membrane-associated molecules and structures, in: "Methodological Surveys in Biochemistry", Vol 9, p. 135, E. Reid, ed., Chichester (1979).
14. A. W. Galston, Polyamines as modulators of plant development, BioScience 33:382 (1983).
15. K. Veluthambi and B. W. Poovaiah, Polyamine-stimulated phosphorylation of proteins from corn (*Zea mays L.*) coleoptiles, Biochem. Biophys. Res. Com. 125:774 (1984).
16. G. D. Kuehn and V. J. Atmar, Posttranslational control of ornithine decarboxylase by polyamine-dependent protein kinase, Federation Proc. 41:3078 (1982).
17. A. Graziana, M. Dillenschneider and R. Ranjeva, A calcium-binding protein is a regulatory subunit of quinate: NAD^+ oxidoreductase from dark-grown carrot cells, Biochem. Biophys. Res. Com. 125:774 (1984).

ESTIMATION OF CYTOPLASMIC CALCIUM

THE DEFINITION AND MEASUREMENT OF INTRACELLULAR FREE Ca

M.V. Thomas

Sittingbourne Research Centre, Shell Research Ltd.

Sittingbourne, Kent ME9 8AG

INTRODUCTION

The measurement of the free Ca concentration in living cells is a challenging problem, and the difficulties are likely to be even greater in plant cells than in animal cells, on which most Ca measurements have been made.

The reasons why knowledge of free intracellular Ca concentration should be of value will not be considered in any detail here, as they are discussed in other presentations. This presentation will concentrate on the concept of free Ca and how it can be measured. It will also discuss briefly the various possible mechanisms for control of the free Ca concentration, in particular the various sources and sinks for free Ca in the cytosol.

The most important points concerning intracellular Ca are that only a small proportion of the total intracellular Ca is present as free ionized Ca in the cytosol, and that the free cytosolic Ca may change as a result of redistribution of the total intracellular Ca rather than as a result of a change in the total Ca content. Therefore measurements of total Ca content, or of Ca uptake or release, cannot be used to estimate the free Ca concentration in the cytosol, and instead this parameter must be measured directly.

An additional complication is that there are two alternative definitions of free ionized Ca, according to whether Ca concentration or Ca thermodynamic activity is used as the unit of measurement. For reasons that will be described below, concentration is the preferred unit of measurement, and this convention will be adopted throughout the present discussion.

Measurements on a variety of different cells have shown that the free Ca concentration in the cytosol is normally on the order of 100nM, and that it may rise into the micromolar range in response to an appropriate stimulus, whereas the total intracellular Ca concentration may be in the millimolar range. These findings have been discussed in my book on intracellular Ca measurement (Thomas, 1982), and in papers and reviews by others working in the field of intracellular Ca measurement (Blinks et al., 1983; Campbell, 1983; Tsien & Rink, 1983). Those publications deal primarily with the measurement of free Ca in animal cells, since hardly any information is yet available for plants (but see Williamson & Ashley, 1982). However, much of the information in them is directly relevant to plants as well.

THE POSSIBLE FORMS OF INTRACELLULAR Ca

Since very little of the total Ca in cells is present as free ionized Ca in the cytosol, most of it must be bound in some way. In principle, this could arise via two different mechanisms, and they both appear to be important. Ca may be chelated to other molecules in the cytosol, which thereby act as Ca buffers, or it may be sequestered inside organelles. The two processes differ in an important respect. Chelated Ca is in thermodynamic equilibrium with the free Ca (except possibly when the free Ca concentration is changing rapidly), whereas sequestration of Ca requires active Ca transport. In view of the fundamentally different nature of these two processes, some care is needed in both the measurement and the discussion of the overall buffering capacity of cytoplasm.

The nature of the molecules that may chelate intracellular Ca and the extent to which this phenomenon occurs are likely to vary significantly between cells, but carboxylate groups on protein molecules probably make an important contribution to the Ca chelating capacity of cytoplasm. The existence of this phenomenon was suggested by a variety of experiments, such as those showing that only a small proportion of the Ca that was introduced into the cytosol of nerve cells appeared as free Ca, even when active Ca sequestration mechanisms were blocked by metabolic inhibitors (Brinley, 1978). In those experiments it was estimated that the chelated Ca was 20-30 times higher than the free Ca.

The interaction of intracellular Ca with various receptors occurs via chelation of Ca to specific receptor sites, so the chelation of and the responses to intracellular Ca may be closely related phenomena. Indeed, where the density of Ca receptor sites is very high (as in striated muscle) they may contribute significantly to the total Ca chelating capacity of cytoplasm. A variety of proteins that can bind Ca have been characterized, and most (if not all) biological phenomena that are controlled by Ca appear to be mediated by binding of Ca to these proteins. Calmodulin is of particular importance in this respect, and many other Ca-binding proteins (such as Troponin C in striated muscle) are structurally related to it (see Van Eldik et al., 1982, and Klee & Vanaman, 1982, for reviews).

Although the chelated Ca will be in equilibrium with the free Ca, it should be emphasised that any factor that is capable of disturbing that equilibrium may consequently change the free Ca concentration. In particular, a change in intracellular pH could have such an effect, just as it affects the apparent affinity of commonly-used Ca buffers such as EGTA.

The major compartments of the cell into which intracellular Ca could be sequestered include organelles such as the endoplasmic reticulum, mitochondria, chloroplasts and the vacuole. The relative importance of intracellular and extracellular sources and sinks for cytosolic Ca varies substantially between different cells. In nerve cells, for example, the increases in intracellular Ca that mediate such phenomena as neurotransmitter release or potasssium ion channel activation arise primarily from Ca influx through the cell membrane, whereas the increases that mediate the activation of the contractile mechanism in skeletal muscle arise primarily from release of Ca from the endoplasmic (sarcoplasmic) reticulum (see Thomas, 1982, for further discussion and references).

The possible role of mitochondria in regulating the free cytosolic Ca concentration is a contentious issue. A number of studies on isolated mitochondria have shown that these organelles can indeed take up Ca. However, such studies have generally employed Ca concentrations in the micromolar range, which we now know to be substantially higher than the normal resting free Ca concentration in the cytosol, and it is not clear to what extent uptake of Ca by mitochondria is physiologically significant. Mitochondria do appear to take up Ca in cells that are subjected to an abnormally high Ca load, such as by Ca injection into insect salivary gland cells (Rose & Loewenstein, 1975), but electron probe analysis of vascular smooth muscle showed that the mitochondria in that tissue did not contain significant amounts of Ca except in cases where the overall ionic balance was disturbed (i.e. elevated intracellular Na concentration and depressed K concentration), suggesting that the cell was in a damaged state (Somlyo, Somlyo & Shuman, 1979).

The driving force for Ca uptake into mitochondria is likely to be the mitochondrial membrane potential. This is normally in the range of 150-200mV, inside negative, and Ca uptake by mitochondria was found to increase as the potential became more negative over this range (Bernadi & Azzone, 1983). If Ca were to distribute passively according to the Nernst equation for a divalent ion (tenfold concentration difference per 29mV potential difference), its concentration within the mitochondria ought to be around 100mM, for a cytosolic concentration of 100nM and a mitochondrial membrane potential of 180mV. This enormous theoretical concentration clearly implies that under normal conditions, mitochondria must actively extrude Ca. Therefore net uptake of Ca into mitochondria, particularly at unphysiologically high cytosolic Ca concentrations, could be a result of overload or failure of the active Ca extrusion mechanism(s). Thus observations that mitochondria can take up Ca in vitro are in no way surprising, and they do not necessarily imply that it occurs to a significant extent in vivo. Very great caution is needed on this point (see Thomas, 1982, for further discussion).

The situation in plant cells is potentially more complicated than in animal cells, since the chloroplasts and the vacuole may also be important in the regulation of free cytosolic Ca.

Chloroplasts appear to behave similarly to mitochondria with respect to Ca. Measurememts of Ca flux rates in spinach chloroplasts showed a similar Ca-dependence to the rates observed in rat liver mitochondria (Kreimer, Melkonian & Latzko, 1985). Uptake of Ca by chloroplasts is also light-dependent, and these authors suggested that the effect occurred as a result of light increasing the chloroplast membrane potential, thereby increasing the driving force for Ca entry, as discussed above for mitochondria. The same caution is needed as for mitochondria in ascribing any physiological significance to this phenomenon, attractive as it may be to do so. Measurements of the cytosolic free Ca in vivo will be necessary to resolve this point. It is also attractive to suggest that changes in the chloroplastic free Ca concentration may modulate various aspects of chloroplast function, but caution has been advised on this point as well (Simon et al., 1984).

The vacuole appears to be of somewhat greater potential importance as a Ca reservoir. The total Ca concentration in the vacuole of Characean cells was found to be 12mM, compared with 8mM in the cytoplasm (Spanswick & Williams, 1965). Using the photoprotein aequorin, Williamson & Ashley (1982) obtained a lower limit of 100uM for the free Ca, compared with around 100nM in the cytoplasm. These figures suggest that Ca is actively transported into the vacuole.

Although intracellular stores of Ca may play an important short-term role in controlling the free cytosolic Ca concentration, they cannot be filled or emptied without limit. On a long-term basis, the free cytosolic Ca concentration must be determined by the Ca permeability and Ca transport properties of the plasma membrane (reviewed by Sulakhe & St. Louis, 1980).

Another point to stress in connection with the chelation and sequestration of intracellular Ca is that both these processes will lower the diffusion coefficient for free Ca in the cytosol, perhaps by two or more orders of magnitude (again see Thomas, 1982). This means that large Ca concentration gradients can occur over relatively small distances within the cytosol, particularly when the average free cytosolic Ca concentration is changing rapidly. Under these conditions the free cytosolic Ca concentration can vary substantially in space as well as in time.

THE ACTIVITY COEFFICIENT OF Ca IN CYTOPLASM

The difference between ionic concentration and ionic activity is a potential source of confusion. Although ionic activity is the preferable measurement on thermodynamic grounds, concentration is the better choice in practice. This is because concentration can be defined exactly in terms of the relative numbers of solute and solvent molecules, whereas activity has to be measured. The basis for measurement of activity is that concentrated solutions behave as if they are less concentrated than expected from extrapolation of the behaviour of very dilute solutions, with respect to such properties as osmotic strength. This diminution is normally expressed in terms of an activity coefficient. For dilute ionic solutions the effect can be described by the Debye-Huckel equation, which ascribes the reduction in activity to electrostatic interactions between the ions in the solution. For more concentrated solutions, as found inside cells, the Debye-Huckel equation has to be modified in a more-or-less empirical way to fit the data.

An additional theoretical complication is that the activity coefficients of individual ions cannot be measured directly; they have to be estimated from the measured activity coefficients of solutions of various ion-pairs, and ultimately an untestable assumption must be made (e.g. that the reference electrode used with an ion-sensitive electrode performs ideally) in order to do this.

The activity coefficient of an ion is affected primarily by the ionic strength of the solution. The way in which the ionic strength is made up (i.e. the nature of the other ions) is only of secondary importance. The effect is clearly different from specific interactions such as those between Ca and EGTA. The individual ions remain fully dissociated, and concentration is therefore an entirely acceptable unit of measurement. Indeed, the uncertainties surrounding ionic strength measurements makes concentration (at a specified overall ionic strength) the only choice for "standards" work such as measurement of buffer affinities.

Measurements of the Ca activity coefficient by various methods have suggested that its value over the range of ionic strengths likely to be encountered inside cells is likely to be between 0.25 and 0.5, with 0.33 being a reasonable approximate estimate for most cells (Thomas, 1982, Figure 2.3). There is normally no need to take the activity coefficient into account when making Ca measurements. All that is required is to calibrate the Ca measurement method _in vitro_ at (approximately) the same ionic strength as the cells, with respect to free Ca concentrations which have been defined by a standard Ca buffer such as EGTA.

The only occasion when the Ca activity coefficient needs to be considered is when considering the movement of Ca between compartments of significantly different ionic strength, since the Ca activity coefficient will differ between the two compartments. Although this situation rarely arises in animal cells, it is of course much more common in plant cells, particularly with respect to Ca movement across the plasma membrane. The intracellular ionic strength is likely to be very much higher than that outside and therefore the Ca activity coefficient will be lower inside (say about 0.33) than outside (may be close to 1.0). This means that the true Ca electrochemical gradient across the membrane will be somewhat higher than calculated from the Ca concentration ratio, and an appropriate correction will need to be made when calculating the gradient.

Ca MEASUREMENT TECHNIQUES

It should be realized that all the currently used methods of measuring free intracellular Ca have been developed for use on animal cells, where some of the requirements (in particular the measurement of rapid transient changes in the free intracellular Ca in nerve and muscle cells) are rather different from those that are likely to be necessary for plant cells. Thus some of the methods that have been particularly useful in animal cells may be rather less so in plants.

The three main methods of measuring the cytosolic free Ca concentration are photoproteins, Ca-sensitive electrodes and dyes. Apart from some of the newer dyes, these methods all require introduction of one or more micropipettes into the cell, either to inject the Ca-sensitive agent (dyes or photoproteins) or to make the measurement (Ca electrodes). These measurement techniques were all pioneered on relatively large animal cells, particularly giant nerve and muscle cells with diameters on the order of 1mm (see the reviews cited in the introduction for further information). Although these methods have been refined to permit their use in small cells, the technical difficulties become significantly greater for cells having a diameter much below 100um. The cell wall of plant cells makes insertion of micropipettes more difficult than in animal cells, and there is also the problem that the pipette tip may reside in the vacuole rather than the cytoplasm if it is inserted too far. Application of these techniques to plant cells is thus a challenging problem, although Williamson and Ashley's (1982) measurements on giant algal cells were obtained in this way, using the photoprotein aequorin.

These techniques are well described in the reviews, and will only be discussed in outline here. The first technique to be applied successfully was the use of the photoprotein aequorin. This substance (which is isolated from jellyfish of the genus _Aequorea_) emits light in a reaction that is catalysed by Ca. The aequorin cannot readily be regenerated _in vitro_, and so great care must be taken to keep it free of Ca before it is introduced into the cell, otherwise its activity will be lost. Another problem is that its rate of reaction varies sigmoidally with the Ca concentration, which complicates calibration. The best feature of the technique is its very high sensitivity, since single photons (representing the reaction of single aequorin molecules) can be detected at high efficiency by photomultiplier techniques.

Measurement of Ca by ion-sensitive electrodes has been possible for some time, but Ca electrodes with sufficient selectivity against other ions to allow their use inside cells are a comparatively recent development. However, there remains a problem with miniaturization of these electrodes. It is difficult to reduce the tip size below about 1 micron without loss of performance, and this makes the technique less suitable for small cells. Another difficulty is that a reference electrode must be inserted into the cell, to allow the membrane potential to be subtracted from the potential recorded by the Ca electrode. A double-barrelled pipette, containing both electrodes, can be used, but this also causes size problems. The other major disadvantage of this technique is that the response time of the electrode can be fairly long (perhaps a second or more) in comparison with the other techniques, although that is not likely to be a particular drawback in most plant cells. There may also be problems of long-term drift. On the other hand, it does not require any specialized equipment apart from a very high-impedance amplifier, which is a possible advantage over the other techniques. Successful use of this technique requires considerable patience, but it ought to be applicable to at least the larger plant cells.

A variety of dye molecules have been used to measure the intracellular free Ca concentration. These molecules undergo a change in optical absorbance and/or fluorescence on chelating Ca, and so they must be used in conjunction with suitable spectrophotometric equipment. The major potential disadvantage of this technique is that chelation of Ca by the intracellular dye molecules may affect the intracellular free Ca concentration. The extent of this disturbance can be assessed by making the measurements over a range of dye concentrations, and the results of such experiments have shown that it is normally possible to obtain satisfactory measurements at dye concentrations below the threshold at which this effect becomes significant.

The dyes generally respond linearly to the free Ca concentration, as long as only a small proportion of the total dye is complexed with Ca. As the Ca concentration increases, the response begins to saturate, and this limits the Ca concentration range over which a dye can be used. For accurate measurement of large transient Ca changes, a dye of relatively low Ca affinity, such as arsenazo III, is most appropriate. On the other hand, such dyes are not sufficiently sensitive to measure resting Ca levels with any accuracy, and for this purpose a dye of much higher affinity, such as quin2, is needed.

Calibration of dye absorbance or fluorescence changes can be difficult. This is particularly true for arsenazo III and related dyes, since the stoicheiometry of their interaction with Ca is complex and other ions, particularly protons and Mg, interfere to a significant extent. Dyes in the quin2 family are much better in this respect.

Another advantage of quin2 and related dyes is that they can be introduced into cells by an ester loading technique. If the four carboxylate groups that bind Ca are converted to acetoxymethyl esters, the resulting uncharged derivative can permeate biological membranes. The ester linkages are hydrolysed by enzymes in the cytosol, resulting in accumulation of the active indicator in the cytosol (Tsien, 1981). This method can be applied to much smaller cells than is possible by the microinjection method, and application of it to plant cells is potentially very attractive. If it proves successful in plants, it would probably be the method of choice.

Quin2 is not a universally applicable Ca indicator (its Ca affinity is rather high and the fluorescence change on binding Ca is relatively low), but Tsien's group is developing analogues with potentially better performance in these respects. In particular, the fura analogues (Grynkiewicz, Poenie & Tsien, 1985) offer a considerable improvement in sensitivity, the fluorescence of fura2 being about 30 times greater than that of quin2. These new dyes could greatly facilitate the measurement of free Ca concentrations in plant cells.

REFERENCES

Bernardi, P., and Azzone, G.F., 1983, Regulation of Ca efflux in rat liver mitochondria. Role of membrane potential, Eur. J. Biochem. 134:377.
Blinks, J.R., Weir, W.G., Hess, P., and Prendergast, F.G., 1983, Measurement of calcium(2+) concentrations in living cells, Prog. Biophys. Mol. Biol. 40:1.
Brinley, F.J., Jr., 1978, Calcium buffering in squid axons, Ann. Rev. Biophys. Bioeng. 7:363.
Campbell, A.K., 1982, "Intracellular Calcium: Its Universal Role as a Regulator," John Wiley, Chichester.
Grynkiewicz, G., Poenie, M., and Tsien, R.Y., 1985, A new generation of Ca indicators with greatly improved fluorescence properties, J. Biol. Chem. 260:3440.
Kreimer, G., Melkonian, M., and Latzko, E., 1985, An electrogenic uniport mediates light-dependent Ca influx into intact spinach chloroplasts, FEBS Lett. 180:253.
Rose, B., and Loewenstein, W.R., 1975, Calcium ion distribution in cytoplasm visualized by aequorin: diffusion in cytosol restricted by energized sequestering, Science, N.Y. 190:1204.
Simon, P., Bonzon, M., Greppin, H., and Marme, D., 1984, Subchloroplastic localization of NAD kinase activity: evidence for a Ca, calmodulin-dependent activity at the envelope and for a Ca, calmodulin-independent activity in the stroma of pea chloroplasts, FEBS Lett. 167:332.
Somlyo, A.P., Somlyo, A.V., and Shuman, H., 1979, Electron probe analysis of vascular smooth muscle: composition of mitochondria, nuclei and cytoplasm, J. Cell. Biol. 81:316.
Spanswick, R.M., and Williams, E.J., 1965, Ca fluxes and membrane potentials in Nitella translucens, J. Exp. Bot. 16:463.
Sulakhe, P.V., and St. Louis, P.J., 1980, Passive and active calcium fluxes across plasma membranes, Prog. Biophys. Mol. Biol. 35:135.
Thomas, M.V., 1982, "Techniques in Calcium Research," Academic Press, London.
Tsien, R.Y., 1981, A non-disruptive technique for loading Ca indicators and buffers into cells, Nature 290:527.
Tsien, R.Y., and Rink, T.J., 1983, Measurement of free calcium(2+) in cytoplasm, Curr. Methods Cell. Neurobiol. 3:249.
Van Eldik, L.J., Zendegui, J.G., Marshak, D.R., and Watterson, D.M., 1982, Calcium-binding proteins and the molecular basis of calcium action, Int. Rev. Cytol. 77:1.
Williamson, R.E., and Ashley, C.C., 1982, Free Ca and cytoplasmic streaming in the alga Chara, Nature 296:647.

MEASUREMENT OF CYTOPLASMIC CALCIUM ACTIVITY

WITH ION-SELECTIVE MICROELECTRODES

Dale Sanders and Anthony J. Miller

Department of Biology
University of York
York YO1 5DD

INTRODUCTION

Evaluation of Methods for Measuring pCa$_c$

Several different techniques are available for measurement of cytoplasmic calcium activity (or its logarithmic equivalent, pCa$_c$). Optical probes involve monitoring changes in one of a number of signals: dye absorbance, fluorescence intensity or bioluminescence. A conceptually different approach stems from the use of Ca^{2+}-selective electrodes in which the sensor is embedded in a liquid membrane plugging the tip of a micropipette.

Each method can be evaluated according to a number of criteria (Blinks et al., 1982): 1. perturbation of pCa$_c$ by the probe itself (potentially a serious problem with many optical indicators, since their utility is based on the fact that Ca^{2+} binding to the ligand in some way affects the optical properties of the ligand); 2. the range of pCa over which the probe responds; 3. the accuracy with which the probe may be calibrated for _in vivo_ use (again, potentially a problem for many of the optical indicators, which are susceptible to interference from Mg^{2+} and H$^+$); 4. whether the cytoplasmic location of the probe can be assured.

Attempts can be made to correct optical signals for some of these potential errors. Nevertheless, microelectrodes made with Simon's neutral carrier resin ETH 1001 (Tsien & Rink, 1980) are not significantly subject to any of the sources of inaccuracy listed above. The ligand is present in insufficient quantity to affect pCa$_c$ directly and shows extremely good selectivity for Ca^{2+} over Mg^{2+} and H$^+$ (see below). Furthermore, the electrodes are responsive over a wide range of pCa, which includes the anticipated physiological range of pCa$_c$ in plants from 5 to 7.

Theoretical Background

For an electrode which exhibits ideal and specific permeability to a given ion A, the electrical potential across it can be described as a function of the ion activities in the internal (i) and external (e) phases by the well-known Nernst equation as

$$E_{ie} = E_o' + \frac{RT}{zF} \ln\left[\frac{[A]_e}{[A]_i}\right], \tag{1}$$

in which the potential difference, E_{ie}, is referenced from the inside to the outside, E_o is a constant reference potential, z is the valence of A, the square brackets designate activity of A and R, T and F have their usual meanings. If the ion activity in the internal phase is maintained constant, Eq. 1 can be rewritten as

$$E_{ie} = E_o' + \frac{2.303RT}{zF} \log[A]_e. \qquad (2)$$

The reference potential, E_o', now subsumes a term for internal activity of A. Substituting in Eq. 2 for a perfectly specific Ca^{2+} electrode at 25 C and expressing the result in millivolts,

$$E_{ie} = E_o' + 29.6\log[Ca^{2+}]_e. \qquad (3)$$

Thus, in principle, the electrode can be calibrated against a range of Ca^{2+}-buffered standards, and the resulting output of the electrode should increase linearly with decreasing pCa, exhibiting a slope of 29.6 mV.

In practice, however, the situation is more complex than this because no ion-selective electrode possesses ideal selectivity for the ion in question. The modified form of Eq. 2 which takes account of the contribution to the overall potential from an interfering ion, B, is the Nicolsky-Eisenman equation:

$$E_{ie} = E_o' + \frac{2.303RT}{z_A F}\log\left[[A]_e + K_{AB}[B]_e^{(z_A/z_B)}\right]. \qquad (4)$$

K_{AB} is the so-called **selectivity coefficient** of the electrode for ion A with respect to ion B. It expresses on a molar basis the relative contributions of ions A and B to the measured potential. Clearly, if cytoplasmic Ca^{2+} is 10^{-6} M or less, it is necessary in the case of an ion such as K^+ (present at 10^{-1} M) for K_{AB} to be extremely low (10^{-5} or less) if the electrode is to be significantly responsive to pCa_c.

Design of Liquid Ion-Selective Electrodes for Ca^{2+}

All liquid ion-selective microelectrodes conform to the same basic design. The membrane across which the potential is sensed consists of a water-immiscible solvent which plugs the tip of a glass micro-pipette. The solvent contains a sensor--either a neutral ligand or a lipophilic electrolyte. In the case of neutral ligands, to be discussed in this paper, the ion is stripped of its hydration shell as it binds tightly and selectively to the ligand and is therefore effectively solubilized in the membrane. Provided the mobility of the ligand-ion complex in the membrane is adequate, a Nernstian potential will develop rapidly across the membrane in accordance with the activity ratio of the ion.

As emphasized above, a Ca^{2+} sensor must exhibit an impressive degree of discimination for Ca^{2+} over other ions. Elegant work by Simon's group has resulted in the design of ligand ETH 1001 (Oehme et al., 1976). The ether oxygens coordinate with Ca^{2+} and hence enable fractional dissipation of the charge on the ion. The arrangement of the coordination sites in a rigid "cage" of suitable dimensions results in a high degree of selectivity for Ca^{2+} over the much smaller (unhydrated) Mg^{2+} ion. ETH 1001 is superior to other Ca^{2+} sensors in its selectivity for Ca^{2+} over interfering ions at physiological concentrations (Blinks et al., 1982).

All intracellular applications of ion-selective electrodes require the positioning of a second, potential-measuring microelectrode in the same physiological compartment. This is because the output from the ion-selective electrode (referenced to ground) consists not only of a potential developed across the liquid membrane, but also of the potential across the biological membrane(s). Passing the signals from the ion-selective and intracellular reference electrodes through a differential amplifier enables resolution of the two components.

MATERIALS AND METHODS

General Principles of Electrode Construction

Liquid ion selective microelectrodes are constructed in a minimum of four stages.

1. Micropipettes are made on a microelectrode puller. The tip diameter of the pipettes should be that which will enable normal membrane potential recordings from the biological preparation when the pipette is filled with KCl i.e. diameter usually between about 0.3 μm and 1 μm.

2. The inside of the pipette must be given a hydrophobic coating to allow good contact between the water-immiscible liquid membrane and the glass. This is important because the membrane is normally of very high resistance, and any aqueous film between the membrane and glass would tend to shunt the potential. The hydrophobic coating also aids mechanical stabilization of the membrane in the pipette tip. To generate the hydrophobic coating, hydroxyl groups on the surface of the glass are covalently bound to silyl compounds in a process known as **silanization.**

3. A small amount of sensor is introduced to the tip of the pipette.

4. The remainder of the pipette is filled with a reference solution. Electrical contact with the external circuitry is obtained by inserting a chloridized Ag wire into the reference solution.

Apparatus

The apparatus required for fabrication and use of Ca^{2+}-selective microelectrodes can be obtained at relatively low cost. We use a Narashige PE-2 electrode puller (supplied by Optical Instrument Services, Croydon) for manufacture of ion-selective and reference pipettes. Unless double-barrelled electrodes are used (see Discussion), two micromanipulators will be required for impalement of electrodes in a cell, and we strongly recommend the Goodfellow-Huxley design (Goodfellow Metals, Cambridge). An amplifier with high input impedance is necessary because ion-selective microelectrodes tend to be of high resistance (10^{11} Ω) and the WPI model FD-223 (Harvard Apparatus, Edenbridge, Kent) has become a standard in the field. This amplifier has an input impedance of 10^{15} Ω, and operates in a differential mode. The high resistance of the electrode also requires that the experimental set-up is shielded from stray capacitance by a Faraday cage, which can be home-built with sheet metal, or wire mesh. Some form of isolation of the preparation from building vibration is also necessary for delicate microelectrode techniques. We use a Physik Instrumente T-251 vibration isolation base (Lambda Photometrics, Harpenden, Herts.) on which is mounted a plate of 3/4" steel to facilitate bolting down of apparatus. [For most purposes, inner tubes from motor cycle tyres can replace a commercial vibration isolation table if they are positioned between the steel plate and a sturdy laboratory bench.] Almost any compound microscope will serve for viewing the preparation, though we have found a 25x long-working-distance objective (Leitz) to be of great utility in allowing freedom of movement around the preparation. Finally, results can be displayed on a two-channel chart recorder. The total cost of the set-up is about £12 000, excluding sales tax ($15 500 at July 1985 exchange rates). Much of this equipment will be present in a well-found electrophysiological laboratory.

Special Considerations for Plant Cells

Two major problems beset any investigator who hopes to make measurements of pCa_c in plant cells with ion-selective electrodes.

The first concerns cell turgor pressure, which tends to force the liquid membrane back into the electrode shank. This generates a shunt between the membrane and the pipette, with concomitant loss of electrode selectivity. Potentially, there are several ways in which this turgor problem might be solved. We have rejected those which involve reduction of cell turgor (e.g. by increase of external osmotic pressure) as causing undesired physiological perturbation, at least in those cells which sense turgor. We have made an attempt to block movement of the liquid membrane by physical means, either by embedding the internal aqueous phase in agar, or by tightly sealing the blunt end of the electrode. However, these methods did not meet with success, as evidenced by the complete failure of ion-selective electrodes to recalibrate when withdrawn from the cell, even though no visually discernable shift in position of the membrane was apparent. Fortunately, a solution exists. Membranes which incorporate poly(vinylchloride) (PVC) can be firmly gelled in the pipette tip. Indeed, in simple calibration solutions, PVC-containing Ca^{2+}-selective electrodes actually exhibit responsivness over a wider range of pCa than their counterparts from which PVC is absent, probably because the polymer reduces electrical shunting through the electrode tip (Tsien & Rink, 1981).

The second problem is that the cytoplasm in most mature plant cells is relatively sparse, most (95-99%) of the intracellular space being occupied by a large vacuole. Although there cannot be much doubt about which of the intracellular compartments a Ca^{2+}-selective electrode is recording from (pCa should be at least 3 times lower in the vacuole than in the cytoplasm), the major task is to ensure a cytoplasmic location of the ion-selective and reference electrodes at the outset. The chances of simultaneous cytoplasmic location of both electrodes can be enhanced considerably by the construction of double-barrelled electrodes (see Discussion). Mechanical devices such as piezoelectric drivers can also help attain cytoplasmic impalement. Alternatively, the investigator can use a well-characterized cytoplasm-rich system such as fungal hyphae, moss protonemata or meristematic cells. We have chosen to initiate our investigations with cytoplasm-rich fragments of the giant-celled alga Chara.

A more minor problem with most electrophysiological studies on non-animal cells is that of exclusion of electrodes from the cell by formation of new plasma membrane over the electrode tip. While this does not prohibit cytoplasmic recording from the vast majority of cell types, it can, nevertheless, result in premature termination of experiments after 15-30 min (Sanders & Slayman, 1982).

Strategy for Obtaining Electrode Measurements of pCa_c

Before directly attempting to obtain accurate measurements of pCa_c with Ca^{2+}-selective microelectrodes, we have undertaken preliminary studies in three stages. Our first objective has been to record intracellularly, without regard to the particular intracellular compartment, with a Cl^--selective electrode. The aim of this study was to overcome the problem of turgor-sensitivity of liquid membrane microelectrodes. Cl^- selective electrodes were chosen because they are better characterized than their Ca^{2+} counterparts. As a result of this study, we satisfied ourselves that PVC-gelled sensors are suitable for intracellular use in plants (Fig. 1).

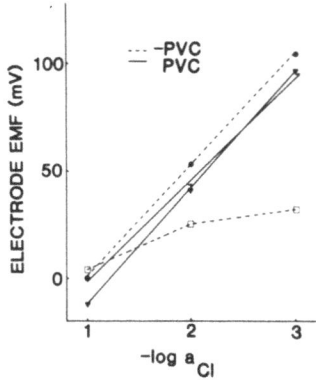

Figure 1. Gelling liquid membrane with PVC
protects ion-selective electrodes against effects of cell turgor. Cl^--selective electrodes
were made either with sensor alone (●,□), or
with sensor + 10% PVC (◄,▼). Response to
calibration solutions before (●,◄) and after
(□,▼) insertion into Chara internodal cell.
Values of electrode e.m.f. before impalement
differed for the two electrode types, and have
been normalized to 0 at log a_{Cl} = -1.

Second, we thought it important to demonstrate that the cells to be used can
be impaled cytoplasmically with conventional potential-measuring electrodes.
Third, we tested Ca^{2+}-selective microelectrodes for their sensitivity to
change in pCa over the physiological range 5-8 in simple calibration solutions. Only when each of these three stages had been successfully completed
did we attempt measurements of pCa_c.

Criteria for Acceptable Measurement

Several facets of the microelectrode method can be used to establish
the reliability of a recording. Ordered insertion of the two electrodes
(reference first) permits any gross membrane damage by the ion-selective
electrode to be detected as membrane depolarization. Furthermore, if the
electrodes are recording from the same compartment, controlled perturbation
of the membrane potential per se should not affect the difference trace,
which is taken to represent pCa. [Large and rapid membrane depolarization
can be expected transiently to manifest itself in the difference trace if
the electrical time constants of the electrodes are significantly different,
but simple methods are available for correction of this artefact (Sanders &
Slayman, 1982).] Finally, the Ca^{2+}-selective electrode should also display
identical calibration curves before and after impalement.

Detailed Methods for Fabrication of Ca^{2+}-Selective Electrodes

In this "cook book" section, we describe how Ca^{2+}-selective microelectrodes are made in our laboratory. Much of the method is based on the
report by Tsien & Rink (1981). Variations on these methods can be found in
Thomas (1978) for liquid membrane microelectrodes generally, and Blinks et
al. (1982) for Ca^{2+}-selective microelectrodes.

An overriding rule for microelectrode manufacture is that work should
be performed in a dry environment. Where exposure to humid air is unavoid-
able, micropipettes should be stored in a dessicator.

Unfilamented borosilicate glass (O.D. = 1 mm) is washed in chromic acid, rinsed well in distilled water, and pulled to give pipette tips of diameter ≃ 1 μm. Micropipettes are positioned, tips up, in a small Al block into which suitably sized holes have been drilled. The block is then placed in an oven for one h at 180 C before introducing 50 μl tri-n-butylchloro-silane (Fluka) in a glass capillary through a port in the top of the oven. After 30 min in the silanizing vapor, the micropipettes are removed from the oven. The thermal capacity of the Al block helps prevent moisture from condensing in the tips of the micropipettes.

1 mg PVC (MW ≃ 200 kDa; BDH Chemicals) is dissolved in 50 μl tetra-hydrofuran (Sigma Chemical) in a Reacti-vial. The solution is kept on ice to prevent evaporation of solvent. Ca^{2+}-sensor cocktail, consisting of 89% w/w 2-nitrophenyl octyl ether (solvent), 10% ETH 1001 and 1% sodium tetra-phenylborate (lipophilic anion, to limit interference from anion-generated diffusion potentials) is purchased from Fluka AG. The PVC solution is mixed with the cocktail to a final dilution (w/w) of 10%. A small quantity (enough to fill the pipette from the tip to the shoulder) of the resulting solution is introduced to the shank of a micropipette using a glass syringe and a 30 G needle. A cat's whisker is then used to move the sensor into the tip of the micropipette and to remove air bubbles. The filled micropipettes are stored tips downward, in a dessicated environment.

After about 12 h, electrodes are back-filled with 0.1 M $CaCl_2$ using a syringe and needle. It is often necessary to remove air bubbles at the gelled membrane/aqueous interface with a cat's whisker. Ag/AgCl wire is inserted into the aqueous phase, and the electrode is sealed. The electrode tips are conditioned in 0.1 M $CaCl_2$ for 10-30 min before use.

Reference electrodes with dimensions identical to the ion-selective electrodes are pulled from filamented borosilicate glass and filled with 100 mM KCl.

The assembled Ca^{2+}-selective electrodes are calibrated in a series of standard solutions from pCa 3 to pCa 8 (Tsien & Rink, 1980). Interference by the ions Mg^{2+}, H^+ and K^+ was estimated by the separate solution method (Lee, 1981). The results are expressed as mV shift in electrode output for changes in ion concentration in the physiological range.

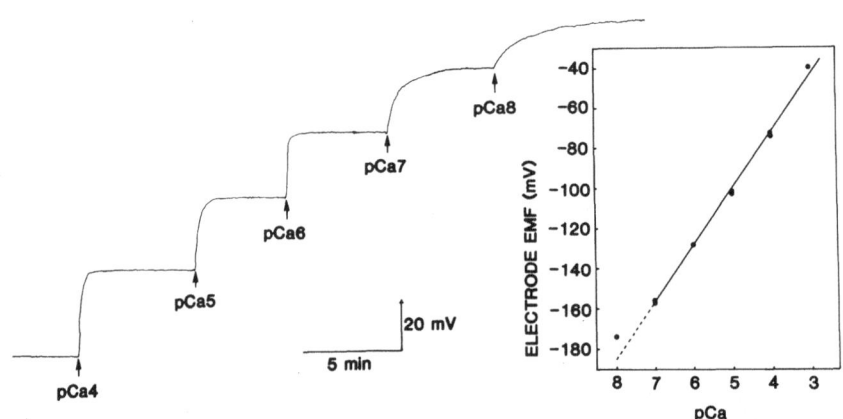

Figure 2. Response of Ca^{2+}-selective microelectrode to changes in pCa (solution exchange time: about 3 s). Composition of standard buffer solutions from Tsien & Rink (1980). Inset: Calibration curve from an ascending and descending series of pCa buffer solutions (two points shown where possible). Slope (pCa 3-7) is 28.3 mV/pCa unit (cf. Eq. 3).

RESULTS

The normal shelf life of the electrodes is 2 days. They generally have a resistance of 50-100 GΩ (compared with 10-50 GΩ for PVC-free electrodes). The electrodes are responsive to changes in pCa over the whole range from 3 to 8 (Fig. 2), though the calibration curve demonstrates some loss of sensitivity above pCa 7.

Over the physiologically important range of pCa 5 to 7, the electrode time constants for response to a change in pCa varied between 6 and 40 s for the example shown in Fig. 2. [We have recently fabricated electrodes with faster time constants: 6 s between pCa 6 and 7.] As reported by Tsien & Rink (1981), these values are considerably higher than the electrical (RC) time constant (usually about 100 ms).

Raising the Mg^{2+} in the calibration buffer from 0 to 5 mM resulted in an average decrease in electrode output (3 electrodes) of 4 mV; pH transition from 7.8 to 6.4, 1 mV; K^+ from 125 to 80 mM, 7 mV, and from 125 to 160 mM, 7 mV. Since the ranges of ion activities tested probably reflect the maximum deviations to which a cytoplasmically-located electrode would be exposed, we conclude with Tsien & Rink (1980; 1981) that interference from these ions will be negligable.

Fig. 3 (left-hand trace) is a recording showing impalement of a cytoplasm-rich fragment of Chara, first by a potential-measuring electrode, then by a Ca^{2+}-selective electrode. A stable value of pCa_c = 6.32 (Ca^{2+} activity = 480 nM) is registered rapidly. This estimate of pCa_c in Chara falls within the range reported by Williamson & Ashley (1982), using aequorin. Impalement of the Ca^{2+}-selective electrode generates little disturbance of membrane potential. The right-hand trace of Fig. 3, shows that darkness causes reversible depolarization of the plasma membrane, but no effect on pCa_c. This might indicate that a change in pCa_c is not responsible for the well-known activation of electrogenic proton pumping in plants by light. More importantly, from a methodological point of view, the absence of response of the pCa trace serves to demonstrate that the electrodes are electrically-coupled i.e. in the same intracellular compartment.

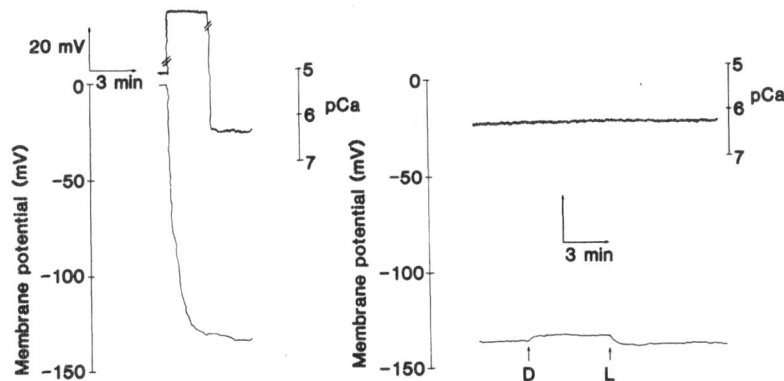

Figure 3. Recordings of pCa_c and membrane potential in cytoplasm-rich fragments of Chara internodal cells. Cells were centrifuged very gently (150 rpm) on a bench centrifuge for 30 min, tied with silk thread 3 mm from the centrifugal end to isolate the cytoplasm and left to recover in pond water overnight. Left: Entry of potential-measuring electrode into fragment, followed after 2.6 min by Ca^{2+}-selective electrode. pCa scale refers only to electrode trace after entry into cell. Right: Effect of light and dark on membrane potential and pCa_c in cell fragment.

DISCUSSION AND CONCLUSIONS

Ca^{2+}-selective electrodes appear to constitute a promising method for measurement of pCa_c in plant cells. They are comparatively easy to fabricate and require no more than conventional electrophysiological apparatus for their deployment. Especially encouraging is the fact that interference by cell turgor pressure can be overcome simply by gelling the liquid membrane with PVC. The major limitations of the electrodes at present are:

1. They respond to changes in pCa in the physiological range over periods of several seconds. Rapid transients (<5 s) in pCa may therefore be undetectable. Although this limitation might preclude their application in the study of pCa_c changes during plant action potentials (Williamson & Ashley, 1982), it is thought, by analogy with animal cells, that the limitation in time resolution is unlikely to prevent detection of "second messenger"-related changes in pCa_c.

2. Cytoplasm-rich cells must be used. It is likely that this restriction can be largely overcome by construction of double-barrelled electrodes, in which the tips of both the Ca^{2+}-selective and the reference electrodes are closely juxtaposed. We are currently making such electrodes.

3. Only single cells in a population can be sampled at any given time.

Against these drawbacks must be set the clear advantages conferred on the electrodes by ETH 1001: negligable Ca^{2+} binding, minimal interference from other ions, and good resolution over a wide range of pCa.

ACKNOWLEDGEMENTS

We are grateful to Drs. H. Behrens (Bonn) & M.R. Blatt (Cambridge) for advice on the fabrication of ion-selective electrodes, and to Dr. M.J. Beilby (Cambridge) for bringing the cytoplasm-rich Chara system to our attention. Financial support was provided by grants from the Nuffield Foundation and the Agricultural and Food Research Council.

REFERENCES

Blinks, J. R., Wier, W. G., Hess, P., Prendergast, F. G., 1982, Measurement of Ca^{2+} concentrations in living cells, Prog. Biophys. Molec. Biol., 40:1.

Lee C. O., 1981, Determination of selectivity coefficients of ion-selective microelectrodes, in: "Ion Selective Microelectrodes and their use in Excitable Tissues," E. Syková, P. Hulk, L. Vyklický, eds., Plenum Press, New York.

Oehme, M., Kessler, M., Simon, W., 1976, Neutral carrier Ca^{2+}-microelectrode, Chimia, 30:204.

Sanders D., Slayman, C. L., 1982, Control of intracellular pH. Predominant role of oxidative metabolism, not proton transport, in the eukaryotic microorganism Neurospora, J. Gen. Physiol., 80:402.

Thomas, R. C., 1978, "Ion-sensitive Intracellular Microelectrodes. How to Make and Use Them," Academic Press, London.

Tsien, R. Y., Rink, T. J., 1980, Neutral carrier ion-selective microelectrodes for measurement of intracellular free calcium, Biochim. Biophys. Acta, 599:623.

Tsien, R. Y., Rink, T. J., 1981, Ca^{2+}-selective electrodes; a novel PVC-gelled neutral carrier mixture compared with other currently available sensors, J. Neurosci. Meth., 4:73.

Willamson, R. E., Ashley, C. C., 1982, Free Ca^{2+} and cytoplasmic streaming in the alga Chara, Nature, 296:647.

NMR AND X-RAY MICROANALYSIS METHODS FOR

MEASUREMENT OF CALCIUM IN PLANT CELLS

W. A. Hughes

Unilever Research Laboratory
Colworth House
Sharnbrook
Bedford, U.K.

1.   Introduction

Calcium plays a vital role in many aspects of plant growth
and metabolism. It is known to regulate cell division, cell wall
expansion, ionic exchange, sugar transport and to be a major
balancing cation in organic acid balance and control of cellular pH.
Accurate measurement of the calcium levels inside cells is
obligatory for any attempt to understand the link between its
activity and function. However the different compartments within
the plant cell hold varying amounts of calcium. The integrity and
shape of the cell is maintained by a cell wall which contains a
large proportion of the cell's calcium bound to pectic substances
(Demarty et al, 1984). The much lower concentration of free
cytoplasmic calcium will be determined by the equilibria with the
larger pools of calcium and the movement of calcium ions across
the plasma membrane. The cytoplasmic calcium is recognised as an
important regulator of a number of enzyme reactions (Dieter,
1984). The calcium levels will change with time as a consequence
of either a direct stimulus (e.g. light) or as a general component
of the ionic net manifested in fluctuation in membrane potentials.
In the latter case these ion levels may reflect an electrical
polarity at a tissue level as described in the early stages of
somatic embryogenesis of carrot (Brawley et al, 1984). Any
measurement of calcium within plant cells must accommodate these
spatial and temporal factors.

The ideal technique to measure calcium is one that is
specific, sensitive and non-invasive. The first two properties
are fairly obvious although the micromolar levels observed for
cytoplasmic calcium in the presence of millimolar magnesium ions
presents a greater problem in determination than that required for
monitoring of cell wall calcium. However the third property - the
need to measure calcium with minimal disturbance to the cell - is
crucial.

Two strategies will be described in this paper to measure
calcium. The first strategy relies on an histochemical approach
with tissue fixation and identification of calcium by x-ray
microanalysis (also called electron probe x-ray microanalysis).

The ultimate insult to the cell of rapid fixation is justified only if every attempt is made to inhibit any redistribution of calcium on preparation. Given that this can be minimized the spatial distribution of calcium within cells and tissues may be studied. However to monitor the lower cytoplasmic calcium over a period of time a specific probe is needed. Introduction of this probe into the plant cell may allow continuous measurement of resting and stimulated calcium levels. However the partition of the probe between the different compartments of the cell may introduce some uncertainty and secondly the probe must not disturb calcium levels or perturb metabolism and induce mitogenic changes. The use of fluorine-labelled chelators as a potential probe with detection by high resolution ^{19}F-NMR spectroscopy will be discussed below.

2. X-ray microanalysis

2.1 Theory

When an electron beam strikes a solid object a number of interactions can occur. When an electron in the high energy beam (20-100 KV) hits an atom it may cause an orbiting electron, from one of the inner shells, to be ejected from the atom. To stabilise the atoms an electron from a higher energy orbital falls into the gap and its excess energy is emitted as an x-ray photon. The gap now left in the higher energy orbital may then be filled with an electron from the next higher energy orbit; this sequence of events may occur several times over, depending on the complexity of the atom, until the atom is stabilised. A large number of x-ray emissions may therefore take place from one ionisation event. (Fig. 1).

These emitted x-rays have a certain energy value and therefore carry information about the atoms in the sample in the region being irradiated by the electron beam. The energy of these x-rays can be measured by using an energy dispersive spectrometer (EDS). The detector is attached to an electron microscope and the

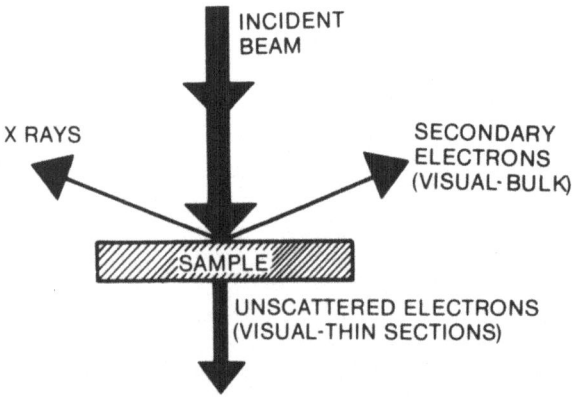

Fig. 1 X-ray microanalysis. Interactions occurring when an electron beam impinges upon a biological sample.

sample can be viewed either in a scanning or conventional transmission manner. Therefore this technique provides a means of correlating morphological appearance with chemical composition.

2.2 Methodology

Samples can be prepared for analysis in many ways either in a resin-embedded, unfixed and dried, frozen-hydrated or frozen-dried form. The redistribution of soluble elements, such as calcium, presents a major problem but can be minimized by use of rapid freezing. The thickness of the sample (either thin-section or bulk) will dictate the method used for analysis. Frozen dried or frozen-hydrated cryosections (0.1-$1.5\mu m$) can be used for determination of intracellular calcium with increased spatial resolution compared with bulk samples but there is relatively a low mass of elements to scan. Bulk specimens on the other hand are much easier to prepare and may be used to localise ions within broader areas of tissue (for discussion see Echlin et al, 1982). There is some some debate on the use of cryoprotectants to preserve morphological appearance (Skaer, 1982) although recently studies on oil palm cells showed little or no deterioration of the structure in the absence of cryoprotectant (Warley et al., 1985).

Most currently used detector systems will identify any element with an atomic number greater than sodium and the results can be displayed on a screen or presented as a colour-contour map of the element in question (for examples see Photo 1).The X-ray map can be compared with a conventional ultrastructural image for exact elemental localisation. Quantitative x-ray microanalysis requires comparison of these spectra with appropriate standards embedded in a gelatin mix. A computer program is available to calculate levels for background radiation, overlapping peaks and variation in section thickness. Knowledge of the protein to dry mass ratio is needed to convert the units of mmoles/kg dry weight to that of mmolar concentration. However this procedure is made more difficult for calcium in biological systems due to the presence of high concentrations of potassium. The potassium $K\beta$ peak partially overlaps with the Ca $K\alpha$ peak.

The detection limits for the lower number atomic elements when using SEM for instance is about 0.1% although this may be lowered under favourable conditions. This represents an estimated lower limit for calcium of 100 μM.

2.3 Applications

X-ray microanalysis has been used to study a number of physiological functions in whole plants (Spurr, 1980). A recent review has identified the specific use of x-ray microanalysis for detection of calcium in animal systems (Somlyo, 1985). An up-to-date survey of the relevant literature has been undertaken and Table 1 contains those reports concerned with both calcium per se or as part of a more general study. The main feature to note is the popularity of calcium localization in discrete bodies such as calcium oxalate crystals. In addition to direct measurement of calcium by x-ray microanalysis an indirect method via precipitation with pyro-antimonate or oxalate has been described (Stockwell and Hanchey, 1982).

Table 1. Application of X-ray microanalysis to detection of calcium

a) General :
Cameron et al, 1984. Onion root meristem.
Dwarte and Ashford, 1982. Protein bodies in celery fruit.
Echlin et al, 1982. Root tips of Lemna minor
Gorton and Satter, 1984. Protoplasts/samanea tree.
Kuo et al, 1982. Seed reserves in several proteaceous species.
Warley et al, 1985. Oil palm suspension cultures.

b) Calcium
(Ando et al, 1981. Products of calcium ATPase activity).
Chino, 1981. Roots of soybean, kidney bean and corn.
Goynes et al, 1984. Calcium oxalate in cotton plants.
Horner and Franceschi, 1978. Calcium oxalate crystals/aquatic plant
Kupila- Ahvenniemi et al, 1973. Cell wall calcium in vicia faba.
Lott et al, 1982. Seed reserves in castor-beans.
Poovaiah and Rasmussen, 1973. Abscission zones of bean leaves
Shibata et al, 1981. Calcium oxalate in rice leaves.
Stockwell, Hanchey, 1982. Bean/use of K-pyroantimonate
Sutherland and Sprent, 1984. Calcium oxalate in legumes

A number of systems have been studied by x-ray microanalysis in the author's laboratory. To investigate the involvement of calcium and other ion fluxes in the endosperm/embryo of oil palm developing fruit, at a particular time of rapid cell division x-ray microanalysis was used in two ways. Both the levels of calcium in freeze-dried endosperm (see spectra in Photo 1) and the locale of calcium by Digimap analysis of frozen-hydrated sections have been determined (see photo 1). It is evident from the latter that calcium is excluded from the embryo at this time - as is the case for potassium and chlorine - whereas phosphorous is included.

The application of x-ray probe microanalysis to subcellular ion localization in oil palm cells has been reported (Warley et al, 1985). The different compartments (cytoplasm + cell wall, nucleus, nucleolus) - as judged by their ultrastructural integrity - contained similar low levels of calcium (12 ± 3 (40 cells) mmol/kg dry weight). Dead cells can be identified by high ratios of phosphorous to potassium. However little change was seen in calcium levels over the growth cycle. A similar approach was used by Cameron et al, 1984 in studying thin cryosections of onion root meristem cells. Recently calcium localisation within the endoplasmic reticulum and mitochondria was reported in rat liver (Somlyo et al, 1985).

The lack of sensitivity of this technique restricts its use to localisation of "bulk calcium". However a sister technique - proton microprobe microanalysis - has been described (Watt et al, 1982). Because the incident high energy protons are much more

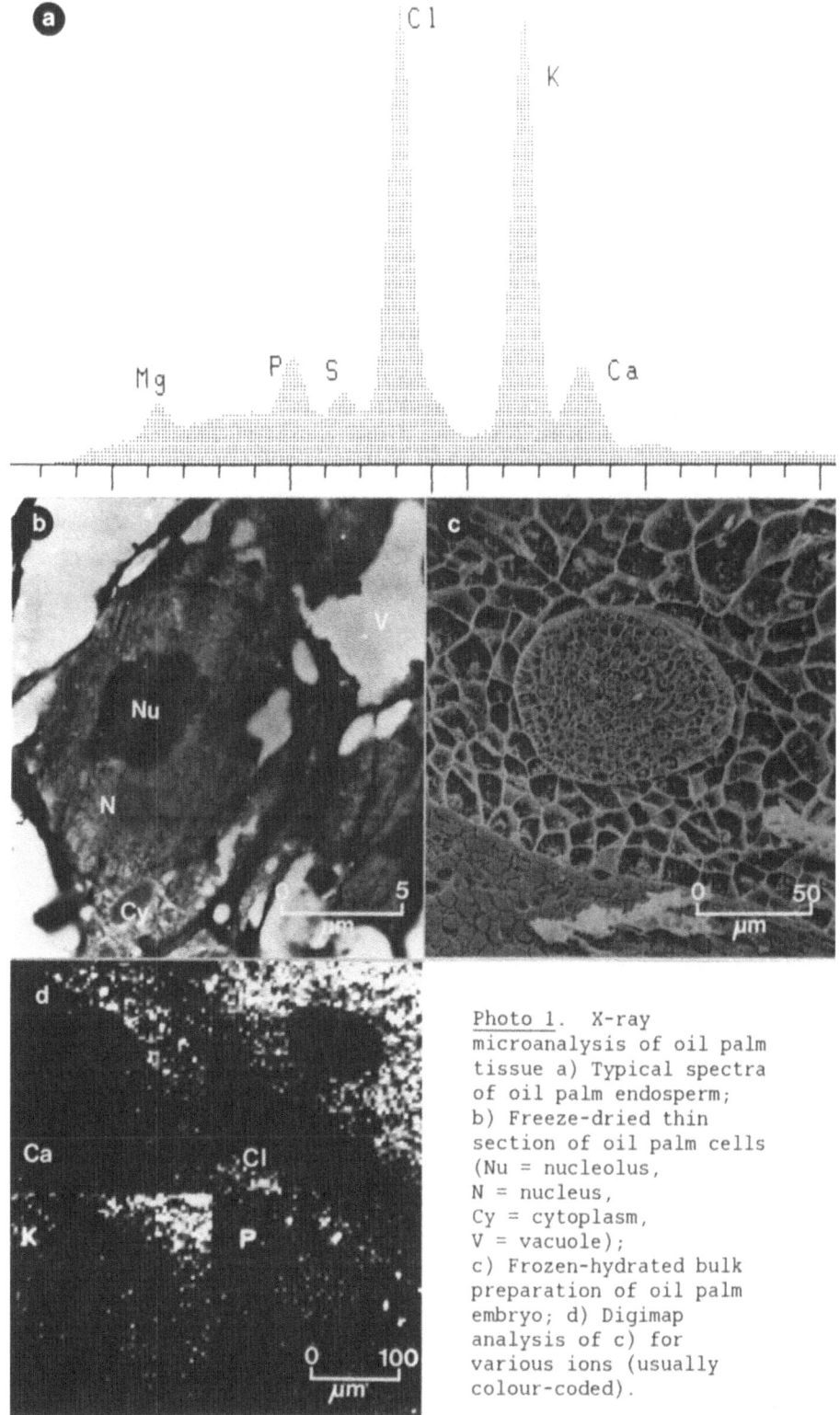

Photo 1. X-ray microanalysis of oil palm tissue a) Typical spectra of oil palm endosperm; b) Freeze-dried thin section of oil palm cells (Nu = nucleolus, N = nucleus, Cy = cytoplasm, V = vacuole); c) Frozen-hydrated bulk preparation of oil palm embryo; d) Digimap analysis of c) for various ions (usually colour-coded).

massive than the electrons they displace, the protons penetrate a
lot further into the sample than the equivalent electron beam.
Combined with a low background PIXE (proton-induced x-ray
emission) offers a possible increase of sensitivity of over 100
fold. Application of this technique to plant systems is still in
its infancy but is reported elsewhere at this conference (Reiss,
et al, 1985).

High Resolution 19-Fluorine NMR

3.1 Theory and Methodology

It has been reported that intracellular calcium levels of
mouse thymocytes can be monitored by ^{19}F-NMR of fluorine-labelled
chelators (Smith et al, 1983). The symmetrically substituted
difluoro derivatives of 1,2-bis (o-amino-phenoxy) - ethane -
N,N,N',N'-tetra acetic acid (nFBAPTA) have ^{19}F- chemical shifts
that are highly sensitive to chelation by calcium unaffected by
the presence of millimolar concentrations of magnesion ion
(Fig. 2). The estimated cytoplasmic calcium of 250 nM can be
compared to the value of 120 nM obtained using quin 2 in the same
cells. A review of this technique has recently been published
(Metcalfe et al, 1985).

3.2 Application

The applications of high resolution NMR to plant cells have
mainly involved the ^{31}P, ^{13}C and ^{15}N nuclei. Several reports have
described ^{31}P-NMR as a useful probe for phosphate metabolism and

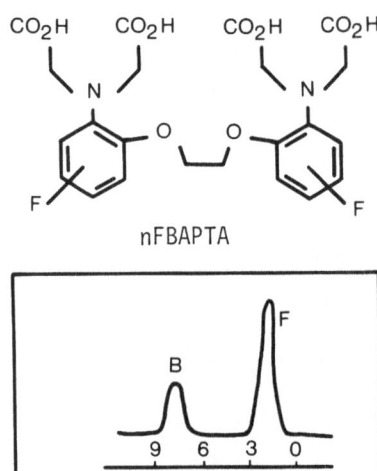

nFBAPTA

Fig. 2 The fluorine-labelled chelator nFBAPTA has a different
chemical shift dependent on whether it is in the free
form (F) or bound with calcium (B). (Metcalfe et al,
1985).

details are available for a perfusion apparatus for oil palm cells (Hughes et al, 1983).

Attempts to couple these systems together have been made in this laboratory but at present have been unsuccessful. The failure of the ester derivatives of nFBAPTA to load in a variety of cell suspensions is not surprising on account of the chemical similarity of these probes to Quin 2 - and the difficulties of using this molecule in plant systems are reported elsewhere (Gilroy, 1985). A number of parameters will be tested such as medium conditions, use of related esters or indeed different delivery mechanisms. Developments in prospect to improve sensitivity of these probes have been described (Metcalfe et al, 1985).

4.   Conclusions

The two methods described have been assessed for their ability to measure calcium in plant tissue. X-ray microanalysis is too insensitive for measurement of the low levels of free cytoplasmic calcium but is useful for localization of larger calcium pools with a spatial resolution of at least 10 nm. Particular attention must be paid to sample preparation and the compromise between preservation of cellular architecture and the redistribution of calcium on preparation. The future success of ^{19}F-NMR as a detection system is dependent on the entry of a suitably labelled probe into the cell but given that continuous monitoring of low levels of calcium over a period of minutes to hours will be possible.

Acknowledgements

I should like to thank my colleagues at Unilever Research Laboratory and especially to thank Dudley Ferdinando. His help in the microanalysis work and in organising the script and photographs has been invaluable.

References

Ando., T., Fujimoto, K., Mayahara, H., Miyajima, H., Ogawa, K., 1981. A new 1 step method for the histochemistry and cytochemistry of calcium ATPase activity. Acta Histochem. Cytochem., 14 ; 705.
Brawley, S. H., Wetherell, D. F., and Robinson, K. R. 1984. Electrical polarity in embryos of wild carrot precedes cotyledon differentiation, Proc. Natl. Acad. Sci., 81 : 6064.
Cameron, I. L., Hunter, K. E., and Smith, N. K. R. 1984, The subcellular concentration of ions and elements in thin cryosections of onion root meristem cells an electron-probe energy dispersive x-ray microanalysis study. J. Cell. Sci., 72 : 295-
Chino, M. 1981. Species differences in calcium and potassium distributions within plant roots. Soil Sci. Plant Nutr. 27 : 487.
Demarty, M., Morvan, C., and Thellier, M., 1984. Calcium and the cell wall, Plant, Cell and Environment, 7 : 441.
Dieter, P. 1984. Calmodulin and calmodulin-mediated processes in plants. Plant, Cell and Environment. 7 : 371.

Dwarte, D. and Ashford, A. E. 1982. The chemistry and microstructure of protein bodies in celery apium-graveolens cultivar South Australian-white endosperm, Bot. Gaz., 143 : 164.

Echlin, P., Lai, C. E., and Hayes, T. L., 1982. Low-temperature x-ray microanalysis of the differentiating vascular tissue in root tips of Lemna Minor L. J. Microsc., 126 ; 285.

Gorton, H. L., Satter, R. L., 1984. Extensor and flexor protoplasts for samnea 2. X-ray analysis of Potassium, chlorine, sulfur, phosphorus and calcium. Plant Physiol. 76 : 685.

Goynes, W. R., Ingber, B. F., and Berni, R. J. 1984. SEM study of calcium oxalate crystals in cotton plants and cotton-related dusts. Scanning Electron Microsc. III : 1443.

Gilroy, S. 1985. Some practical aspects of the application of Quin-2 to plant systems. See this volume.

Horner, H. T. Jr., Franceschi, V. R., 1978. Calcium oxalate crystal formation in the air spaces of the stem of myriophyllum. Scanning Electron Microsc. II : 69.

Kuo, J., Hocking, P. J., and Pate, J. S. 1982. Nutrient reserves in seeds of selected proteaceous species from South Western Australia. Aus. J. Bot., 30 : 231-

Kupila-Ahvenniemi, S., Karjalahti, L. and Kauppi, A., 1973. Electron probe x-ray microanalyzer studies on cell wall calcium in wounded vicia-faba internodes. Ann. Bot. Fenn., 10 : 97.

Lott, J. N. A., Greenwood, J. S., Vollmer, C. M. 1982. Mineral reserves in castor-beans ricinus-communis the dry seed. Plant Physiol., 69 : 829-

Metcalfe, J. C., Hesketh, T. R., and Smith, G. A. 1985. Free cytosolic Ca^{2+} measurements with fluorine labelled indicators using ^{19}F NMR. Cell Calcium, 6 : 183.

Pooviaiah, B. W. and Rasmussen, H. P., 1973. Calcium distribution in the abscission zone of bean leaves. Plant Physiol., 52 : 683.

Reiss, H-D., Herth, W., and Schnepf, E., 1985. Calcium and polarity in tip-growing plant cells. See this volume.

Shibata, S., Sato, K., and Hoshikawa, K. 1981. Calcium oxalate crystals in rice oryza-sativa cultivar Sasanishiki leaves. Jpn. J. Crop Sci., 50 :210.

Skaer, H. le B. 1982. Chemical cryoprotection for structural studies. J. Microsc. 125 : 137.

Smith, G. A., Hesketh, R. T., Metcalfe, J. C., Feeney, J. and Morris, P. G. 1983. Intracellular calcium measurements by ^{19}F-NMR of fluorine-labelled chelators. Proc. Natl. Acad. Sci., 80 : 7178.

Somlyo, A. P. 1985. Cell calcium measurement with electron probe and electron energy loss analysis. Cell Calcium 6 : 197.

Somlyo, A. P., Bond, M., and Somlyo, A. V. 1985. Calcium content of mitochondria and endoplasmic reticulum in liver frozen rapidly in vivo. Nature 314 : 622.

Spurr, A. R. 1980. Application of x-ray microanalysis in botany. Scanning Electron Microscopy II : 535.

Stockwell, V., and Hanchey, P., 1982. Cytohistochemical techniques for calcium localization and their applications to diseased plants. Plant Physiol., 70 : 244.

Sutherland, J. M. and Sprent, J. I. 1984. Calcium oxalate crystals and crystal cells in determinate root nodules of legumes Planta, 161 : 193.

Warley, A., Ferdinando, D., and Hughes, W. A. 1985. Application of X-ray microanalysis to cell suspensions of oil palm (Elaeis Guineensis). Planta. In press.

Watt, F., Grime, G. W., Blower, G. D., Takacs, J. and Vaux, D. J. T. 1982. The Oxford 1μm proton microprobe. Nucl. Instr. and Meth., 197 : 65.

PHYSIOLOGICAL AND CELLULAR RESPONSES TO CALCIUM

LONG DISTANCE TRANSPORT OF CALCIUM

CALCIUM REGULATION OF MITOSIS: THE METAPHASE/ANAPHASE TRANSITION

Peter K. Hepler

Department of Botany
University of Massachusetts
Amherst, MA. 01003

INTRODUCTION

Calcium is commonly thought to be a regulator of mitosis. Fluxes in the concentration of calcium ions ([Ca]) might stimulate one or more processes that participate in controlling the formation and function of the mitotic apparatus (MA) (Hepler and Wolniak, 1984). The discovery several years ago that an elevated [Ca] caused microtubules (MTs) to depolymerize has greatly influenced our thinking about ion regulation of the MA. Specifically it has seemed attractive that during prophase the [Ca] would be low (0.1 uM) to allow formation of the MA but that during late metaphase or at the onset of anaphase the [Ca] would rise and thereby activate events including MT depolymerization that are involved in chromosome transport (Hepler and Wolniak, 1984).

Support for the idea that Ca regulates mitosis comes from several different studies. Elevated levels of Ca, for example, cause spindle MTs to disassemble in vitro (Salmon and Segall, 1980) and in vivo (Kiehart, 1981). The Ca regulatory components including calmodulin (Welsh et al., 1978; Zavortink et al., 1983), Ca -ATPase (Petzelt, 1984) and an extensive membrane system (Hepler and Wolniak, 1984), which is capable of sequestrating the ion (Silver et al., 1980), are present within the MA. It has also been reported that elevated levels of Ca facilitate anaphase movement of chromosomes (Cande, 1984; Hauser and Beier, 1980). Although these studies provide circumstantial evidence that Ca regulates mitosis they do not tell us if the [Ca] inside the cell actually changes, when these postulated fluxes occur, and which components and processes are modulated.

For a variety of reasons it seems that the metaphase/anaphase transition is a time and place where Ca fluxes might occur. The onset of anaphase is an abrupt event in which, for the cell types under discussion, all the chromosomes separate at the same instant, suggesting that the process is pan-cellular and is under ionic-electronic control. Moreover, at the onset of anaphase chromosomes begin movement to the spindle poles and spindle fibers shorten, an event that might entail a Ca-dependent depolymerization of MTs. If fluxes in [Ca] occur during mitosis, in all probability, one of these fluxes would be coupled with the metaphase/anaphase transition. For these reasons it seems worthwhile to direct attention at this moment during mitosis in the hopes of better understanding the broader aspects of Ca regulation.

The essay that follows includes a brief discussion of recent work that focuses on the involvement of Ca during the metaphase/anaphase transition. In the concluding section a model for Ca regulation is put forth with the intent of bringing together diverse evidence and of stimulating new experimentation on the vexying question of Ca and mitosis.

MEASUREMENT OF Ca AND OTHER IONS

That Ca and other ions change during the metaphase/anaphase transition is supported by recent studies in which we analyzed the fluorescence from dividing endosperm cells of Haemanthus that had been labelled variously with different permeant dyes (Wolniak et al., 1983). Chlorotetracycline (CTC) was used to monitor changes in "membrane-associated" Ca . Quantitative measurements over regions of the MA revealed that the fluorescence declined a few minutes prior to the onset of anaphase. These results are consistent with a release of Ca from membrane storage sites such as the endoplasmic reticulum and a concomitant increase in the free cytoplasmic [Ca]. Two other dyes, anilino naphthalene sulfonate and dioxacarbocyanine , which report on changes in membrane potential, showed marked increases in fluorescence that were spatially localized over the MA and temporally correlated with the onset of anaphase (Wolniak et al., 1983). Taken together these results indicate that pronounced changes in Ca and possibily other ions occur in the MA at the time of the metaphase/anaphase transition. The reduction in CTC and thus in membrane associated Ca , however, only indirectly suggests that the cytoplasmic free [Ca] may have increased.

Recent results on sea urchin zygotes provide the first indication that the free cytoplasmic [Ca] increases during metaphase/anaphase and at other times in the division cycle. Poenie et al. (1985) loaded Lytechinus pictus eggs with fura-2, a newly developed fluorescent indicator for free Ca , and observed a strong fluorescence increase at fertilization and lesser, but measurable, increases at periodic intervals thereafter correlating approximately to pronuclear migration, streak stage, nuclear envelope breakdown, metaphase/anaphase and cleavage. The study on Lytechinus marks a major step forward in our quest to decipher the role of Ca during mitosis. However it is essential that it is repeated in other systems. Furthermore it will be important to correlate precisely the timing of the Ca pulse with the particular event in question. For example, at the metaphase/anaphase boundary does the Ca pulse precede or follow the separation of the chromosomes? Accordingly, in collaboration with M. Poenie and R.Y. Tsien, I have attempted to observe changes in [Ca] using dividing stamen hair cells of Tradescantia in which the microscopic delineation of the phases of mitosis is much easier than with Lytechinus zygotes. Free fura-2 anion was injected into dividing cells and changes in fluorescence measured continuously thereafter as the cell progressed through mitosis. However, owing to the fact that the dye moved from the cytoplasm to the vacuole it was not possible to obtain reliable information about [Ca] changes in the MA.

There have been several recent attempts to measure Ca fluxes during mitosis using quin-2 (Keith et al., 1985; Fannin and Sisken, 1984; Wolniak et al., 1984). The difficulties inherent in the use of this probe, most notably its excitation at short wavelengths (339 nm), which are transmitted poorly by glass, and its relatively poor quantum yield (Tsien et al., 1982) cause one to view these studies with caution. The report by Keith et al. (1985) showing a gradual reduction in [Ca] in cultured Pt K2 cells at metaphase/anaphase using quin-2 presents results that are markedly different from those of Poenie et al. (1985) on sea urchins using fura-2. The apparent superiority of fura-2 over quin-2 and the compelling nature of the observations on Lytechinus including a clear demonstration of the well known Ca spike at fertilization lend credibility to the report by Poenie et al. (1985). However it is quite evident that much more needs to be done. These different results will have to be sorted out and, importantly, the existence of Ca fluxes established in different systems.

PHIYSIOLOGICAL STUDIES ON Ca

Further evidence that Ca is involved in the metaphase/anaphase transition comes from different physiological studies. Izant (1983), using PtK cells, has reported that a lowered intracellular [Ca] brought about by microinjection of a Ca-EGTA buffer (pCa 7.3) retards the onset of anaphase. The calmodulin antagonists chloropromazine (Boder et al., 1983), W-7 calmidazolium, and trifluoperizine (Keith et al., 1983) also retard the onset of anaphase or arrest cells in metaphase. These studies suggest that lowering intracellular [Ca] or blocking calmodulin activity prolong metaphase. Contrasting views have been presented, however, by Chai and Sandberg (1983) who report that restriction of extracellular calcium with EGTA or lanthanum (La) promotes the progression of Chinese hamster DON cells from metaphase to telophase, and by Chafouleas et al. (1982) who note that the calmodulin antagonist W-13 has no effect on mitosis in Chinese hamster ovary cells.

Attempts to modulate mitosis with elevated levels of Ca have also yielded conflicting results. An excess of Ca in the medium, for example, has been reported to inhibit Chinese hamster DON cell progression from metaphase to telophase (Chai and Sandberg, 1983). Similarly, the application of the Ca ionophore, A23187, prolongs metaphase in HeLa cells (Sisken, 1980). However, A23187 plus exogenous Ca , under conditions that promote hydrogen ion efflux and hence Ca influx, have no effect on mitosis in spermatogenous cells of Marsilea (Wick, 1978). A more direct approach to the effect of high intracellular Ca on mitosis has come from

the recent study of Izant (1983) who has microinjected
citrate-glutamate-buffered solutions of Ca into PtK cells and has
reported that 1.0 uM free Ca hastens the onset of anaphase.
Unfortunately, it is not known how cells injected with a resting level of
Ca (0.1 uM) behave, and thus there is a question of whether the cells
with 1.0 uM calcium are hastened through metaphase or are simply
proceeding at a normal rate.

The role of calcium during mitosis is thus far from settled.
Depending on the particular study, increasing or decreasing the [Ca] can
either retard or advance the onset of anaphase. To further pursue and
clarify these questions I have examined the effect of agents that restrict
extracellular Ca on the ability of stamen hair cells of Tradescantia to
progress through mitosis (Hepler, 1985). Stamen hair cells are well suited
for these investigations since they can be cultured for several hours in
simple, minimal media (Hepes; KCl) and since their mitotic events occur
with considerable regularity. Control cells, for example, progress from
nuclear envelope breakdown to the onset of anaphase (metaphase transit
time) in 32.5 $\pm$ 3.9 min, initiate a cell plate in an additional 18.9 $\pm$
1.1. min, and move their chromosome at a rate of 1.45 $\pm$ 0.043 um/min
(Hepler, 1985).

Using agents that lower extracellular [Ca] , Ca-EGTA buffers (pCa
8-pCa 5), or block transport of the ion into the cell, La and
methoxy-verapamil (D-600), the three events listed above, namely metaphase
transit time, cell plate initiation time, and rate of chromosome motion,
have been measured. The results show that low extracellular Ca (pCa 8)
greatly extends metaphase transit time (32.5 vs 52.6 min), and in some
instances permanently arrests cells in metaphase. pCa 7 or higher Ca ,
however causes no prolongation of metaphase (Hepler, 1985). Blockade of Ca
entry with La or D-600 also markedly lengthens metaphase transit time. In
La this period lasted 50.3 min while in D-600 almost all cells under
examination become permanently arrested. If the D-600 treated cells were
also cultured with an elevated level of Ca (1 mM) then the cells entered
anaphase at the control time. Light microscopic examination of Ca
stressed cells indicated that chromosome condensation progressed normally
and that the cells appeared to become arrested at the moment just prior to
chromosome separation (Hepler, 1985).

In contrast to their effect on metaphase transit time these
Ca-perturbing agents exert little or no effect on the time of cell plate
initiation or on the rate of chromosome motion. For cells treated with
pCa 8 Ca-EGTA buffer the results are especially clear; the same cells that
had been prolonged in metaphase, when they entered anaphase, initiated
their cell plate at the control time and moved their chromosomes at the
control rate (Hepler, 1985).

These studies taken together indicate that the metaphase/anaphase
transition in stamen hair cells of Tradescantia is regulated by a Ca
influx at the plasma membrane (PM). Although the agents used to restrict
Ca are different in their chemical properties and mode of action they all
appear to act in the extracellular wall space or at the PM. The Ca

chelator, EGTA, for example, because of its size, charge, and hydrophilicity will not cross the PM and is thus confined to the extracellular space where it will reduce Ca and thus the availability of the ion. La achieves its effect by competing for Ca binding sites (Martin and Richardson, 1979; dos Remedios, 1981). It also does not cross the PM, as has been shown directly by electron microscopy (Thomson et al., 1973). By tightly binding to Ca transport complexes and displacing the ion without itself being transported La creates a condition in which Ca would fail to flow into the cell. D-600 inhibits the slow, voltage-dependent Ca channel (Janis and Triggle, 1983). In cardiac tissue it acts on the sarcolemma rather than the sarcoplasmic reticulum. Its mode of action, however, appears to involve sites on the inside of the PM (Hescheler et al., 1982). Thus by different mechanisms the various agents appear to restrict or block Ca influx from the extracellular space. These conclusions cause us to direct attention to the PM as the cellular component that regulates the Ca influx which is required for the initiation of anaphase.

CONCLUSIONS

In previous discussions of Ca regulation of mitosis emphasis had been placed on the endomembranes (Hepler and Wolniak, 1984). Their intimate association with the spindle fibers and their Ca content make the elements of endoplasmic reticulum likely candidates for the release and sequestration of Ca . The current results do not diminish the importance of endomembranes rather they add for our consideration the PM as an important controlling factor in mitosis. The results suggest that the PM by itself or together with the endomembranes controls the influx of Ca into the MA and thus the stimulation of sister chromosome separation that marks the onset of anaphase.

Transport of Ca down its concentration gradient occurs via voltage-dependent, ion selective channels (Reuter, 1983). In Tradescantia the specific inhibition of the metaphase/anaphase transition by D-600 lends support to the idea that voltage-gated Ca channels participate in the regulation process. Thus a cell close to the metaphase/anaphase boundary may initially experience a depolarization of the membrane potential. This event causes the voltage-gated Ca channels to open and allows the ion to flow down its concentration gradient into the cell raising the cytoplasmic [Ca] . One of two events seems possible at this point: either the increased Ca directly stimulates the chromosomes to separate and begin anaphase, or the increased Ca stimulates a further, more massive, release of Ca from the MA-associated endomembrane system and this secondary event activates anaphase onset (Fig. 1).

The latter system stated above is known as "Ca -induced, Ca -release" and occurs in cardiac muscle (Fabiato, 1983) and non muscle cells (Ridgway et al., 1977). Its possible operation in stamen hair cells of Tradescantia may allow one to account more readily for Ca regulation of the events of mitosis besides the metaphase/anaphase transition. If the PM alone regulated Ca within the MA then why is anaphase motion not sensitive to Ca   deprivation? Does this mean that Ca   does not contribute to the control of MT depolymerization? Possibly the PM alone can generate a sufficient pulse of Ca   to carry the cell through anaphase but it seems more likely that the endomembranes also contribute to an ion flux. Regardless of the specific mechanism, when a cell reaches a threshold in the Ca   anaphase is initiated and the subsequent mitotic events are completed without further Ca   influx at the PM.

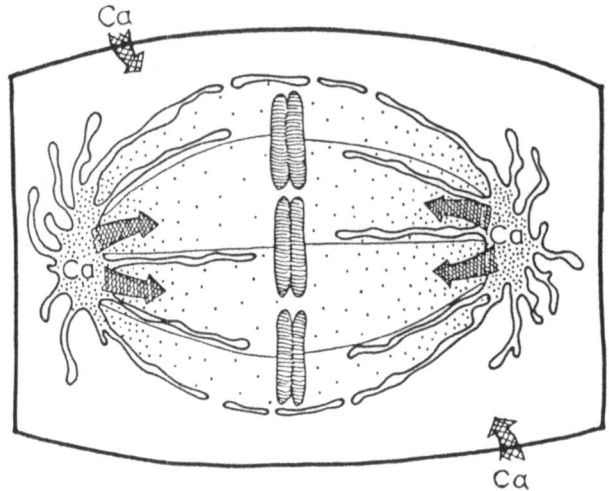

FIG. 1. A diagramatic representation of a cell in late metaphase. The suggestion is made that a depolarization of the membrane potential triggers the opening of voltage-gated Ca channels on the PM. Ca flows into the cell and regulates the onset of anaphase. Either the Ca influx directly stimulates the separation of chromosomes, or it induces a further release of this ion from the spindle-associated endomembrane system, which then initiates anaphase (Hepler, 1985).

The conclusions stated above raise questions and suggest experiments for future studies on the regulation of mitosis. One specific prediction is that Ca   entry at the PM is activated by a depolarization of the membrane potential. This prediction can be tested in different ways, either through direct measurements by intracellular recording or by experimentally depolarizing the electrical potential and monitoring its effect on mitosis. The conclusions also predict that the cytoplasmic [Ca] increases just prior to metaphase/anaphase. It is imperative that these measurements be made in many different systems so that the "Ca   theory" of mitosis can be put on a firm foundation.

ACKNOWLEDGEMENTS
     This work has been supported by grant number   GM 25120 from the National Institutes of Health.

REFERENCES

Boder, G.B., Paul,D.C., and Williams, D.C., 1983, Chloropromazine inhibits mitosis of mammalian cells. Eur. J. Cell Biol., 31: 399-353.

Cande, W.Z., 1981, Physiology of chromosome movement in lysed cell models. In "International Cell Biology" 1980-1981, H.G. Schweiger, ed., Springer-Verlag, Berlin 382-391.

Chafouleas, J.G., Bolton, W.E., Hidaka, H., Boyd III, A.E. and Means, A.R., 1982, Calmodulin and the cell cycle: involvement in regulation of cell cycle progression, Cell, 28: 41-50.

Chai, L.S., and Sandberg, A.A., 1983, Effect of divalent cations and chelators on metaphase to telophase progression and nuclear envelope formation in Chinese hamster cells, Cell Calcium, 4: 237-252.

Fabiato, A., 1983, Calcium-induced release of calcium from the cardiac sarcoplasmic reticulum, Am.J.Physiol., 245: C1-C14.

Fannin, F.F., and Sisken, J.E., 1984, Quin 2 fluorescence in HeLa cells during the course of mitosis, J. Cell. Biol., 99 (5, Pt.2): 428a.

Hauser, M., and Beier, A.M., 1980, Caffeine and Ca accelerate chromosome movement in locust meiosis, Eur. J. Cell Biol., 22: 313.

Hepler, P.K., 1985, Calcium restriction prolongs metaphase in dividing Tradescantia stamen hair cells, J. Cell Biol., 100: 1363-1368.

Hepler, P.K., and Wolniak, S.M., 1984, Membranes in the mitotic apparatus: their structure and function, Int. Rev. Cytol., 90: 169-238.

Hescheler, J., Pelzer, D., Trube, G., and Trautwein, W., 1982, Does the organic calcium channel blocker D600 act from inside or outside the cardiac cell membrane? Pfluegers Arch. Gesamte Physiol.Menschen Tiere, 393: 287-291.

Izant, J.G., 1983, The role of calcium ions during mitosis. Calcium participates in the anaphase trigger, Chromosoma (Berl.) 88: 1-10.

Janis, R.A., and Triggle, D.J., 1983, New developments in Ca channel antagonists, J.Med. Chem., 26: 775-785.

Keith, C.H., Maxfield, F.R., and Shelanski, M.L., 1985, Intracellular free calcium levels are reduced in mitotic Pt K2 epithelial cells, Proc. Natl. Acad. Sci. USA, 82: 800-804.

Keihart, D.P., 1981, Studies on the in vivo sensitivity of spindle microtubules to calcium ions and evidence for a vescicular calcium-sequestrating system, J.Cell Biol., 88: 604-617.

Martin, R.B., and Richardson, F.S., 1979, Lanthanides as probes for calcium in biological systems, Q.Rev. Biophys., 12: 181-209.

Petzelt, C., 1979, Biochemistry of the mitotic spindle, Int. Rev. Cytol., 60: 53-92.

Petzelt, C., 1984, Localization of an intracellular membrane-bound Ca ATPase in PtK-cells using immunofluorescence techniques, Eur. J. Cell Biol., 33: 55-59.

Poenie, M., Alderton, J., Tsien, R.Y., and Steinhardt, R.A., 1985, Changes in free calcium levels with stages of the cell division cycle, Nature, 315: 147-149.

dos Remedios, C.G., 1981, Lanthanide ion probes of calcium binding sites on cellular membranes, Cell Calcium, 2: 29-51.

Reuter, H., 1983, Calcium channel modulation by neurotransmitters, enzymes and drugs, Nature, 301: 569-574.

Ridgway, E.B., Gilkey, J.C., and Jaffe, L.F., 1977, Free calcium increases explosively in activating medaka eggs, Proc. Natl. Acad. Sci. USA, 74: 623-627.

Salmon, E.D., and Segall, R.R., 1980, Calcium-labile mitotic spindles isolated from sea urchin eggs (Lytechinus variegatus), J. Cell Biol., 86: 355-365.

Silver, R.B., Cole, R.D., and Cande, W.Z., 1980, Isolation of mitotic apparatus containing vesicles with calcium sequestering activity, Cell, 19: 505-516.

Sisken, J., 1980, The significance and regulation of calcium during mitotic events. In "Nuclear Cytoplasmic Interactions in the Cell Cycle", G.L. Whitson, ed., Academic Press, New York. 271-292.

Thomson, W.W., Platt, K.A., and Campbell, N., 1973, The use of lanthanum to delineate the apoplastic continuum in plants, Cytobios, 8: 57-62.

Tsien, R.Y., Pozzan, T., and Rink, T.J., 1982, Calcium homeostasis in intact lymphocytes: cytoplasmic free calcium monitored with a new, intracellularly trapped fluorescent indicator, J. Cell Biol., 94: 325-334.

Welsh, M.J., Dedman, J.R., Brinkley, B.R., and Means, A.R., 1978, Calcium-dependent regulator protein: localization in the mitotic apparatus of eukaryotic cells, Proc. Natl. Acad. Sci. USA, 75: 1867-1871.

Wick, S.M., 1978, Ionophore A23187 stimulates H+ release from developing plant spores, J. Cell Biol., 79 (2, Pt.2): 172a.

Wolniak, S.M., Bart, K.M., and Hepler, P.K., 1984, A change in the intracellular free calcium concentration accompanies the onset of anaphase, J. Cell Biol., 99 (5, Pt. 2): 429a.

Wolniak, S.M., Hepler, P.K., and Jackson, W.T., 1983, Ionic changes in the mitotic apparatus at the metaphase/anaphase transition, J. Cell Biol., 96: 598-605.

Zavortink, M., Welsh, M.F., and McIntosh, J.R., 1983, The distribution of calmodulin in living mitotic cells, Exp. Cell Res., 149: 375-385.

CALCIUM AND CALMODULIN AS REGULATORS OF CHROMOSOME MOVEMENT DURING MITOSIS IN HIGHER PLANTS

Anne-Marie Lambert and Marylin Vantard

Laboratoire de Biologie Cellulaire Végétale
Université Louis Pasteur, Institut de Botanique
67083 Strasbourg Cedex, France

INTRODUCTION

Microtubules represent the main component of the mitotic apparatus. It is well-established that the movement of chromosomes is directly related to microtubule assembly-disassembly (for reviews: Bajer and Molè-Bajer, 1972; Inoué and Ritter, 1975). However, the mitotic force production and the control of spindle function in the living cell remain unknown. These processes may be due to a complex cascade of events.

Data obtained on the in vitro microtubule assembly suggests that calcium and Calmodulin are candidates as regulators of these mechanisms.

Since the Weisenberg experiments (1972), numerous results indicate that brain microtubule polymerization-depolymerization in vitro is Ca^{2+} sensitive and that this Ca^{2+} sensitivity may vary from the micromolar (Nishida et al., 1979) to millimolar (Olmsted and Borisy, 1975) range depending on the purity of the microtubule protein extracts. Such Ca^{2+} concentrations in vitro are greater than the general level of free Ca^{2+} found in the living cytoplasm ($10^{-7}M$). Therefore, Ca^{2+} regulation of microtubule assembly in vivo will be dependent on local and sudden free Ca^{2+} efflux/influx from Ca^{2+} sequestering compartments. The signals that regulate such events are not understood, particularly in plant cells where such ion storage systems as vacuoles, mitochondria and plastids are important.

Calmodulin, like Troponin C, enhances highly, Ca^{2+} sensitivity of brain microtubule disassembly (Marcum et al., 1978). However, the molecular mechanisms of Ca^{2+} - Calmodulin interactions with microtubules are not yet clarified. Calmodulin binding to Tubulin (Kumagaï et al., 1982), to Tau factors (Sobue et al., 1981), or to other associated proteins like MAPs (Rebhun et al., 1980) or to STOPs (Stable-tubule only proteins) on cold-stable microtubules (Job et al., 1982) is now well-documented but often remains controversial.

Little is known of Ca^{2+} sensitivity of microtubules polymerized from non-neural sources. Some data indicate that microtubule assembly is possible in the absence of Ca^{2+} chelators (Doenges, 1978).

Data on plant microtubule assembly in vitro is still rare and restricted to a few examples (Morejohn and Fosket, 1982; Mizuno, 1985) such as endosperm cells (Picquot and Lambert, 1985). The effects of Ca^{2+} - Calmodulin on plant microtubule protein are not yet documented in vitro. The role of Ca^{2+} ions during mitosis, in higher plant cells has been studied. The experiments suggest that Ca^{2+} acts as the anaphase trigger (Izant, 1983) and regulates the metaphase-anaphase transition (Wolniak et al., 1983; Hepler, in this book). However, little is known about the potential regulation of the mitotic process by Ca^{2+} and Calmodulin in plant cells.

Immunocytochemical procedures (Welsh et al., 1978; De Mey et al., 1980) as well as microinjection (Zavortink et al., 1983) demonstrated the presence of Calmodulin within the mitotic apparatus of various animal cell lines. It was shown that Calmodulin is associated with the kinetochore fibers and the polar regions where centrioles are present.

We identified Calmodulin in the mitotic spindle of higher plant endosperm cells, which are characterized by the lack of centrioles (Lambert et al., 1983). Correlative studies of Tubulin and Calmodulin distribution indicate that Calmodulin localization changes during mitosis parallel the shortening of kinetochore microtubules in anaphase. Calmodulin is also located on aster-like microtubule centers that establish the mitotic polarity in plant cells (Vantard et al., 1985). Recently, Calmodulin was identified in root tip cells and a particular distribution in the phragmoplast was observed (Wick et al., 1985).

Whether Calmodulin regulates microtubule dynamics in plant cells is not known. In this context, endosperm cells of Haemanthus and Clivia appear to be suitable models for the study of Ca^{2+} - Calmodulin activity in plant mitosis (Vantard et al., 1985). Precise analysis of Calmodulin distribution at different and well-controlled stages of mitosis permits us to propose a hypothesis concerning the potential and selective role of Ca^{2+} - Calmodulin as regulators of kinetochore microtubule disassembly in the acentriolar mitotic apparatus of higher plants. No mitotic enzymes have yet been identified during microtubule dynamics, and the extracellular and intracellular signals that may induce sudden Ca^{2+} efflux of sequestring systems are entirely unknown.

MATERIALS AND METHODS

Endosperm cells of the Monocotyledons Haemanthus Katherinae BaK. and Clivia nobilis Lindl. have been chosen due to their characteristics. These cells lack a cellulose wall and can therefore be considered as native protoplasts. In vivo visibility of chromosome movements is exceptional, as well as the size of the mitotic spindle which is 80 μm in length. These cells were extensively used to study mitosis (Bajer, 1958, review; Bajer and Molè-Bajer, 1972) and are particularly suitable for immunocytochemical studies as no pre-treatment with wall enzymes is needed. Immediate access to the cell membrane is possible and facilitates the penetration of antibodies.

Living Cells

Preparations of mitotic cells were made as described (Molè Bajer and Bajer, 1968; Lambert and Bajer, 1977). Colchicine and Cold treatments were done as described (Vantard et al., 1985).

Immunocytochemical Procedures

Rabbit antibodies against spinach Calmodulin were kindly provided by L. Van Eldik; they were produced as described for vertebrate Calmodulin (Van Eldik and Watterson, 1981). Affinity purified rabbit antibodies to dog

brain Calmodulin was a generous gift of J. De Mey (De Mey et al., 1980).
Tubulin antibodies were obtained either using dog brain Tubulin (De Mey et
al., 1982) or pig brain Tubulin. For light- and electron microscopic localization of Calmodulin and Tubulin, indirect immunofluorescence (Schmit et al.,
1983) and immuno-gold-staining (De Mey et al., 1982) were done. Fixation and
permeabilization of the cells were developed as described (Schmit et al.,
1983; Vantard et al., 1985). Preparations were observed with a Leitz Ortho-
plan microscope, in epifluorescence using HBO 100/W_2 lamp and FITC Ploemopack
filters, and a 63X oil fluorescence objective, numerical aperture 1.30
(Leitz). Details are given elsewhere (Vantard et al., 1985).

Cell Models

Cell preparations were flushed by an extraction buffer containing 65 mM
PIPES, 25 mM HEPES, 1 mM EGTA, 1 mM $MgSO_4$ and 0.1 mM GTP at pH 6.94. Cell
lysis was obtained by addition of Brij 58 at 0.08% for 2 to 3 minutes
followed by a quick wash in the extraction buffer to remove the detergent.
Precise concentration of free Ca^{2+} added to the models is maintained by a
Ca^{2+}-EGTA buffer system, at precise pH, as estimated by computer program
(Cande, 1980).

Calmodulin Inhibitors

Calmidazolium (R 24571), Janssen Life Sciences Products, was used at
different concentrations in the range of 1.5×10^{-5}M to 1.5×10^{-4}M in the
culture medium. Calmidazolium solutions were perfused in cell preparations
at precise stage of chromosome movement.

Calcium Effect on Cytoplasmic and Spindle Microtubules

Precise concentrations of free Ca^{2+} were added to cell models using
Ca^{2+}-EGTA buffers at different stages of the cell cycle and during mitosis.
These data were compared to results obtained by perfusion of Ca^{2+} in presence of A 23187 ionophore (0.5 µg/ml) to the living cells at selective steps
of chromosome migrations. Ca^{2+} effects on the different classes of microtubules were estimated with indirect immunofluorescence as described above.

RESULTS

Detailed analysis of Calmodulin distribution during mitosis of endosperm cells was described recently (Vantard et al., 1985). We summarize
here the most essential observations that suggest the potential role of
Calmodulin in the regulation of kinetochore microtubule assembly-disassembly.
These microtubules represent the selective and structural support of chromosome movement.

Changes in Calmodulin Distribution from Interphase to Mitosis

During interphase Calmodulin exhibits a diffuse localization and no
particular domains could be detected (Fig. 1). In such interphasic stage,
microtubules form a regular meshwork in the cytoplasm, as previously
described (Schmit et al., 1983).

At the onset of prophase, when chromosomes begin to condense in the
nucleus, the microtubule pattern is entirely and progressively rearranged.
Close to the nucleus, microtubule density increases and microtubules tend
to align parallel to the future spindle axis. At this time, transient microtubule converging centers are detected and resemble the asters of animal
cells, in spite of the lack of centrioles in these higher plant cells.
During these complex processes of microtubule reorientation and rearrangement

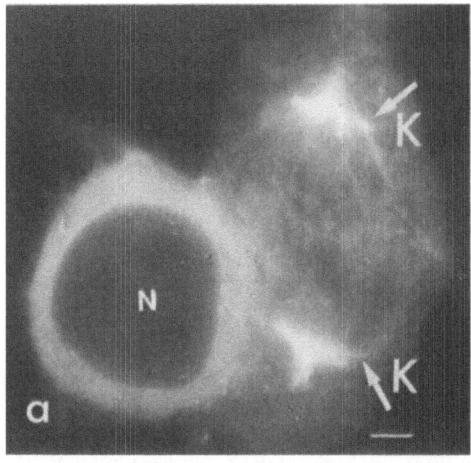
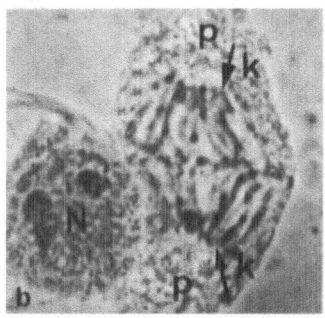

Fig. 1. Two neighbour cells treated with spinach Calmodulin antibodies.
a: In Interphase, Calmodulin staining is diffuse, while during
mitosis in anaphase, Calmodulin exhibits a cone-like distribution
between kinetochores and the poles. No Calmodulin is detected in
the interzone.
b: Same preparation in phase contrast. K: kinetochore location.
Comparison with Fig. 3 indicates that Calmodulin is selectively
associated with the kinetochore microtubules. Bar: 10 μm.
N: nucleus.

Calmodulin distribution becomes concentrated at these regions where microtubule converging centers are formed (Vantard et al., 1985).

After nuclear envelope breakdown, when spindle fibers form the bipolar
mitotic apparatus, as well-documented before, Calmodulin remains associated
with the polar regions, even when microtubule aster-like centers are no
longer detected in metaphase and early anaphase. Detailed correlations
between Tubulin and Calmodulin patterns indicate that Calmodulin localization is restricted to the distribution of kinetochore fibers. No Calmodulin
is detected in the interzone during anaphase (Fig. 1).

Association of Calmodulin with Kinetochore Microtubules

Detailed analysis of mitotic cells showed that the immunofluorescence
pattern of Calmodulin, with spinach or brain antibodies, was restricted to
the distribution of kinetochore fibers. Both antibodies gave similar pictures. Ultrastructural studies using antibodies labelled with 5 nm immunogold permitted also the detection of Calmodulin localization at the close
vicinity of microtubules (Vantard et al., 1985).

To determine more precisely the localization of Calmodulin we studied
anaphase cells concurrently with Tubulin and Calmodulin antibodies. Normal
cells and cells pre-treated with Colchicine (8×10^{-5}M) or Cold (3°C) for
2 to 5 minutes were compared. Both Colchicine and Cold induce rapid and
selective disassembly of non-kinetochore microtubules (Lambert and Bajer,
1977) while kinetochore microtubules are resistant. Therefore, the use of
drug or cold treated cells permitted us to analyse Calmodulin distribution
in a mitotic spindle depleted of all interpolar and interzonal microtubules.
Figure 2 illustrates this study. It demonstrates that, in spite of such
drastic changes in the Tubulin pattern in Colchicine treated cell, Calmodulin
localization remains quasi identical to the control cell (comparison between
Figs 2a and d, and also Fig. 3).

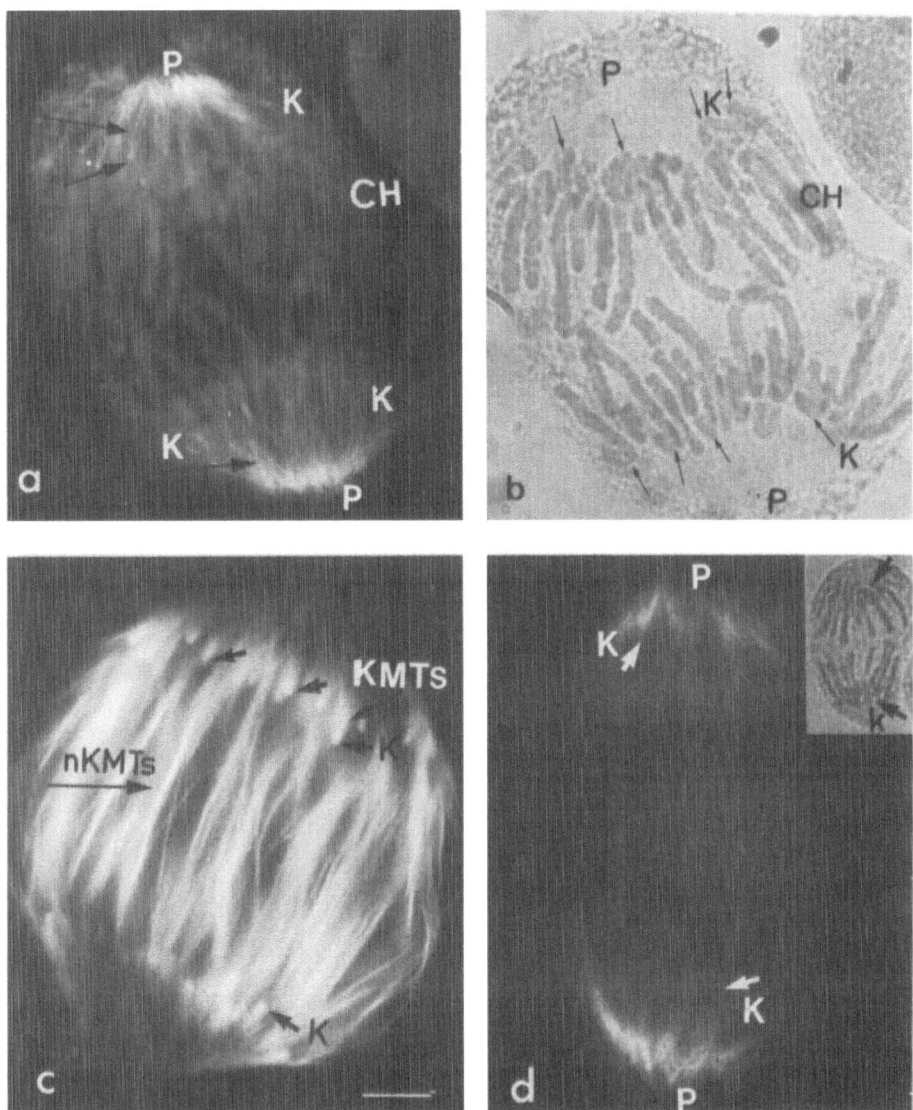

Fig. 2. Calmodulin distribution is compared with Tubulin pattern in anaphase,
using indirect immunofluorescence. a: Calmodulin staining in normal
cell (b: the same cell with stained chromosomes): Calmodulin fluorescence is detected, as thin stripes, along kinetochore microtubules. No fluorescence in the interzone where numberous bundles of
non-kinetochore microtubules are present as seen in c with Tubulin
antibodies. In colchicine treated cell (8.10⁻⁵M), in d, Calmodulin
distribution is comparable to the control cell, although non-
kinetochore microtubules are disassembled with such treatment, and
kinetochore microtubules remain only. (cf. Fig. 3). Bar: 10 μm.
KMTs: Kinetochore microtubules. nKMTs: non-KMTs. P: pole.

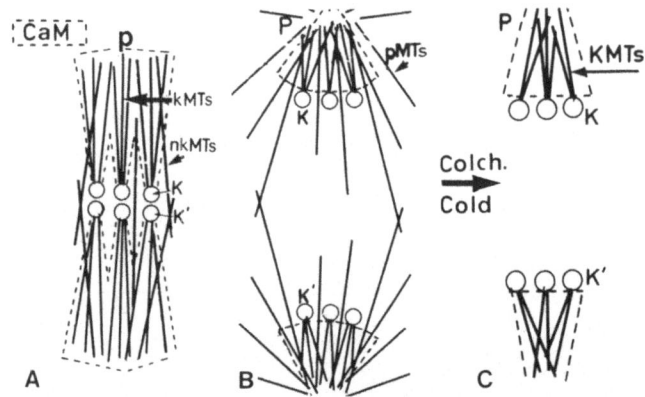

Calmodulin localization marked by dotted lines.

Fig. 3. Dynamics of Calmodulin (CaM) distribution from Metaphase (A) to
Anaphase (B: normal cell, C: treated cell with Colchicine, or cold
Temperature). Abbreviations as in Fig. 2, pMTs: polar MTs assembled
during anaphase. CaM immunofluorescent pattern is restricted to the
distribution of Kinetochore-MTs and follows their disassembly in
anaphase (comparison: A-B). When the spindle is depleateated of
non-KMTs by cold or colchicin, CaM distribution remains comparable
to control cell (comparison between C and B).

Differential Calcium Sensitivity of Microtubule Populations in Lysed Cell Models

Detergent extracted endosperm cells were used as a model system to
analyse the Ca^{2+} sensitivity of in vivo polymerized microtubules under in
vitro conditions. Ca^{2+}-EGTA buffers were used to monitor Ca^{2+} concentration
as described in methods. In such models, cytoplasmic microtubules in interphase were disassembled at μM range of Ca^{2+}, i.e. close to the physiological
cytosolic concentrations in vivo. However, in the same experimental conditions, spinkle microtubules showed high resistance to calcium, up to 100 μm
concentration. Kinetochore microtubules were the most resistant. These data
were in agreement with the experiments on living cells perfused with Ca^{2+}
associated to A 23187 ionophore. Interphase microtubules were always more
sensitive while spindle microtubule resistance was variable.

Effects of Calmidazolium (Calmodulin inhibitor) in Metaphase

Calmidazolium was perfused to living endosperm metaphase cells, at
1.5×10^{-5}M concentration. Kinetochore splitting was inhibited in most
cells. Results were variable when the inhibitor was given in late metaphase, suggesting a point of "no return", in relation to the anaphase
trigger.

DISCUSSION

Our data on mitotic endosperm cells permitted us to follow the dynamics of Calmodulin distribution in an acentriolar spindle. The results we obtained are summarized in the schematic interpretation of Fig. 3.

Our data suggest, therefore, that Calmodulin is directly involved in the regulation of kinetochore microtubule dynamics and consequently regulates chromosome movement during mitosis.

Immunocytochemical evidence of Calmodulin along kinetochore microtubule length (cf. Fig. 2a), does not, however, prove its activity. Calmodulin and the complex Ca^{2+}-CaM cannot be distinguished yet, by normal techniques. The potential distribution of CaM molecules on kinetochore microtubules, from the pole to the kinetochore itself, is compatible with tubulin treadmilling. CaM may be activated only in particular domains of the kinetochore fiber, at the polar end (- end), or at different assembly-disassembly sites along the microtubule. Such activation would be related to free Ca^{2+} changes.

Calmodulin-Tubulin binding (Kumagai et al. 1982) remains controversial. Tubule-Calmodulin association through Mt-associated protein is a better candidate. Surprising similarities between in vitro cold-stable MTs (Job et al. 1981) and in vivo cold-resistant Kinetochore-MTs (Lambert et al. 1977, cf. Fig. 3c) suggest comparable molecular properties. In vitro STOPs proteins (Stable Tubule Only Proteins) are responsible for Cold and Ca^{2+} resistance which is compensated by Calmodulin. Kinetochore-MTs also exhibit Ca^{2+} resistance in lysed cell models, and may therefore possess similar associated protein. Figure 4 shows how Ca^{2+}, as a second messenger, may activate CaM associated to Kinetochore-MTs, inducing activation of the STOP-like P-associated protein. Phosphorylation of this protein induces tubulin activation (disassembly related to GDP - GTP regulation). STOP-like protein may bind CaM at physiological Ca^{2+} concentrations or in a Ca^{2+}-independent manner. A sudden free Ca^{2+} increase (efflux from Ca^{2+} sequestering systems)

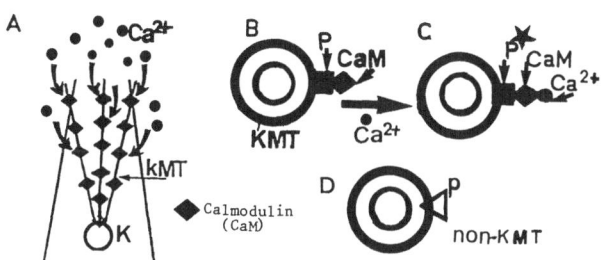

Fig. 4. Hypothetical representation of Ca^{2+}-Calmodulin regulation of kinetochore microtubule (KMT) disassembly. A: one anaphasic kinetochore fiber surrounded by 2 non-kinetochore MTs Calmodulin (CaM) is selectively bounded to KMTs and is activated by free Ca^{2+} (arrows). B,C: cross-section of one kinetochore MT, B: inactive form (permanent CaM binding), C: activated form after Ca^{2+}-binding and subsequent activation of the P protein (phosphorylation: ✹) followed by tubulin activation (disassembly). D: cross-section of non-KMT with different (p) protein that does not bind CaM.

(Keith et al. 1985), could induce punctual CaM-P activation and regulate MT disassembly through a Ca^{2+} gradient along MT length. Regulation of Ca^{2+} influx-efflux would result in the control of tubulin assembly-disassembly at the pole. Ca^{2+}-CaM dependent enzymes are known in plants (Marmé et al., 1984; Ranjeva et al.,1985; Trewavas, 1976). Kinetochore-MT-associated protein may function as a protein-kinase; however, such enzymes have not yet been found during mitotic chromosome movement.

ACKNOWLEDGEMENTS

We are grateful to Dr L. Van Eldik and Dr J. De Mey for providing purified Calmodulin antibodies and for helpful discussions.

REFERENCES

Bajer, A., 1958, Cine-micrographic studies on endosperm mitosis, Exp. Cell Res., 14.

Bajer, A. and Molè-Bajer, 1972, "Spindle Dynamics and Chromosome Movements", Acad. Press, New York.

Cande, W.Z., 1980, A permeabilized cell model for studying cytokinesis using mammalian tissue culture cells, J. Cell Biol. 87:326.

Deery, W., Means, A.R. and Brinkley, B.R., 1984, Calmodulin-microtubule association in cultured mammalian cells, J. Cell Biol., 98:904.

De Mey, J., Moeremans, M., Guens, G., Nuydens, R., Van Belle, H. and De Brabander, M., 1980, Immunocytochemical evidence for the association of Calmodulin with microtubules of the mitotic apparatus, in "Microtubules and MT Inhibitors", M. De Brabander and J. De Mey, eds, Elsevier-Amsterdam.

De Mey, J., Lambert, A.M., Bajer, A., Moeremans, M. and De Brabander, M., 1982, Visualization of microtubules in interphase and mitotic plant calls of Haemanthus with Immuno-Gold-Technique, Proc. Natl Acad. Sci. USA, 79:1898.

Doenges, K., 1978, Assembly of non-neural microtubules with Ca^{2+}, FEBS Lett., 89.

Inoué, A. and Ritter, H., 1975, Dynamics of mitotic spindle organization and function, in "Molecules and Cell Development", S. Inoué and Stephens, eds, Raven Press, New York.

Izant, J.G., 1983, The role of Ca^{2+} ions during mitosis, Chromosoma, 88:1.

Job, D., Fischer, E.H. and Margolis, R.L., 1981, Rapid disassembly of cold-stable microtubules by Calmodulin, Proc. Natl Acad. Sci. USA, 78: 4679.

Keith, C., Ratan, R., Maxfield, F., Bajer, A. and Shelanski, M., 1985, Nature, in press.

Kumagaï, H., Nishida, E. and Sakai, H., 1982, The interaction between Calmodulin and microtubule proteins, J. Biochem., 91:1329.

Lambert, A.M., 1980, The role of chromosomes in anaphase trigger, Chromosoma, 76:101.

Lambert, A.M. and Bajer, A., 1977, Microtubule distribution and reversible arrest of chromosome movements by low temperature, Cytobiologie, 15:1.

Lambert, A.M., Vantard, M., Van Eldik, L. and De Mey, J., 1983, Immunolocalization of Calmodulin in higher plant endosperm cells during mitosis, J. Cell Biol., 97:40a.

Marcum, J.M., Dedman, J.R., Brinkley, B.R. and Means, A.R., 1978, Control of microtubule assembly-disassembly by calcium dependent regulator protein, Proc. Natl Acad. Sci., USA, 75:3771.

Marmé, D. and Dieter, P., 1984, The role of Ca^{2+} and Calmodulin in plants, in "Calcium and Cell Function", IV, Cheung, ed., 263, Acad. Press, New York.

Mizuno, K., 1985, *In vitro* assembly of microtubules from tubulins of
    several higher plants, Cell Biol. Inter. Rep., 9:13.
Molè-Bajer, J. and Bajer, A., 1968, Studies of selected endosperm cells
    with light and electron microscope, La Cellule, 67:257.
Morejohn, L.C. and Fosket, D., 1982, Higher plant tubulin identified by
    self-assembly into microtubules *in vitro*, Nature, 297:426.
Nishida, E., Kumagaï, H., Ohtsuki, I. and Sakaï, H., 1979, The interactions
    between calcium dependent regulator protein in cyclic phosphodiester-
    ase and microtubule proteins. J. Biochem., 85:1257.
Olmsted, J. and Borisy, G.G., 1975, Ionic and nucleotide requirements for
    microtubule polymerization *in vitro*, Biochemistry, 14:2996.
Picquot, P. and Lambert, A.M., 1985, Electrophoretic properties and *in
    vitro* self microtubule assembly of plant endosperm tubulin extrac-
    ted from Haemanthus cells, Biol. Cell, 53:39.
Ranjeva, R., Graziana, A., Dillenschneider, M., Charpenteau, M. and Boudet,
    A., 1985, A novel plant calciprotein as transient subunit of
    enzymes, in this book.
Rebhun, L., Jemiolo, D., Keller, T., Burgess, W. and Kretsinger, R., 1980,
    Calcium-Calmodulin and control assembly of brain and spindle micro-
    tubules, *in* "Microtubules and MT inhibitors", De Brabander, ed.,
    Elsevier.
Schmit, A.C., Vantard, M., De Mey, J. and Lambert, A.M., 1983, Aster-like
    microtubule centers establish spindle polarity in higher plant
    cells, Plant Cell Rep., 2²285.
Sobue, K., Morimato, K., Inu, M., Kanda, K. and Kakiuchi, S., 1982, Control
    of actin myosin interaction by Calmodulin, Biomed. Res., 3:188.
Trewavas, A., 1976, Post-translational modification of proteins by phosphory-
    lation, Ann. Rev. Plant Physiol, 27:349.
Van Eldik, L.J. and Watterson, D.M., 1981, Reproductible production of anti-
    serum against vertebrate calmodulin and determination of the immuno-
    reactive site, J. Biol. Chem., 256:4205.
Vantard, M., Lambert, A.M., De Mey, J. and Van Elkik, L.J., 1985, Charac-
    terization and immunocytochemical distribution of Calmodulin in
    higher plant endosperm cells: localization in the mitotic apparatus,
    J. Cell Biol., 101, in press.
Weisenberg, R.C., 1972, Microtubule formation *in vitro* in solutions con-
    taining low calcium concentrations, Science, 177:1104.
Welsh, M.J., Dedman, J.R., Brinkley, B. and Means, A., 1978, Calcium-depen-
    dent regulator protein: localization in the mitotic apparatus of
    eukaryotic cells, Proc. Natl Acad. Sci., USA, 75:1867.
Wick, S., Muto, S., Duniec, J., 1985, Double immunofluorescence labelling
    of Calmodulin and tubulin in dividing plant cells, Protoplasma,
    126:207.
Wolniak, S.M., Helper, P. and Jackson, W., 1983, Ionic changes in the
    mitotic apparatus at metaphase-anaphase transition, J. Cell Biol.,
    96:598.
Zavortink, M., Welsh, M. and McIntosh, J.R., 1983, Exp. Cell Res. 149:375.

CALCIUM MEDIATION OF CYTOKININ-INDUCED CELL DIVISION

Mary Jane Saunders

Botany Department
Louisiana State University
Baton Rouge, LA 70803 USA

INTRODUCTION

Cytokinins are a class of plant hormones that are defined by their ability, in the presence of optimal auxin, to initiate cell division in tobacco pith culture (Skoog et al., 1965). For a variety of reasons it seems plausible that cytokinins exert at least part of their effect through modulation of intracellular calcium ion concentration ($[Ca^{2+}]_i$). An argument in favor of this hypothesis can be made by integrating the literature on cytokinin stimulation of cell division, mitotic regulation by Ca^{2+}, and cytokinin modulation of Ca^{2+} uptake. In addition, there is strong evidence that cytokinins stimulate an asymmetrical division in <u>Funaria</u> by increasing $[Ca^{2+}]_i$ (Saunders and Hepler, 1981, 1982, 1983).

Cytokinins induce cell division in a variety of tissue culture systems including pea root callus, tobacco pith, cocklebur stem callus, carrot root callus, and soybean cotyledon callus. The production of cytokinins in callus fluctuates; high levels of cytokinin are correlated with levels of mitotic activity. In intact plants, rich sources of cytokinins are found in rapidly dividing and meristematic tissues such as endosperm, fruitlets, cambium, and root tips (see Letham, 1978). Tobacco protoplasts require cytokinins to divide and can be synchronized by hormone addition indicating a control point in the cell cycle (Meyer and Cooke, 1979). Research on effects of cytokinins on division has fallen off in recent years and research on whole plants has been hampered by uncertainty about interaction with endogenous levels. However, the necessity for exogenous cytokinins for growth of tissue cultures that are not cytokinin-producing is unquestioned.

There is increasing evidence that the induction and regulation of cell division in both plants and animals is mediated by Ca^{2+} (see Jaffe, 1980; Hesketh et al., 1985). Induction of meiosis in starfish oocytes by 1-methyladenine (1-MeAde) is a well characterized system in which transient increases in $[Ca^{2+}]_i$ have been implicated (Moreau et al., 1978). This reaction is especially interesting because of the structural similarity between 1-MeAde and cytokinins, although they do not cross-react. Mammalian lymphocytes also require an increase in $[Ca^{2+}]_i$ to trigger mitosis (Hesketh et al., 1985). Yeast cells have similar control points in the cell cycle; they can be induced to divide by A23187 + Ca^{2+}; without Ca^{2+} they are blocked in G_2 (Duffus and Patterson, 1974). The mitotic apparatus itself and in particular spindle microtubules appear to be regulated by Ca^{2+} that is sequestered and released by membranes associated with the mitotic apparatus (see Hepler, 1985). In mitotic Pt K2 epithelial cells, a drop in free Ca^{2+} has been measured during mitosis using quin-2 implicating a Ca^{2+} sequestration mechanism (Keith et al., 1985). Calcium fluxes across the plasma membrane during mitosis have been measured in sea urchin zygotes (Clothier and Timourian, 1972) and Physarum (Holmes and Stewart, 1977). Triggering the common developmental event of mitosis occurs in some cells as the result of a Ca^{2+} transient; in others a sustained Ca^{2+} elevation is necessary. The effects of calcium upon a cell may be dependent on the state of the cell and/or the duration of the signal.

There is little information about the effect of cytokinins on Ca^{2+} uptake or $[Ca^{2+}]_i$. The literature that does exist, however, argues that cytokinins exert at least part of their effect through the modulation of $[Ca^{2+}]_i$. In Achlya, cytokinins stimulate Ca^{2+} release from a glycoprotein localized on the surface and enhance uptake of the ion into the cells (LeJohn and Cameron, 1973). Kinetin and Ca^{2+} have synergistic effects on ethylene production by mung bean hypocotyl segments (Lau and Yang, 1975) and deferral of sensecence in Zea leaf discs (Poovaiah and Leopold, 1975). In Xanthium cotyledons, dry weight gain is stimulated two-fold by cytokinin and three fold by cytokinin plus CaCl (Leopold et al., 1973). Ralph et al. (1976) found that Ca^{2+} will substitute for cytokinin in a leaf disc expansion assay and that cytokinins plus Ca^{2+} affect membrane protein phosphorylation. Recently Elliott (1983) reported that calmodulin inhibitors block cytokinin-induced betacyanin synthesis in Amaranthus cotyledons and postulates that a rise in calcium is an intermediary in this process. Finally, a plasmalemma-enriched membrane preparation from soybean hypocotyls has been isolated that contains an ATP-dependent Ca^{2+} pump that is cytokinin modulated (Kubowicz et al., 1982).

Accepting the premise that cytokinin is intimately involved in the induction of cell division, and the premise that the regulation and/or induction of cell division is mediated by Ca^{2+}, it seems logical to hypothesize that the cytokinin stimulation of cell division is mediated by calcium. This hypothesis has been tested by studies on cytokinin-induction of bud formation in the protonema of the moss Funaria hygrometrica. Filamentous protonemata, consisting of chloronema cells with large chloroplasts, give rise after 8-10 days of growth to caulonema cells with small chloroplasts, target cells for cytokinin action. The first detectable morphological event occurs approximately 10 hr after cytokinin treatment and consists of a bulge on the distal end of target cells. The nucleus migrates to this region and by 24 hr the first asymmetric division has taken place, cutting off a small lens-shaped bud initial cell. In the continued presence of cytokinin the bud initial divides in three planes forming the bud, a mass of cells with a tetrahedral apical cell.

Jaffe (1980) has proposed three conditions that should be met before one can claim that Ca^{2+} is acting as a second messenger. These conditions have been met and substantiate the hypothesis that cytokinin-induced mitosis in Funaria hygrometrica is Ca^{2+}-mediated. More specifically:
1. Increases in membrane-associated Ca^{2+} after cytokinin treatment have been detected using chlorotetracycline (CTC) and correlated with the migrating nucleus, the site of the initial asymmetric division and the rapidly dividing cells of the bud (Saunders and Hepler, 1981).
2. Artificial induction of a rise in free calcium has been accomplished using the Ca^{2+} ionophore A23187 which stimulates the initial division in the absence of cytokinin (Saunders and Hepler, 1982).
3. Prevention of a natural rise in $[Ca^{2+}]$ after hormone treatment has been explored using Ca^{2+} antagonists in the presence of cytokinin. The Ca^{2+} transport inhibitors La^{3+}, D 600 and verapamil and the intracellular Ca^{2+} antagonist TMB-8 blocked bud formation; extracellular Ca^{2+} is essential

because blocking uptake with Ca^{2+}-transport inhibitors blocks both nuclear migration and subsequent division. In addition, TMB-8 distorts cell plate formation. The transition of the nucleus from G_2 to mitosis is sensitive to calmodulin inhibitors (Saunders and Hepler, 1983).

It seems plausible to suggest that cytokinin may exert its effect on asymmetrical cell division in Funaria by activating and or concentrating Ca^{2+} channels at the presumptive bud site. This would have localized effects on cytoplasmic free Ca^{2+} concentration and would create a microdomain of higher free Ca^{2+} at distal ends of target cells. To test this hypothesis I examined changes in the distribution and activation of plasma membrane ion channels after cytokinin treatment in caulonema cells of Funaria using a non-intrusive vibrating microelectrode (Jaffe and Nuccitelli, 1974). The role of the cytoskeleton in determination of the bud site was investigated using the microtubule depolymeizer, colchicine, the microtubule stabilizer taxol and the microfilament inhibitors cytochalasin B and cytochalasin D (see Williamson, 1984). Changes in free Ca^{2+} levels after cytokinin treatment during establishment of the presumptive bud site and in dividing bud cells were measured using a fluorescent Ca^{2+} indicator fura-2 AM (Grynkiewicz et al., 1985).

MATERIALS AND METHODS

Funaria hygrometrica was cultured as described in Saunders and Hepler (1981) using Laetsch's (1967) medium. Vibrating microelectrode studies were conducted at the National Vibrating Probe Facility, Marine Biological Laboratories, Woods Hole, MA, USA. The microelectrode used was a metal filled glass micropipette with an electroplated platinum ball (10 μm diam.) on its tip, attached to a piezoelectric element and oscillated at a constant frequency of 200 cps with a tip excursion of 20 μm. Potential differences measured by the electrode tip at the extremes of its excursion were converted to current density. A 546 nm narrow-pass green filter (Carl Zeiss Inc.) was used during all measurements. Cytokinin was added to the medium during microelectrode measurements by adding benzyladenine (BA) (Sigma Chemical Co., St. Louis, MO, USA) to a final concentration of 1 μM. For Ca^{2+}-uptake inhibitor experiments gadolinium nitrate (Sigma) was added during measurements to a final concentration of 1 μM.

Colchicine (Sigma) was dissolved in Laetsch's medium and used at concentrations between 10 μm and 1 mM. Taxol (gift of National Cancer Institute, Bethesda, MD, USA) was dissolved in DMSO (stock concentration 10 mM) and used at final concentrations of 1 μM to 10 μM. Cytochalasin B and cytochalasin D (Sigma) were dissolved in DMSO (stock concentration 5 mg/ml) and used at final concentrations of 1 μg/ml to 40 μg/ml. The fluorescent free Ca^{2+} indicator fura-2 AM (Molecular Probes, Junction City, OR, USA) was dissolved in DMSO stock concentration 1 mM) and used at a final concentration of 10 μM. Cells were loaded with fura-2 AM by incubating for 1 hr in the probe followed by a 30 min rinse in Laetsch's medium.

Measurements of free Ca^{2+} were performed on a Leitz Ortholux II microscope (Ernest Leitz, Wetzlau, Germany) equipped with an MPV photometer. For fura-2 AM measurements, 340 nm or 380 nm narrow pass filters (Ditric Optics, Marlboro, MA, USA) were placed in the excitation path, Leitz filter cube A (340-380 nm excitation filter, 430 nm barrier filter) was used with a 510 nm narrow pass filter (Ditric) in the photometer. The aperture of the photometer was adjusted to read entire cells or selected areas within one cell.

RESULTS

　　Electrical current (by convention, flow of positive charge) was mapped around target caulonema cells to establish distribution of ion channels before and after cytokinin treatment (Table 1). In all caulonema cells measured the greatest inward current (0.5 $\mu A/cm^2$) was detected at the nuclear zone before cytokinin treatment. Inward current away from the nuclear zone was ca. 0.25 $\mu A/cm^2$ and homogenous; no differences were seen between distal and proximal ends of the cell. Immediately after addition of benzyladenine (BA) (1 μM final concentration) an increase in inward current was detected along the length of target cells. Current at the distal end continued to rise while current at the proximal end and nuclear zone fell back to resting levels. After 1 hr current at distal ends of target cells reached a maximum while the current at the nuclear zones and proximal ends fell to resting levels. Ten hr after BA treatment a slight bulge and accumulation of vesicles can be seen at distal ends of target cells, defining the presumptive bud site. Current influx at the presumptive bud site was lower than immediately after BA treatment and continued to fall as the new growth zone became established. Current around cells of a developing bud is low (0.25 $\mu A/cm^2$) and homogeneous. The addition of the lanthanide gadolinium nitrate (1 μM), a competitive Ca^{2+} uptake inhibitor (dos Remedios, 1981), reduced inward current to zero both before and after BA treatment.

　　The microtubule inhibitor colchicine had no effect on the position of bud formation along the length of target cells at any concentration tested (10μM–1mM). At concentrations from 0.7 mM to 1 mM, the nucleus migrated to the presumptive division site, swelling of the wall occurred but mitosis was blocked. At lower concentrations (70μM–100μM) mitosis occurred but morphologically distorted buds formed. The microtubule stabilizer taxol had no effect on nuclear migration, mitosis or placement of the bud on target

Table 1. Current density at the surface of <u>Funaria</u> caulonema cells

Electrode position	Treatment	Inward Current ($\mu A/cm^2$)	
Distal end	–	0.24 ± 0.06	n=48
(5–50μm from distal	+1μM Gd^{3+}	0	n=3
cross wall)	+1μM BA		
	5 min	0.64 ± 0.12	n=6
	30 min.	1.1 ± 0.2	n=5
	1 h	1.2 ± 0.2	n=5
	10 h	$.67 \pm 0.3$	n=4
	24 h	$.25 \pm 0.01$	n=12
	+1μm BA;15 min;1μM Gd^{3+}	0	n=2
Proximal end	–	0.25 ± 0.04	n=40
(5–50μm from proximal	+1μM Gd^{3+}	0	n=2
cross wall)	+1μM BA		
	30 min	0.23 ± 0.02	n=3
	1 h	0.25 ± 0.02	n=3
	10 h	0.24 ± 0.03	n=3
Nuclear zone	–	0.51 ± 0.03	n=40
(approximately halfway	+1μMGd^{3+}	0	n=2
along cell)	+1μM BA		
	5 min	0.90 ± 0.1	n=6
	30 min	0.56 ± 0.2	n=5
	1 h	0.28 ± 0.1	n=4
	10 h	0.26 ± 0.06	n=4
	+1μM BA;15 min;1μM Gd^{3+}	0	n=2

cells at any concentration tested. In contrast, the microfilament inhibitors cytochalasin B (25 μg/ml) and cytochalsin D (15 μg/ml) displaced the site of bud formation from the distal end of cells to halfway along the length of the cell, over the nucleus. Higher concentrations (40μM CB, 25μM CD) also displaced buds to over the nucleus and the buds that formed were abnormal.

Target cells preloaded with the fluorescent Ca^{2+} indicator fura-2 AM show an immediate increase in fluorescence at 510 nm with excitation at 340 nm (compared to excitation at 380 nm) after BA treatment. The increase is detected homogeneously within the cell. However, cells pretreated with cytokinin for 10-15 hrs that have a bulging growth zone at distal ends have greater fluorescence (excitation 340 nm) at that end of the cell compared to proximal ends. Fluorescence of fura-2 AM in bud cells show an opposite trend indicating a decrease in free Ca^{2+} in these mitotically active cells.

DISCUSSION

These results indicate 1) untreated caulonema cells have maximum inward current at the nuclear region, 2) addition of cytokinin induces an increase in inward current along the length of target cells that subsequently shifts to distal ends (presumptive bud site) and 3) after establishment of the growth zone at the presumptive bud site inward current falls. This current has a Ca^{2+} component since current falls to zero after treatment with Gd^{3+}. Determination of the presumptive bud site is microfilament and not microtubule dependent. Free Ca^{2+} rises homogenously within target cells immediately after cytokinin treatment but after the establishment of a bulging growth zone, free Ca^{2+} is spatially non-uniform (higher at the presumptive bud site) and decreases in cells of the developing bud. These observations lend support to my hypothesis that an initial mode of action of cytokinin is activation of plasma membrane ion channels that subsequently creates a microdomain of high $[Ca^{2+}]_i$ at the presumptive bud site.

Incorporating these results with previous research I propose the following model for cytokinin-induced cytomorphogenesis in Funaria.
A. In the absence of cytokinin, protonemata of Funaria grow as filamentous mats consisting of three cell types - elongating and actively dividing tip cells, green nondividing cells (chloronemata) and cytokinin-target cells (caulonemata). Because of the tip-growing, filamentous nature of the system a molecular polarity must exist within the caulonema cells. Localization of cytokinin receptors (✱) on the plasma membrane may be determined by this gradient and be clustered at the distal end. Open Ca^{2+}-channels (◖) show maximal concentration over the nucleus which lies midway along the long axis, next to the side wall, anchored by a web of microtubules and arrested in G_2 of the cell cycle. We can envision the caulonema cell before cytokinin treatment as partially polarized but in quiescent state.
B. Addition of cytokinin (●) presumably results in its binding to hormone receptors (⊙✱). This affects Ca^{2+}-channel proteins and opens closed channels (◖) to O) along the length of the cell. The membrane permeability to Ca^{2+} (✱) increases and the ion moves passively and homogeneously into the cell down its steep concentration gradient.
C. Calcium influx becomes localized at distal ends of target cells as channels accumulate there by a microfilament based process. Ca^{2+} influx would thus be localized to the distal end whereas the efflux pumps may be either randomized or concentrated at the proximal end of the cell. Ca^{2+} permeability may also affect other ion transport and membrane potential. Intracellular $[Ca^{2+}]$ would now increase within the cell in a gradient that is highest at the distal end. Because of active sequestering by mitochondria and endoplasmic reticulum and active extrusion at the plasma

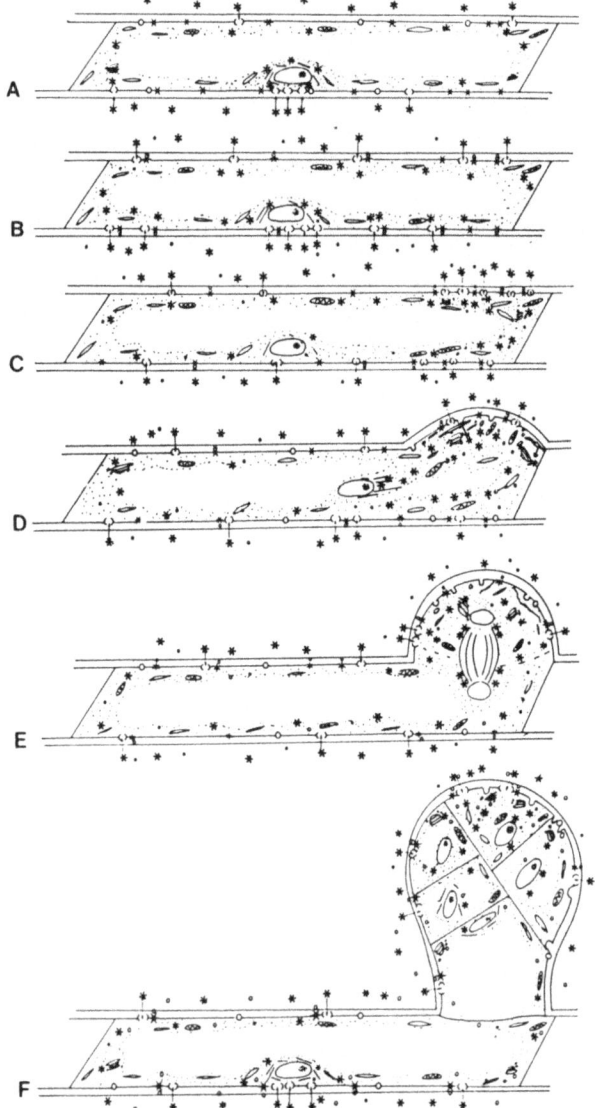

Figure 1. Model for Cytokinin-Induced Bud Formation in Funaria

membrane, the ability of Ca^{2+} to diffuse through the cytoplasm is limited and one could imagine that initially, high $[Ca^{2+}]_i$ would be localized near the surface at the distal end of the target cell.[1] Supporting evidence comes from the fact that the first morphological changes detectable are localized wall expansion and an increase in vesicles next to the plasma membrane in that region of the cell (Conrad and Hepler, 1985). Local high $[Ca^{2+}]_i$ at the plasma membrane would facilitate vesicle fusion to the plasma membrane at the new growth zone, the presumptive bud site.

D. The zone of high $[Ca^{2+}]$ is first created within the swollen presumptive bud site and later extends further down the long axis of the cell. At this point Ca^{2+} channels are no longer activated or concentrated at the distal end and the source of Ca^{2+} is from internal stores within endomembranes at the distal end. Calcium concentrations, therefore, vary both temporally and spatially within the cytoplasm of the target cell. There are multiple receptors for calcium within the cell and a wide variety of molecular changes may occur in response to the regionally elevated $[Ca^{2+}]$. These include: a) activation of calmodulin which in turn can activate a variety of enzymes, b) increased microtubule depolymerization releasing the nucleus from its anchorage on the side wall, c) stimulation of an acto-myosin microfilament system that directs membrane flow and/or organelle movement, d) conversion of the cytoplasmic matrix from a gel to a sol that might bring about a regulated concentration of the cytoskeleton to the point of the highest $[Ca^{2+}]$, the presumptive bud site. The contraction-collapse of a filamentous lattice could provide the motive force for the migration of the nucleus to that region. It is attractive to imagine that several of these processes work together to cause the profound cytoplasmic rearrangements that occur following cytokinin induction.

E. When the nucleus reaches the presumptive bud site, mitosis is triggered. The mitotic apparatus creates its own microdomain and ionic fluxes are be regulated within it. After mitosis, the newly formed cell plate fuses with caulonema walls at a position which may be determined by concentration of Ca^{2+} channels, producing a small bud initial cell.

F. Since the initial division is asymmetrical, the partitioning of membrane components between the caulonema cell and the bud initial cell is unequal. The plasma membrane of the initial cell may be enriched in cytokinin receptors and thus continued stimulation by cytokinin affects only the initial cell; the caulonema cell reverts to its nondividing state. The small volume and large concentration of organelles and endomembranes of the initial cell would lead to highly regulated sequestration and release of Ca^{2+}. The relationship between the division rate and the growth rate of the cells may provide geometrical constraints that determine the pattern of divisions and give rise to the organized mass of cells termed a bud.

REFERENCES

Conrad, P. A., and Hepler, P. K., 1984, Ultrastructural changes associated with cytokinin induced bud formation in the moss Funaria hygrometrica. J. Cell Biol. 99: 244a.

Clothier, G., and Timourian, H., 1972, Calcium uptake and release by dividing sea urchin eggs. Exp. Cell Res. 75:105.

Duffus, J. H., and Patterson, L. J., 1974, Control of cell division in yeast using the ionophore A23187 with calcium and magnesium. Nature 251: 626.

dos Remedios, C. G., 1981, Lanthanide ion probes of calcium-binding sites on cellular membranes. Cell Calcium 2: 29.

Elliott, D. C., 1983, Inhibition of cytokinin-mediated responses by calmodulin binding compounds. Plant Physiol. 72: 215.

Grynkiewicz, G., Poenie, M., and Tsien, R. Y., 1985, A new generation of Ca^{2+} indicators with greatly improved fluorescence properties. J. Biol. Chem. 260: 3440.

Hepler, P. K., 1985, Calcium restriction prolongs metaphase in dividing *Tradescantia* stamen hair cells. J. Cell Biol. 100: 1363.

Hesketh, T. R., Moore, J. P., Morris, J. D. H., Taylor, M. V., Roberts, J., Smith, G. A., and Metcalfe, J. C., 1985, A common sequence of calcium and pH signals in the mitogenic stimulation of eukaryotic cells. Nature 313: 481.

Holmes, R. P., and Stewart, P. R., 1977, Calcium uptake during mitosis in the myxomycete *Physarum polycephalum*. Nature 269: 592.

Jaffe, L. F., 1980, Calcium explosions as triggers of development. Ann. New York Acad. Sci. 339: 86.

Jaffe, L. F., and Nuccitelli, R., 1974, An ultrasensitive vibrating electrode for measuring steady extracellular currents. J. Cell Biol. 63: 614.

Keith, C. H., Maxfield, F. R., and Shelanski, M. L., 1985, Intracellular free calcium levels are reduced in mitotic Pt K2 epithelial cells, Proc. Natl. Acad. Sci. USA 82: 800.

Kubowicz, B. D., Vanderhoef, L. N., and Hanson, J. B., 1982, ATP-dependent calcium transport in plasmalema preparations from soybean hypocotyls. Plant Physiol. 69: 187.

Laetsch, W. M., 1967, Ferns. in: "Methods in developmental biology," Wilt, F. H., Wessells, N. K., eds., Thomas Y. Crowell Co., New York.

Lau, O.-L., and Yang, S. F., 1975, Interaction of kinetin and calcium in relation to their effect on stimulation of ethylene production. Plant Physiol. 55: 738.

LeJohn, H. B., and Cameron, L. E., 1973, Cytokinins regulate calcium binding to a glycoprotein from fungal cells. Biochem. Biophys. Res. Commun. 54: 1053.

Leopold, A. C., Poovaiah, B. W., de la Fuente, R. C., and Williams, R. J., 1973, Regulation of growth with inorganic solutes. in: "Plant Growth Substances," Hirokawa Pub. Co., Tokyo.

Letham, D. S., 1978, Cytokinins. in: "Phytohormones and related compounds - a comprehensive treatise, Vol. I, The biochemistry of phytohormones and related compounds," Letham, D. S., Goodwin, P. B., Higgins, T. J. V., eds., Elsevier/North-Holland Biomedical Press, Amsterdam, Oxford, New York.

Meyer, Y., and Cooke, R., 1979, Time course of hormonal control of the first mitosis in tobacco mesophyll protoplasts cultivated in vitro. Planta 147: 181.

Moreau, M., Guerrier, P., Doree, M., and Ashley, C. C., 1978, Hormone-induced release of intracellular Ca^{2+} triggers meiosis in starfish oocytes. Nature 272: 251.

Poovaiah, B. W., and Leopold, A. C., 1973, Deferral of leaf senescence with calcium. Plant Physiol. 52: 236.

Ralph, R. K., Buillivant, S., and Wojcik, S. J., 1976, Effects of kinetin on phosphorylation of leaf membrane proteins. Biochem. Biophys. Acta 421: 319.

Saunders, M. J., and Hepler, P. K., 1981, Localization of membrane-associated calcium following cytokinin treatment in *Funaria* using chlorotetracycline. Planta 152: 272.

Saunders, M. J., and Hepler, P. K., 1982, Ca^{2+} inophore A23187 stimulates cytokinin-like mitosis in *Funaria*. Science 217: 943.

Saunders, M. J., and Hepler, P. K., 1983, Calcium antagonists and calmodulin inhibitors block cytokinin-induced bud formation in *Funaria*. Develop. Biol. 99: 41.

Skoog, F., Strong, F. M., and Miller, C. O., 1965, Cytokinins. Science 148: 532.

Williamson, R. E., 1984, Calcium and the plant cytoskeleton. Plant, Cell and Envir. 7: 432.

This research sponsored in part by NSF grant PCM 84-08496.

UPTAKE AND RELEASE OF CA^{2+} IN THE GREEN ALGAE
MOUGEOTIA AND MESOTAENIUM

M.H. Weisenseel

Botanisches Institut
Universität Karlsruhe (TH)
Kaiserstr. 12, FRG

SUMMARY

Mougeotia cells were immersed in unbuffered artificial
pond water with low calcium concentration and irradiated with
far-red light, followed by red light. During irradiations the
calcium ion concentration and the pH of the medium were
monitored with extracellular electrodes. The measurements show
that in far-red light the calcium ion concentration of the
medium remained constant or increased slowly. During red light
irradiation, the pH of the medium increased by about two units,
the cells took up calcium ions from the medium quite rapidly,
and then released them slowly back into the medium. The effect
of red light on the calcium ion uptake could be simulated
partly by increasing the pH of the medium artificially with
NaOH.

Protoplasts freshly prepared from Mougeotia cells by
enzymatic digestion of the cell walls showed calcium ion
uptake, but no release, during red light irradiation. Red light
irradiation of protoplasts also caused the extracellular pH
to increase and the chloroplasts to move. Thus, the red light-
mediated uptake of calcium ions by intact Mougeotia cells is
probably membrane controlled and not caused by binding of
calcium to the cell walls. The results moreover indicate that
a pH-gradient at the plasmalemma is involved in the uptake
of calcium ions.

In Mesotaenium cells, surrounded by well buffered media,
red light caused a decrease in the surface charge of the cells,
as measured by cell electrophoresis. This decrease is dependent
on the uptake of calcium ions. Addition of EDTA or Verapamil
to the medium prevents the red light-induced decrease in
surface charge. We assume that the decrease in surface charge
is, in part, caused by an efflux of mucus and calcium from
the cells.

INTRODUCTION

Calcium ions are known to couple stimulus and response in

many animal cells. In such responses, the primary stimulus interacts with a receptor in the plasma membrane and causes a transient increase in the intracellular calcium ion concentration from 0.1 - 1 to 10 uM (e.g. Campbell 1983). In plant cells, too, there is increasing evidence for calcium-coupling of stimulus and response. For instance, light-mediated uptake and/or release of calcium ions have been demonstrated in Nitella internodes (Weisenseel and Ruppert 1977), Mougeotia cells (Dreyer and Weisenseel 1979, Roux 1983), oat cell protoplasts (Hale and Roux 1980), Phormidium trichomes (Häder 1984) and Onoclea spores (Wayne and Hepler 1985).

Such observations stimulate further questions and experiments that can help to elucidate the mechanism of signal transduction in plant cells. For instance, what controls the calcium-flux across the cell membrane? Is it the receptor itself, or the membrane potential, or a pH gradient across the membrane? What are the consequences of an elevated intracellular calcium ion concentration? Does it activate or inactivate genes, or stimulate exocytosis of cell wall material, or induce intracellular electrophoresis? How does the cell maintain calcium homeostasis? We have started to tackle some of these questions by measuring the uptake and release of calcium ions from Mougeotia cells and protoplasts with extracellular electrodes, and we began to investigate the effect of calcium ions on the zeta potential of Mesotaenium cells with cell electrophoresis.

UPTAKE AND RELEASE OF CA^{2+} IN MOUGEOTIA CELLS

Mougeotia cells were grown for 3 - 4 weeks under 12 h white light (ca. 150 lux) daily at 16 - 18°C in a medium containing 90 mg K_2HPO_4, 90 mg $NaNO_3$ and 100 ml extract from rotten leaves in 1800 ml water. For each experiment about 200 mg of algae (i.e. ca. 1.5 - 2.0.10^6 cells) were preincubated in 100 ml of experimental medium and kept in darkness for 24 h.

The experimental medium contained 1 mM NaCl, 1 mM $NaHCO_3$, 0.2 mM KCl and 1 uM $CaCl_2$, pH 7.3 - 7.6. After preincubation, the cells were transferred to 11 ml of fresh medium and into a plexiglass vessel surrounded by water of 20 - 0.1°C. The cells were then irradiated with continuous far-red light (15-25 Wm^{-2}, glass filter RGN9/3mm, Schott a. Gen.) or continuous red light (150-200 Wm^{-2}, glass filter RG 610/2 mm, Schott a. Gen.). The activity of calcium ions and protons in the medium was monitored during the irradiations with a calcium-sensitive electrode (model 93-20, Orion) and a small combination pH electrode (Lot 405-M3, Ingold). The Ag-AgCl electrode of the pH electrode was used as a common reference for both electrodes to minimize noise and drift. The output of the electrodes was recorded with electrometers (model 910C, Keithley) and a two channel chart recorder.

When Mougeotia cells were irradiated for 20 min with far-red light followed by 10 min darkness, the calcium ion concentration of the medium remained constant or increased slightly. This slight increase is probably caused by leakage of calcium from the electrode and loss of calcium from the cells. During red irradiation the cells took up calcium ions

from the medium quite rapidly. Then they released it back
into the medium more slowly. The measurable decrease of
external calcium ions began with a lag-time of 4 - 5 min and
reached its maximum 15 - 20 min after the onset of red light.
Figure 1 A shows a representative example of the changes in
calcium ion concentration during irradiation. This type of
kinetic was found in most experiments; in rare cases the cells
showed only uptake of calcium during red light irradiation.
Figure 1 B summarizes 11 experiments.

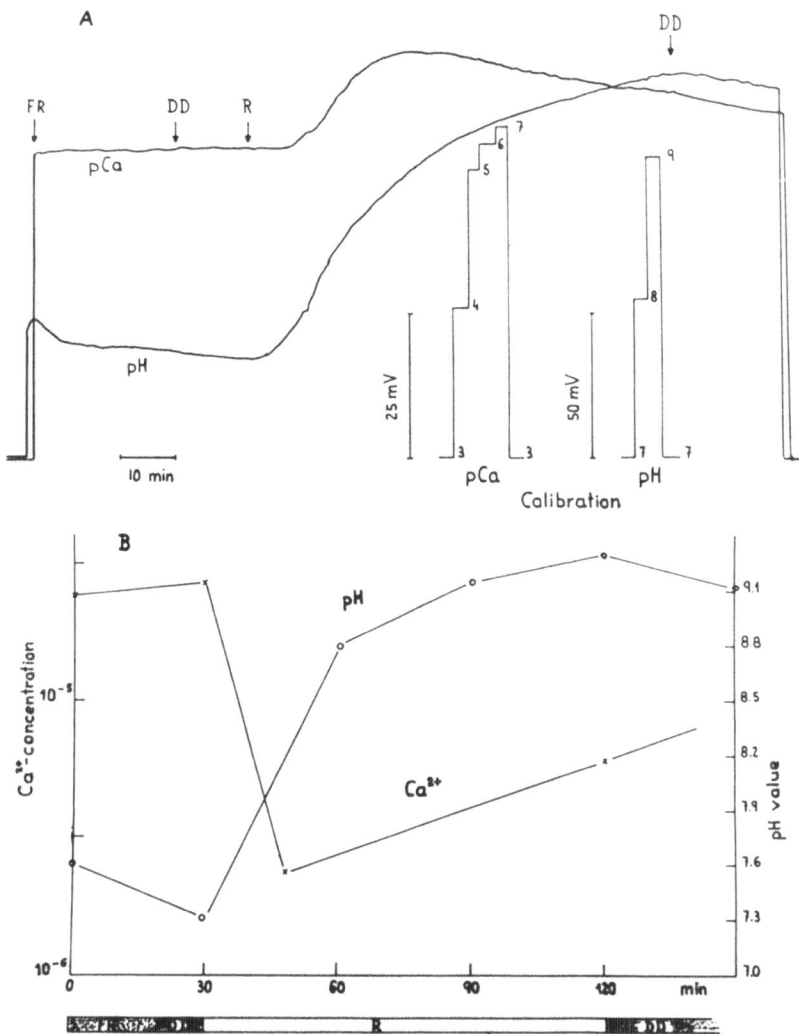

Fig. 1: The effect of red light (R), far-red light (FR) and
        darkness (DD) on the changes of calcium ions and
        protons in the medium surrounding Mougeotia cells.
        A: Representative example of the effect. B: Average
        values of 11 experiments (Ca^{2+}-concentration in
        $mol.l^{-1}$). The calibration for calcium ions gives
        nominal values; the actual concentrations at 10^{-6} M
        and 10^{-7} M are most likely higher.

During red irradiation, the pH of the medium increased
rapidly, attaining values of 9.3 and higher (Fig. 1).

Experiments with pH-buffers in the medium indicate that this pH-increase is essential for the uptake of calcium. Simulating the red light-induced pH-increase in the dark by adding 0.05 N NaOH to the medium caused uptake and release of calcium in Mougeotia cells similar to red light.

The observed calcium depletion in the medium during red light irradiation is most likely a response of the living cells. It can neither be explained by simple precipitation of $CaCO_3$, nor by binding of calcium ions to the cell walls. The maximum amount of $CaCO_3$ in the medium was 10 x less than the soluable amount. A similar calcium ion uptake was recorded when $NaHCO_3$ was replaced by NaCl, and an uptake of calcium during red light irradiation was also found with protoplasts.

UPTAKE OF CA^{2+} IN MOUGEOTIA PROTOPLASTS

One to two grams of Mougeotia were used to prepare protoplasts. The cell walls were digested in a medium containing 500 or 700 mM mannitol, 0.3 % cellulysin (Calbiochem), 0.1 % macerase (Calbiochem) and 0.1 % pectinase Rohament P-5 (Serva), pH 4.8 - 5.2. After about 4 h at room temperature, the released protoplasts were washed free of enzymes and collected on a discontinuous sucrose-mannitol gradient. Five ml of protoplasts were placed on a pad of 5 ml 550 mM, respectively 750 mM sucrose in a narrow plexiglass cylinder and both electrodes were brought into contact with the protoplast layer. The pad of sucrose medium was used to minimize destructive contact of protoplasts with the bottom of the vessel and to allow chloroplasts from burst protoplasts to sediment to the bottom. After each experiment the concentration of protoplasts was determined with a haemocytometer.

During far-red irradiation and in darkness, the concentration of calcium ions in the medium around protoplasts increased slightly and the pH decreased (Fig. 2). When the protoplasts were irradiated with red light, the calcium concentration of the medium decreased for at least 60 min, i.e. during the whole time tested so far. No release of calcium ions from the protoplasts was observed during this period. The pH in the protoplast layer increased by about one unit, in some cases by 2 units, during red light irradiation. When the red light was turned off, a rapid release of calcium started. The viability of the protoplasts could be demonstrated with FDA-fluorescence and by chloroplast movements.

LIGHT AND CA^{2+}-UPTAKE AFFECT THE SURFACE CHARGE OF MESOTAENIUM CELLS

Mesotaenium caldariorum is an unicellular green alga, very similar in morphology and physiology to Mougeotia. It is well suited to study the effect of stimuli and calcium ions on the surface charge. Mesotaenium was cultured in Petri dishes at a temperature of 16 - 18°C under white light of about 100 lx illuminance. The daily photoperiod was 12 h. The culture medium was the same as described by Gärtner (1970). All measurements were carried out with about four-weeks-old cultures and cell densities of about 10^5 cells ml^{-1}. The following medium was used for the measurements: 1 mM NaCl, 1 mM KCl, 1 mM $CaCl_2$, 4 mM MES (Morpholinoethane sulfonic acid) and 4 mM TRIS (Tris(hydroxymethyl) aminomethane).

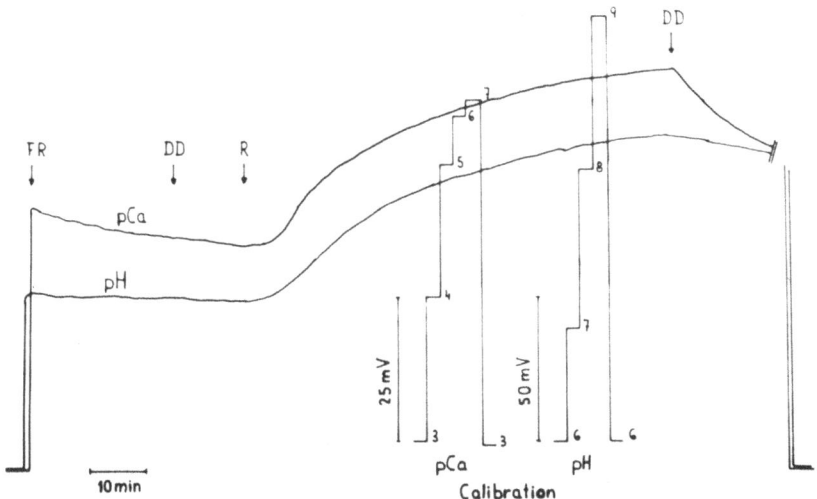

Fig. 2: Representative example of the effect of red light (R)
far-red light (FR) and darkness (DD) on the changes of
calcium ions and protons in the stagnant medium
surrounding Mougeotia protoplasts (ca. $170 \cdot 10^3$
protoplasts/ml). (About calcium calibration see Fig. 1.)

The electrophoretic measurements were carried out using a
Zeiss cytopherometer equipped with a flat chamber of 0.75 mm
depth, 14 mm height and 35 mm length. A current flow of
1.5 mA ± 0.5 % through the chamber was established by a
stabilized current source, yielding an electric field strength
of 22 V cm^{-1} within the stationary plane of the viewing field.
To determine the electrophoretic mobility u = v/E of the cells
(v = cell migration velocity, E = electric field strength),
the time required for passing a distance of 150 µm was measured.
From the measured values for cell mobility, the zeta potentials
were calculated according to Smoluchowski's equation. This
equation for calculating zeta potentials applies to charged
particles of the size of Mesotaenium cells.

Thirty minutes irradiation of Mesotaenium cells with red
light or far-red light of 9 Wm^{-2} each resulted in highly
different zeta potentials of the cells (Fig. 3). In the pH
range investigated, i.e. between pH 5 and pH 9, the zeta
potentials were distinctly higher after far-red light exposure
than after red light irradiation. The differences were largest
at pH 7 and decreased significantly at pH 6.

Remembering our hypothesis that phytochrome mediates
calcium ion fluxes across the cell membrane (Haupt and Weisenseel 1976), we investigated the effect of calcium ions on the
zeta potentials of Mesotaenium cells. For this purpose, the
chelating agent EGTA (Ethylene glycol- bis (ß-aminoethyl ether)
N, N', N'-tetraacetic acid) was added to the medium to reduce
the external calcium concentration. The results for different
EGTA concentrations and incubation periods show that the
difference between red light and far-red light irradiation
became negligible when EGTA of 1 mM was present for 1 h or
longer. Experiments using the calcium channel blocker
Verapamil (Sigma) showed that 10 µM Verapamil in the medium
inhibited the effect of red light exposure on the zeta potential.

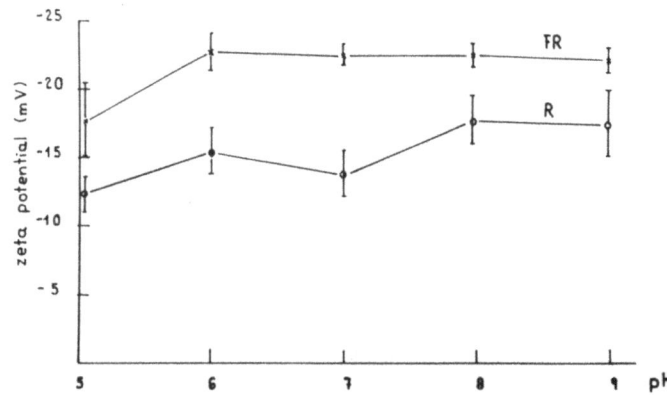

Fig. 3: Zeta potentials of Mesotaenium cells after red light
(R) or far-red light (FR) irradiation in media with
different pH values. The data are mean values ± S.E.
of 10 - 20 different cells of a representative
experiment. (Data from H.-G. Stenz and M.H. Weisenseel,
submitted to the "Journal of Plant Physiology".)

To explain the red light-induced decrease of the surface
charge of Mesotaenium cells we assume that the first step is a
red light-mediated influx of calcium into the cells. The influx
of calcium ions might then cause a reduction of the surface
charge by stimulating exocytosis of less negatively or
positively charged material (possibly including calcium ions)
to the cell surface. Our first EM pictures of Mesotaenium cells
after red, respectively far-red irradiation give support to
this possibility.

Final remark: Most of the data and conclusions described
above should be considered preliminary. It was my intention to
present results from very recent experiments and work in
progress to satisfy the demands of a workshop, i.e. to discuss
new findings and to exchange ideas. We are aware that many more
experiments and controls have to be carried out to elucidate
the true mechanism and function of light and calcium fluxes in
Mougeotia and Mesotaenium cells.

Acknowledgement: The financial support of our research by
the Deutsche Forschungsgemeinschaft is gratefully acknowledged.

REFERENCES

Campbell, A.K., 1983, "Intracellular Calcium: Its universal
role as regulator", John Wiley and Sons, New York.

Dreyer, E.M., and Weisenseel, M.H., 1979, Phytochrome-mediated
uptake of calcium in Mougeotia cells, Planta, 146: 31 - 39.

Gärtner, R., 1970, Die Bewegung des Mesotaenium-Chloroplasten
im Starklichtbereich, Z. Pflanzenphysiol., 63: 147 - 161.

Häder, D.P., 1984, Wie orientieren sich Cyanobakterien im Licht, *Biologie i. u. Zeit,* 3: 78 - 83.

Hale, C.C., and Roux, S.J., 1980, Photoreversible calcium fluxes induced by phytochrome in oat coleoptile cells, *Plant Physiol.,* 65: 658 - 662.

Haupt, W., and Weisenseel, M.H., 1976, Physiological evidence and some thoughts on localized responses, intracellular localisation and action of phytochrome, *in:* "Light and Plant Development", H. Smith, ed., Butterworths, Boston, pp 63 - 74.

Roux, S.J., 1983, A possible role for Ca^{2+} in mediating phytochrome responses, *in:* "The Biology of Photoreception", Symposium of the Society for Experimental Biology, Number 36, Cambridge University Press, Cambridge, London, New York, pp 561 - 580.

Wayne, R., and Hepler, P.K., 1985, Red light stimulates an increase in intracellular calcium in the spores of Onoclea sensibilis, *Plant Physiol.,* 77: 8 - 11.

Weisenseel, M.H., and Ruppert, H.K., 1977, Phytochrome and calcium ions are involved in light-induced membrane depolarization in Nitella, *Planta,* 137: 225 - 229.

CALCIUM POOLS, CALMODULIN AND LIGHT-REGULATED CHLOROPLAST MOVEMENTS IN MOUGEOTIA AND MESOTAENIUM

Sigrid Jacobshagen, Doris Altmüller,
Franz Grolig and Gottfried Wagner

Botanisches Institut I
der Justus-Liebig-Universität
Senckenbergstrasse 17-21
D-6300 Giessen, FR Germany

INTRODUCTION

Mesotaenium often is understood as an unicellular substitute of Mougeotia, particularly in terms of blue and of red light-mediated chloroplast reorientational response. A major difference in the sensory transduction chain of these two species of the order Conjugate, however, is easily seen: For performance of the red light-mediated low irradiance response, a single flash of red light, followed by complete darkness, suffices in Mougeotia (Haupt and Wagner, 1984). The "memory" here for the perceived reorientational stimulus declines by a half-life of 90 min, compared with a time period of 40 min to perform the chloroplast reorientational response (Wagner and Klein, 1981). In contrary, Mesotaenium either needs continuous red irradiation or repetitive pulses of red light. The threshold length of dark intervals intermittent with the pulses of red light here is about 10 min, while the reorientational response needs 40 min, similarly as in Mougeotia (Haupt, 1982). Thus, "memory" in Mesotaenium appears shorter than the time period of response, while "memory" in Mougeotia appears longer.

This significant difference in "memory" may reside with the photoreceptor (Haupt, 1982). Additionally, major parts of the sensory transduction chain, including calcium compartmentation, calmodulin as a predominant calcium target and the actomyosin motor apparatus including anchorage proteins to actin may significantly be different in Mesotaenium compared to Mougeotia. Therefore, Mesotaenium was grown in a mass culture to allow calmodulin purification. Also, calcium vesicles have been localized by the in vivo-calcium stain chlorotetracycline and compared to similar data from Mougeotia, Spirogyra and Zygnema (Rossbacher et al., 1984).

An intriguing red/far red-reversible calcium vesicular chlorotetracycline fluorescence is reported here for Mougeotia, possibly reflecting a phytochrome-regulated

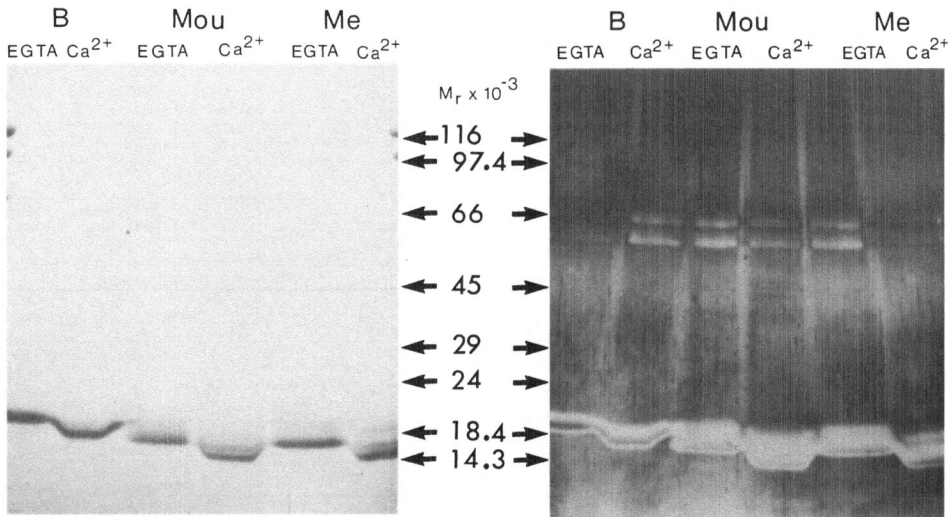

Figure 1.
SDS-PAGE of purified calmodulin from bovine brain (B), from Mougeotia (Mou) or from Mesotaenium (Me) in absence (EGTA, 10 mmol/l) or in presence of calcium (Ca^{2+}, 25 mmol/l). Relative molecular mass (M_r) of coelectrophoresed standard proteins is also shown. The gel was stained by Coomassie brillant blue G 250 (left), then stained by silver (right).

vesicular calcium release, as is calcium-dependent fast protein phosphorylation.

MATERIAL AND METHODS

Mesotaenium caldariorum, strain 648-1 from the Algal Collection at Göttingen, FRG, was grown in a synthetic medium (Stabenau, 1978) as described for Mougeotia sp. (Wagner et al., 1984). Bovine brain was fresh from a nearby slaughter house. For isolation of Mesotaenium calmodulin, 150 g fresh weight were sonified as described (Wagner et al., 1984) in 600 ml of ice-cooled extration buffer ((3-(N-morpholino)-propanesulfonic acid) (MOPS) 50 mmol/l, pH=7.5; $MgCl_2$ 2 mmol/l; ethylene diamine tetraacetic acid (EDTA) 3 mmol/l; dithiothreitol (DTT) 10 mmol/l; ascorbic acid 20 mmol/l; Polyclar AT insoluble 48 g, phenylmethylsulfonyl fluoride (PMSF) 1 mmol/l; isopropanol 3% (v/v)). The cell homogenate was squeezed through miracloth (Serva, Heidelberg, FRG) and centrifuged for 30 min 35,000 g at 4°C. The supernatant was heat precipitated by 5 min at 80°C, centrifuged for 30 min at 35,000 g at 4°C and recentrifuged after addition of 5 mmol/l $CaCl_2$ as described above.

Mesotaenium calmodulin was purified as described for Mougeotia and bovine brain calmodulin (Wagner et al., 1984). Samples of Mesotaenium supernatant protein in extration buffer plus 5 mmol/l $CaCl_2$ were applied to 6 ml (10 mm inner diameter, 76 mm length) column of chlorpromazine-sepharose 4 B (CAPP; Jamieson and Vanaman, 1979) equilibrated with equilibration buffer (MOPS 50 mmol/l, pH=7.5; NaCl 0.1 mol/l; $CaCl_2$ 5 mmol/l; $MgCl_2$ 1 mmol/l; DTT 1 mmol/l).

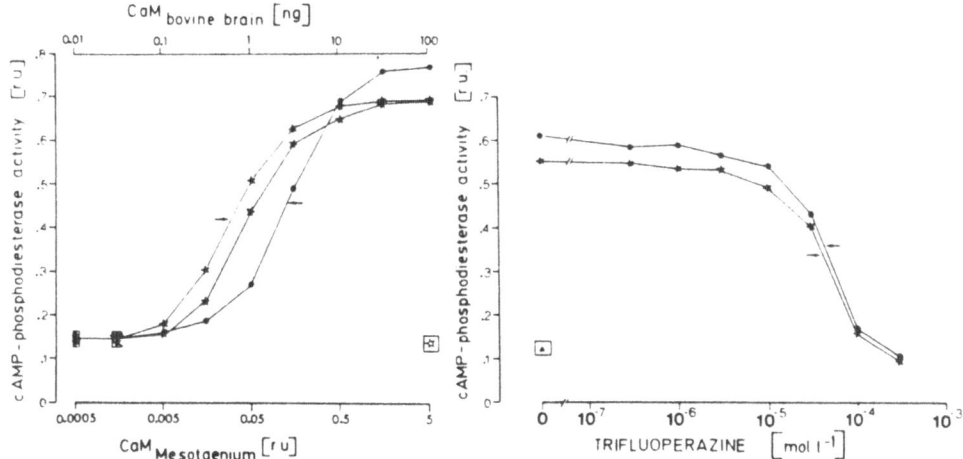

Figure 2.
Activity of bovine heart cAMP-phosphodiesterase in presence
of various amounts of bovine brain calmodulin (●) or
Mesotaenium calmodulin (★). Increase of the cAMP-phosphodiesterase activity by Mesotaenium calmodulin was
abolished in presence of 5 mmol EGTA (✩). Arrows indicate
half-saturation of cAMP-phosphodiesterase by calmodulin
(EC_{50}). Relative units (r.u.)

Figure 3.
Activity of bovine heart PDE in presence of 21 nmol/l bovine
brain calmodulin (●) or 21 nmol/l Mesotaenium calmodulin (★
) inhibited by various amounts of trifluoperazine. cAMP-phosphodiesterase activity in absence of calmodulin and of inhibitor is also shown (control, ▲). Arrows indicate half-
saturation of calmodulin-activated cAMP-phosphodiesterase by
inhibitor (IC_{50}=42 μmol/l TFP). Relative units (r.u.).

The column was then washed with 0.5 mol/l NaCl in equilibration buffer until the absorbance at 280 nm wavelength dropped
to base-line level. Bound calmodulin was then eluted,and
fractions (1 ml) collected at a rate of 12 ml/h as described
(Wagner et al., 1984). Calmodulin from bovine brain and
Mougeotia was purified accordingly.

    Gel electrophoresis on 10% ,or 6% to 20% gradient SDS-
polyacrylamide gels was performed as described by Laemmli
(1970). Separated proteins were stained either with Coomassie
brillant blue G 250 or with silver according to Nielson and
Brown (1984). For autoradiography dried gels were laid onto a
Kodak X-omat XAR 5 film and exposed in a black box at -24°C.
Exposure time was 7-9 days. Cyclic nucleotide phosphodiesterase test was as described by Jacobshagen et al., this
volume. Trifluoperazine was kindly provided by Röhm-Pharma,
Weiterstadt, FRG and dissolved in cAMP-phosphodiesterase test
medium.

    Mougeotia was homogenized (1 g fresh weight per 0.8 ml
buffer) by sonification for 7 min in HEPES buffer (50
mmol/l, pH=8.2, containing: $MgCl_2$ 2 mmol/l; $CaCl_2$ 0.2 mmol/l;

Table 1.
Relative molecular mass of calmodulin in absence (EGTA) and
presence (Ca^{2+}) of calcium, and percentage of calmodulin per
extracted soluble protein from bovine brain, Mougeotia or
Mesotaenium.

	Relative molecular mass (M_r)		Percentage of calmodulin in soluble protein
	EGTA	Ca^{2+}	
Bovine brain	19,200	16,800	0.66
Mougeotia	16,300	14,300	0.28
Mesotaenium	16,300	14,300	0.15

EDTA 3 mmol/l; DTT 20 mmol/l; $Na_2SO_3/Na_2S_2O_5$ 5 mmol/l; Polyclar AT insoluble of 2.25 times the amount of the fresh
weight; 0.1% (w/v) BSA and 10% (w/v) sucrose (Fig. 6) or 20%
(v/v) glycerine (Fig.7). After filtration through nylon mesh,
the crude extract was centrifuged for 10 min 500 g to remove
cellular debris. 0.1%-0.2% (w/v) Triton X-100 was added to
the supernatant which was left on ice for about 15 min and
then immediately used for the phosphorylation assay. The
freshly prepared homogenate was transferred into a waterbath
of 24°C and the reaction started with the addition of 296 to
370 KBq $[\gamma^{32}P]$ -ATP per 200 µl aliquot for different times.
The reaction was stopped in aliquots of 200 µl by addition of
equal volume of sample buffer on ice (Laemmli, 1970). The
samples were boiled for 5 min, and 100 µl each were loaded
onto the tracks of the gel.

For in vivo-fluorescence measurements, Mougeotia filaments were incubated in 2 mmol/l chlorotetracycline (CTC) for
5 min, rinsed three times in MXS-medium (Wagner and Rossbacher, 1980), allowed to adapt to the CTC-treatment for 3 h
in darkness and transferred onto microscopic slides under dim
green safelight (Wagner et al., 1984). Single spot fluorescence, stained by CTC in cells of face on-oriented chloroplast, was monitored by a microfluorimetric set-up using a
Zeiss universal microscope with epifluorescence as described
and provided by Dr. E. Hartmann, Mainz, FRG (Hartmann; 1984).

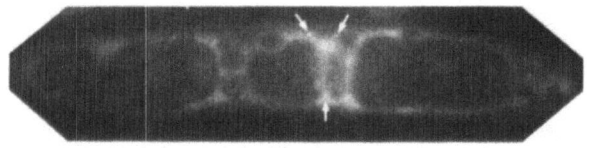

Figure 4.
Micrograph of a fluorescent Mesotaenium cell, stained by
chlorotetracycline. Calcium-CTC-stained vesicles, far less
abundant than in Mougeotia, are seen as white dots (arrows;
for details, see text). Low CTC-fluorescence from the ground
cytoplasm is also seen. x 700.

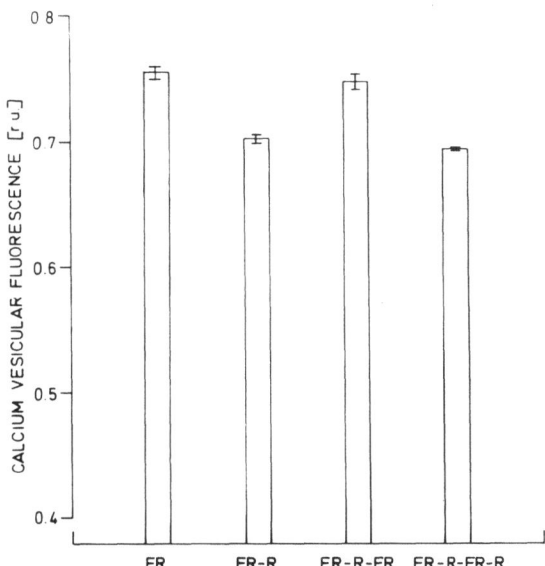

Figure 5.
Calcium vesicular fluorescence in <u>Mougeotia</u>, stained <u>in vivo</u>
by chlorotetracycline, as a function of far-red/red pre-
irradiation. Fluorescence emitted from individual calcium
vesicles was quantified in relative units (r.u.) by a micro-
fluorimeter after 5 min-pulses each of far red (FR) or red
(R) light (see MATERIALS and METHODS). The bars indicate
standard error of the mean (SE). Photon fluence rates: 8.5
μmol m^{-2} x s^{-1} from DAL 663 nm; 10 μmol m^{-2} x s^{-1} from DAL
730nm.

The photomultiplier tube was shielded against chlorophyll
fluorescence by red absorbing glass filters and a diaphragm
matched to the calcium vesicular diameter. Photomultiplier
current was converted into voltage and read in relative units
through a thermoprinter. Actinic light was provided by the
microscopic bulb through interference filters (DAL 663 nm;
DAL 730 nm; Schott, Mainz, FRG) and transmitted through the
condensor lens onto the microscopic slide. To avoid fading of
CTC fluorescence, excitation was kept to a minimum by a
mechanic shutter, and registration of fluorescence output was
within seconds. Fluorescence micrograph of CTC-stained
<u>Mesotaenium</u> was taken on Kodak Plux X pan film.

RESULTS

    The theoretical total binding capacity of the CAPP af-
finity column used for the calmodulin isolations was calculated from the absorption difference ($\Delta(\Delta OD)$) of the
CAPP-solution before and after coupling to CNBr-activated
sepharose 4B. The concentration of ligand then was 4.3 μmol
per ml swollen gel with a capacity of the column, ready for
use, of 216 mg calmodulin. Calcium/EGTA reversibility of
calmodulin-binding was tested by means of bovine brain calmodulin from supernatant of crude homogenate after two-step
heat precipitation. In the elution diagram both, peak ultra-
violet absorption and peak calmodulin content matched each

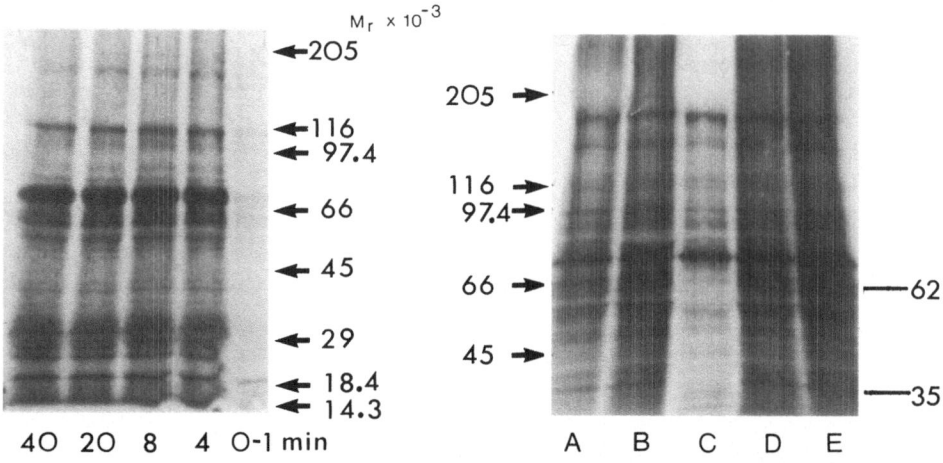

$M_r \times 10^{-3}$

40 20 8 4 0-1 min

A B C D E

Figure 6.
In vitro-phosphorylation of Mougeotia proteins as a function
of time. The autoradiograph shows phosphorylated Mougeotia
proteins in a large range of relative molecular mass (M_r).
The maximum level in phosphorylation appears 4 min after
incubation with $[\gamma\ ^{32}P]$-ATP. Mougeotia cells were
harvested from a slowly growing culture.

Figure 7.
In vitro-phosphorylation of Mougeotia proteins in presence of
EGTA (4 mmol/l; lane C) or in presence of calcium ($CaCl_2 \geqslant$
0.2 mmol/l; lanes A, B, D, E). Incubation with $[\gamma\ ^{32}P]$-ATP
was 4 min in lane (A) and 20 min in lanes (B), (C), (D),
(E). Mougeotia cells were harvested from a fast growing
culture.

other exactly as did the shape of the curves (Jacobshagen et
al., this volume). The eluted protein was a highly purified
calmodulin, but a tiny contamination is seen by SDS-PAGE
after silver staining (Fig. 1). By the identical technique,
calmodulin was isolated and purified out of heat-precipitated
supernatant from Mesotaenium and from Mougeotia homogenate.
For Mesotaenium the identity of calmodulin being eluted from
the CAPP-affinity column upon addition of 10 ml EGTA was
proved by PDE-test of two peak absorbing fractions of the
eluate and quantified via standard concentrations of bovine
brain calmodulin (Fig. 2). Half saturation of PDE activation
is found here by 2.6 ng bovine brain calmodulin which is up
to a factor of four less than that found before (Wagner et
al., 1984). This reflects the increased sensitivity of the
PDE test as used for these determinations (Jacobshagen et
al., this volume).

Percentage of calmodulin in extracted soluble protein
(%w/w) was calculated for each homogenate via protein tests
(Bradford, 1976) and gravimetric determinations. The data are
shown in Table 1 together with the calculated relative molecular mass of calmodulin from bovine brain, from Mesotaenium
and from Mougeotia, determined by SDS-PAGE in presence and in

absence of calcium. Despite the observed difference in molecular mass and in the capability to activate PDE (Table 1; Fig. 2), bovine brain calmodulin and Mesotaenium calmodulin turned out identical in terms of sensitivity to the inhibitor TFP (Fig. 3). The half saturating concentration (IC_{50}) was 42 µmol/l TFP in contact either with 21 nmol/l bovine brain calmodulin or with 21 nmol/l Mesotaenium calmodulin. At the highest trifluoperazine concentration tested, i.e. at 0.3 mmol TFP/l, a side effect of TFP is seen with inhibition of PDE to an activity less than that observed in absence of calmodulin.

The major part of calcium, residing in intracellular stores of cytoplasmic vesicles in Mougeotia, in Spirogyra and in Zygnema (Rossbacher et al., 1984), was found also in Mesotaenium. The calcium vesicles in this alga turned out similar to those in Spirogyra and Zygnema, but smaller and less abundant than those in Mougeotia (Fig. 4). Therefore, Mougeotia was used first for detection of red/far red-reversible vesicular calcium fluxes as monitored by the in vivo-calcium stain chlorotetracycline (Fig. 5). A light regime of 5 min far-red followed by 5 min red resulted in an accompanying decrease of the relative output of calcium vesicular fluorescence by 7%. Repetition of the far-red/red light regime for one more cycle proved the consistent changes in vesicular fluorescence (Fig. 5).

Verbelen and coworkers localized phytochrome in Mougeotia by immunocytochemical techniques (Verbelen et al., 1983). During the preliminary work, they found phytochrome concentrated in discrete areas at the edge of the Mougeotia chloroplast the side being similar to that of the calcium vesicular distribution . The hypothetical chain of events giving raise to a precisely controlled reorientational movement of the single chloroplast both in Mougeotia and in Mesotaenium could lead from light absorption via phytochrome or blue light receptor to calcium as the signal and than to calmodulin which upon calcium activation regulates actin-myosin interaction in these green algae. In Mougeotia, blue light-mediated low irradiance response (Haupt, 1982) turned out sensitive to calmodulin inhibitors (Wagner, unpubl.), as does red light-mediated response (Wagner et al., 1984; Serlin and Roux, 1984).

For the molecular action of Ca^{2+}-calmodulin in these reorientational responses, a probable target would be the motor apparatus or parts of it, e.g. myosin kinase. To screen for possible protein kinase activities in Mougeotia, the time sequence of protein phosphorylations was followed in cell homogenates after addition of [γ 32P]-ATP and is shown in Fig. 6. Starting with 1 to 2 min after [γ 32P]-ATP incubation, protein phosphorylations have reached maximum level within 4 min. With extended periods of incubation, significant dephosphorylations are not seen, possibly due to rapid protein denaturation by high concentration of endogenous polyphenolic compounds. Despite these cell-specific complications (Altmüller et al., 1984), phosphorylation of two proteins (Mr 62,000; 35,000) turned out significantly dependent on Ca^{2+} (Fig. 7).

DISCUSSION

Within the range of accuracy of the method used, relative molecular mass of calmodulin from Mesotaenium turned out identical to that of Mougeotia, but measurably smaller than that from bovine brain. This difference should be expected from the phylogenetic distance of bovine and Mesotaenium. Percentage of calmodulin in soluble protein turned out smaller, however, in Mesotaenium compared to closely related Mougeotia. The improved extraction buffer used in Mesotaenium but not in Mougeotia (see MATERIALS and METHODS) may partly explain the difference. A species-specific difference in the fractional amount of calmodulin, however, could also be suggested, consistent with the small size and low population of calcium vesicles in Mesotaenium compared to Mougeotia. Thus, species-specific differences in photosensory "memory" of Mougeotia and Mesotaenium (see INTRODUCTION) may partly relay on calcium and calmodulin. Mesotaenium calmodulin is inhibited by trifluoperazine in a stoichiometry exactly that found for bovine brain calmodulin. Besides capability to activate cAMP-phosphodiesterase, this gives additional proof of the isolated Mesotaenium protein being calmodulin; an unspecific side effect of trifluoperazine at high concentrations is also seen.

A major function of calcium-calmodulin in Mougeotia and in Mesotaenium appears regulation of actin-myosin interaction possibly via protein phosphorylation (Wagner et al., 1984). Rapid in vitro-protein phosphorylation have been reported here in Mougeotia with two proteins being significantly phosphorylated as a function of calcium. The proteins of M_r 62,000 dalton and 35,000 dalton have not yet been identified neither was involvement of calmodulin. In vitro-phosphorylation of rabbit myosin light chains, however, has been observed catalyzed by Mougeotia cell homogenate only in presence of calcium (D. Altmüller, unpubl.). This may indicate a calcium-regulated myosin kinase in Mougeotia (Heigl et al., 1985).

A most intriguing result reported here is red/far-red reversible change in fluorescence of the calcium vesicles in Mougeotia. The in vivo-calcium stain used, i.e. trifluoperazine, mainly reflects membrane-bound calcium (Schneider et al., 1983). If this can be taken for granted in Mougeotia and if phytochrome is involved in this process (Verbelen et al., 1983), a red light-mediated fluorescence decrease possibly reflects P_{fr}-induced release of membrane-bound calcium, which is P_r-reversible. For the molecular mechanism of phytochrome-regulated calcium release, the immunocytochemical data of phytochrome localization suggest an ion exchange mechanism depending on the membrane binding state of phytochrome (Saunders et al., 1983). Thus, the cytoplasmic concentration of calcium would increase upon membrane-binding of soluble phytochrome in its P_{fr}-state, and decrease upon phytochrome release in its P_r-state. Obviously, this fraction of soluble phytochrome must be different from the other fraction permanently bound at or within the Mougeotia plasmalemma which in a tetrapolar gradient possibly controls the membrane anchorage sites to actin (Haupt and Wagner, 1984; Wagner and Grolig, 1985).

Acknowledgements. We thank Dr. Elmar Hartmann (Institut für Allgemeine Botanik der Universität Mainz, FRG) for giving access to his microfluorimetric equipment. Financial support from the Deutsche Forschungsgemeinschaft is gratefully acknowledged.

REFERENCES

Altmüller, D., Grolig, F., and Wagner, G., 1984, Europ. J. Cell Biol., 33, S 5: 3.
Bradford, M. M., 1976, Anal. Biochem., 72: 248.
Hartmann, E., 1984, J. Hattori Bot. Lab., 55: 87.
Haupt, W., 1982, Annu. Rev. Plant Physiol., 33: 205.
Haupt, W., and Wagner, G., 1984, Chloroplast movement, in: "Membranes and Sensory Transduction", pp. 331-375, Colombetti, G., Lenci, F., eds. Plenum, New York.
Heigl, M., Altmüller, D., Grolig, F., and Wagner, G., 1985, Europ. J. Cell Biol., 36, S 7: 24.
Jamieson, G. A., Jr., and Vanaman, T. C., 1979, Biochem. Biophys. Res. Commun., 90: 1048.
Laemmli, U. K., 1970, Nature, 227:680.
Nielsen, B. L., and Brown, L. R., 1984, Anal. Biochem., 141: 311.
Rossbacher, R., Wagner, G., and Pallaghy, Ch. K., 1984, Nucl. Inst. Meth. Physics Res., 231: 664.
Saunders, M. J., Cordonnier, M.-M., Palevitz, B. A., and Pratt, L. H., 1983, Planta, 159: 545.
Schneider, A. S., Herz, R., and Sonenberg, M., 1983, Biochemistry, 22: 1680.
Serlin, B. S., and Roux, St. J., 1984, Proc. Natl. Acad. Sci. (U.S.A.), 81: 6368.
Stabenau, H., 1978, Ber. Dtsch. Bot. Ges., 91:251.
Verbelen, J.-P., DeGreef, J. A., Hartmann, E., and Wagner, G., 1983, Arch. Inst. Physiol. Biochim., 91: B46.
Wagner, G., and Grolig, F., 1985, Molecular mechanisms of photoinduced chloroplast movements, in: " Sensory Perception and Transduction in Aneural Organisms", in press, Song, P. S., Lenci, F., Colombetti, G., eds. Plenum, New York.
Wagner, G., and Klein, K., 1981, Protoplasma, 109: 169.
Wagner, G., and Rossbacher, R., 1980, Planta, 149: 298.
Wagner, G., Valentin, P., Dieter, P., and Marmé, D., 1984, Planta, 162: 62.

CALCIUM AND POLARITY IN TIP GROWING PLANT CELLS

H.-D. Reiss, W. Herth and E. Schnepf

Zellenlehre
Universität Heidelberg
F.R.G.

Polarity of various plant cells is manifested most obviously by tip growth. Tip growth occurs, e.g. in some algae, fungal hyphae, moss and fern protonemata, root hairs and pollen tubes. These tubular cells are also characterized by a polar distribution of cell organelles. Golgi vesicles, produced in a certain distance from the tip, are transported vectorially into the cell apex. They accumulate there to form a "tip body" or "clear cap" and fuse with the plasmalemma, extruding matrix substances of the cell wall. Polar growth implied additionally the polar formation of cellulose microfibrils (Reiss et al., 1984b). Exogenous factors like gravity and light as well as endogenous factors, e.g. growth substances, turgor pressure, ion currents, the plasma membrane, the nucleus and cytoskeletal elements may be involved in inducing polarity and controlling tip growth. Polarity itself is based on the interrelationship of these factors (for review see Sievers and Schneft, 1981).

During the last years we investigated the roles of calcium in tip growth. Generally, ionic currents traverse tip growing plant cells and are involved in cell morphogenesis. Calcium ions are one of the components of these currents (review see Weisenseel and Kicherer, 1981).

All tip growing cells revealed a tip-to-base calcium gradient after application of the fluorescent antibiotic chlorotetracycline (CTC; Reiss and Herth, 1978, 1979b). Because CTC fluoresces remarkably only as a membrane bound calcium complex one could suspect that the CTC fluorescence indicates the actual membrane distribution. This is not correct. Changes of CTC fluorescence do not correspond to shifts of membrane distributions during pollen tube germination. The density of membranes in the pollen tube tip scarcely differs from that in more basal regions. Drastical disturbance of the tip region of moss protonemata by heat-shock or UV-light do not influence the CTC fluorescence (Reiss et al., 1985b; Schmitt, 1982; Fig. 1).

With another technique, the analysis by proton-induced X-ray emission (PIXE), it was possible to confirm the existence of the calcium gradient. PIXE is in principle similar to the commonly used electron microprobes. It has a better elemental resolution limit, also in thick samples and allows to detect free ions, if a corresponding preparation technique is available (Bosch et al., 1980). PIXE analyses of pollen tubes and moss protonema cells revealed a similar calcium distribution as visualized by CTC fluorescence (Reiss et al., 1983, 1984a). Chemically fixed and shock frozen pollen tubes

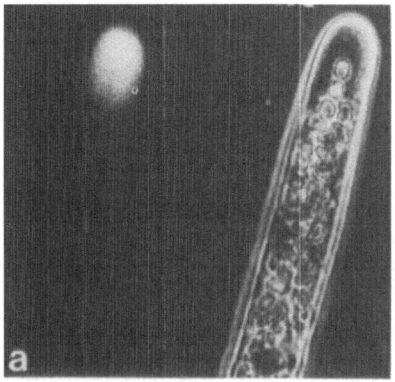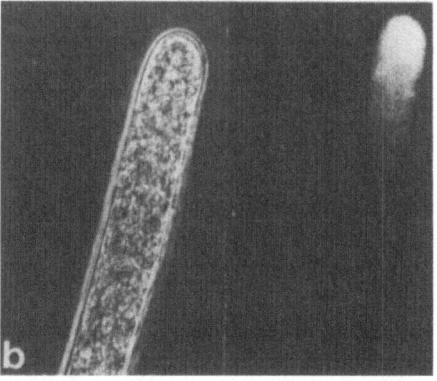

Fig. 1. Phase contrast and CTC (10^{-4}M) fluorescence of a moss protonema
tip cell after few seconds (a) and 2 min (b) of UV-irradiation.

differ only in the shape of the gradient which is somewhat steeper in the
chemically fixed tubes (Reiss et al., 1983). Recently it was possible to
display two-dimensional elemental maps by PIXE (Reiss et al., 1985a;
Fig. 2).

The concentration of detectable intracellular calcium is drastically
enhanced after 10 min treatment with CTC. Obviously CTC fixes additionally
free calcium to membranes (Reiss et al., 1983). Therefore, CTC fluorescence
may reflect the actual in vivo calcium distribution only then, when the
photograph was taken directly after CTC application.

To clarify whether the calcium gradient is the cause or the consequence
of an organelle gradient, we applied various inhibitors which might specifically or unspecifically disturb the calcium distribution, mostly combined
with the CTC and/or PIXE techniques. Some inhibitors stop tip growth, but do
not influence the CTC fluorescence, as cytochalasin B (Fig. 3). Obviously
the CTC gradient is not absolutely coupled with the tip growth rate.

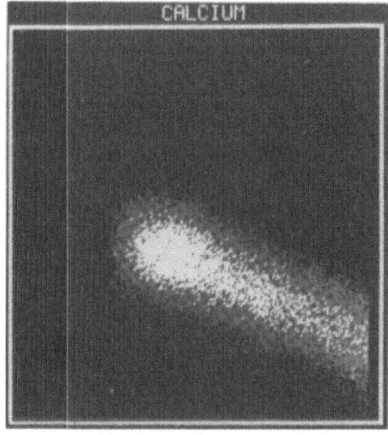

Fig. 2. Two-dimensional calcium distribution in a chemically fixed pollen
tube tip, detected with the Oxford proton microprobe.

Fig. 3. CTC fluorescence of a pollen tube, treated for 30 min with cytochalasin B (0.1 µg/ml).

Ionophores which are not specific for calcium, stop tip growth and cause other effects, e.g. Golgi swelling, but they do not destroy the polar arrangement of cell organelles (Reiss, 1980; Reiss and Herth, 1980). Drugs which disturb the calcium distribution within the cell, influence in contrast also the intracellular organelle arrangement. For instance, the calcium ionophore A-23187 equilibrates the gradient, as measured also by PIXE. Finally, the CTC fluorescence in pollen tubes is lost (Reiss and Herth, 1978; Reiss et al., 1983). Increasing with time also the polar distribution of cell organelles is affected. Golgi vesicles discharge their contents at the tube flanks (Reiss and Herth, 1979a).

After longer CTC treatment the pollen tubes grow irregularly and the CTC fluorescence becomes irregularly distributed along the whole pollen tube (Fig. 4).

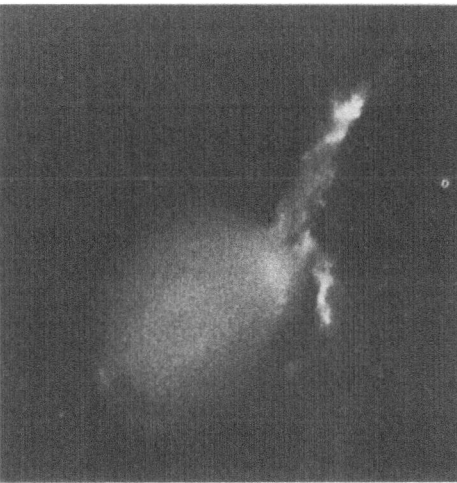

Fig. 4. CTC fluorescence of a pollen tube grown in a CTC (10^{-4}M) containing medium.

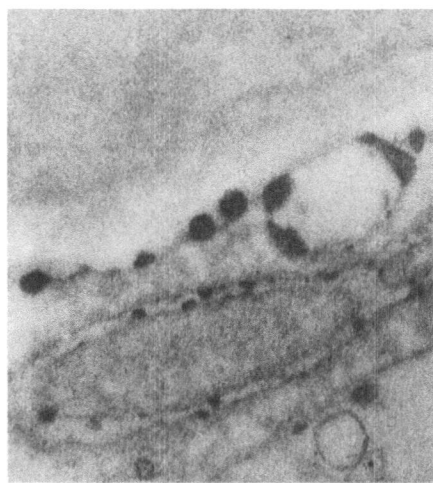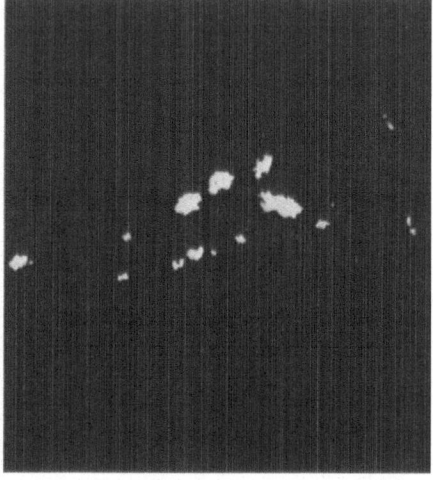

Fig. 5. Electron-opaque deposits in the pollen tube, yielded by the
Belitser technique and analysed with ESI for calcium.

Ultrastructural studies revealed that vesicle accumulation and extrusion occurs at many distinct areas along the whole tube. In consequence the
cell wall is irregularly thickened (Reiss and Herth, 1982).

Both experiments indicate that the equilibration of the CTC gradient
and the artificial binding of extensive high amounts of calcium during long
CTC treatment lead to the loss of the organelle gradient and the development
of new gradients on "wrong" places respectively. The calcium gradient seems
to be somehow translated into polar organelle distribution, perhaps by an
electrophoretical force as discussed by electrophysiologists (see Weisenseel
and Kicherer, 1981) or via the actin system (Brawley and Robinson, 1985).

The CTC as well as the PIXE technique is not suitable to detect calcium-
binding sites within the cell. The reliability of the commonly used precipitation techniques is disputed (Schnepf et al., 1982). We applied the technique
of Belitser (Belitser et al., 1982). The cells were fixed in the presence
of extensive calcium, and analysed light and electron microscopically to
detect the possible calcium-binding sites. It could be shown with electron
spectroscopic imaging (ESI), a newly developed method of Zeiss (Egle et al.,
1984), that the electron-opaque deposits, yielded by the Belitser technique
are rich in calcium (Fig. 5). They are mainly located on the plasma membrane
at the ER and at mitochondria, exactly the organelles which have been suggested to be involved in regulation of the free calcium concentration in
the cytosol. However, it was not possible to distinguish between the natural
calcium-binding sites and such membranes on which calcium was artificially
bound by the technique used. In any case, a gradient in calcium-binding
sites could not be detected (Herth et al., 1985).

Also in tip growing plant cells free calcium ions might control specific enzyme activities via calmodulin. With the fluorescing calmodulin-
binding phenothiazines it could be shown that the calmodulin distribution
does generally not coincide with the calcium distribution (Hausser et al.,
1984). But in some cases, as in the moss protonema cell shown in Figure 6,
a polar distribution of calmodulin was detected. Further experiments with
immunofluorescence methods are necessary to verify or falsify this spatial
relationship.

Fig. 6. Phenothiazine (10^{-6}M) fluorescence in a moss protonema tip cell.

Cell polarization and the establishment of calcium gradients in tip growing cells seem to be self-enhancing processes. From various experiments it is highly probable that external calcium ions enter the growing tip region via polarly distributed calcium channels in the plama membrane (for discussion see Weisenseel and Kicherer, 1981; Brawley and Robinson, 1985). To test whether calcium channel blockers have any influence on the calcium gradient in pollen tubes, we used the organic agent, nifedipine, and in comparison cobalt ions, as a possible anorganic blocking agent. Nifedipine causes disturbances and loss of the CTC gradient and other structural effects. Germinating pollen grains no longer form a tube. Instead they protrude a large, monstrous outgrowth along the whole colpus (Fig. 7).

Growing tubes form multiple tips or branch (Reiss and Herth, 1985). Similar effects are obtained by cobalt ions. PIXE analyses demonstrated that cobalt is similarly but not completely identically distributed as calcium (Reiss and Traxel, 1985).

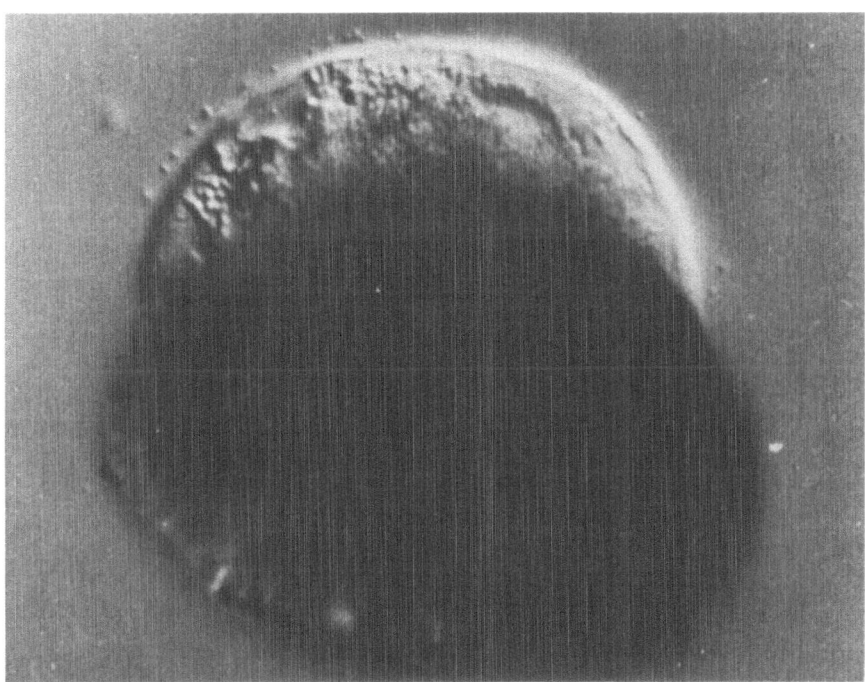

Fig. 7. Germinating pollen grain in the presence of 10^{-5}M nifedipine.

REFERENCES

Belitser, N.V., Zaalishvili, G.V. and Sytnianskaja, N.P., 1982, Ca^{2+}-binding sites and Ca^{2+}-ATPase activity in barley root hairs, Protoplasma, 111:63.

Bosch, F., El Goresy, A., Herth, W., Martin, B., Nobiling, R., Povh, B. Reiss, H.-D. and Traxel, K., 1980, The Heidelberg proton microprobe, Nucl. Sci. Appl., 1:33.

Brawley, S.H. and Robinson, K.R., 1985, Cytochalasin treatment disrupts the endogenous current associated with cell polarization in fucoid zygotes: studies of the role of F-actin in embryogenesis, J. Cell Biol., 100:1173.

Egle, W., Kurz, D. and Rilk, A., 1984, The EM 902, a new analytical TEM for ESI and EELS, Magazine for Electron Microscopists, 3:4.

Hausser, I., Herth, W. and Reiss, H.-D., 1984, Calmodulin in tip growing plant cells, visualized by fluorescing phenothiazines, Planta, 162:33.

Herth, W., Reiss, H.-D., Hertler, B., Bauer, R., Traxel, K. and Ender, C., 1985, Localization of potential Ca^{2+}-binding sites in lily pollen tubes and maize calyptra cells: transmission electron microscopy, proton microprobe (PIXE) analysis and electron spectroscopic imaging (ESI), J. Ultrastruct. Res., submitted.

Reiss, H.-D., 1980, Calcium Gradienten und Spitzenwachstum bei Pollenschläuchen von Lilium longiflorum, Doctoral Thesis, Universität Heidelberg.

Reiss, H.-D. and Herth, W., 1978, Visualization of the Ca^{2+}-gradient in growing pollen tubes of Lilium longiflorum with chlorotetracycline fluorescence, Protoplasma, 97:373.

Reiss, H.-D. and Herth, W., 1979a, Calcium ionophore A-23187 affects localized wall secretion in the tip region of pollen tubes of Lilium longiflorum, Planta, 145:225.

Reiss, H.-D. and Herth, W., 1979b, Calcium gradients in tip growing plant cells visualized by chlorotetracycline fluorescence, Planta, 146:615.

Reiss, H.-D. and Herth, W., 1980, Effects of the broad-range ionophore X-537A on pollen tubes of Lilium longiflorum, Planta, 147:295.

Reiss, H.-D. and Herth, W., 1982, Disoriented growth of pollen tubes of Lilium longiflorum Thunb. induced by prolonged treatment with the calcium-chelating antibiotic, chlorotetracycline, Planta, 156:218.

Reiss, H.-D. and Herth, W., 1985, Nifedipine-sensitive calcium channels are involved in polar growth of lily pollen tubes, J. Cell Sci., (in press).

Reiss, H.-D. and Traxel, K., 1985, Hint of polar distribution of calcium channels under PIXE analysis, Biol. Trace Element Res., (in press).

Reiss, H.-D., Herth, W., Schnepf, E. and Nobiling, R., 1983, The tip-to-base calcium gradient in pollen tubes of Lilium longiflorum measured by proton-induced X-ray emission (PIXE), Protoplasma, 115:153.

Reiss, H.-D., Nobiling, R. and Ender, C., 1984a, Tip-to-base calcium gradients in tip growing plant cells measured by PIXE, Nucl. Instr. Meth., B3:660.

Reiss, H.-D., Schnepf, E. and Herth, W., 1984b, The plasma membrane of the Funaria caulonema tip cell: morphology and distribution of particle rosettes and the kinetics of cellulose synthesis, Planta, 160:428.

Reiss, H.-D., Grime, G.W., Li, M.Q., Takacs, J. and Watt, F., 1985a, Distribution of elements in the lily pollen tube tip, determined with the Oxford scanning proton microprobe, Protoplasma, 126:147.

Reiss, H.-D., Herth, W. and Nobiling, R., 1985b, Development of membrane- and calcium-gradients during pollen germination of Lilium longiflorum, Planta, 163:84.

Schmitt, U., 1982, Licht-und elektronenmikroskopische Untersuchungen über die cytologischen Auswirkungen von Hitzeschocks bei Poterioochromonas malhamensis und Funaria caulonemen, Doctoral Thesis, Universität Heidelberg.

Schnepf, E., Hausmann, K. and Herth, W., 1982, The osmium tetroxide-
potassium ferrocyanide (OsFeCN) staining technique for electron
microscopy: a critical evaluation using ciliates, algae, mosses
and higher plants, Histochemistry, 76:261.

Sievers, A. and Schnepf, E., 1981, Morphogenesis and polarity of tubular
cells with tip growth, in: "Cell Biology Monographs", vol.8, p.265,
Springer, Wien New York.

Weisenseel, M.H. and Kicherer, R., 1981, Ionic currents as control mechanism in cytomorphogenesis, in: "Cell Biology Monographs", vol.8,
p.379, Springer, Wien New York.

TRANSDUCTION OF THE GRAVITY STIMULUS IN CRESS ROOTS:

A POSSIBLE ROLE OF AN ER-LOCALIZED Ca^{2+} PUMP

A. Sievers

Botanisches Institut der Universität

Venusbergweg 22, D - 5300 Bonn 1, FRG

INTRODUCTION

The high sensitivity of plants against a gravistimulus can be deduced from the lowermost limits found for the gravireaction, e. g. the presentation time of the cress root is only 12 s, and the perception time of some plants is only 0,5 s. A deviation from the plumb line of only 0,5° is perceived by the plant as a stimulus. If sedimentable amyloplasts function as statoliths their displacement in the cell cannot play the major role in graviperception. One extreme example may visualize that: during the perception time amyloplasts with an average sedimentation rate of 1 µm/min are displaced by only about 0.3% of their diameter. Evidently the natural substratum for the amyloplasts of a vertically growing organ should be investigated for having a role in the transduction of the physical stimulus into a physiological signal. Roots are especially suitable for the investigation of this signal transduction because here the site of perception of the stimulus and the site of the response to the stimulus are spatially separated and situated in different tissues. The cells specialized for graviperception are in the center of the calyptra and the responding root movement has its origin in the elongation zone.

In the preceding years several reviews about graviperception in plants have been written (Audus, 1979; Volkmann and Sievers, 1979; Sievers and Hensel, 1982; Hensel and Sievers, 1983; Sievers, 1984). Therefore, only the most recent papers on this subject will be cited below.

THE DISTAL ENDOPLASMIC RETICULUM: A CLEAR INDICATION OF THE POLARITY OF STATOCYTES

The typical radially symmetrical structure of the calyptra and the typical structural polarity of the statocytes was the underlying reason for the model of graviperception postulated by Sievers and Volkmann (1972). Any deviation from the gravity vector results in differential contacts between the amyloplasts and the distal ER complexes. This is especially obvious after tilting the root into the horizontal position.

In the prospective statocytes of the embryonic radicle this structural polarity does not yet exist. It develops during seed germination and is fully expressed prior to any first gravitropic bending. Since even roots grown on a horizontally rotating axis of the clinostat develop the structural polarity of their statocytes one can draw the conclusion that it is endogenous.

Recent investigations provide first insights into the mechanism of the development of this structural polarity (Hensel, 1986). If cytochalasin B is applied to young roots they synthesize more ER in the proximal cell pole than in the distal one. In older roots the distal ER complex is found as in control roots but in addition to this ER is also found proximally near the nucleus. If roots are kept permanently in the drug the typical structural polarity is completely missing: the nucleus is situtated in the distal and the ER almost exclusively in the proximal cell pole. These effects of cytochalasin B on statocyte development allow the conclusion that microfilaments play an essential role in directing the transport of ER cisternae and thereby in the expression of the statocyte polarity.

The stability of this polarity can be influenced by a number of very different exogenous factors. Many such experiments show that polarity is a precondition for the functioning of graviperception.

The distal position of the ER complex is stabilized by microtubuli which form a network between ER cisternae and the plasmalemma. After application of colchicin and subsequent inversion of the root the distal ER complex moves down to the center of the cell (Hensel, 1984). This finding could be a basis to explain an often reported observation, namely, that the ER complexes in the statocytes of many plants have been found not to be localized in the distal cell poles: could it be that this results from an artifact due to destruction of microtubules during the fixation process for electron microscopy?

Weak electric fields exogenously applied to roots turn the endogenous cell polarity into the direction of the electric field vector. The ER cisternae are moved towards the direction of the cathode and the nuclei to that of the anode (Behrens et al., 1985a). Since this movement is reversible one can draw the conclusion that the structural polarity of the statocyte is a consequence of the natural endogenous electric properties of the roots. Furthermore, statocytes show their typical gravielectric response (a change in membrane potential) only if the structural polarity is intact (Behrens et al., 1985b); if this polarity is disturbed they exhibit a constant resting potential after gravistimulation.

When roots are pregrown under normal culture conditions and then become rotated on the horizontal clinostat (20 h at 2 rpm) this treatment leads to destruction of the polarity due to autolysis in the statocytes. The steady change of the gravitational vector apparently is sensed as a too strong overstimulation. The roots respond to this loss of statocytes by shortening the time of the mitotic cycles in the meristem.

After centrifugation of the roots the cell organelles of the statocytes become stratified according to their density. Within a few minutes the original polarity is restituted. The ER complex moves back from the cell flanks to the distal cell pole so that the amyloplasts sediment onto it again. This movement of the ER complex back to its previous position also is caused by microfilaments: after application of cytochalasin B the physical stratification is no longer restituted (Wendt and Sievers, 1985).

It is especially autolysis and stratification of the statocytes which influence the gravitropic bending of the roots. About 20% of the roots with damaged statocytes do not longer react gravitropically. This phenomenon very clearly demonstrates the role of statocytes as stimulus-perceiving cells, especially, since autolysis is restricted to statocytes alone and does not occur in other root cells. Centrifuged roots with stratified statocytes, however, show the same kinetics of bending as the controls but the latency time is extended by several minutes. This is exactly the time which is needed to restitute after stratification the original polarity in the distal part of the cell. If the restitution of polarity is prevented by the application of cytochalasin B such roots do not bend gravitropically in their normal way. Thus, we obtained clearcut correlations of intact structural polarity and graviperception.

PROPERTIES OF THE ENDOPLASMIC RETICULUM IN THE ROOT TIP

Since, according to the perception model of Sievers and Volkmann (1972), the ER complex functions as gravisensitive structure, its molecular features should be investigated. Whether the ER complex is involved in protein synthesis is not known although attached ribosomes indicate that this is the case. Protein synthesis, however, as a relatively time consuming process should be of lesser importance for the rapid steps of graviperception.

Membranes of the ER complex exhibit lectin-binding sites (concanavalin A) as shown in situ after glutaraldehyde fixation of roots and confirmed in vitro on membrane-fractions in our laboratory. The use of non-ionic detergents in the in-situ labeling experiments indicates that the concanavalin A receptors are glycoproteins of the ER membranes. - With the freeze-etch technique it was shown that another distinct feature of the ER in statocytes is that the number of intramembraneous particles is smaller than that of ER in cells of the dermatocalyptrogen and of the outer secretion layer (D. Volkmann, unpublished).

By means of centrifugation intact root tips (calyptra and approx. half of the apical meristem) can be separated from the root bases. After homogenization and sucrose density-gradient centrifugation, membrane-fractions of each of the two batches were analyzed further. The cytochrome c reductase of root tip ER is characterized by a shoulder at 1.16 g/cm^3 (in low Mg^{2+} gradients) which is lacking in basal root ER.

Whether this is a characteristic feature of statocyte ER could not be determined as yet, because statocyte ER makes up only 3.2% of the root tip ER which makes further biochemical analysis difficult.

A well-known function of ER in animal cells is to store Ca^{2+} ions with a calcium pump. Since the regulatory role of Ca^{2+} ions in many cellular processes in animals and plants is well established it was an obvious objective to show the presence of such a Ca^{2+} pump in ER membranes of roots. In the statocytes very fast alterations of contacts between amyloplasts and ER probably are primary events in graviperception which could exert a regulatory influence on such a presumably important function as a Ca^{2+} pump. The isolation of a highly purified fraction of RER (80% as shown by morphometry) has been demonstrated (Buckhout, 1984). Ca^{2+} transport is strictly ATP-dependent with a narrow pH optimum at pH 7.5. The K_m for Ca^{2+} is 2.5 μM and for ATP 73 μM which indicates physiological relevance. This Ca^{2+} transport is not inhibited by CCCP or by mitochondrial Ca^{2+} transport inhibitors (ruthenium red, La Cl_3, oligomycin, azide) and contributions by tonoplast vesicles is ruled out by morphometry which showed only 12% contamination by unidentified smooth membranes. Interestingly, Ca^{2+} transport is inhibited by phenothiazine inhibitors (chlorpromazine, fluphenazine, trifluoperazine) but neither exogenously added nor endogenous calmodulin showed any influence on the in vitro activity. Possibly, these drugs can react unspecifically in vitro with the ER membrane. Knowing these in vitro properties of the Ca^{2+} pump in ER should be extremely helpful in relating these to regulatory functions of such a pump in the ER in the primary steps of graviperception.

CONCLUDING REMARKS

We do not yet know whether the Ca^{2+} pump found in the ER of the root can be regulated in its activity by a change in the contacts (pressure?) between ER and sedimenting amyloplasts. Perhaps the application of drugs and ionophores which specifically influence ion gradients will be helpful in the future. The importance of changes of ion fluxes across biomembranes of root statocytes has been demonstrated by various electrophysiological methods (Behrens et al, 1985b). These changes represent the most rapid events in graviperception which have been found so far (see Lühring et al., this Volume).

ACKNOWLEDGEMENT. I thank the Deutsche Forschungsgemeinschaft for the generous support of our experimental work.

REFERENCES

Audus, L. J., 1979, Plant geosensors, J. Exper. Bot. 30: 1051
Behrens, H. M., Baumann, M., and Sievers, A., 1985a, The polar organization of cress root statocytes is altered by the application of weak external electrical fields, Europ. J. Cell Biol. 36, Suppl. 7: 7

Behrens, H. M., Gradmann, D., and Sievers, A., 1985b, Membrane-
    potential responses following gravistimulation in roots
    of Lepidium sativum L., Planta 163: 463
Buckhout, T. J., 1984, Characterization of Ca^{2+} transport in
    purified endoplasmic reticulum membrane vesicles from
    Lepidium sativum L. roots, Plant Physiol. 76: 962
Hensel, W., 1984, A role of microtubules in the polarity of
    statocytes from roots of Lepidium sativum L., Planta
    162: 404
Hensel, W., 1986, Cytochalasin B affects the structural polarity
    of statocytes from cress roots (Lepidium sativum L.),
    Protoplasma: in press
Hensel, W., and Sievers, A., 1983, Graviperception in plant
    cells, The Physiologist 26, Suppl.: S-60
Sievers, A., 1984, "Sinneswahrnehmung bei Pflanzen: Graviper-
    zeption," Westdeutscher Verlag, Opladen
Sievers, A., Hensel, W., 1982, The nature of graviperception,
    in: "Plant Growth Substances 1982," P. F. Wareing, ed.,
    Academic Press, London New York
Sievers, A., Volkmann, D., 1972, Verursacht differentieller
    Druck der Amyloplasten auf ein komplexes Endomembransystem
    die Geoperzeption in Wurzeln? Planta 102: 160
Volkmann, D., Sievers, A., 1979, Graviperception in multicel-
    lular organs, in: "Encyclopedia of Plant Physiology. N. S.,
    vol. 7: Physiology of movements," W. Haupt, and Feinleib,
    M. E., eds., Springer, Berlin Heidelberg New York
Wendt, M., Sievers, A., 1985, Restitution of polarity in stato-
    cytes from roots of Lepidium sativum L. after centri-
    centrifugation, Europ. J. Cell Biol. 36, Suppl. 7: 72

ELECTRICAL EVENTS IN GROWING LEPIDIUM ROOT TIPS

SEEM TO BE CORRELATED WITH GRAVITROPIC DYNAMICS

H. Lühring, H. M. Behrens and A. Sievers

Botanisches Institut der Universität

Venusbergweg 22, D - 5300 Bonn 1, FRG

INTRODUCTION

In the root of a higher plant, the mechanism to perceive gravity, and transduce it into a physiological signal, is located in the statocytes of the calyptra. A prerequisite for the induction of the graviresponse of the Lepidium root is the integrity of structural polarity of the statocytes (ref. see Sievers, this Vol.). Recently, electrophysiological experiments have been carried out to approach the molecular mechanism of graviperception and transduction (Behrens et al., 1985b, Behrens et al., 1982, Behrens and Gradmann, 1985). To discuss electrical phenomena at the plasma membrane of root cells, the contribution of different ions to electrogenesis was investigated. Association of these results with data yielded from gravitropic reaction kinetics (Behrens et al., 1985c), and structural changes after external electrical field application (Behrens et al., 1985a) may hold for the idea that transduction of a gravistimulus primarily consists of electrical events, which even might be the signal transmitted from the statenchyma to target cells which trigger graviresponse.

MATERIAL AND METHODS

Seeds of Lepidium sativum were soaked for 30 to 60 min in distilled or tap water. They were germinated vertically in a moist chamber at 23° in the dark for 20 to 24 h, thereafter, single seedlings transferred to the experimental setup. Except the measurements of membrane voltage during gravistimulation, the roots were kept in artificial pond water of known composition. It contained (in mM): 0.01 KCl, 0.99 NaCl, or 1 KCl, zero NaCl, respectively, 0.1 $CaCl_2$, 5 HEPES, pH adjusted with Tris to the desired value, or a sodium phosphate buffer. For measurements of time-dependent curvature of the root, the media contained KCl and NaCl, 10 mM in all, 0.1 mM $CaCl_2$, 5 mM HEPES. External currents were measured in medium containing (in mM): 0.1 KCl, 1 NaCl, 0.1 $CaCl_2$, 1 Tris. Four methods have been applied, which are briefly outlined in the following, details are given in the references cited.

Vibrating probe (Behrens et al., 1982).

Natural electric current surrounding the root was measured by a vibrating metal-filled micropipet. The local voltage difference (ΔV) over the solution's resistivity ($\rho \cdot d$), where d is the amplitude of vibration, gives the local current $I = \Delta V / \rho \cdot d$. The electrode was stepwise positioned along the the root surface with a micromanipulator.

Recording of membrane voltage and input resistance

A platform which could be tilted into a 45° postion carried the root in a small moist chamber and three micromanipulators. Conventional glass-microelectrodes were inserted from both sides into the root cells, and intracellular potential recorded versus a reference electrode potential in the apoplast (Fig. 1A). Rectangular current pulses were supplied by a function generator, fed into a cell via the voltage recording electrode and measured via the reference electrode by a current voltage transducer. For experiments dealing with the influence of different ions on membrane voltage and input resistances of root cells, a cuvette was used which was permanently perfused with bathing medium (Fig. 1B). Current was applied through one barrel of a double-barreled microelectrode. Current voltage curves were obtained by application of low frequency triangle- shaped current.

External electrical field (Behrens et al., 1985b).

Weak external electrical fields (1 V/cm, 9 μA/mm^2) were applied for about 30 min in transversal and longitudinal direction to the root axis via a conductive bathing medium. Structural changes in the statocytes were investigated by electron microscopy.

K$^+$ dependent curvature (Behrens et al., 1985c).

Roots were preincubated with droplets of control solution at their tip (in mM:1 KCl, 9 NaCl, 0.1 CaCl$_2$, 5 HEPES); at the beginning of the experiment, the drop was exchanged to one of certain composition, after 30 min it was removed, and the assembly containing 10 seedlings as a whole tilted in a 90° position. Photographs were taken every 10 min, the growth increment of lower and upper flank morphometrically determined.

RESULTS

Natural Electric Currents

The surface of the Lepidium root was scanned with a vibrating microelectrode, and external currents could be detected, which formed a symmetric pattern around the tissue, entering at the tip and leaving it at the basal elongation zone (Fig.2A). After gravistimulation, current left the root cap at the upper flank, but with an increased density entered at the lower flank (Fig. 2B). Bromocresol purple containing agar is stained by an embedded root, according to the above mentioned currents, thus pointing at H$^+$ as a major current component.

Membrane Voltage (V_m) and Input Resistance

Fig. 3 shows the result of a simultaneous recording of membrane voltage immediately after tilting the root into a 45° position. The electrical response of the statocytes started only several seconds after stimulation, with a depolarization of the lower flank cells, and a hyperpolarization of upper flank statocytes. By insertion of two microelectrodes into cells of opposite flanks, or in some distance at the same flank, and applying current pulses through one of the electrodes, symplasmic electrical coupling was evaluated in terms of the length constant, i.e. the distance where the signal is reduced to 1/e of its original amplitude. The ratio of the length constant in longitudinal to that in transversal direction was calculated to be 3 : 1, i.e. intercellular communication is predominant along the root axis. The membrane resistance was estimated to be about o.2 $\Omega \cdot m^2$. To evaluate the contribution of different ions to membrane voltage, roots were measured in a permanently perfused bathing medium. The resting membrane voltage was about -200 mV, and the cells reacting very sensitive to external changes in K^+ concentration (Fig. 4A). If the control medium contained only 10 µM KCl (low K^+), there was no effect of external pH variation on V_m, even in the presence of 2 µM CCCP which should act as a protonphore (Fig. 4B). Otherwise, if 1 mM K^+ is provided in the control medium, V_m is strongly depolarized by lowering the external pH (Fig. 5A). DNP in low K^+ medium hyperpolarized the membrane (Fig. 5B), but depolarized it in high K^+ medium (Fig. 5A). Fig. 6 shows current voltage curves of root cells under control conditions, and after application of DCCD and W-7, respectively. DCCD depolarizes the cells and increases the input resistance. W-7, a calmodulin antagonist, also depolarizes and increases resistance after about 5 min of application.

Time-dependent Curvature

The kinetics of graviresponse dependent on external K^+ concentration was investigated. Fig. 7 displays the time-dependent length difference between lower and upper flank upon gravistimulation, after preincubation in control medium and high K^+ concentration. Comparative measurements with mannitol did not reveal any inhibition, but W-7 had a strong inhibitory effect.

External Field Application

An external electrical field of 1 V/cm and 9 µA/mm² current density displaced the apical ER in the statocytes to the cathodal side of the cell, the nucleus moved in direction of the anode. This behaviour could be verified for electrical fields transversal to the root axis and in parallel, as well. After stopping the treatment, both, the ER and the nucleus moved into their original position (Fig. 8).

DISCUSSION

The fact that after gravistimulation electrical events appear a considerable time befor curvature occurs, shows that they are no side effects of this complex reaction but might be of importance in some early stage of the signal chain. Actually, changes in external current patterns could be detected 30 s after stimulation (not earlier for technical reasons). The consequence - or cause - was recorded in the same range: Deflections of membrane voltage in the perceiving statenchyma within 10 s. Since from previous studies it was obvious that the particular structural polarity of the stato-cytes plays a distinct role in graviperception, the attention is turned to the distal ER complex. Biochemical investigation showed that the ER is a Ca^{2+} store, the gradient across the ER membrane maintained by an active ATP-consuming transport mechanism (Sievers et al., 1984). Can Ca^{2+} be the mediator of transduction? To have a basis, the influence of different ions on the electrogenesis of the root cells must be examined. Evidently, the plasma membrane is highly permeable for K^+, since external K^+ largely affects membrane voltage. On the other hand, one would expect this for H^+, too, because of the external currents measured with the vibrating probe. Indeed, there is a significant effect of external pH changes on V_m, provided, the K^+ concentration in the medium is high enough to set the K^+ equilibrium potential more positive than V_m at rest; otherwise, H^+ fluxes may easily be compensated by K^+ fluxes. There exists an active transport mechanism in the plasma membrane, becoming visible in current voltage curves after application of DCCD. There is another component which is neither congruent with a passive transport system, nor it is with the DCCD-inhibited H^+ pump. This mechanism significantly contributes to electrogenesis and is inhibited by W-7. W-7 is known as a potent calmodulin antagonist (Hidaka and Tanaka, 1983). If W-7 does penetrate the membrane without disturbing effect on the established transport units, there will be good reason to assume large scale regulation of membrane characteristics by Ca^{2+}-calmodulin complex.

Results from graviresponse kinetics suggest that K^+ plays an important role in the transduction-transmission chain. High K^+ concentrations impede the kinetic, and the morphometrical results show that in contrast to control experiments the growth rate of the upper flank is depressed, whereas the lower flank will not particularly respond, its rate is slowed as in control. Valinomycin, a K^+ ionophore, has no influence on the kinetics of curvature. Obviously, K^+ conductivity is very high in any case, second, the cell must be in a hyperpolarized state for an optimal response, and third, that apoplastic K^+ concentration is rather low, since otherwise occuring H^+ influx will depolarize the plasma membrane. This could be true for the upper flank of the stimulated root. At the lower flank one has to expect an opposite behaviour. As it cannot be the same mechanism, there might be involved an endogeneous process. The displacement of amyloplasts could play a role in the behaviour of the lower flank cells. From electrical field applications we have concluded that an intracellular electrical axis exists, aligned with the structural polarity. The endogeneous electrical field might be altered by amyloplast sedimentation, i. e. they would function as dipoles which are displaced in a homogeneous electrical field. The effect on transport characteristics of the plasma membrane, or Ca^{2+} release from ER cisternae might be the reason for the depolarization as the primary event. Future work will comprise experiments employing ion-sensitive microelectrodes to detect ion fluxes across the plasma membrane, and the patch clamp technique to characterize the activity and regulation of distinct carriers.

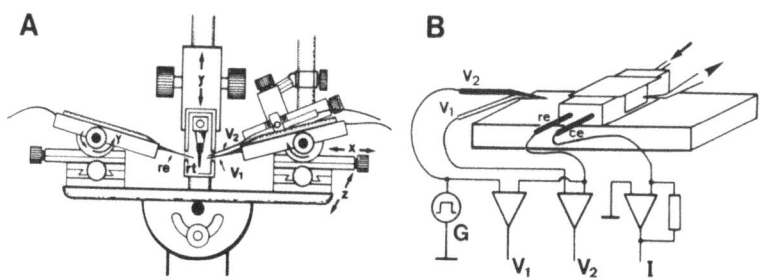

Fig. 1. Experimental arrangements for electrophysiological measurements. A: Tiltable platform with the root (rt) placed in a moist chamber, the reference electrode (re) joined the apoplast, two microelectrodes (V_1, V_2) could be inserted into the tissue. Electrodes could be adjusted in x-, y-, and z-direction. B: Membrane voltage measurement under permanent perfusion of test solution with double-barreled electrodes (V_1, V_2), reference (re) and current electrode (ce). Arrow and arrowhead indicate medium supply and withdrawal, resp. Current was supplied by a function generator (G), and measured by a current voltage transducer (I).

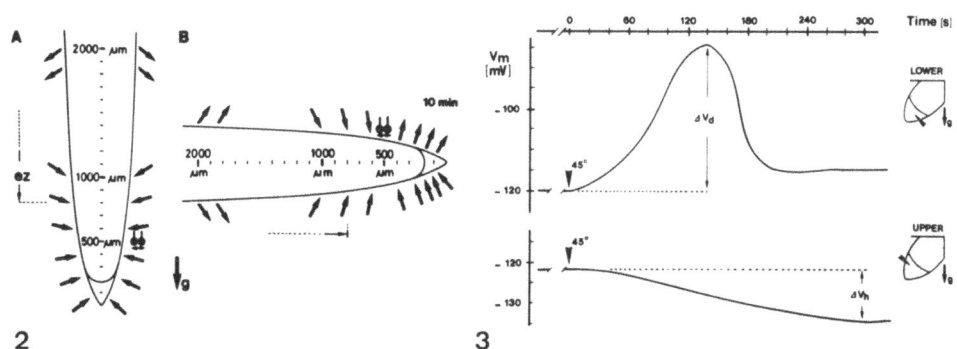

Fig. 2. External current measured with the vibrating probe (⚫⚫). Arrows show direction of cation flux, the gravity vector is displayed. A: Current surrounding the vertically growing root. B: Current pattern 10 min after positioning the root horizontally.

Fig. 3. Membrane voltage changes in statocytes of lower (top) and upper (bottom) flank after tilting the platform into 45° position. Insets give the resp. electrode locations.

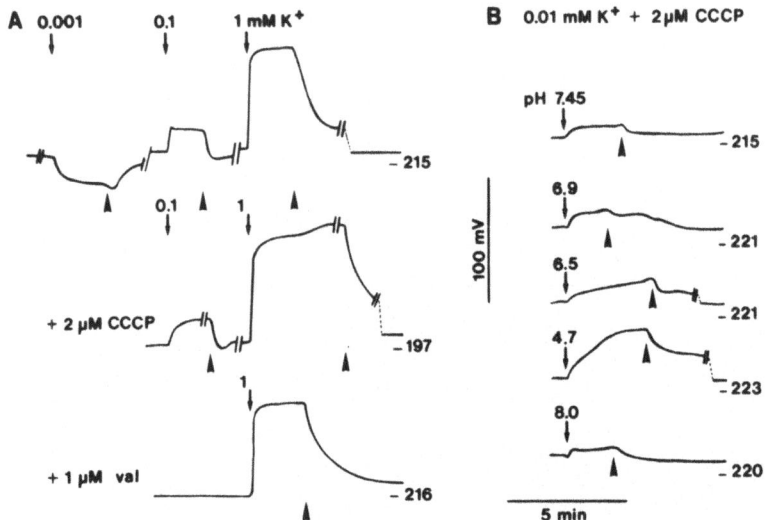

Fig. 4. A: Effect of external K$^+$ on V_m of subepi-
dermal cells of the elongation zone.
B: Effect of changes in external pH in
the presence of CCCP. - Resting V_m is
denoted right-hand side of each trace,
upward arrowheads indicate wash-out
with control medium (in mM: 0.01 KCl, 0.99
NaCl, 0.1 CaCl$_2$, 2 Na$_2$H/NaH$_2$PO$_4$, pH 7.45).

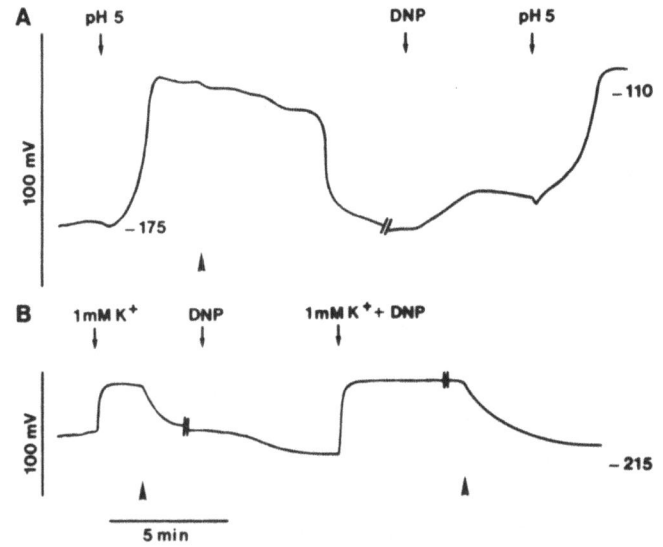

Fig. 5. Effect of DNP, K$^+$, and pH on V_m of high
(A) and low (B) K$^+$ preincubated roots.
Arrowheads indicate wash-out with control
medium, resting V_m is denoted at each
trace. Control solution in mM: A: 1 KCl,
2 NaCl, 0.1 CaCl$_2$, 5 HEPES, pH 7.
B: 0.01 KCl, 0.99 NaCl, 0.1 CaCl$_2$, 2 Na$_2$H/
NaH$_2$PO$_4$, pH 7.45.

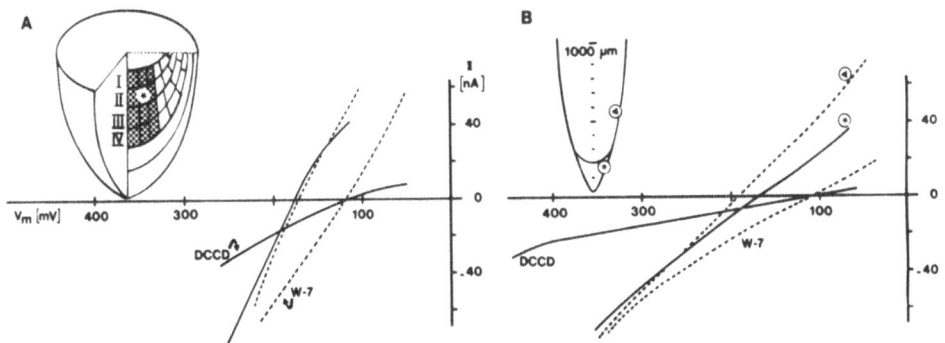

Fig. 6. Current voltage curves of root cells in control medium
and after application of DCCD or W-7. A: Statocytes
of 2nd story. B: Cells of the peripheric calyptra
and meristematic zone. Control solution in mM: 1 KCl
2 NaCl, 0.1 CaCl$_2$, 5 HEPES, pH 7.5.

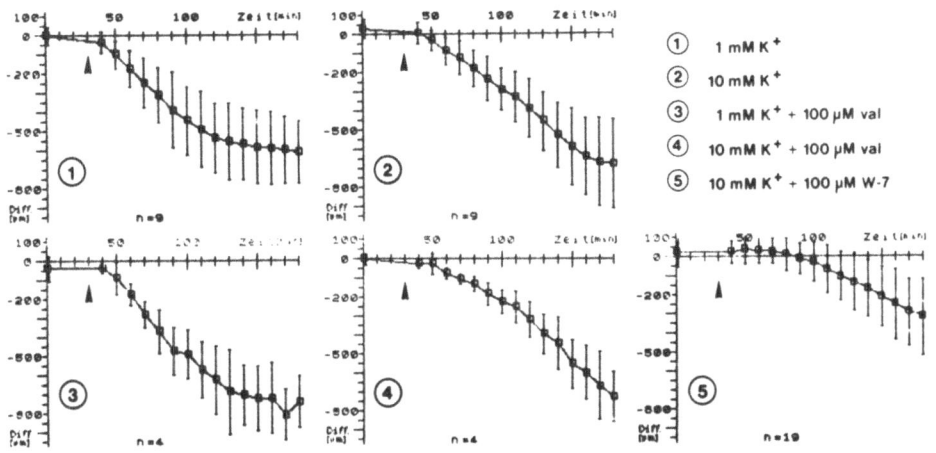

① 1 mM K$^+$

② 10 mM K$^+$

③ 1 mM K$^+$ + 100 μM val

④ 10 mM K$^+$ + 100 μM val

⑤ 10 mM K$^+$ + 100 μM W-7

Fig. 7. Differences of length between lower and upper flank
of gravistimulated roots, dependent on external K$^+$
and on W-7, resp. Arrowheads indicate beginning of
gravistimulation. Control medium in mM: 1 KCl, 9 NaCl,
0.1 CaCl$_2$, 5 HEPES, pH 7.5.

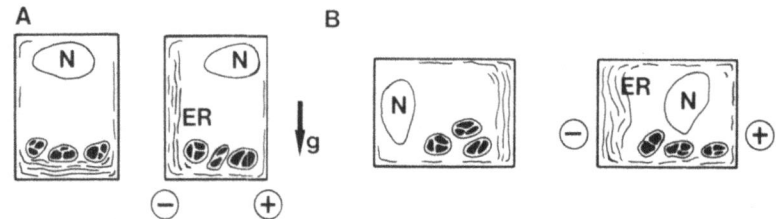

Fig. 8. Displacement of endoplasmic reticulum (ER) and nucleus
(N) in statocytes after 30 min application of an
external DC-field (1V/cm, 9 µA/mm^2). A: Transversal
electrical field across a vertically growing root tip.
B: Electrical field along the root axis. - Polarity of
the electrical field is shown by ⊕ and ⊖ .Bathing
medium in mM: 3.1 KCl, 1 NaCl, 0.1 CaCl$_2$, 1 MES.

ACKNOWLEDGEMENT

For contribution of graviresponse kinetics and electrical
field application results, we thank U. Kehren and M. Baumann.
This work was supported by the Deutsche Forschungsgemeinschaft.

REFERENCES

Behrens, H.M. and Gradmann, D., 1985, Electrical properties
of the vertically growing root tip of Lepidium sativum L.
Planta, 163:453.
Behrens, H.M., Weisenseel, M.H. and Sievers, A., 1982, Rapid
changes in the pattern of electric current around the
root tip of Lepidium sativum L. followin gravistimulation, Plant Physiol., 70:1079.
Behrens, H.M., Baumann, M. and Sievers, A., 1985a, The polar
organisation of cress root statocytes is altered by the
application of weak external electrical fields, Eur.
J. Cell Biol., Suppl. 7, 36: 7.
Behrens, H.M., Gradmann, D. and Sievers, A., 1985b, Membrane-
potential responses following gravistimulation in roots
of Lepidium sativum L., Planta, 163:463.
Behrens, H.M., Kehren, U. and Sievers, A., 1985c, Effect
of ionophores on growth and on the gravitropic reaction
of cress roots (Lepidium sativum L.), Eur. J. Cell Biol.,
Suppl. 7, 36:7.
Hidaka, H. and Tanaka, T., 1983, Naphthalene sulfonamides
as calmodulin antagonists, Meth. Enzymol., 102 G:185.
Sievers, A., Behrens, H.M., Buckhout, T.M. and Gradmann, D.,
1984, Can a Ca^{2+} pump in the endoplasmic reticulum of
the Lepidium root be the trigger for rapid changes in
membrane potential after gravistimulation?, Z. Pflanzenphysiol., 114:195.

LOCALIZATION OF CALCIUM IN THE SENSORY CELLS OF THE DIONAEA TRIGGER HAIR BY LASER MICRO-MASS ANALYSIS (LAMMA)

B. Buchen and W.H. Schröder[*]

Botanisches Institut der Universität, Venusbergweg 22
D-5300 Bonn 1, FRG
[*]Institut für Neurobiologie, KFA Jülich, D-5170 Jülich, FRG

INTRODUCTION

In <u>Dionaea</u>, mechanical bending of the trigger hair induces action potentials which spread over the trap lobes to the motor cells (review Bentrup 1979). The perception of the stimulus and its transformation into a physiological signal occurs in a ring of specialized epidermal cells in the indentation zone of the trigger hair. These sensory cells (Haberlandt 1906) are characterized by a highly evolved ER complex at the apical and the basal cell pole. The ER surrounds several vacuoles containing polyphenols (Buchen et al. 1983). In order to study the function of these cell structures in sensory transduction, we examined the development of the trigger hair (Casser et al. 1985). During its development, a change in the membrane potential could be measured for the first time when the structural polarity of the sensory cell was established. Yet the short action potentials which are necessary for trap closure were fired only if the typical ER complex in the cell poles was visible. Since membrane potential changes are mediated by ions, we tried to identify and to localize ions possibly involved in these processes. Here we present the first results.

MATERIAL AND METHODS

<u>Dionaea muscipula</u> Ellis ex L. was cultured in a green-house under natural light conditions. All experiments were done after mechanical stimulation of the trigger hairs. Conventional fixation and embedding procedures were described by Buchen et al. (1983).

Cytochemistry. The potassium-pyroantimonate technique for the precipitation of cations was carried out after Walz (1979). Prefixation in 2% glutaraldehyde was followed by postfixation in 2.5% potassium-pyroantimonate in 1% OsO_4 in 0.05 M phosphate buffer (pH 7.4; 2 h; room temperature). Serial sections incubated in 5mM EGTA (ethylene glycol bis(ß-amino-ethyl-ether)-N,N'-tetraacetic acid) for 60 min at 60°C to remove Ca^{2+}-pyroantimonate were compared to control sections (Walz 1979).

Freeze substitution. A small piece of the leaf with a trigger hair was frozen by rapid immersion in liquified propane, transferred into a vial containing 1% OsO_4 in dry acetone (solid at -194°C). The sub-

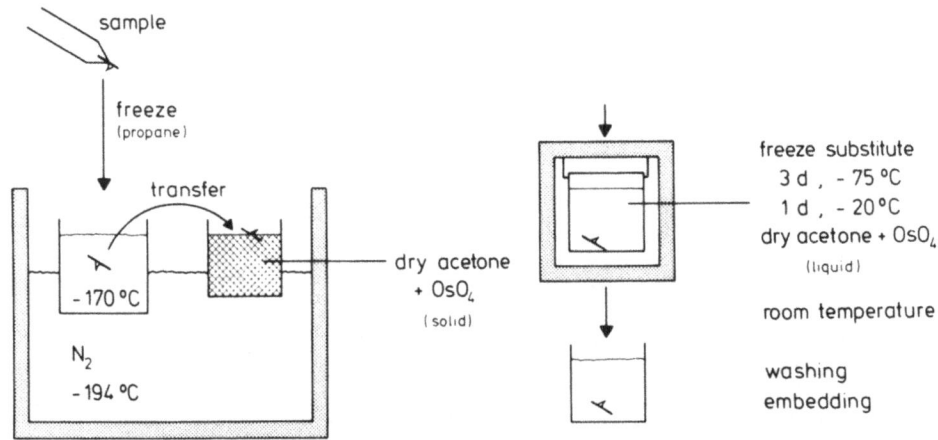

Fig. 1. Cryofixation and freeze-substitution procedure

stitution occurred at −75°C for 3 d, and at −25°C for 1 d (Fig. 1). After washing in dry acetone, the specimens were infiltrated and embedded at room temperature as usually. Sections of 0.5 µm in thickness were dry cut on a Reichert OM U_2 ultramicrotome (Reichert, Wien, Austria) and transferred to a copper grid without support film.

X-ray and laser micro-mass analysis. Only freeze-substituted and dry-cut specimens were used. Energy-dispersive X-ray analysis (EDAX) was done with a Philipps EM 400 electron microscope combined with an EDAX 9100 (Phillips, Eindhoven, The Netherlands) using a voltage of 120 kV, a spot size of about 1 µm in diameter, and a time of 100 s for analysis. The LAMMA method with a LAMMA 500 (Leybold-Heraeus, Cologne, FRG) was performed as described in Schröder and Fain (1984; detailed in Fain and Schröder 1985).

RESULTS

(i) After postfixation in the mixture of OsO_4 and potassium-pyroantimonate, sensory cells of Dionaea showed electron-dense precipitates in the vacuoles, the ER cisternae, the mitochondria and at the plasma membrane. The staining was selective, since for example lipid droplets showed no precipitates. Yet the staining intensity varied from cell to cell . Outside the plasma membrane of the podium cells, the apoplastic space showed dark precipitates. If sections were incubated in EGTA, the dark precipitates were no longer detectable or, as in the vacuoles, the contrast was reduced. However, the use of this precipitation method does not preclude ion redistributions during fixation.

(ii) The main problem for ion localization was to preserve the ultrastructure of the cells without using an aequeous fixative that might cause uncontrollable ion movements. Unexspectedly, the freeze-substituted cells of the trigger hair showed a good structural preservation (Fig. 2), although a high number of samples could not be used because of damage by ice crystals, by poor infiltration and embedding. Well preserved sensory cells, however, showed the polar arrangement of ER cisternae and vacuoles, and the nucleus and mitochondria in the cell center. Often, the plasma membrane was slightly detached from the walls of the sensory cells. The membrane systems and compartments were intact and could easily be identified (Fig. 3). Besides ER cisternae in concentric arrangement,

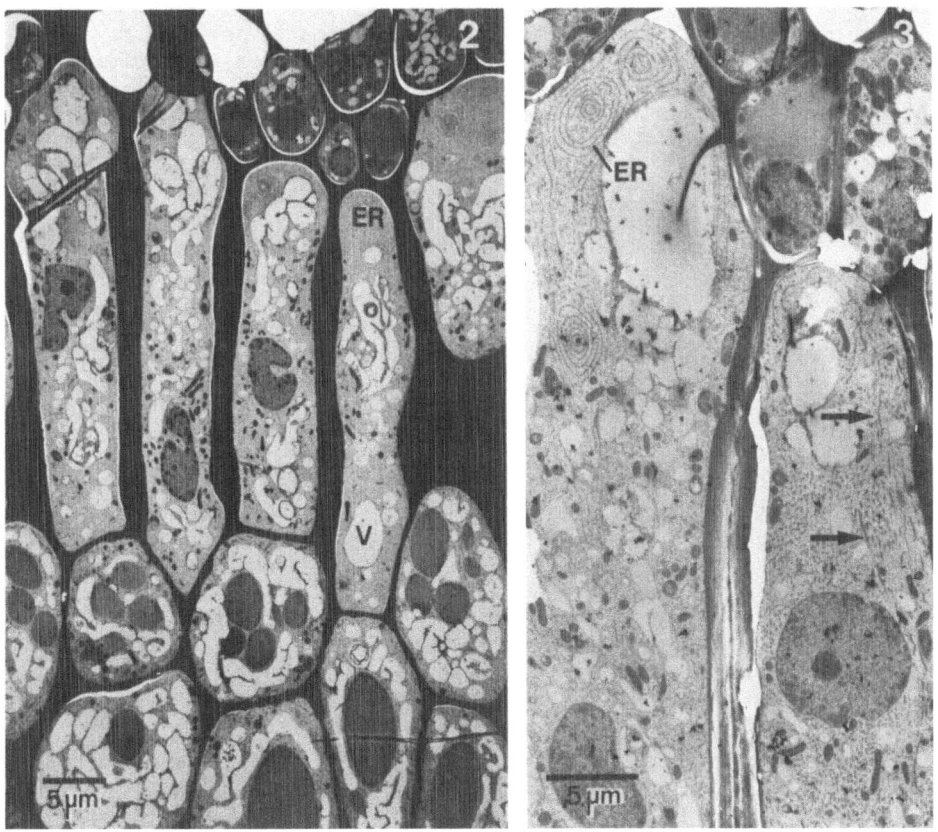

Fig. 2. Freeze-substituted <u>Dionaea</u> sensory cells showing the polar arrangement of ER and vacuoles (V). The cell structures are well preserved.

Fig. 3. Apical part of freeze-substituted sensory cells. Besides ER cisternae (ER), thicker fibrillar elements traverse the cell (arrows).

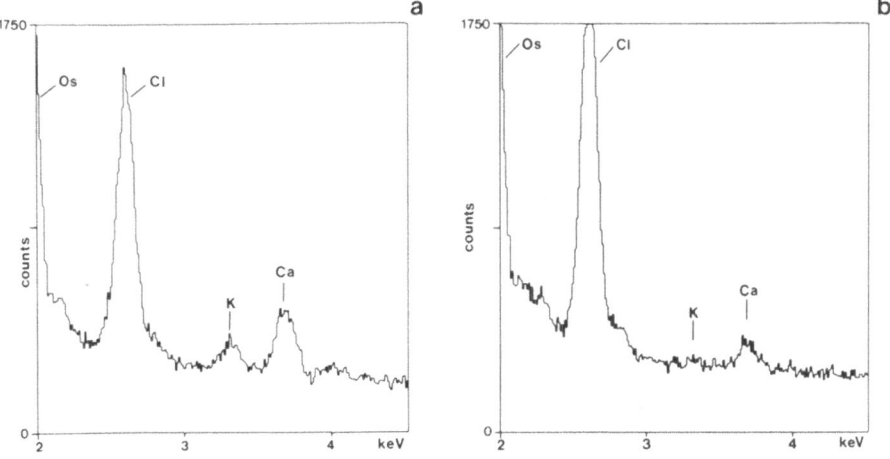

Figs. 4a, b. EDAX spectra of the cell wall (a) and cytoplasm (b) of a freeze-substituted sensory cell.

a unique structural component was found in the sensory cells (Fig. 3,➡).
Long elements of medium electron density traverse the cells, preferentially in the direction of the long axis. Near the nucleus or at the
cell poles, they are curved. The diameter of these elements is 120 nm
and more. A distinct membrane surrounding the elements could not be dis-
cerned, but fine fibrillar structures could be seen within them. The fine
structure and identification of these elements is under investigation;
possibly a highly developed filament system necessary for the bending
and re-bending of the trigger hair has been discovered.

(iii) Using dry sections of a freeze-substituted trigger hair,
we compared the X-ray energy spectra from different areas of the tissue.
Control spectra from pure embedding medium showed a high amount of
Cl, but no K or Ca. In the cell wall between two sensory cells, the high
peaks for K (K_α = 3.31 KeV) and Ca (K_α = 3.69 KeV) were well above the
background (Fig. 4a). In the cytoplasm of the sensory cell (below the
nucleus), a high amount of K and Ca was measured, too (Fig. 4b). Comparing
both energy spectra, we found less K and Ca within the sensory cell,
but relative more Ca than K. The absolute values for both cations
were not determined. With the procedure used we can only draw conclusions
on the distribution of Ca, as loss and redistribution of monovalent
ions (e.g. K) are not to be excluded.

(iv) In the LAMMA probe, an area of 1 - 5 μm in diamter of a 0.5 μm
thick section was evaporated and ionised by a high-power laser. The ions
were analysed in a time of flight mass spectrometer. The measurements are
based on the different flight times of ions with different m/e values.

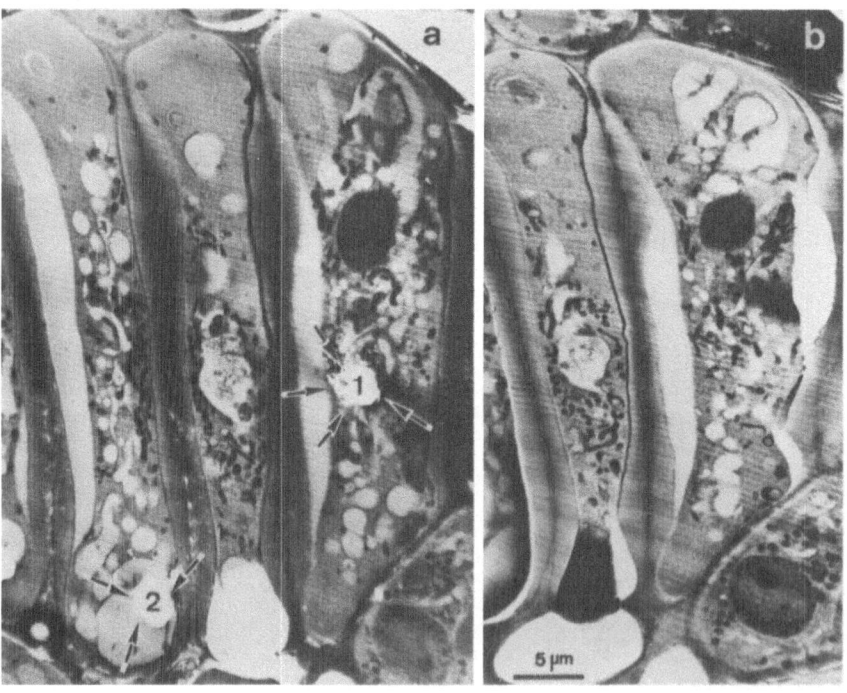

Figs. 5a, b. Serial sections (0.5 μm thick) of freeze-substituted sensory cells. a: Two holes (arrows, 1 and 2) produced by LAMMA.
b: The structural details of the vaporized area 1.

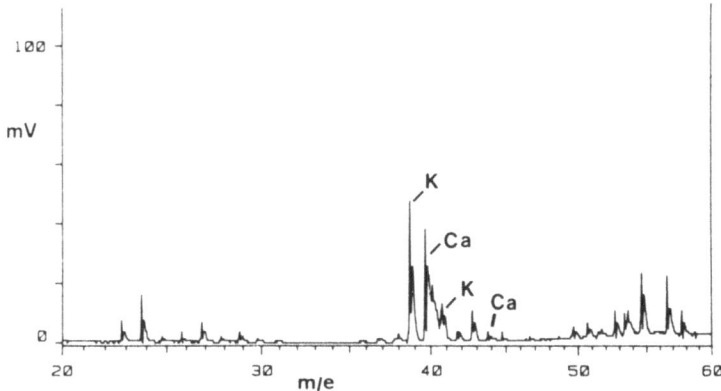

Fig. 5c. LAMMA spectrum of the area 1 marked in Fig. 5a
by arrows. Note the prominent ^{40}Ca peak (area integral;
rel. %: 57.4 for ^{40}Ca compared to 35.3 for ^{39}K; Table 1)

Different regions within the trigger hair and the trap lobe were
investigated with the LAMMA probe. It was possible, to direct the
laser beam very well under light microscopical observation and to
control and measure the analysed area later in the electron microscope.
Limitations for the method were the difficulties to produce sections
(dry-cut) of similar relative thickness with the sensory cells in
one plane. As vacuoles and ER cisternae were in close contact, normally
a hole shot by the laser beam covered more than one compartment. Thus,
the spectra obtained represent data on relatively large regions.

Fig. 5a shows holes produced by the laser beam in the sensory
cells (arrows). For comparison, the structural details of the analysed
area beneath the nucleus are given in a serial section (Fig. 5b).
The mass spectrum of this area showed high peaks for Mg, K and Ca
(Fig. 5c). The Ca peak is higher than that one of K. The same result
was obtained for other samples and at different sites within the sensory
cells. For example, the area 3 in the apical part mainly included
cytoplasm and small portion of vacuoles. The corresponding spectra showed
high peaks for Ca, K, and Mg; Ca being the highest one.

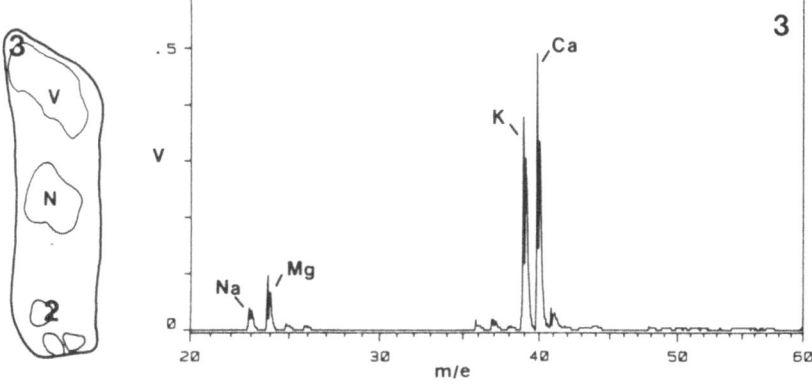

Fig. 6. LAMMA spectrum of area 3 marked in the schema of
the sensory cell.

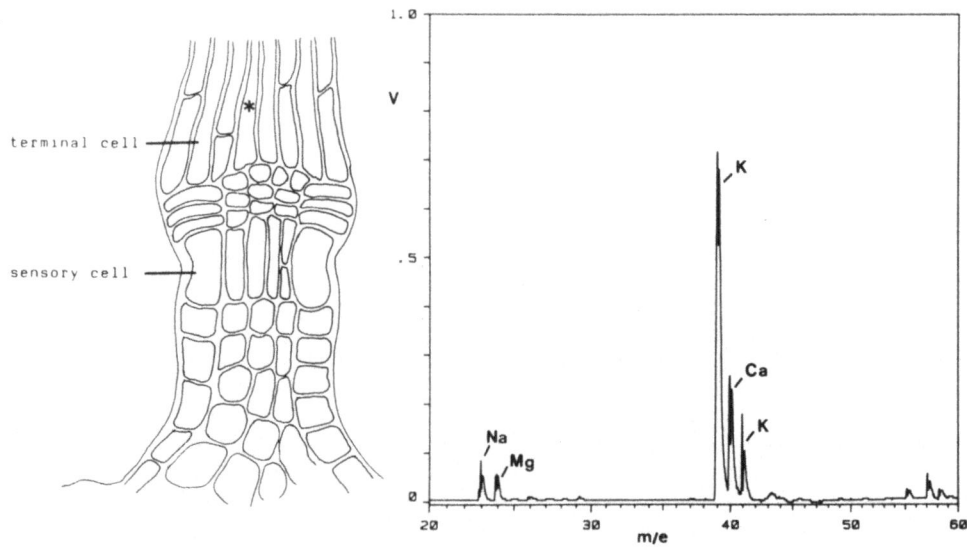

Fig. 7. LAMMA spectrum of the vacuole in a cell of the terminal part of the trigger hair (✻). Note the high peak for ^{39}K

In contrast, for the vacuoles of the terminal cells of the trigger hair (Fig. 7) as well as for the vacuoles of the epidermal and hypodermal cells of the trap lobes (Fig. 8) we obtained spectra with a predominant K peak. Thus, comparing the intensity of the element signals at the different sites (Table 1) we found a higher amount of Ca as compared to K in the sensory cells. In the cells of the terminal part of the trigger hair or in the trap lobe epidermis, however, the K peaks are higher than the Ca ones as would be expected for plant cells. The absolute Ca concentration can be estimated as ranging in mM in the sensory cells. This estimated value is based on the comparison with data obtained with standards (Schröder and Fain 1984; Fain and Schröder 1985).

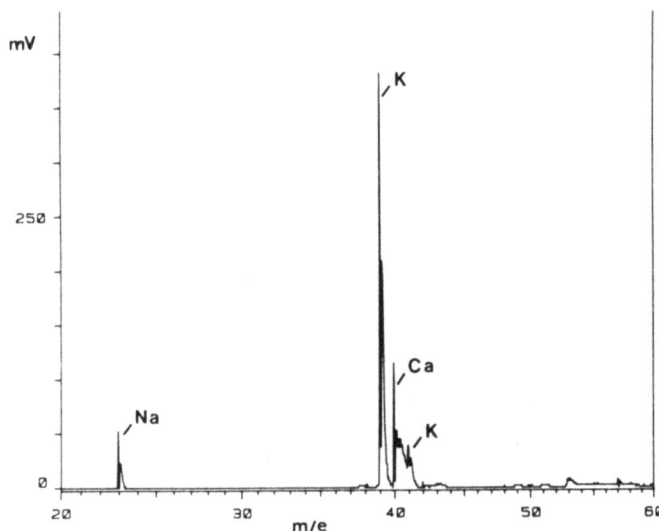

Fig. 8. LAMMA spectrum of a vacuole of a hypodermal cell (trap lobe)

Table 1. Relative units and % of cations from different areas of
freeze-substituted Dionaea cells measured by LAMMA
(taken from individual spectra)

cell (area)	^{23}Na	^{24}Mg	^{39}K	^{40}Ca
cell wall	40	105	1700	2085
	1.0%	2.7%	43.3%	53.1%
epidermal leaf cell (vacuole)	26	0	379	150
	4.7%	0.0%	68.3%	27.0%
terminal cell (vacuole)	187	144	2260	813
	5.5%	4.2%	66.4%	23.9%
sensory cell (basal vacuole)	305	630	908	1157
	10.2%	21.0%	30.3%	38.6%
sensory cell (apical vacuole)	451	541	1069	1212
	13.8%	16.5%	32.7%	37.0%
sensory cell (apical cytoplasm)	97	191	907	978
	4.5%	8.8%	41.7%	45.0%
sensory cell (cytoplasm, center)	3	7	48	78
	2.2%	5.2%	35.3%	57.4%

DISCUSSION

In Dionaea, the membrane resting potential is potassium-mediated
and the action potential induced by bending of the trigger hairs seems
to be calcium-mediated (Jacobson 1974). For characterization and localization of ions on the electron microscopic level, many ion trapping
methods have been employed, especially in animal tissue. In plant cells,
the use of these methods is hampered by cuticules and cell walls as
diffusion barriers and by large vacuoles. Furthermore, most of the precipitation methods are not specific for a certain ion and need carefully
be controlled by other methods (critical evalution of K-pyroantimonate
for example Simson and Spencer 1975). Hence, we take the precipitates by
K-pyroantimonate and their solution by EGTA in the ER and the vacuoles
of the Dionaea sensory cells only as a hint for the presence of high
concentrations of calcium and other cations, but not as a proof.

One precondition for ion localization is to avoid any aqueous prefixation that causes redistribution. A method of choice is shock-freezing
followed by freeze-substitution (e.g. Läuchli et al. 1970; Harvey et al.
1976; Fain and Schröder 1985). In spite of a thick cuticle and thick
cell walls, the ultrastructure of sensory cells of Dionaea was well
preserved by freeze-substitution (Figs. 2,3, 5). Since no prefixation
or cryoprotection substances were applied prior to freezing, this good
structural preservation leads us to the conclusion that high intracellular ion concentrations might be in part responsible for this fact by
serving as natural cryoprotectives.

The results of the energy-dispersive X-ray analysis as well as the
LAMMA measurements showed high peaks for Ca and K in the sensory cells
(Figs. 4, 6). An internal control is provided by the vacuoles of the
leaf epidermis or of the terminal cells which are characterized by
relatively higher levels of K and lower levels of Ca (Figs. 7, 8).

We estimate the Ca concentration in the sensory cells to be in the range of mM. This range is higher than compared to the neighbouring cells of the trigger hair and the trap lobes. We assume that the calcium is compartmentalized in the sensory cells. An analysis of the cation content of a single compartment in the sensory cell (ER or vacuole), however, was as yet not successfull. The area of a section vaporized by the laser beam normally included not only vacuoles, but also ER and some other organelles. The proportions of the compartments in the measured areas varied. Therefore, our next aim is to study the cation content in individual compartments of the Dionaea sensory cells.

Acknowledgements. The encouragement and helpful discussions during this study by Prof. Dr. A. Sievers, Botanical Institute, University of Bonn, are gratefully acknowledged. The cooperation with M. Naib-Majani, Institute of Cytology, University of Bonn, in performance of freeze-substitution methods deserves thanks. H.-J. Ensikat and A. Klein are thanked for careful technical assistance. Supported by SFB 160.

REFERENCES

Bentrup, F.W., 1979, Reception and transduction of electrical and mechanical stimuli, in: "Encyclopedia of plant physiology, N.S., vol.7: Physiology of movements," W. Haupt, and Feinleib, M.E., eds., Springer, Berlin Heidelberg New York

Buchen, B., Hensel, D., and Sievers, A., 1983, Polarity in mechanoreceptor cells of trigger hairs of Dionaea muscipula Ellis, Planta 158: 458

Casser, M., Hodick, D., Buchen, B., and Sievers, A., 1985, Correlation of excitability and bipolar arrangement of endoplasmic reticulum during the development of sensory cells in trigger hairs of Dionaea muscipula Ellis, Europ. J. Cell Biol. 36: Suppl. 7, 12

Fain, G.L., and Schröder, W.H., 1985, Ca content and Ca exchange in dark-adapted toad rods, J. Physiol., in press

Haberlandt, G., 1906, "Sinnesorgane im Pflanzenreich zur Perzeption mechanischer Reize," 2nd edn. Engelmann, Leipzig

Harvey, D.M.R., Hall, J.L., and Flowers, T.J., 1976, The use of freeze-substitution in the preparation of plant tissue for ion localisation studies, J. Microsc. 107: 189

Jacobson, S.L., 1974, Effect of ionic environment on the response of the sensory hair of Venus's-flytrap, Can. J. Bot. 52: 1293

Läuchli, A., Spurr, A.R., and Wittkopp, R.W., 1970, Electron probe analysis of freeze-substituted, epoxy resin embedded tissue for ion transport studies in plants, Planta 95: 341

Schröder, W.H., and Fain, G.L., 1984, Light-dependent calcium release from photoreceptors measured by laser micro-mass analysis, Nature 309: 268

Simson, J.A.V., and Spicer, S.S., 1975, Selective subcellular localization of cations with variants of the potassium pyroantimonate technique, J. Histochem. Cytochem. 23: 575

Walz, B., 1979, Subcellular calcium localization and ATP-dependent Ca^{2+}-uptake by smooth endoplasmic reticulum in an invertebrate photoreceptor cell. An ultrastructural, cytochemical and X-ray microanalytical study, Europ. J. Cell Biol. 20: 83

LONG DISTANCE TRANSPORT OF CALCIUM

John A. Raven

Department of Biological Sciences
University of Dundee
Dundee, DD1 4HN
Scotland

SUMMARY

The low free Ca^{2+} concentration in the cytosol restricts the symplastic flux of Ca^{2+} over long distances by either diffusive or mass flow processes. Restriction on Ca^{2+} supply to Ca^{2+}-requiring parts of plants due to the restricted mobility of Ca^{2+} in the symplast is best exemplified by the shoots of terrestrial vascular plants. Here an adequate supply of Ca^{2+} to the shoot as a whole is provided by apoplastic mass flow of the xylem solution. However, most of the xylem water passes to mature photosynthetic organs (the main sites of transpiration), while most of the Ca^{2+} is needed in growing regions with low transpiration rates. Ca^{2+} movement to growing regions probably involves a mixture of (axial) mass flow in the phloem and the xylem, the latter aided by radial and tangential Ca^{2+} fluxes, <u>via</u> parenchyma cells, which transfer Ca^{2+} from xylem streams destined for mature photosynthetic regions to those passing to growing regions. Ca^{2+} transport processes which do not involve Ca^{2+} mass flow in xylem or phloem are unlikely to be quantitatively significant in mediating <u>axial</u> Ca^{2+} fluxes over more than a few mm.

INTRODUCTION

Calcium is needed within, and in the immediate environment of, all plant cells (Hanson, 1984). However, not all cells in a multicellular plant may have access to environmental Ca, a circumstance which is most obvious for aerial portions of terrestrial plants. Such restricted access to external Ca means that long-distance (in excess of a few mm) transport of Ca is required. This paper sets out to quantify the long-distance transport processes in plants, and the limitations which they impose on Ca availability to different plant parts. Recent reviews of long-distance Ca transport include Clarkson (1984), Ferguson (1979), Hanger (1979), Hanson (1984) and Raven (1977).

POTENTIAL PATHWAYS AND MECHANISMS OF LONG-DISTANCE Ca TRANSPORT

Spatially, we distinguish movement of Ca in the apoplast (outside the plasmalemma, but within the confines of the plant body) from that in the symplast (inside the plasmalemma: strictly the cytosol of individual cells and of the plasmodesmata connecting them); see chapter 9 of Raven (1984a).

Mechanistically, Ca movement in aqueous apoplastic or symplastic phases can be by diffusion through static solutions or by mass flow of solution. It is important to note that diffusion is, by its nature, decremental: solute flux is caused by a solute concentration difference so that, for a given concentration difference ($C_{source} - C_{sink}$, where concentrations are in mol m^{-3}) and solute diffusion coefficient (D, m^2s^{-1}), the flux J(m^2s^{-1}) is inversely proportional to the pathlength l(m):

$$J = \frac{D(C_{source} - C_{sink})}{l} \qquad (1)$$

By contrast, mass flow is not intrinsically decremental with respect to solute concentration as a function of movement from source to sink:

$$J = \bar{v} \, C \qquad (2)$$

where J is the solute flux (mol $m^{-2}s^{-1}$), $\bar{v}$ is the mean velocity of solution (m s^{-1}) and C is the concentration of the moving solution (mol m^{-3}).

It is important to note that these equations apply to steady state fluxes with no exchange of the moving species (chemical Ca or tracer Ca) with other compartments en route. Much confusion has been generated by miss-interpretation of transients after addition of ^{45}Ca to Ca-transporting systems (see analysis by Safford and Bassingthwaite, 1977).

QUANTITATION OF Ca FLUXES BY VARIOUS PATHWAYS AND MECHANISMS

Objectives

Our aim here is to set approximate upper limits on fluxes by various mechanisms in the multitude of Ca transport pathways and mechanisms. In each case the flux will be initially expressed as mol Ca (m^2 transverse area of transport pathway)$^{-1}s^{-1}$, and then scaled to give mol Ca (m^2 transverse area of plant axis)$^{-1}s^{-1}$ using the data in Table 2 of Raven and Rubery (1982). This provides a basis for comparison with Ca transport requirements (next Section).

Diffusion in the apoplast: Cell Walls

With a mean cell wall fixed negative charge of 500 mol m^{-3} (Demarty, Morvan and Thellier, 1984) the upper limit of Ca^{2+} in this phase is 250 mol Ca^{2+} m^{-3}. For the extreme case of C_{source} = 250 mol Ca^{2+} m^{-3} and C_{sink} = 0 mol Ca^{2+} m^{-3} (Nakajami et al, 1981), with l = 10 mm and $D_{Ca^{2+}}$ = $5.10^{-11}m^2s^{-1}$ (based on the self-diffusion coefficient for Sr^{2+} in cell walls at 2°C, scaled to 25°C assuming a Q_{10} of 1.2: Briggs, Hope and Robertson, 1961), equation (1) gives $J_{Ca^{2+}}$ of 1.25 μmol (m^2 transverse area of cell wall)$^{-1}s^{-1}$ or (cell walls = 0.05 of axis area) 62.5 nmol (m^2 transverse area of plant axis)$^{-1}s^{-1}$. This Ca^{2+} flux is in exchange for some other cation. The computed value is clearly an upper limit as can be seen from the negligible apoplastic backflux in polar 'cell to cell' transport systems in higher plants where there are no obvious constraints (such as casparian strips) to such an apoplastic flux (Vaughan, Evans and Hutchin, 1967; Raven, 1979).

Diffusion in the Apoplast: Xylem Conduits

For the extreme case of a non-moving xylem solution, and assuming a C source of 10 mmol Ca m^{-3} (e.g. Ferguson, 1980), a C_{sink} of zero (complete removal of Ca and its accompanying anion), $D_{Ca^{2+}} = 8.10^{-10}$ m^2s^{-1} (Nye and Tinker, 1977, Table 4.2) and, as in (6) $l = 10$ mm. These values give (equation (1)) a $J_{Ca^{2+}}$ of 0.8 μmol $m^{-2}s^{-1}$ on the basis of xylem area (0.05 of axis area) or 40 nmol $m^{-2}s^{-1}$ on the basis of axis area.

We note that the upper limit of total Ca concentration in the xylem solution is likely to be constrained by the (pH –dependent) solubility of calcium phosphate and the (pH –independent) solubility of calcium sulphate, as well as by Ca competition for chelators which also bind trace metals (White et al, 1981). Accordingly, the regulation of pH (Clarkson, Williams and Hauson, 1984) and chelator concentration (White et al, 1981) in the xylem is likely to be crucial in permitting the simultaneous transport (by mass flow or diffusion) of Ca, P, S and trace metals.

Mass flow in the apoplast: xylem conduits

Assuming 10 mol Ca per m^3 of xylem solution independent of $\bar{v}$ (in practice the value of C in equation (2) falls with increasing $\bar{v}$) and a high value for $\bar{v}$ in transpiration (0.8 m s^{-1}: Passioura, 1972) we compute an upper limit on Ca flux of 8 mol Ca $m^{-2}s^{-1}$ (xylem area basis) or 0.4 mol Ca (m^2 axis area)$^{-1}s^{-1}$. A more 'typical' transpirational $\bar{v}$ value (17 mm s^{-1}; Passioura, 1972) yields a Ca flux of 8.5 mmol Ca (m^2 axis area)$^{-1}s^{-1}$, while a root pressure Ca flux of 70 $\mu mol/m^2$ axis area)$^{-1}s^{-1}$ can be deduced from the data of Pitman and Welfare (1978) assuming a mean root length of 10 mm. It is important to emphasise that these computations assume no net flux of Ca between the xylem solution and the surrounding tissues; even if such net fluxes occur, leading to an increment or decrement in Ca concentration with distance along the xylem, there are no grounds for regarding the lateral fluxes as essential to the movement of Ca with the rest of the xylem solution. 'Exchange Transport' of Ca (Bell and Biddulph, 1963; Hanger, 1979; van de Geijn and Smuelders, 1981; Clarkson, 1984) in the xylem is not a quantitatively verified mechanism of transport. Evidence which seems to favour this mechanism involves a missinterpretation of the relationship between net and tracer Ca 'fluxes' (see Safford and Bassingthwaite, 1977), or a dismissal of the possibility of varying Ca concentration in xylem sap (lower concentrations at higher water flow rates) in interpretting data in which Ca transfer to the shoot is not proportional to the rate of transpiration.

Diffusion through the symplast

The eukaryotic cytosol has ~ 0.1 mmol m^{-3} of free Ca^{2+} (pCa ~ 7), 10 mmol m^{-3} of Ca bound to low Mr compounds, and $\leqslant 1$ mol m^{-3} of Ca bound to proteins (Baker, 1972; Kretsinger, 1977). The most likely low Mr chelator is citrate^{3-}, in view of its relatively high pK_D (analagous to pK_a for H^+) for Ca^{2+} of 4.8 (Williams, 1977). The citrate^{3-} pK_D for Mg is 3.2 so, with pMg = 4 and pCa = 7 (Tsien, 1983), our ~ 10 mmol m^{-3} of Ca-citrate$^-$ requires the simultaneous presence of ~ 250 mmol m^{-3} of Mg-citrate$^-$ and ~ 1.6 mol m^{-3} of citrate^{3-}. Since proteins cannot pass through plasmodesmata (Goodwin, 1983) the Ca species involved in the plasmodesmata portion of the pathway are free Ca^{2+} and Ca-citrate$^-$. If we permit free $[Ca^{2+}]$ to vary between 1.1 x 0.9 x the mean 0.1 mmol m^{-3}, the maximum gradient of Ca^{2+} and Ca citrate$^-$ is some 2 mmol m^{-3}. We further assume that diffusion of Ca^{2+} and Ca citrate$^-$ through plasmodesmata is rate-limiting for the symplastic mobility of Ca (see Tyree, 1970; and below). Gunning and Robards (1976, Table 15.1) quote solute fluxes through a plasmodesma of $< 10^{-18}$ mol s^{-1} with concentration differences across plasmodesmata of 10 mol m^{-3}. If Ca^{2+} + Ca-citrate$^-$ behave like other solutes, the maximum concentration difference of 2 mmol m^{-3} gives a flux per plasmodesma of 2.10^{-22} mol Ca s^{-1}. For a 10 mm pathlength with cells 20 μm long our maximum driving force is distributed over 500 arrays of plasmodesmata in series so the flux through an individual plasmodesma is reduced to 4.10^{-25} mol s^{-1}. With a maximum areal density of plasmodesmata of 2.10^{13} m^{-2} (Gunning and Robards, 1976) the maximum Ca flux is 8.10^{-12} mol Ca $(m^2$ parenchyma$)^{-1}$ s^{-1} or, with parenchyma occupying 0.84 of the axis area, 6.72 pmol Ca $(m^2$ axis area$)^{-1}$ s^{-1}. Overall, Wolfe and Gunning (1982) opine that the desmotubule "lumen" is not a pathway for transport between cells; even if it were, and the larger Ca concentration gradients possible in this endoplasmic reticulum derivative allowed fluxes 10^3 times those through the annulus of cytosol, the Ca flux over 10 mm could still not exceed 10 nmol Ca $(m^2$ axis$)^{-1}$ s^{-1}.

Diffusion Across the Symplast of Individual Cells with Transmembrane Fluxes into and Out of Cells

This 'cell-to-cell' pathway differs from that above in that passive Ca entry into a cell at the 'source' end and its active efflux from the cell at the 'sink' end (with diffusion in the symplastic and apoplastic aqueous phases) permits the full permissible difference in cytosol diffusible [Ca] between 'source' and 'sink' portions of cytosol to be developed across a single cell. Since Ca-binding proteins, as well as free Ca^{2+} and Ca-citrate$^-$, can diffuse through cytosol, the flux of Ca across 20 μm of cytosol with a concentration difference of 2 mmol m^{-3} for Ca^{2+} plus Ca-citrate$^-$ (for which D is assumed to be 8.10^{-10} m^2 s^{-1}) and of 200 mmol m^{-3} for Ca-protein (assumed D of 5.10^{-11} m^2 s^{-1}; Table 4.3 of Nye and Tinker, 1977) is 580 nmol m^{-2} s^{-1} on a cytosol area basis, i.e. 487 nmol $(m^2$ axis area$)^{-1}$ s^{-1} for non-vacuolate cytosol occupying 0.84 of the axis area. For vacuolate cells with no vacuolar contribution to the transcellular Ca flux and 0.05 of the axis area occupied by cytoplasm this becomes 29 nmol $(m^2$ axis area$)^{-1}$ s^{-1}. If the transcellular pathway involves a cytosol-vacuole-cytosol transition in a vacuolate cell, then the cytosol (C_{source}- C_{sink}) for the various Ca forms can develop over the 0.25 μm of cytosol on either side of the vacuole, giving a flux between plasmalemma and tonoplast of 36.7 μmol m^{-2} s^{-1} on the basis of axis area, 0.79 of which is vacuole. For the remaining 19.5 μm of the intraplasmalemma pathway (i.e. through the vacuole) such a flux would need a (C_{source}-C_{sink}) of 0.89 mol Ca m^{-3}; this is acceptable in the context of a mean vacuolar [Ca] of say 10 mol m^{-3}. Diffusion through cell walls (see (b) above) could support the highest computed fluxes computed above without the need to invoke excessive Ca concentration differences across the apoplast.

We now consider constraints on this mechanism of Ca transport which may be imposed by the potential magnitude of transmembrane Ca fluxes. Active Ca^{2+} flux from cytosol across plasmalemma or tonoplast is unlikely to exceed 1 μmol Ca^{2+} $(m^2$ membrane$)^{-1}$ s^{-1} (primary active Ca^{2+} transporters, moving 500 μmol Ca^{2+} per g protein per second at 25°C occupying 1/5 of the total protein in an 8 nm thick membrane, density 1.2 Mg m^{-3}, 70% by weight protein: MacLennan et al., 1971); the corresponding passive influx to the cytosol across tonoplast or plasmalemma would need only 10^{-4} of the membrane protein to be a passive Ca^{2+} uniporter, $M_r = 10^5$, of the kind described by Reuter (1983). Amplification of the active fluxes on a projected cell area basis by membrane folding is more plausible for plasmalemma ('transfer cells' with wall invaginations) than for tonoplast but is only, alas, essential (to balance the aqueous phase fluxes) in vacuolate cells where it would be needed at the plasmalemma and the tonoplast!

Turning to estimates of transcellular fluxes of Ca in plants, Clarkson (1984) quotes values of up to 79 nmol m^{-2} s^{-1} for Zea and Cucumis endodermis; Vaughan, Evans and Hutchin (1867) find 180 nmol m^{-2} s^{-1} for basipetal Ca flux in *Zea* root segments (assuming 3 mm^2 cross sectional area per root) and De Guzman and De La Fuente (1984) find 150 nmol m^{-2} s^{-1} for acropetal Ca flux in *Helianthus* hypocotyl segments (assuming 4 mm^2 cross sectional area per hypocotyl). Similar maximal values are found for vertebrate epithelial transcellular Ca fluxes (Terepka *et al.*, 1976).

The totality of the data discussed here suggests that an upper limit on transcellular Ca fluxes in plant cells is 1 μmol Ca (m^2 cell area)$^{-1}$ s^{-1}. We may note that this can only be achieved at the cost of a very substantially higher respiration rate than is commonly found (Penning de Vries, 1975) in non-growing higher plant tissues, with ATP/O$_2$ = 6, Ca^{2+}/ATP = 1.0 at each membrane, and a fresh weight/dry weight ratio of 5. The cost of this cell to cell transport is ∿ 1000 times that of Ca transport by mass flow in xylem or phloem (see Raven, 1984a).

It is important to emphasise that the steady-state Ca gradient across the cytosol implicit in such transcellular Ca fluxes is not necessarily inconsistent with the localized 'informational' role of Ca^{2+} in the cytosol (Lowenstein, 1975).

Mass Flow in Streaming Cytoplasm

Although cytoplasmic streaming may be important in Ca distribution in giant-celled algae (Raven, 1984a) and in root hairs and mycorrhizas (Harley and Smith, 1983), it is unlikely to make a major contribution to the transcellular fluxes in cells 20 μm long considered above. Tyree (1970) and Nobel (1983) discuss the relative importance of diffusion and cytoplasmic streaming in solute movement across small cells; their conclusion that cytoplasmic streaming is relatively unimportant may be rather less true for Ca than for other solutes in view of the potential importance of Ca transport in protein complexes of low diffusion coefficient, and in organelles entrained in moving cytosol. More data on exchange rates of free Ca with proteins and organelles are needed to quantify the role of such transport.

Mass Flow in the Phloem

The total Ca concentration in sieve-tube sap obtained by cutting or *via* excised stylets of phloem-feeding hemipterans is usually in the range 0.3-3.0 mol m^{-3} (Jeschke, Atkins and Pate, 1985; Raven, 1977; Richardson, Baker and Ho, 1982; Ziegler, 1975). While this is in general accord with the notion that sieve-tube sap is a derivative of the cytosol it is likely that free [Ca^{2+}] is higher than in 'normal' eukaryotic cytosol. Thus the best low M$_r$ chelator in sieve-tube sap (citrate^{3-}; pK$_{D(Ca)}$ ≅ 4.8) is only present at about twice the total Ca concentration (Richardson, Baker and Ho, 1982; Hayashi and Chino, 1985), so that only about half of the Ca would be Ca-citrate$^-$. There is no apparent correlation between the Ca content of sieve-tube sap and that of proteins, making major, general Ca binding by proteins unlikely (see McEuen, 1979; McEuen, Hart and Sabuis, 1981); the extremes of Ca:protein ratios reported for sieve-tube sap are 1 mol Ca per 188 g protein (total Ca^{2+} = 1.65 mol m^{-3}) in *Tilia* and 1 mol Ca per 112,000 g protein (total Ca^{2+} = 0.9 mol m^{-3}) in *Cucurbita* (Ziegler, 1975; Richardson, Ho and Baker, 1982). A substantial chelation of Ca^{2+} in phloem sap must occur in at least some plants if the solubility product of calcium and phosphate ions is not to be exceeded (Fukumorita *et al.*, 1983).

The other requirement for an estimate of Ca flux in the phloem is one for $\bar{v}$ (equation (2)). Passioura and Ashford (1874) quote a $\bar{v}$ of 1.7 mm s^{-1} for the remaining seminal root of root pruned *Triticum* while Kallarackal and Milburn (1984) found 2.3 mm s^{-1} for *Ricinus* exuding from a cut. With a Ca concentration of 3 mol m^{-3} we thus have a Ca flux of 6.3 mmol Ca m^{-2} s^{-1} on the basis of sieve-tube area or, if sieve-tubes occupy 0.01 of the axis cross-section, 63 μmol Ca (m^2 axis area)$^{-1}$ s^{-1}; a more 'usual' phloem velocity of 0.5 mm s^{-1} gives a Ca flux of 14 μmol (m^2 axis area)$^{-1}$ s^{-1}.

The negligible phloem mobility of Ca indicated by many [45]Ca studies is probably a result of competition for [45]Ca between phloem loading and other processes, and of [45]Ca exchange *en route* across the sieve-tube plasmalemma, and with immobile Ca-binding sites inside the sieve-tube (Hanger, 1979; Safford and Bassingthwaite, 1977).

Ca FLUXES BY VARIOUS PATHWAYS AND MECHANISMS IN RELATION TO PLANT Ca REQUIREMENTS

The Flux of Ca from Below-ground to Above-ground Portions of Terrestrial Vascular Plants

The adequacy of mass flow in the xylem to account for Ca fluxes from below-ground to above-ground portions of vascular land plants can be tested by promiscuously combining the highest observed Ca concentration in xylem sap (10 mol m^{-3}) and the smallest observed water flux to the shoot during growth (in CAM plants: 28 g H_2O per g dry gain by the shoot, made up of 18 g H_2O lost in transpiration (Kluge and Ting, 1978) 1.5 g used in net photosynthesis, and 8.5 g retained as tissue water; cf. Allaway, 1976). This yields a Ca content of shoot dry matter of 0.88 mmol per g dry weight or 22 mg Ca per g dry weight which is well above the critical Ca concentration (i.e. the lowest Ca content which yields maximum growth rate) in higher plants (Hanson, 1984).

Not only is xylem mass flow adequate to account for Ca transport to vascular land plant shoots, it is the only mechanism which can yield the required Ca fluxes. A herb with a relative growth rate of 0.1 d^{-1}, with a shoot dry weight of 10 g and 20 mg Ca per g dry weight needs a mean Ca flux to the shoot of 5.8 nmol s^{-1}. With a hypocotyl area of 20 mm^2 the required Ca flux is 29 mmol (m^2 axis)$^{-1}$ s^{-1}; only xylem mass flow can support such a flux (see previous section).

The Distribution of Ca Among Above-ground Organs of Terrestrial Vascular Plants

Most of the H_2O passing in the xylem evaporates from mature photosynthetic tissue whose only net Ca requirement relates to Ca oxalate precipitation as part of a 'biochemical pH stat' in certain plants with NO_3^- as their N source. Growing aerial tissues are frequently not interpolated between the below-ground portions of the plant and the main transpiring surfaces; examples are apical meristems, and fruits. A study of the carbon, water and Ca budget of growing fruits, together with the Ca and organic C concentrations in sieve-tube saps, of a number of fruits (original data, and literature citations, in Hocking, 1982; Pate *et al.*, 1985; Peoples *et al.*, 1985; and Raven, 1977) permits determination of the contribution of phloem transport of Ca to the total Ca flux to the fruit. Frequently the phloem only provides 0.05-0.20 of the total Ca; with a phloem Ca flux of 14 μmol/m^2 peduncle area^{-1} s^{-1}; the remaining Ca flux amounts (see above) to some 50 μmol Ca (m^2 peduncle area)$^{-1}$ s^{-1}. It is clear (see above) that none of the Ca transport mechanisms, other than mass flow in the xylem, have a capacity large enough to account for this Ca flux, even when distances of only 10 nm are considered; and that the highest capacity non-mass flow mechanism, i.e. cell to cell transport, has a very high energy cost per unit Ca moved relative to the mass flow mechanisms (∿ 1000 fold for a 10 mm path length) in xylem or phloem (see above, and Raven, 1984a).

However, there are a number of fruits for which the net H$_2$O flux along the peduncular xylem is out of the fruit (e.g. *Vigna unguiculata*: Peoples *et al.*, 1985), i.e. the phloem overprovides the fruit with water relative to organic C. However, Pate *et al.* (1985) showed that, during the dark period of the diel cycle, the H$_2$O balance of the *Vigna* fruit required a net influx of water in the xylem, with a larger net efflux *via* the xylem in the light period. Thus, even in *Vigna*, there is the possibility of net Ca influx to the fruit *via* the xylem. However, much more work is needed before Ca supply to aerial fruits, and other growing regions, can be considered to be quantitatively explained in terms of Ca influx *via* phloem and xylem. Estimates of xylem solution influx to, and efflux from growing regions are required (Pate *et al.*, 1985), as are values of Ca concentration in these xylem streams. It is possible that there is a transfer of Ca from xylem streams destined for mature photosynthetic organs to those streams destined for growing regions in the manner shown for organic N compounds (Pate, 1980). We note that transfer of a substantial fraction of any xylem-borne solutes between xylem streams in the shoot requires, in view of restrictions on the magnitude of transcellular ion fluxes, plasmalemma area abutting on shoot xylem of a similar magnitude to that involved in xylem loading in roots.

CONCLUSIONS: THE PHYLOGENY OF LONG-DISTANCE TRANSPORT OF Ca

A low concentration of free Ca^{2+} in the cytosol was probably an early evolutionary acquisition (Kretsinger, 1977). However, it was not until terrestrial macrophytes combined the spatial separation of growing and photosynthetic regions, and the large cell wall Ca^{2+} requirement (both features in common with many aquatic macrophytes: Raven, 1984a,b) with the absence of an extracellular Ca^{2+} source for much of the plant surface, that restrictions on symplastic Ca^{2+} transport became significant for plant nutrition. Delivery of Ca^{2+} to transpirational termini by apoplastic mass flow, with Ca^{2+} demand by growing regions with low rates of transpiration, involves symplastic transport in the shoot; the two most significant are probably the (restricted) Ca^{2+} mass flow in the phloem, and xylem-to-xylem transport *via* xylem parenchyma cells. This latter mechanism may underlie the increased area of xylem element-parenchyma contact observed in the fossil record of vascular plants which, at least in part, parallels the extent of differentiation of mature photosynthetic from vegetative growing regions and reproductive regions in the aerial shoot (Bower, 1930; Raven, 1984b; Niklas, 1984).

ACKNOWLEDGEMENTS

J.J. MacFarlane contributed useful discussions on xylem-to-xylem transfer; B.A. Osborne made useful suggestions throughout

REFERENCES

Allaway, W. G., 1976, Influence of stomatal behaviour on long distance transport, in: "Transport and Transfer Processes in Plants," I. F. Wardlaw and J. B. Passioura, eds., Academic Press, New York.

Baker, P. F., 1972, Transport and metabolism of calcium ions in nerve, *Progr. Biophys. Mol. Biol.*, 24:177.

Bell, C. W. and Biddulph, O., 1963, Translocation of Ca: Exchange vs. mass flow, *Pl. Physiol.*, 38:610.

Bower, F. O., 1930, "Size and Form in Plants, With Special Reference to the Primary Conducting Tracts,", MacMillan, London.

Briggs, G. E., Hope, A. B., and Robertson, R. N., 1961, "Electrolytes and Plant Cells," Blackwell Scientific Publications, Oxford.

Clarkson, D. T., 1984, Calcium transport between tissues and its distribtuion in the plant, *Pl. Cell Environ.*, 7:449.

Clarkson, D. T., Williams, L., Hanson, J. B., 1984, Perfusion of onion root xylem vessels: a method and some evidence of control of the pH of the xylem sap, *Planta*, 162:361.

De Guzman, C. C., and De La Fuente, R. K., 1984, Polar calcium flux in sunflower hypocotyl segments. I. The effect of auxin, *Pl. Physiol.*, 76:347-352.

Demarty, M., Morvan, C., and Thellier, M., 1984, Calcium and the cell wall, *Pl. Cell Environ.*, 7:441.

Ferguson, A. R., 1980, Xylem sap from *Actinidia chinensis*: apparent differences in sap composition arising from the method of collection, *Ann. Bot.*, 46:791.

Ferguson, I.B., 1979, The movement of calcium in non-vascular tissue of plants, *Commun. Soil Sci. Pl. Anal.*, 10:217.

Fukumorita, T., Noziri, Y., Haraguchi, H., and Chino, M., 1983, Inorganic content of rice phloem sap, *Soil Sci. Pl. Nutr.*, 29:185.

Goodwin, P. B., 1983, Molecular size limit for movement in the symplast of *Elodea* leaf, *Planta*, 157:124.

Gunning, B. E. S., and Robards, A. W., 1976, Plasmodesmata: current knowledge and outstanding problems, in:"Intercellular Communication in Plants: Studies on Plasmodesmata," B. E. S. Gunning and A. W. Robards, eds., Springer Verlag, Berlin.

Hanger, B. C., 1979, The movement of calcium in plants, *Commun. Soil Sci. Pl. Anal.*, 10:171.

Hanson, J. B., 1984, The functions of calcium in plant nutrition, *Advances in Plant Nutrition*, 1:149.

Harley, J. L., and Smith, S. E., 1983, "Mycorrhizal Symbiosis", Academic Press, London.

Hayashi, H., and Chino, M., 1985, Nitrate and other anions in rice phloem sap, *Pl. Cell Physiol.*, 26:325.

Hocking, P.J., 1982, Accumulation and distribution of nutrients in the fruits of castor bean (*Ricinus communis* L.), *Ann. Bot.*, 49:51.

Jeschke, W. D., Atkins, C. A., and Pate, J. S., 1985, Ion circulation via phloem and xylem between root and shoot of nodulated white lupin, *J. Pl. Physiol.*, 117:319-330.

Kallarackal, J., and Milburn, J. A., 1984, Specific mass transfer and sink-controlled phloem translocation in castor bean, *Austr. J. Pl. Physiol.* 11:483.

Kluge, M., and Ting, I. P., 1978, "Crassulacean Acid Metabolism: An Analysis of an Ecological Adaptation," Springer-Verlag, Berlin.

Kretsinger, R. H., 1977, Evolution of the informational role of calcium in eukaryotes, in: "Calcium-binding Proteins and Calcium Function," R. H. Wasserman *et al.*, eds., North-Holland, New York.

Loewenstein, W. R., 1975, Permeable junctions, *Cold Spring Harbor Symp. Quant. Biol.*, 40:49.

McEuen, A. R., 1979, Studies on calcium and a calcium-binding protein in the sieve-tube sap of *Cucurbita maxima* and related species. Ph.D. Thesis, University of Aberdeen.

McEuen, A. R., Hart, J. W., and Sabnis, D. D., 1981, Calcium-binding protein in sieve-tube exudate, *Planta*, 151:531.

MacLennan, D. H., Seeman, P., Iles, G. H., and Yip, C. C., 1971, Membrane formation by the ATPase of sarcoplasmic reticulum. *J. Biol. Chem.*, 296:2702.

Nakajami, N., Morikawa, H., Igarushi, S., and Senda, M., 1981,
Differential effect of calcium and magnesium on mechanical
properties of pea cell walls, *Pl. Cell Physiol.*, 22:1305.

Niklas, K. J., 1984, Size-related changes in the primary axis anatomy of
some early tracheophytes, *Palaeobiol.*, 10:487.

Nobel, P. S., 1983, "Biophysical Plant Physiology and Ecology,"
W. H. Freeman and Co., San Francisco.

Nye, P. H., and Tinker, B. B., 1977, "Solute Movement in the Soil-Root
System," Blackwell Scientific Publications, Oxford.

Overall, R. L., Wolfe, J., and Gunning, B. E. S., 1982, Intercellular
communication in *Azolla* roots: I. Ultrastructure of plasmodesmata,
Protoplasma, 111:134.

Passioura, J. B., 1972, The effect of root geometry on the yield of spring
wheat growing on stored water, *Austr. J. Agric. Res.*, 23:745.

Passioura, J. B., and Ashford, A. E., 1974, Rapid translocation in the
phloem of wheat plants, *Austr. J. Pl. Physiol.*, 1:521.

Pate, J. S., 1980, Transport and partitioning of nitrogenous solutes,
Ann. Rev. Pl. Physiol., 31:313.

Pate, J. S., Peoples, M. B., van Bel, A. J. E., Kuo, J., and
Atkins, C. A., 1985, Diurnal water balance of the cowpea fruit,
Pl. Physiol., 77: 148.

Penning de Vries, F. W. T., 1975, The cost of maintenance processes in
plants, *Ann. Bot.*, 39:77.

Peoples, M. B., Pate, J. S., Atkins, C. A., and Murray, D. R., 1985,
Economy of water, carbon and nitrogen in the developing cowpea
fruit, *Pl. Physiol.*, 77:142.

Pitman, M. G., and, Welfare, D., 1978, Inhibition of ion transport in
excised barley roots by abscisic acid; relation to water
permeability of the roots, *J. exp. Bot.*, 29: 1125.

Raven, J. A., 1977, H^+ and Ca^{2+} in phloem and symplast:relation of
relative immobility of the ions to the cytoplasmic nature of the
transport paths, *New Phytol.*, 79:465.

Raven, J. A., 1979, The possible role of membrane electrophoresis in the
polar transport of IAA and other solutes in plant tissues,
New Phytol., 83:299.

Raven, J. A., 1984a, "Energetics and Transport in Aquatic Plants,"
A. R. Liss, New York.

Raven, J. A., 1984b, Physiological correlates of the morphology of
early vascular plants, *Bot. J. Linn. Soc.*, 88:105.

Raven, J. A., and Rubery, P. H., 1982, Co-ordination of development:
hormone receptors, hormone action and hormone transport, in:
"The Molecular Biology of Plant Development," H. Smith and
D. Grierson, eds., Blackwell Scientific Publications, Oxford.

Reuter, H., 1983, Calcium channel modulation by neurotransmitters,
enzymes and drugs, *Nature*, 301:569.

Richardson, P. T., Baker, D. A., and Ho, L. C., 1982, The chemical
composition of cucurbit vascular exudates, *J. exp. Bot.*, 33:
1239.

Safford, R. E., and Bassingthwaite, J. B., 1977, Calcium diffusion in
transient and steady states in muscle, *Biophys. J.* 20:113.

Terepka, A. R., Coleman, J. R., Armbrecht, H. J., and Gunter, T. E.,
1976, Transcellular transport of calcium, *Symp. Soc. Exp. Biol.*,
30:117.

Tsien, R. Y., 1983, Intracellular measurements of ion activites,
Ann. Rev. Biophys. Bioeng., 12:91.

Tyree, M. T., 1970, The symplast concept. A general theory of
symplastic transport according to the thermodynamics of
irreversible processes, *J. Theor. Biol.*, 26:181.

Van de Geijn, S. C., Smeulders, F., 1981, Diurnal changes in the flux of calcium toward meristems and transpiring leaves in tomato and maize plants, *Planta*, 151:265.

Vaughan, B. E., Evans, E. C. III and Hutchin, M. E., 1967, Polar transport characteristics of radiostrontium and radiocalcium in isolated corn root segments, *Pl. Physiol.*, 42, 747-750.

White, M. C., Baker, D. A., Chaney, R. L., and Decker, A. M., 1981, Metal complexation in xylem fluid. II. Theoretical equilibrium model and computational computer programme, *Pl. Physiol.*, 67: 301.

Williams, R. J. P., 1977, Calcium chemistry and its relation to protein binding, in: "Calcium-binding Proteins and Calcium Function," R. H. Wasserman *et al.*, eds., North-Holland, New York.

Ziegler, H., 1975, Nature of transported substances, in: "Encyclopedia of Plant Physiology, New Series, Volume 1," M. H. Zimmermann and J. A. Milburn, eds., Springer Verlag, Berlin.

REGULATION OF FREE CYTOSOL CALCIUM ENTRY

CONTROLS ON CALCIUM INFLUX IN CORN ROOT CELLS

John B. Hanson, Magaly Rincon and Sharon A. Rogers

Department of Plant Biology
University of Illinois
Urbana, IL 61801, USA

INTRODUCTION

Our laboratory has been investigating why injury or shock to corn roots causes the plasmamembrane to become "leaky", inhibits H^+-pumping and active ion influx, and depolarizes the cell (Gronewald et al 1979, Cheeseman and Hanson 1980, Gronewald and Hanson 1980, 1982, Chastain et al 1981, Chastain and Hanson 1982). In recent years this investigation has come to focus on a role for Ca^{2+} as the agent for signaling injury (Zocchi and Hanson 1982, Zocchi et al 1983, Hanson 1984). Our working hypothesis has been that schematically illustrated in Fig. 1.

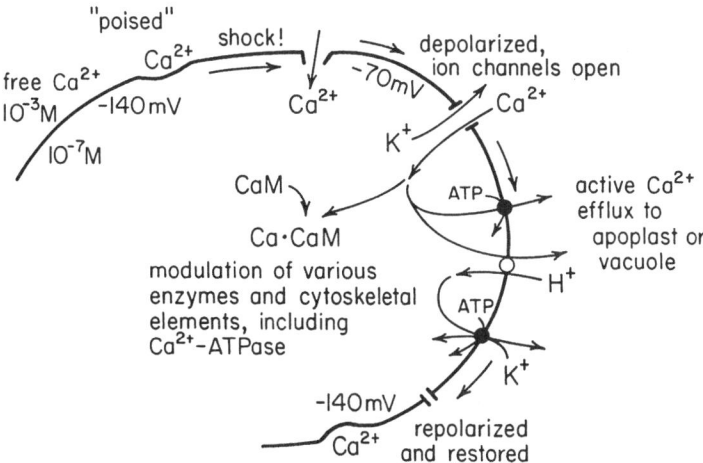

Fig. 1. Schematic diagram of Ca^{2+} fluxes in response to injury and during subsequent recovery.

Stability of the plasmalemma requires millimolar external Ca^{2+} (Hanson 1984). We visualize the binding potential for Ca^{2+} at stabilizing sites on the plasmalemma surface to be poised with the large potential for Ca^{2+} influx. The nature of the sites is unknown, but the stability is reflected in less passive and more active ion transport (ie: due to "tight" membranes and higher $\Delta\bar{\mu}_{H^+}$ -- Hanson 1984). On receipt of an injury or shock the poise is upset and a very small amount of Ca^{2+} is driven in, depolarizing the membrane and opening ion channels (a single non-specific channel is shown

253

for convenience only). The leakage of K^+ from injured plant tissue is well known (Van Steveninck 1975) and where investigated depolarization is found [e.g., corn cells are depolarized by wounding (Gronewald et al 1979) and cold shock (Nelles and Laske 1982)]. We have found $^{45}Ca^{2+}$ influx increased by cold shock (Zocchi and Hanson, 1982), and on the basis of increased membrane protein phosphorylation the Ca^{2+} is assumed to enter the cytoplasm (Zocchi et al 1983).

As indicated in Fig. 1, among the responses to increased cytosol free Ca^{2+} is the activation of Ca^{2+}-ATPase through the Ca·CaM complex (Dieter and Marme 1980). Also, a number of laboratories have offered evidence for a Ca^{2+}/H^+ exchange driven by $\Delta\bar{\mu}_H+$ (Hanson 1984), but there is no evidence on its activation. The relative contribution of these two mechanisms to Ca^{2+} homeostasis is unknown. However, homeostasis must be attained because with "washing" or "aging" there is recovery of membrane stability and transport functions. Recovery is in two phases: an inductive phase, lasting about an hour, in which H^+-pumping is largely restored and "leaks" sealed, and a developmental phase in which induced biosyntheses are expressed.

We report here additional studies on the influx of Ca^{2+} into corn root tissue in response to injury or shock.

METHODS AND MATERIALS

Techniques for raising corn seedlings (Zea mays L., Crow's Hybrid 430) with 0.1 mM $CaSO_4$, excising non-growing root segments 0.5-2.5 cm from the tip, and washing to restore active ion transport have been described (Leonard and Hanson 1972, Gronewald et al 1979, Chastain et al 1981), as have determinations of net H^+ efflux and $^{45}Ca^{2+}$ influx (Chastain and Hanson 1982, Zocchi and Hanson 1982). The standard medium for washing was aerated 0.1 mM $CaSO_4$ + 0.2 mM KH_2PO_4 (pH 6.0 with 1 mM Mes-Tris) at 30°C, and the same medium labeled with ^{86}Rb or ^{45}Ca was used for ion influx. Removal of exchangeable $^{45}Ca^{2+}$ was in ice cold 10 mM $CaCl_2$ + 5 mM KCl for 30 min, at which time loss was essentially complete. Cold shocking was in distilled H_2O chilled with ice (4 to 5°C) or in a freezer (< 0.5°). Additional details and variations are given with the data.

Microsomes were isolated as the 13,000-80,000 xg fraction of freshly excised root tip homogenates as described by Zocchi et al (1983) except that grinding was in a cold mortar and fusicoccin was omitted. The pellet was separated on dextran cushions essentially by the method of Churchill et al (1983), collecting the 0-7% and 7-15% dextran interface layers. The 7% fraction has the characteristics of tonoplast enrichment and the 15% of plasmalemma enrichment as determined by sensitivity of membrane ATPase to chloride, nitrate, K^+, vanadate and DCCD (Uri Ladror, unpublished research). Ca^{2+} uptake was as described (Zocchi and Hanson, 1982). Pretreatment of the root tissue was for 15 min (1 h in a few experiments) in the grinding medium at room temperature. IAA was maintained at 10 µM during vesicle isolation, but was not present in the dextran cushions or in the Ca^{2+} uptake assays.

RESULTS

Table 1 shows that shocks and injuries in addition to cold will increase the rate of $^{45}Ca^{2+}$ influx. All these shocks block net H^+ pumping in a fashion relieved by fusicoccin (FC) (Chastain and Hanson 1981, 1982; Rincon unpublished), and to the extent examined cause depolarization and K^+ leakage. Respiratory inhibitors and uncouplers known to depolarize (Lin and Hanson 1976) likewise increase Ca^{2+} influx. Fusicoccin (FC), which activates H^+ pumping and hyperpolarizes both freshly cut and washed root segments (Gronewald et al 1979), reduces Ca^{2+} influx.

Table 1. Non-exchangeable $^{45}Ca^{2+}$ influx to washed root segments in 15 min at 30°C as affected by various treatments. Data from a large group of experiments normalized with untreated controls at 100%.

Treatment		$^{45}Ca^{2+}$ influx (% of control)
Rubbing	(2 min pretrt.)	121
Cutting	(in half)	126
40°C	(5 min pretrt.)	149
4°C	(5 min pretrt.)	172
pH 3	(5 min pretrt.)	179
KCN	(50 μM during uptake)	155
FCCP	(10 μM during uptake)	199
FC	(10 μM during uptake)	69

We chose cold shocking as a standard injury for subsequent experiments.

Injured root cells repolarize with washing (Lin and Hanson 1974) and there should be a decline in Ca^{2+} influx if it is voltage regulated. Figure 2 plots Ca^{2+} influx after different periods of washing excised root segments. For comparison earlier data on active K influx and net H^+ efflux are plotted (Gronewald and Hanson 1980). Net K^+ leakage and loss of adenine nucleotides stops at the end of the inductive period of about 1 hour (Gronewald and Hanson 1982). In this first hour there is also a decline in $^{45}Ca^{2+}$ influx, but not thereafter. The anomalous "blip" in Ca^{2+} influx at 30 min can be attributed to spontaneous acidification of the apoplast, producing an "acid shock" (Table 1). The acidification arises from the more rapid recovery in K^+ influx than in anion influx (Gronewald and Hanson 1980).

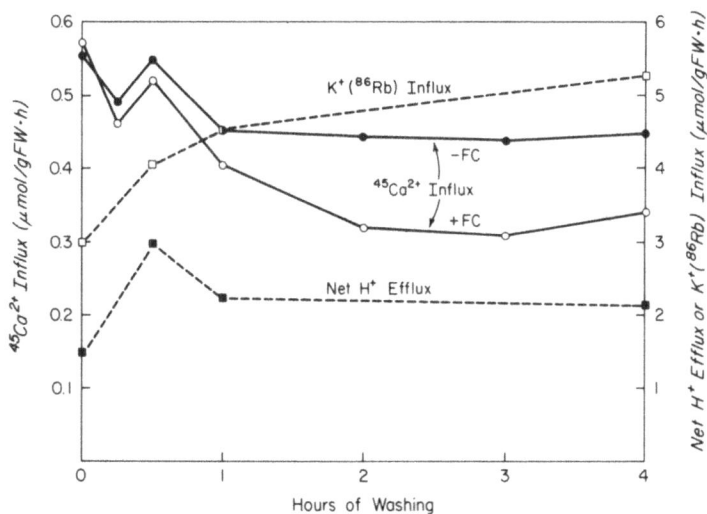

Fig. 2. $^{45}Ca^{2+}$ influx after different periods of washing corn root segments. Fusicoccin (10 μM) added during $^{45}Ca^{2+}$ influx assays. Means from 3 experiments.

Fusicoccin does not suppress $^{45}Ca^{2+}$ uptake in freshly excised (and very leaky) segments, but does so starting at about 30 min of washing (Fig. 2).

Earlier work showed FC to produce pronounced increases in net H^+ efflux, K^+ influx and electrogenic cell potential in freshly excised tissue (Gronewald et al 1979). It appears that the initial shock of excision must not only open Ca^{2+} channels, but must largely keep them open despite increased H^+-pumping and cell potential. If this is the case, the FC-induced hyperpolarization should cause a greater Ca^{2+} influx, and indeed the experiments suggest this (zero time, Fig. 2).

To gain more information on this point we produced different levels of injury by cold shocking for 0.5 and 5 min, judging the level of injury by determining net K^+ leakage (Table 2). Ca^{2+} influx was followed in parallel experiments (Table 2).

Table 2. Effect of fusicoccin on net K^+ flux and $^{45}Ca^{2+}$ influx in cold shocked tissue. Washed root segments were placed in distilled water at 0.5°C as indicated, and transferred to std. solution at 30°. Change in K^+ content over 30 min was determined. $^{45}Ca^{2+}$ uptake was for 15 min (Table 1). 10 µM FC added as indicated. Means from 3 experiments.

| Cold Shock | Ion influx (µ mol/g FW · h) | | | |
| | net K^+ | | $^{45}Ca^{2+}$ | |
	-FC	+FC	-FC	+FC
none	3.58	6.20	0.50	0.34
0.5 min	-0.25	1.29	0.84	0.68
5 min	-1.18	0.24	0.86	1.03

As shown by the greater leakage of K^+ there is a cumulative aspect to cold shock such that a 5 min exposure is more effective than a 0.5 min. With the longer exposure to cold the proportion of K^+ channels open (or held open) may increase, permitting passive efflux. However, this seems inconsistent with Ca^{2+} influx being increased to the same level by both cold exposures. The response to FC was also different -- with both exposures net K^+ influx was restored to about the same extent, while FC reduced Ca^{2+} influx with the 0.5 min shock and increased it with the 5 mm shock. The explanation is that under these conditions K^+ uptake is active, probably linked to the H^+-ATPase (Cheeseman and Hanson 1979), and cold shock blocks H^+ pumping (Chastain and Hanson 1982) and K^+ influx (Gronewald and Hanson 1980). Fusicoccin by activating H^+-ATPase restores active K^+ transport. Thus the cold shock affects not only K^+ channel opening, but also active influx, and the data largely reflect the latter. On the other hand, passive Ca^{2+} influx is only indirectly affected by H^+-pumping. The Ca^{2+} channels open with the brief shock, but leave voltage control functional; prolonged cold gives the same channel opening but damages the voltage control, permitting hyperpolarization to increase Ca^{2+} influx.

We tried to determine the minimum time for expression of the cold shock. Five seconds, the shortest time for making the transfers, was equally or more effective than 0.5 or 5 min in opening the Ca^{2+} channels. With up to 1 min of 0.5°C exposure FC could retard Ca^{2+} influx; beyond 1 min FC increased Ca^{2+} influx. Hence the opening of the Ca^{2+} channels is very rapid, but if the cold persists the voltage control is lost.

But is it voltage control? Table 3 gives results from including nigericin or tripropyltin in the Ca^{2+} uptake medium. Nigericin carries out electroneutral H^+/K^+ exchange, and in these washed segments inhibits active K^+ influx (Chastain et al 1981). Tripropyltin carries out OH^-/Cl^- exchange, and strongly inhibits K^+ influx (Chastain and Hanson 1981). However, neither ionophore affected passive Ca^{2+} influx. Since neutral exchange should have collapsed Δ pH without lowering ΔΨ, control of Ca^{2+} influx is indicated to reside with voltage.

Table 3. Effect of ionophores on $^{45}Ca^{2+}$ influx into washed corn root segments. Shock at 0.5°C for 30 sec.

Ionophore	$^{45}Ca^{2+}$ influx - μ mol/g FW · h	
	Control	Shocked
None	0.44	0.87
Nigericin (5 μM)	0.47	0.87
Tripropyltin (5 μM)	0.44	0.80

We checked the sensitivity of the cells to cold shock during the course of washing, and were surprised to find peak sensitivity at 1 hour (Fig. 3), the point at which "leaks" are resealed. In muscle membranes phosphorylation of Ca^{2+} channels is proposed to increase the probability of their opening upon membrane depolarization (Reuter 1983). Something similar may occur in plants because ^{32}Pi incorporation by the tissue into microsomal membranes also peaks at 1 hour (Fig. 3). The plasmalemma-enriched fraction of the microsomes has 2.5-fold more active Ca^{2+}-stimulated protein kinase than the tonoplast, and possesses a bound phosphoprotein phosphatase (Ladror and Hanson 1984). Thus the potential exists for control of plasmalemma Ca^{2+} channels by reversible phosphorylation, but we have no firm evidence for such control.

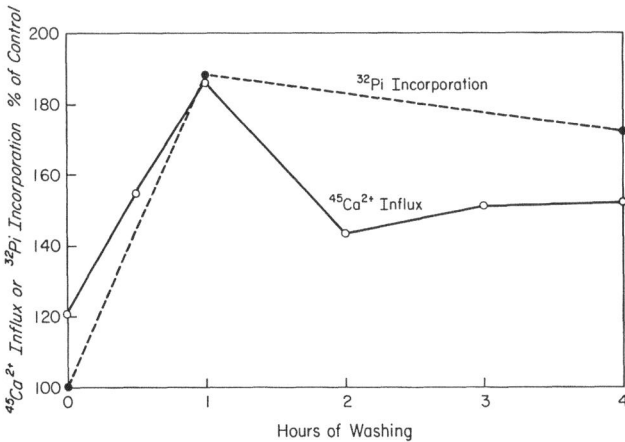

Fig. 3. $^{45}Ca^{2+}$ influx in response to cold shock (0.5°C for 5 min) and ^{32}Pi incorporation by the segments into microsomal membrane protein (no shock) after various periods of washing. ^{32}Pi incorporation for 10 min from 5 mM KH_2PO_4 + 0.5 mM $CaCl_2$ (pH 6) essentially by method of Zocchi et al (1983).

Verapamil, which blocks Ca^{2+} channels in animal membranes (Reuter 1983) was partially inhibitory to Ca^{2+} influx, but only at millimolar concentrations. However, with our standard assay La^{3+} was quite effective in blocking Ca^{2+} influx with an I_{50} of 70-80 μM for both control and cold shocked tissue. The same I_{50} suggests that shock simply opens additional channels of the kind carrying out control influx.

In millimolar concentrations, Ca^{2+} and La^{3+} reduced the effects of cold shock. Furthermore, 5 mM Ca^{2+} protected the voltage control mechanism from the damaging effect of 5 min cold shock as shown by FC reducing, not increasing, $^{45}Ca^{2+}$ influx (Table 4, c.f. Table 2).

Table 4. Protective effect of 5 mM Ca^{2+} against cold shock. Root segments were washed for 110 min in standard medium and then transferred for 10 min to std. medium with 5 mM $CaCl_2$ substituting for 0.1 mM as shown. Cold shock was at 0.5°C for 5 min in distilled water (std. procedure) or 5 mM $CaCl_2$. Influx determined as in preceding experiments.

$CaCl_2$	Control		Cold Shock	
	-FC	+FC	-FC	+FC
mM		μ mol $^{45}Ca^{2+}$/g FW · h		
0.1	0.41	0.29	0.67	0.72
5.0	0.35	0.32	0.61	0.46

Another action of FC is shown in Figure 4. Part of the $^{45}Ca^{2+}$ accumulated by freshly cut root segments is pumped out into the medium, and this is stimulated by FC in a fashion subject to uncoupling (i.e., Ca^{2+}/nH^+

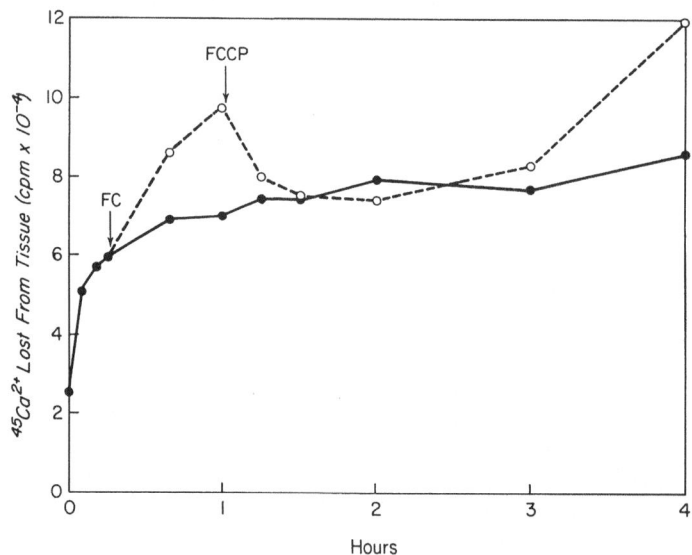

Fig. 4. Stimulation of $^{45}Ca^{2+}$ efflux by FC. Freshly cut segments were preloaded in 0.1 mM $^{45}CaCl_2$ for 30 min at 30°C, exchanged against 10 mM $CaCl_2$ in the cold for 30 min, and transferred to 0.1 mM K_2SO_4 to follow $^{45}Ca^{2+}$ efflux.

exchange driven by $\Delta\mu_{H+}$). The rapid Ca^{2+} efflux after 2 hours in FCCP is due to cell deterioration. Hence, soon after cutting (and in this case chilling during exchange), when FC-stimulated H^+-pumping is not effective in reducing Ca^{2+} influx (Fig. 2), it is effective in driving Ca^{2+} out (Fig. 4).

Oddly enough in washed root tissue FC does not increase Ca^{2+} efflux pumping (Quintero and Hanson 1984). As the plasmalemma recovers voltage control of the Ca^{2+} channels it loses its FC-stimulated Ca^{2+} efflux.

Earlier work had shown FC-stimulated, FCCP-uncoupled Ca^{2+} transport in microsomal vesicles from corn roots (Zocchi & Hanson 1983). Also, we had discovered that pretreatment of the extension zone of soybean hypocotyls with IAA would increase Ca^{2+} uptake by subsequently isolated microsomal vesicles (Kubowicz 1982). Auxin can promote net H^+-pumping in growing corn roots (Mulkey et al 1982). Thus the possibility arises that auxin might be involved in recovery of injured tissue through Ca^{2+}/H^+ exchange activity. Table 5 gives data from a group of exploratory experiments in which excised 0-3 cm root tips were pretreated ± IAA, and subsequently isolated tonoplast and plasmalemma enriched fractions were tested for Ca^{2+} influx.

Table 5. Increased uncoupler sensitive $^{45}Ca^{2+}$ uptake by tonoplast and plasmalemma enriched membrane preparations from auxin-pretreated tissue. (See Methods & Materials). FCCP = 10 µM. *P < 0.05 **P < 0.01

Membranes	$^{45}Ca^{2+}$ uptake (n mol/mg protein·h)		
	-FCCP	+FCCP	ΔFCCP
Tonoplast			
Control	14.38	10.52	-3.86**
IAA pretrt.	17.79	14.02	-3.77**
Plasmalemma			
Control	7.23	6.61	-0.62
IAA pretrt.	9.16	8.09	-1.07*

The tonoplast is a much more stable membrane than the plasmalemma, and comes through isolation with up to 50% uncoupler promoted H^+-ATPase activity. This is reflected in the decline in active Ca^{2+} uptake with the addition of FCCP. On the other hand the excitable plasmalemma suffers considerable damage, and it is difficult to obtain even 20% uncoupling of ATPase actively. However, the pretreatment and isolation with 10 µM IAA provides either some protection or some restoration of uncoupler-sensitive Ca^{2+} transport. We have not followed up this finding other than to establish that the IAA pretreatment also increases uncoupler-promoted ATPase activity in the plasmalemma, but not the tonoplast. Pretreatment is required; IAA added directly to the ATPase assay medium has no effect over a wide range of concentration.

DISCUSSION

Present evidence indicates that there are voltage-controlled Ca^{2+} channels which open in response to actions or agents which depolarize. Conversely, actions (washing) or agents (fusicoccin) which repolarize tend to close the channels. Channel opening can be reduced by high external Ca^{2+}, and may be increased by protein phosphorylation. The plasmamembrane appears to be an excitable membrane, but there is no evidence as yet that excitability rests with destabilization of bound Ca^{2+} (Fig. 1). Severe injury (eg: prolonged cold shock of a chilling-sensitive plant) appears to damage the membranes such that voltage control is lost, and can only be restored by the metabolism which accompanies washing. However, protection against damage is given by 5 mM external Ca^{2+}, a concentration in the normal range for fertile soils.

REFERENCES

Chastain, C. J., Hanson, J. B, 1982, Control of proton efflux from corn root tissue by an injury-sensing mechanism. Plant Sci. Lett., 24:97-104.

Chastain, C. J., LaFayette, P. R. and Hanson, J. B., 1981, Action of protein synthesis inhibitors in blocking electrogenic H^+ efflux from corn roots. Plant Physiol., 67:832-835.

Cheeseman, J. M. and Hanson, J. B., 1979, Energy-linked potassium influx as related to cell potential in corn roots. Plant Physiol., 64:842-845.

Churchill, K. A., Holaway, B., Sze, H., 1983, Separation of two types of electrogenic H^+-pumping ATPases from oat roots. Plant Physiol., 73:921-928.

Dieter, P. and Marme, D., 1980, Calmodulin activation of plant microsomal Ca^{2+} uptake. Proc. Nat. Acad. Sci. USA, 77:7311-7314.

Gronewald, J. W. and Hanson, J. B., 1980, Sensitivity of the proton and ion transport mechanisms of corn roots to injury. Plant Sci. Lett., 18:143-150.

Gronewald, J. W. and Hanson, J. B., 1982, Adenine nucleotide content of corn roots as affected by injury and subsequent washing. Plant Physiol., 69:1252-1256.

Gronewald, J. W., Cheeseman, J. M., and Hanson, J. B., 1979, Comparison of the responses of corn root tissue to fusicoccin and washing. Plant Physiol., 63:255-259.

Hanson, J. B., 1984, The functions of calcium in plant nutrition. Adv. Plant Nutr., 1:149-208.

Kubowicz, B. D., Vanderhoef, L. N. and Hanson, J. B., 1982, ATP-dependent calcium transport in plasmalemma preparations from soybean hypocotyls. Plant Physiol., 69:187-191.

Ladror, U. and Hanson, J. B., 1984, Intrinsic protein kinase activity in corn microsomal membranes. Plant Physiol., 75-S:134.

Leonard, R. T. and Hanson, J. B., 1972, Induction and development of increased ion absorption in corn root tissue. Plant Physiol., 49:430-435.

Lin, W. and Hanson, J. B., 1974, Increase in electrogenic membrane potential with washing of corn root tissue. Plant Physiol., 54:799-801.

Lin, W. and Hanson, J. B., 1976, Cell potentials, cell resistance and proton fluxes in corn root tissue. Effects of dithioerythritol. Plant Physiol., 58:276-282.

Mulkey, T. J., Kuzmanoff, K. M., Evans, M. L., 1982, Promotion of growth and hydrogen ion efflux by auxin in roots of maize pretreated with ethylene biosynthesis inhibitors. Plant Physiol., 70:186-188.

Nelles, A. and Laske, E., 1982, Temperature dependency of the membrane potential of corn coleoptile cells. Biochem. Physiol. Pflanzen, 177:107-113.

Reuter, H., 1983, Calcium channel modulation by neurotransmitters, enzymes and drugs. Nature, 301:569-574.

Van Steveninck, R. M. F., 1975, The "washing" or "aging" phenomenon in plant tissues. Ann. Rev. Plant Physiol., 26:237-258.

Zocchi, G. and Hanson, J. B., 1982, Calcium influx into corn roots as a result of cold shock. Plant Physiol., 70:318-319.

Zocchi, G., Rogers, S. A. and Hanson, J. B., 1983, Inhibition of proton pumping in corn roots is associated with increased phosphorylation of membrane proteins. Plant Sci. Lett., 31:215-221.

ATPases AND MEMBRANE PROPERTIES IN RELATION TO ECOLOGICAL DIFFERENCES

Anders Kylin and Marianne Sommarin

Dept. of Plant Physiology, University of Lund
Box 7007, S-220 07 Lund, Sweden

INTRODUCTION

Animal cells are surrounded by comparatively well regulated body fluids. In contrast, plant cells - and particularly the root cells - must actively be able to maintain their cytoplasm and organelles in a state that is sufficiently constant to allow the biochemical processes to proceed within the limits of life, and this in spite of the often quite considerable short-time fluctuations that occur in water and ion status and in temperature of the surrounding soil and air. The regulation of ion transport through the plasmalemma should thus have a key role for the internal balance of the plant cell.

The practical starting-points for this general statement were summarized by Kähr et al. (1977). ATPases synergistically activated by $(Na^+ + K^+)$ were found in plants by application of the essentially ecological consideration that there may be a selection pressure on halophytes to develop a biochemical system for Na^+/ K^+ exchange of this type (Kylin 1972); and they have so far not been described from non-halophytic species of higher plants. Conversely, a general comparison between oat and wheat and their ATPases as activated by Mg^{2+}, Mn^{2+} and Ca^{2+} led to "rediscovery" of the fact that oat gives a good crop on acid soils, whereas wheat needs a good supply of calcium. Furthermore, the ATPases of the microsomal fractions from roots of oat and wheat respond also to ionic strength in the root medium and to growth temperature in a way that appears understandable by parallelization with the field properties of the two species.

With particular regard to calcium a great amount of ecophysiological data, thinking and terminology exists (see for instance Kinzel 1982). As an example one may evoke the experience of Ingestad (1974), who discerns between the grass type, which "protects" itself against calcium already at the root level; whereas the cucumber type takes up calcium freely and transports it in great quantities to the shoot, where calcium is then "detoxified" by organic counterions. The acid-soil type (for instance blueberries, lingonberries and Rhododendron), finally, grows only on soils that are low in calcium, and is "burnt" to death by excess calcium since the plants take up and transport the element freely as the cucumber type but their production of organic acids is too low to balance excess cations.

In a long series of investigations, French workers have compared

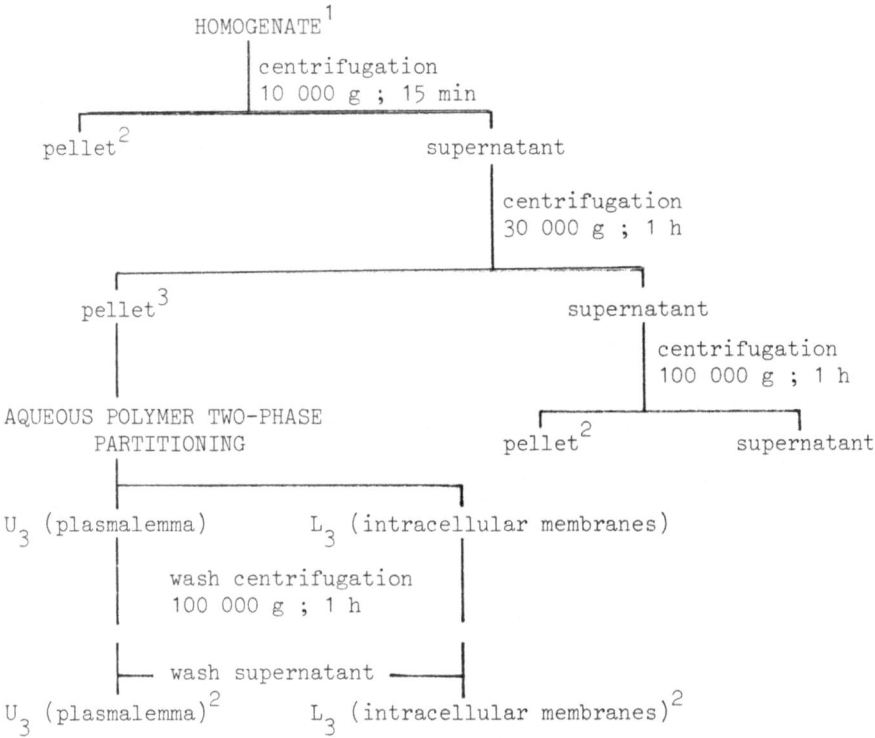

```
                        HOMOGENATE¹
                            │centrifugation
                            │10 000 g ; 15 min
          ┌─────────────────┴─────────────────┐
     pellet²                              supernatant
                                              │centrifugation
                                              │30 000 g ; 1 h
       ┌──────────────────────────────────────┴───────┐
    pellet³                                      supernatant
                                                      │centrifugation
                                                      │100 000 g ; 1 h
    AQUEOUS POLYMER TWO-PHASE              ┌───────────┴───────────┐
         PARTITIONING                  pellet²              supernatant

  U₃ (plasmalemma)        L₃ (intracellular membranes)
              │   wash centrifugation          │
              │   100 000 g ; 1 h              │
              ├─── wash supernatant ───────────┤
  U₃ (plasmalemma)²       L₃ (intracellular membranes)²
```

1. Roots were homogenized in 10 mM Tris-HCl, pH 7.5, 0.25 M sucrose and 1 mM EDTA (5 ml per g fresh weight) in a Sorvall Omnimixer for 30 sec. The homogenate was filtered through nylon cloth (60 μm).

2. Suspension in 10 mM Tris-HCl, pH 7.5 and 0.25 M sucrose.

3. Suspension in 5 mM potassium phosphate, pH 7.8, 0.25 M sucrose and 4 mM KCl.

Figure 1. Flow Scheme

calcifuge (acid-soil) _Lupinus_ and calcicole _Vicia_ with special regard to membrane reactions, role of membrane lipids and regulation of cytoplasmic calcium(list of references in Lamant,1980). One of the most remarkable findings is that a high supply of Ca^{2+} in the medium leads to a high binding of Ca^{2+} to the different membrane systems in cells of _Lupinus_, whereas the membranes of _Vicia_ are comparatively unaffected. This is connected with a higher level of acid phospholipids in _Lupinus_ membranes than in membranes from _Vicia_; and the effects on membrane fluidity also lead to disturbed permeability and function of the membranes. A specific, technical problem arises from the high absorption of Ca^{2+}, since this changes the specific density of the membranes and thus changes their position during differential or gradient centrifugation.

Interpretation of biochemical data obtained with microsomal fractions is uncertain when correlation with physiological behaviour is desired since the preparations contain a mixture of different membrane types. For this reason, members of our laboratory set out to apply the method of aqueous polymer two-phase partitioning for preparations of pure plasmalemma (Lundborg et al. 1981, Widell et al. 1982). We will give data for preparations from roots of acid-soil oat and calcareous-soil wheat with regard to ATPases activated by Ca^{2+}, Mn^{2+} and Mg^{2+} and discuss them against the general background given, and with regard to the new uncertainties that are met with.

MATERIALS AND METHODS

Spring wheat (Triticum aestivum L., cv. Drabant) and oat (Avena sativa L., cv. Brighton) were cultivated and the roots homogenized as described by Sommarin et al. (1985). The flow scheme (Fig. 1) shows the centrifuge fractions obtained. The microsomal fraction (30 000 g pellet) was further separated by the aqueous polymer two-phase technique as applied by Lundborg et al. (1981) and Widell et al. (1982), where the U_3 fraction yields highly purified plasmalemma vesicles. The ATPase analyses were performed by standard procedures, and non-specific activity was always deducted.

RESULTS

Our first comparison between plasmalemma preparations from oat and wheat concerned activation by Mg^{2+} and its modulation by K^+ (Sommarin et al. 1985). The oat plasmalemma behaves as an almost ideal biochemical preparation as regards dependence on pH and temperature as well as with respect to specificity for ATP. The same is true for the K^+ effect on the ATPase activity of plasmalemma from wheat roots, but the effect of Mg^{2+} is in this case less specific, and the dependencies on pH and temperature are complicated. Biologically, and with regard to the complex functions of the plasmalemma, the wheat preparations appear more "natural" than the preparations from oat. A possible explanation - besides "impurities", which can never be excluded but appear less likely due to the data of Lundborg et al. (1981) and Widell et al. (1982) - would be the presence of biochemical units that

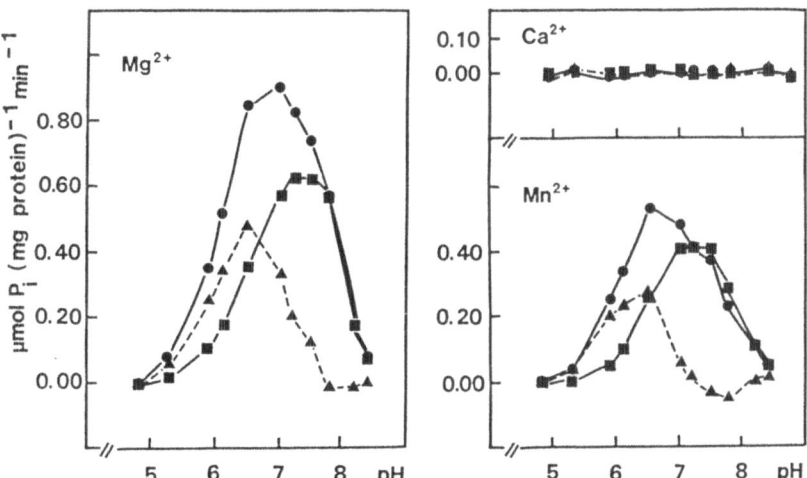

Figure 2. Effect of K^+ on the ATPase activities in oat plasmalemma as dependent on pH. ■: 1 mM Mg^{2+} or Mn^{2+} or 2 mM Ca^{2+} as indicated. ●: Divalent + 25 mM KCl. ▲: Effect of K^+.

Table 1. Effect of K^+ on ATPases in wheat
plasmalemma at pH 6.5. Activities
are expressed in μmol/mg min.

Divalent	Alone	+ K^+	Δ K
Mg^{2+}	0.186	0.432	0.246
Mn^{2+}	0.208	0.323	0.115
Ca^{2+}	0.104	0.095	-0.009

are more easily removed from oat than from wheat plasmalemma. In barley, an increase of pH in the preparation medium has led to the separation of Ca^{2+}-phosphatase from Mg^{2+}-ATPase in the microsomal fraction (DuPont and Hurkman 1985).

The next step has been a comparative investigation over the plasma-lemma ATPases as stimulated by Mg^{2+}, Mn^{2+} and Ca^{2+} (Sommarin et al., this volume). There is virtually no Ca^{2+}-ATPase in plasmalemma from oat, although the preparation shows a weak Ca^{2+}-ITPase. Mn^{2+} and even more Mg^{2+} activate oat ATPase with good specificity. Wheat root plasmalemma shows Ca^{2+}-ATPase, but the specificity is so low that one is rather inclined to regard the activity as NTPase – but not as a general phosphatase since AMP, PNPP and PP_i are not attacked. Both Mg^{2+} and Mn^{2+} activate the wheat system – and Mg^{2+} more so than Mn^{2+} – with reasonable specificity for ATP although not as complete as in the case of oat. Both the Mg^{2+} and Mn^{2+} activities are further stimulated by K^+ in both oat (Fig. 2) and wheat (Table 1), but there is no effect of K^+ on the Ca^{2+}-NTPase. Also, there is no inhibition by nitrate in our preparation (data not shown), so that contamination by tonoplast fragments should be negligible.

As compared to the microsomal fraction (see Kähr et al. 1977), the plasmalemma activity is low and the order between events is changed: both so that oat plasmalemma is more active than plasmalemma from wheat as against the reverse in the microsomes, and so that the divalents activate wheat plasmalemma in the order Mg^{2+} > Mn^{2+} > Ca^{2+} as compared to Ca^{2+} > Mn^{2+} > Mg^{2+} in the microsomal fraction of wheat. In order to get a better understanding of these differences, we have analyzed all fractions obtained by the fractionation procedure (Fig. 1, Table 2). In oat all pellets contain ATPase activated by divalents, with particularly high specific activity in the plasmalemma although it represents a minor part of the total membrane-bound ATPase (probably in the order of 25%, if the losses during phase partitioning are considered). Only unspecific background activity could be demonstrated in the original homogenate or in the supernatants of oat, so that the presence of low-molecular, soluble inhibitors may be indicated. Ca^{2+}-ATPase is weak and ATPase activated by Mg^{2+} and/or Mn^{2+} is dominating in all oat membranes. Contrariwise, Ca^{2+}-ATPase is the highest of the three investigated ones in all fractions from wheat root except in the plasmalemma. The plasmalemma fraction recovered represents even much less of the total membrane activity than in oat – in the order of 1 to 5% depending upon the ion regarded and with a rough estimate of the preparation losses. The highest specific activities for divalent activation occurs in the 100 000 g pellet from wheat, but in this case there is no stimulation by potassium (data not shown), so that plasmalemma fragments are not indicated. Furthermore, all supernatants show high divalent-induced ATPase activity, and the collected membranes contain only in the

Table 2. ATPase activities stimulated by Mg^{2+}, Mn^{2+} and Ca^{2+} in the fractions obtained according to the flow sheme (Fig.1). Total ATPase activity is expressed in μmol/min and the specific ATPase activity in nmol/mg min.

Fraction	Total ATPase activity			Specific ATPase activity		
	Mg^{2+}	Mn^{2+}	Ca^{2+}	Mg^{2+}	Mn^{2+}	Ca^{2+}
WHEAT						
Homogenate	1053	1461	1879	603	837	1076
10 000 g pellet	43	62	71	416	606	693
10 000 g supernatant	955	1308	1945	627	859	1277
30 000 g pellet	66	97	116	682	997	1194
U_3 (plasmalemma)	2.6	2.3	0.8	437	380	130
L_3	15	21	29	577	805	1103
30 000 g supernatant	810	954	1463	609	717	1100
100 000 g pellet	97	144	204	1969	2914	4129
100 000 g supernatant	706	834	1298	545	644	1002
OAT						
Homogenate	–	–	–	–	–	–
10 000 g pellet	15	8.8	1.7	170	102	20
10 000 g supernatant	–	–	–	–	–	–
30 000 g pellet	36	29	3.0	528	420	44
U_3 (plasmalemma)	8.1	5.3	0.2	1155	754	33
L_3	10	11	2.3	354	381	80
30 000 g supernatant	–	–	–	–	–	–
100 000 g pellet	7.9	10	3.4	157	206	67
100 000 g supernatant	–	–	–	–	–	–

order of 20% of the total. It should, however, be noted that we have so far tested a representative selection of substrates only for the plasmalemma and for the microsomal fraction, so that the proportions given are preliminary and intended to show "the state of the art".

With regard to the main starting-point of our research, it is interesting to note that the plasmalemma Ca^{2+}-ATPase in wheat roots is affected by aluminium (C. Lind and T. Lundborg, oral communication; Table 3). In a resistant cultivar, 0.11 mM Al in the medium increases the Ca^{2+}-ATPase, whereas 0.22 mM Al takes the activity back to the control level. Aluminium in the assay medium is inhibitory. In an Al-sensitive cultivar, aluminium decreases the Ca^{2+}-ATPase of the plasmalemma all over. As for Ca^{2+}-ATPase in the microsomal fraction and for Mg^{2+}-ATPase in both cases there may be a slight positive effect (10 – 30% increase) on the resistant cultivar by 0.11 mM Al in the culture medium, but otherwise only negative effects were noted. An ability of the resistant cultivar to respond to aluminium by adaptation of the ion transport mechanism may be indicated.

Table 3. Influence of aluminium on Ca^{2+}-ATPase in plasmalemma of an Al-resistant and an Al-sensitive cultivar of wheat. Data (% of specific activity of control) by courtesy of C. Lind and T. Lundborg.

Cultivar	Al (0.3 mM) in assay	Al (mM) in culture medium		
		0	0.11	0.22
Kadett (resistant)	−	100	349	90
	+	85	233	86
WW 20299 (sensitive)	−	100	86	79
	+	74	55	43

DISCUSSION

As stated in the Introduction, work with the microsomal fraction gave a biochemical parallel to the field properties of acid soil oat and calcium-dependent wheat (Kähr et al. 1977). The further stimulation of Mg^{2+}-ATPase and Mn^{2+}-ATPase by K^+ is strong in oat and was undetectable in the microsomal fraction of wheat, and this has also been discussed in relation to the higher tolerance for acid soils by oat as compared to wheat (Wignarajah et al. 1983). Ca^{2+}-ATPase was not modulated by K^+, but at high pH an influence by Cl^- and HCO_3^- was noted (Björkman et al. 1980).

The plasmalemma contains only a minor part of the total ATPase activity, but since the plasmalemma is the first to meet fluctuations on the outside of the cell, it is of interest to have the situation analyzed for this particular system. The difference in the reaction of a sensitive and a resistant cultivar towards Al in the culture medium (Table 3) is an indication that the plasmalemma responds to external influences; at the same time as it touches upon aluminium toxicity, an ecological problem of high actuality.

With regard to the main theme of the present conference, the absent (oat) or low (wheat) activity of Ca^{2+}-ATPase in the plasmalemma appears of interest. One normally regards a Ca^{2+} outpump coupled to a Ca^{2+}-ATPase in the plasmalemma as one of the regulatory mechanisms that keep the level of Ca^{2+} low in the cytoplasm (Penniston 1983). Our rather negative finding underlines the possibility that other membrane systems and compartments are important for this purpose in at least some plants. Endoplasmic reticulum, mitochondria and vacuoles have been suggested in addition to the plasmalemma (Marmé and Dieter 1983); and the perspective that we would like to propose is that the balance of Ca^{2+}-transport through different membrane systems may be of interest in relation to the reaction of a plant species towards calcium in the soil.

With a new type of plasmalemma preparation in our hands, a few words should be devoted to its properties and problems. The primary preparation appears to consist of sealed, right-side-out vesicles (Larsson et al. 1984). This opens the question on how to energize the ATPase sites. In our case we have worked with frozen and thawed preparations and without sucrose/osmoticum in the assay medium, and the vesicles should reasonable be bursted open. However, there are effects on the activity by both sucrose (diminishing) and detergents like Triton X-100 (increasing) in themselves (A. Bérczi and I. M. Møller, oral communication), so that it is not yet clear on what quantitative assumptions we should correlate transport data and data on ATPase kinetics. There is also the apparent conflict between "biochemical purity" (oat) and "biological likelihood" (wheat – see the points made in the first paragraph of Results). Should one regard the "simplicity" of the oat plasmalemma and the "complexity" of the wheat pre-parations as due to the couple purity/impurity; or are the differences an expression of different conditions for desaggregation/aggregation between the basal membrane and smaller subunits? Carbohydrates will interact with phospholipid monolayers (Crowe et al. 1984) and desaggregation/aggregation could certainly change the functional properties of a membrane system – that is, the phenomena that today are biochemical problems to be overcome may at the same time be properties that are significant for the adaptation of a plant to fluctuations in the surroundings.

Acknowledgements

Our thanks are due to Mrs Inger Rohdin for skilfull technical assis-tance and to Dr. Tomas Lundborg for helpful discussions. Support from the Swedish Natural Science Research Council is gratefully acknowledged.

REFERENCES

Björkman, T., Lundborg, T., Grønbaek, O. & Kylin, A. 1980, Inhibition by bicarbonate of divalent cation activated adenosine triphosphatase from the microsomal fraction of wheat roots, Acta Chem. Scand. B 34:451-452.

Crowe, J. H., Whittam, M. A., Chapman, D. & Crowe, L. M. 1984, Interactions of phospholipid monolayers with carbohydrates, Biochim. Biophys. Acta 769:151-159.

DuPont, F. M. & Hurkman, W. J. 1985, Separation of the Mg^{2+}-ATPases from the Ca^{2+}-phosphatase activity of microsomal membranes prepared from barley roots, Plant Physiol. 77:857-862.

Ingestad, T. 1974, Towards optimum fertilization, Ambio 3:49-54.

Kähr, M., Bervaes, J., Kylin, A. & Kuiper, P. J. C. 1977, Influence of mineral nutrition on ATPase activities and relation of saturated to unsaturated fatty acids in roots of wheat and oats, in: "Trans-membrane Ionic Exchanges in Plants", M. Thellier et al. eds., pp. 213-217. ISBN 2-222-02021-2.

Kinzel, H. 1982, Pflanzenökologie und Mineralstoffwechsel, Verlag Eugen Ulmer, Stuttgart. ISBN 3-8001-3427-6.

Kylin, A. 1973, Adenosine triphosphatases stimulated by (sodium + potassium): Biochemistry and possible significance for salt resistance, in: "Ion Transport in Plants", W. P. Anderson ed., pp. 369-377. Academic Press, London. ISBN 0-12-058250-3.

Lamant, M. 1980, Calcium nutrition in plants. Cytological and biochemical aspects (localization in cell membranes), in: Abstracts of Lec-tures and Poster Demonstrations, IInd Congress of the FESPP, E. Vieitez ed., pp. 70-71, Santiago de Compostela, Spain.

Larsson, C., Kjellbom, P., Widell, S. & Lundborg, T. 1984, Sidedness of plant plasma membrane vesicles purified by partition in aqueous two-phase systems, FEBS Lett. 171:271-276.

Lundborg, T., Widell, S. & Larsson, C. 1981, Distribution of ATPases in wheat root membranes, Physiol. Plant. 52:89-95.

Marmé, D. & Dieter, P. 1983, Role of Ca^{2+} and calmodulin in plants, in: "Calcium and Cell Function", Wai Yiu Cheung ed., pp. 263-311. ISBN 0-12-171404-7.

Penniston, J. T. 1983, Plasma membrane Ca^{2+}-ATPases as active Ca^{2+} pumps, in: "Calcium and Cell Function", Wai Yiu Cheung ed., pp. 99-149. ISBN 0-12-171404-7.

Sommarin, M., Lundborg, T. & Kylin, A. 1985, Comparison of K,MgATPases in purified plasmalemma from wheat and oat - substrate specificities and effects of pH, temperature and inhibitors, Physiol. Plant. 65, in press.

Sommarin, M., Lundborg, T. & Kylin, A. This volume, Properties and distribution of ATPases in oat and wheat with special reference to the plasmalemma.

Widell, S., Lundborg, T. & Larsson, C. 1982, Plasma membranes from oats prepared by partition in an aqueous polymer two-phase system, Plant Physiol. 70:1429-1435.

Wignarajah, K., Lundborg, T., Björkman, T. & Kylin, A. 1983, Field properties of ion uptake of wheat and oat: are they expressed in the modulatory influences of ions on membrane ATPases? Oikos 40:6-13.

FUNCTION OF CALCIUM-CALMODULIN IN CHLOROPLASTS

F. L. Crane and R. Barr

Department of Biological Sciences
Purdue University
West Lafayette, IN 47907 U.S.A.

Chloroplasts contain a large amount of calcium, up to 25 mM, but only 0.5 mM may be unbound.[1,2] Much of this calcium is in the thylakoid lumen while the stroma has a low calcium content.[2] Calcium pumps in chloroplast membranes and H^+/Ca^{++} antiports have been described[2] which can energize the accumulation of calcium in the chloroplast. Calcium uptake is also stimulated by light.[3] It is now clear that this calcium is important in chloroplast function.

Calcium and Photosynthetic CO_2 Fixation

Calcium in the stroma may activate at low concentration or inhibit at high levels both fructose 1,6-bisphosphatase and sedoheptulose-1,7-bisphosphatase which are key regulatory enzymes in CO_2 fixation. A low level (1 μm) of calcium in the stroma will favor photosynthesis[2] (Table I).

NAD kinase in chloroplasts is regulated by a Ca^{++} dependent protein similar to calmodulin and the enzyme is bound to the outer chloroplast membrane.[4] This enzyme would provide the NADP needed for photosynthesis.

A Role For Calcium In Photosystem II

Several laboratories have developed very good evidence for a calcium role in photosystem II (PSII). Various treatments have been used to remove calcium or modify the exposure at a calcium site so that activity is lost which can be restored specifically with added calcium ions and not by other divalent cations. This has been demonstrated in broken chloroplasts, detergent-prepared PSII enriched membrane fragments and cells or thylakoids from blue green algae. Much of this work has been reviewed.[2,5,6]

TABLE I

Control By Calcium-Calmodulin For Stroma Enzymes

Enzyme	Effect	Reference
Fructose-1,6-bis phosphatase	Ca^{++} inhibition	2
Sedoheptulose-1,7-bisphosphatase	Ca^{++} inhibition	2
NAD kinase	Ca^{++} (calmodulin) stimulation	2,4

TABLE II

Restoration Of Photosystem II Activity By Calcium

Particle type	Extraction Procedure	Acceptor	Photosystem II Activity μmoles mg chl^{-1} hr^{-1}				Ref.
			Untreated		Extracted		
			-Ca^{++}	+Ca^{++}	-Ca^{++}	+Ca^{++}	
PSII	NaCl	C$_{12}$BQ	780		196	616	32
PSII	NaCl	Ph BQ	340	370	160	290	26
PSII	NaCl	2,5 DMBQ	270		7	38	27
Thylakoid	hypotonic acid	2,5 DMBQ	372	361	90	192	12
Thylakoid	acid EGTA	DCIP	817	896	158	271	13
PSII	trypsin	Fe(CN)$_6$			137	394	29
PSII	CaCl$_2$	Ph BQ	530		100	190	30
Mangrove	none	Fe(CN)$_6$	27	58			31
Synechoccus	PSII	Fe(CN)$_6$	342		2	62	16
Phormidium	French press	Fe(CN)$_6$			0.6	5.6	17

C$_{12}$BQ, dichlorobenzoquinone; Ph BQ, phenylbenzoquinone; DMBQ, dimethylbenzoquinone; DCIP, dichloroindiphenol.

The type of calcium restoration effects which have been observed are illustrated in Table II. Salt treatment of inverted thylakoids and detergent treated membrane fragments (PSII particles) remove three extrinsic proteins, 18 kD, 23 kD and 33 kD with activity loss. Readdition of the 24 kD and 33 kD proteins or high concentrations (5-50 mM) of calcium salts can restore activity. The actual release of calcium during salt treatment apparently requires exposure of the particles to light.[7,8,9] Light also increases the response to calcium in cells of Anacystis.[6] The conclusions which may be drawn as a result of the work from several laboratories are as follows:

1. Calcium is bound in a light sensitive association with the intrinsic protein components of PSII on the lumen side of the thylakoid membrane.[7,8]

2. The 33 kD, 23 kD and 18 kD extrinsic proteins protect the calcium and manganese sites and removal of these proteins can induce a partial high concentration calcium restoration of PSII.[5]

TABLE III

Effect Of Calcium Chelators On Photosystems I and II In Chloroplasts

Photoreaction		Percent of control activity in presence of:			
		TMB8	500 μm	Murexide	200 μm
H$_2$O → MV	PS I and II		56		46
H$_2$O → DMBQ (DBMIB)	PS II		96		94
H$_2$O → Fe(CN)$_6$ pH 6.0	PS II		107		95
H$_2$O → Fe(CN)$_6$ (DBMIB)	PS II		82		
H$_2$O → SM (DCMU)	PS II		25		94
asc. TMPD → MV	PS I		105		71

EGTA at pH 7.2 and sodium oxalate have no effect on PSII reactions. Barr and Troxel unpublished. MV, methylviologen; TMPD, tetramethylphenylenediamine; SM, silicomolybdate.

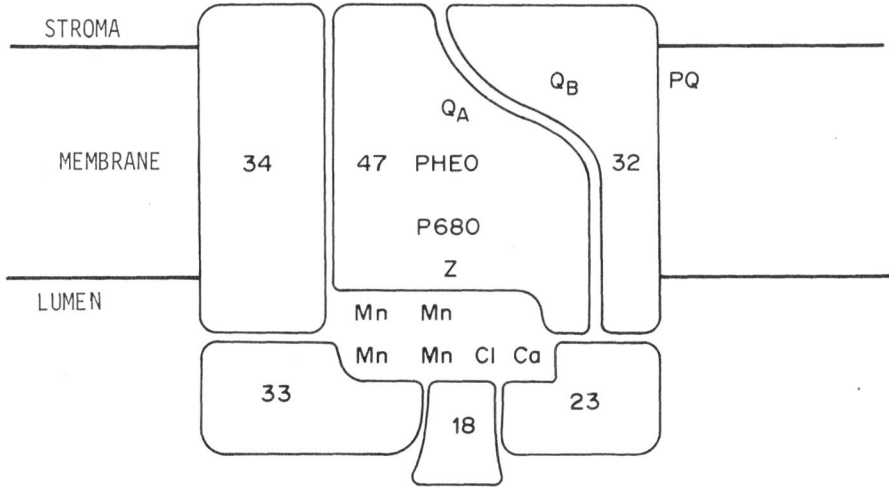

Fig. 1. Association of 33, 23, and 16 kD proteins with the Mn complex and Ca and Cl sites in PSII.[5]

3. After protein removal and light induced calcium removal the reduction of Z^+ is inhibited. This is consistent with a calcium requirement for interaction between the manganese cluster and the primary oxidant Z^+.[8,9,10,20]

The protected nature of the calcium site is shown by failure of the calcium chelator EGTA to inhibit PSII activity in untreated photosystem particles even though EGTA will prevent calcium restoration of extracted particles.[10] Permeable calcium chelators like TMB-8 also do not inhibit PSII activity in broken chloroplasts (Table III). The relation of the calcium site to the components of PSII is shown by Fig. 1.

Extraction of thylakoids with acidic EGTA or acetic acid at pH 4 induces a specific calcium requirement in thylakoids for the restoration of overall electron transport through PS I and II as well as on various partial reactions of PS I and II[11,12] (Table IV). All of the assays used for PSII show loss of activity with extraction and partial reactivation by calcium.

The loss of activity and restoration by calcium when diphenylcarbazide (DPC) is substituted for water as the electron donor for PSII[13] is consistent with a role for calcium after S_1-S_4 states of manganese and near the primary oxidant Z^+ (P680)[28] (Table IV). Calcium has been shown to accelerate the reduction of Z^+ by ascorbate[14] with specific effects over Mg^{++} at low concentrations (< 10 mM). EDTA treatment of inverted thylakoids did not inhibit the DPC → DCIP reaction whereas it inhibited water → DCIP activity and this activity could be restored by Mn, Mg and Ca^{++} (4 mM).[15] EDTA treatment of cyanobacteria thylakoids does induce a calcium requirement for PSII[16] and increases calcium stimulation in sonicated _Chlorella_ chloroplasts.[17]

Evidence For Calcium Function In Photosystem I

Tamura et al.[18] found that cations may non-specifically control the binding of plastocyanin in the lumen of the thylakoid. Treatment of chloroplasts with hypotonic acidic EGTA causes a loss of PSI activity as assayed using ascorbate TMPD (in presence of DCMU) as electron donor and methylviologen as acceptor.[11,13] There is no effect when ascorbate DCIP is used as electron donor. Since ascorbate TMPD donates electrons at the level of

TABLE IV

Calcium Restoration Effects With Different Assays
For Photosystem I and II Activity

| | | | Rates of O_2 Release | |
Function	Assay	Control	Extracted	Extracted +30 mM $CaCl_2$
PS II	$H_2O \rightarrow$ dichloroindophenol	817	158	277
PS II	$H_2O \rightarrow$ silicomolhybdate ph 6.0	271	62	165
PS II	$H_2O \rightarrow$ ferricyanide pH 6.0	344	62	137
PS II	$H_2O \rightarrow$ dimethylbenzoquinone	817	158	271
PS II	DPC $\rightarrow$ dichloroindophenol*	130(170)	43	90
PS I & II	$H_2O \rightarrow$ methylviologen	1139	158	190
PS I	Ascorbate TMPD $\rightarrow$ methylviologen	1137	658	974**
PS I	Ascorbate DCIP $\rightarrow$ methylviologen	1297	1116	1149

Data from Barr et al.[13] and Troxel et al.[11]

Extracted with EGTA under acidic conditions (pH 4.0) and low
  osmolarity. Treatment with acetic acid at pH 4 also can be used.

*For DPC $\rightarrow$ dichloroindophenol calcium gives a large increase in the con-
  trol (unextracted) rate in contrast to small effects in the other
  assays.

**Restoration with 15 mM Ca^{++} usually requires added plastocyanin.

plastocyanin (Fig. 2)[19,20] whereas the reduced indophenol is oxidized after
plastocyanin the inhibition is near plastocyanin. Addition of plastocyanin
and calcium ions fully restores PSI activity (Table V). Magnesium ions do
not replace calcium in this restoration. On this basis we have indicated a
site for calcium action near plastocyanin in PSI. The chelator inhibition
studies in Table III indicate that the calcium at this site is not available
to permeable chelators like the refractory site in PSII. The restoration is
not a non-specific ion effect on plastocyanin binding since magnesium ions
do not replace calcium.

Calmodulin and Photoreactions

Calmodulin is present in chloroplasts.[21,22] There is no positive
evidence for a calmodulin requirement in any photoreaction. Some evidence

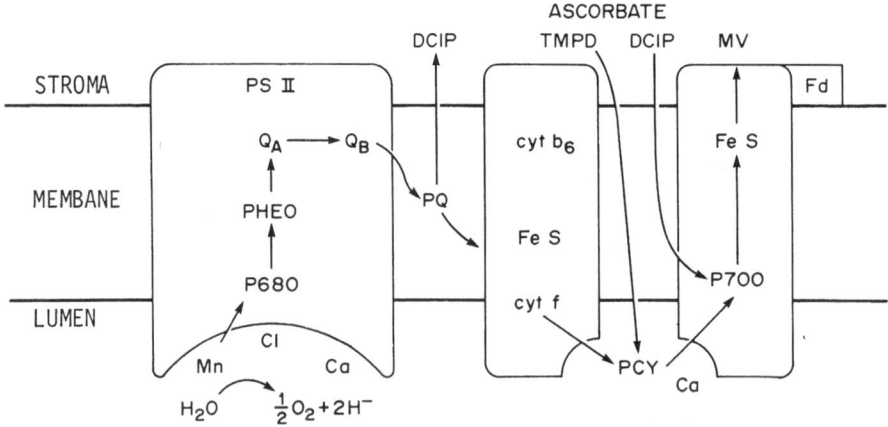

Fig. 2. Relation of PSI and PSII in thylakoids and the TMPD or $DCIPH_2$
  reaction with PSI.

TABLE V

Restoration of Photosystem I Activity with Calcium and
Plastocyanin After Osmotic Shock with Acidic EGTA

	Ascorbate-TMPD methyl viologen
	μmole/mg Chl/hr
Untreated chloroplasts	1131
EGTA treated chloroplasts	612
EGTA treated plus 20 mM Ca^{++}	630
EGTA treated plus plastocyanin 0.06 mg/ml	848
EGTA treated plus Ca^{++} and plastocyanin	1301

Data based in part of Troxel et al.[11] There is no corresponding
stimulation of the ascorbate DCIP methylviologen PSI assay.

has been presented for calmodulin in PSII particles in addition to the 33,
23 and 18 kD extrinsic proteins.[23] Addition of calmodulin has not restored
any activity.

On the other hand, several compounds which can have anticalmodulin
effects inhibit photoreactions in chloroplasts and algae.[33] The phenothiazines clearly inhibit PSII[24] but Nakatani[28] has shown that the S states of
manganese may be affected rather than the reduction of Z^+ which requires
calcium. The effect of a complete series of calmodulin antagonists on PSII
is shown in Fig. 3. The inhibition pattern does not correspond to effects
expected for calmodulin.[25] Likewise, effects of these agents on PSI do not
correspond to calmodulin sensitivity (Table VI). On the other hand, the
series of inhibitors show effects expected for calmodulin when they inhibit
the combined PSI and II (Fig. 4). One must ask if there is another protein

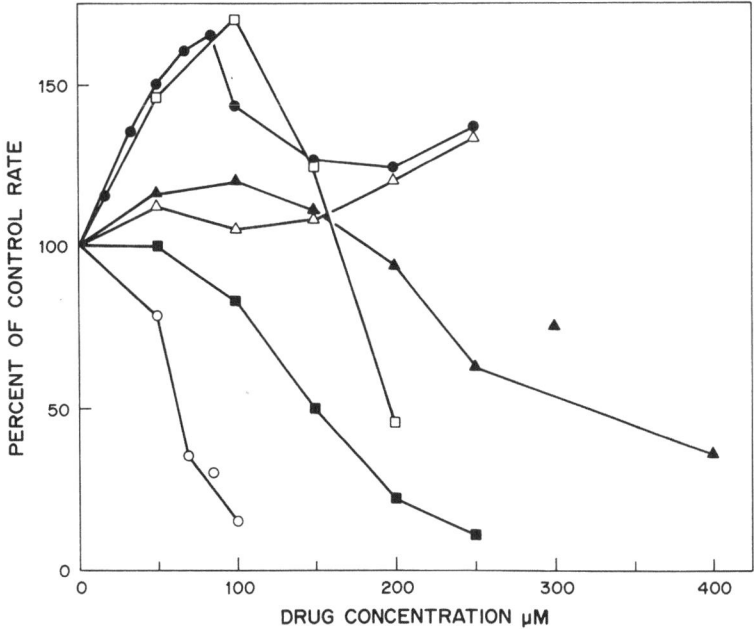

Fig. 3. Effects of calmodulin antagonists on PSII assayed with indophenol
($H_2O \rightarrow DCIP$).[20] Pimozide ●. Calmidazolium ○. W7△. W5◓.
Trifluoperazine ■. Chlorpromazine ▲. Chlorprothixine □.

TABLE VI

Effect of Calmodulin Antagonists on Photosystem I Activity in Chloroplasts

Addition	Ascorbate-TMPD methylviologen μmole mg chl^{-1} hr^{-1}
None	1254 $\pm$ 125
50 μm calmidazolium	1567
100 μm pimozide	1866
250 μm chlorprothixene	2148
200 μm W7	1354
100 μm trifluoperazine	1292

resembling calmodulin which functions between the photosystems and is not detected in partial assays.

In conclusion, one can consider two regulator pools of calcium in the chloroplast. In the stroma an increase in calcium may inhibit CO_2 fixation reactions. The NADPH supply may be controlled at the outer membrane or peripheral space. Concentration of calcium in the thylakoid lumen will stimulate photoreactions at two and possibly three sites. A photoactivated pump may control the calcium level in these compartments.[34] There is no clear evidence yet for calmodulin function at the interior sites even though typical antagonist effects suggest a role for calmodulin-like proteins.

Supported by grants from NSF and NIH. We appreciate gifts of calmodulin antagonists from the following: W-5, Diversified Biotech, Boston, R24571, Janssen Pharmaceutica, fluphenazine, Serva, Pimozide, McNeil Pharmaceutical and chlorprothixene, Hoffman-La Roche.

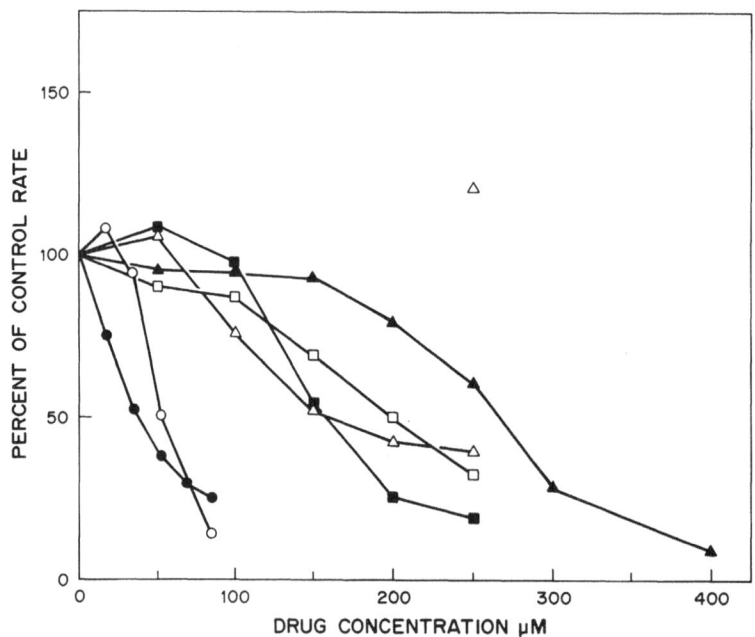

Fig. 4. Effect of calmodulin antagonists on PS I and II. Oxygen release in presence of methylviologen ($H_2O \rightarrow MV$).[20] Symbols as for Fig. 3.

References

1. J. P. Grouzis, Fixation du calcium par les thylakoids isoles de Lupin jeune et de Fauverole, Physiol. Veg. 16:593 (1978).
2. J. J. Brand and D. W. Becker, Evidence for direct roles of calcium in photosynthesis, J. Bioenerg. Biomemb. 16:239 (1984).
3. G. Kriemer, M. Melkonian, and E. Latzko, An electrogenic uniport mediates light dependent Ca^{++} influx into intact spinach chloroplasts, FEBS Lett. 180:253 (1985).
4. P. Simon, M. Bonzon, H. Greppin, and D. Marmé, Subchloroplast localization of NAD kinase activity, FEBS Lett. 167:332 (1984).
5. N. Murata and M. Miyao, Extrinsic membrane proteins in the photosynthetic oxygen evolving complex, TIBS 10:122 (1985).
6. J. J. Brand, P. Mohanty, and D. C. Fork, Reversible inhibition of photochemistry of photosystem II by calcium removal from intact cells of Anacystis nidulans, FEBS Lett. 155:120 (1983).
7. M Miyao and N. Murata, Partial disintegration and reconstitution of the photosynthetic oxygen evolution system, Biochim. Biophys. Acta 725: 87 (1983).
8. J. P. Dekker, D. F. Ghanotakis, J. J. Plijter, H. J. Van Gorkom, and G. T. Babcock, Kinetics of the oxygen evolving complex in salt-washed photosystem II preparations, Biochim. Biophys. Acta 767:515 (1984).
9. T. Ono and Y. Inoue, Ca^{++} dependent restoration of O_2-evolving activity in $CaCl_2$ washed PSII particles depleted of 33, 24 and 16 kDa proteins, FEBS Lett. 168:281 (1984).
10. D. F. Ghanotakis, J. N. Topper, G. T. Babcock, and C. F. Yocum, Water soluble 17 and 23 kDa polypeptides restore oxygen evolution activity by creating a high affinity binding site for Ca^{++} on the oxidizing side of PSII, FEBS Lett. 170:169 (1984).
11. K. S. Troxel, R. Barr, and F. L. Crane, The role of Ca^{2+} in electron transport of spinach chloroplasts, Proc. Indiana Acad. Sci. 89:343 (1980).
12. R. Barr, K. S. Troxel, and F. L. Crane, A calcium selective site in photosystem II of spinach chloroplasts, Plant Physiol. 73:309 (1983).
13. R. Barr, K. S. Troxel, and F. L. Crane, EGTA, a calcium chelator, inhibits electron transport in photosystem of spinach chloroplasts at two different sites, Biochem. Biophys. Res. Communs. 92:206 (1980).
14. C. T. Yerkes and G. T. Babcock, Surface charge asymmetry and a specific calcium effect in chloroplast photosystem II, Biochim. Biophys. Acta 634:19 (1981).
15. N. K. Packham and J. Barker, Stimulation by Mn^{++} and other divalent cations of electron donation reactions to PS II, Biochim. Biophys. Acta 764:17 (1984).
16. K. Satoh and S. Satoh, Inhibition by EDTA and restoration by Mn^{2+} and Ca^{2+} of oxygen evolving activity in photosystem II preparations from the thermophilic cyanobacterium, Synechococcus sp., Biochim. Biophys. Acta 806:221 (1985).
17. R. G. Piccioni and D. C. Marzerall, Calcium and photosynthetic oxygen evolution in cyanobacteria, Biochim. Biophys. Acta 504:384 (1978).
18. N. Tamura, S. Itoh, Y. Yamamoto, and M. Nishimura, Electrostatic interaction between plastocyanin and P700 in the electron transfer reaction of photosystem I enriched particles, Plant Cell Physiol. 22:603 (1981).
19. A. Trebst, Plastoquinone in photosynthetic electron flow, in: "Coenzyme Q," pp. 257-284, G. Lenaz, ed., Wiley, London (1985).
20. R. Barr, R. Melhem, A. L. Lezotte, and F. L. Crane, Stimulation of electron transport from photosystem II to photosystem I in spinach chloroplasts, J. Bionerg. Biomemb. 12:197 (1980).

21. H. W. Jarett, C. J. Brown, C. C. Black, and M. J. Cormier, Evidence that calmodulin is in chloroplasts of peas and serves a regulatory role in photosynthesis, J. Biol. Chem. 257:13795 (1982).
22. S. Muto, Distribution of calmodulin within wheat leaf cells, FEBS Lett. 147:161 (1982).
23. R. W. Sparrow and R. R. England, Isolation of a calcium binding protein from an oxygen evolving PSII preparation, FEBS Lett. 177:95 (1984).
24. R. Barr, K. S. Troxel, and F. L. Crane, Calmodulin antagonists inhibit electron transport in photosystem II of spinach chloroplasts, Biochem. Biophys. Res. Communs. 104:1182 (1982).
25. G. A. Nelson, M. L. Andrews, and M. J. Karnovsky, Control of erythrocyte shape by calmodulin, J. Cell. Biol. 96:730 (1983).
26. M Miyao and N. Murata, Calcium ions can substitute for the 24 kDa polypeptide in photosynthetic oxygen evolution, FEBS Lett. 168:118 (1984).
27. T. Ono and Y. Inoue, Ca^{++}-dependent restoration of O_2-evolving activity in $CaCl_2$ washed PSII particles depleted of 33, 24 and 16 kD proteins, FEBS Lett. 168:281 (1984).
28. H. Y. Nakatani, Inhibition of photosynthetic oxygen evolution by calmodulin-type inhibitors and other calcium antagonists, Biochem. Biophys. Res. Communs., 121:626 (1984).
29. M. Völker, T. Ono, Y. Inoue, and G. Renger, Effect of trypsin on PSII particles. Correlation between hill activity, Mn abundance and peptide pattern, Biochim. Biophys. Acta 806:25 (1985).
30. T. Kuwabara, M. Miyao, T. Murata, and N. Murata, The function of .33 kDa protein in the photosynthetic oxygen-evolution system studied by reconstitution experiments, Biochim. Biophys. Acta 806:283 (1985).
31. C. Critchley, I. C. Baiann, Govindjee, and H. S. Gutowsky, The role of chloride in O_2 evolution by thylakoids from salt-tolerant higher plants, Biochim. Biophys. Acta 682:436 (1982).
32. D. T. Ghanotakis, G. T. Babcock, and C. F. Yocum, Calcium reconstitutes high rates of oxygen evolution in polypeptide depleted photosystem II preparations, FEBS Lett. 167:127 (1984).
33. J. E. Burris and C. C. Black, Jr., Inhibition of coral and algal photosynthesis by Ca^{++} antagonist phenothiazine drugs, Plant Physiol. 71: 712 (1983).
34. S. Muto, S. Izawa and S. Miyachi, Light induced Ca^{++} uptake by intact chloroplasts, FEBS Lett. 139:250 (1982).

THE ROLE OF INTRACELLULAR ORGANELLES IN THE REGULATION OF CYTOSOLIC

CALCIUM LEVELS

Anthony L. Moore*, Michael O. Proudlove*, and Karl E.O. Ackerman**

*Biochemistry Dept., University of Susses
 Brighton BN1 9QG, UK
**Dept. of Biochemistry
 Abo Akadami, 20500 Turku 50, Finland

INTRODUCTION

Within animal cells, the role of calcium as a second messenger of such diverse activities as cell mobility, elongation and division in addition to its role in neurotransmitter and hormone secretion and its effect on the rate of catabolism is well established (Åkerman, 1982; Åkerman and Nicholls, 1983). The cytosolic [Ca^{2+}] in animal cells is in the order of 10^{-7} M (Åkerman and Nicholls, 1983) i.e. orders of magnitude lower than its extracellular concentration. Activation by various excitatory or stimulatory signals results in an increase in the internal free Ca^{2+} concentration by one or two orders of magnitude. This rise in cytosolic Ca^{2+} triggers the above mentioned activities via Ca^{2+}-sensitive enzymes located in the cell. Low cytosolic [Ca^{2+}] is maintained by active extrusion driven by a calmodulin-dependent $Ca^{2+}(Ca^{2+}/Mg)$-ATPase located in the plasma-membrane. It is buffered by transport systems residing in mitochondria, endoplasmic reticulum, and other organelles (Åkerman, 1981; Borle, 1981). Whilst the basic principles of cellular Ca^{2+} regulation as well as Ca^{2+}-dependent enzyme activation have been well documented in animal cells, information on analogous systems in plants is poor. In general, cytosolic [Ca^{2+}] in plants is considered to be low (although no measurements have been reported but see Åkerman et al., 1983) and to be controlled both by active extrusion out of the cell via plasma membrane Ca^{2+}-translocating APTases and by Ca^{2+} sequestering by mitochondria, vacuoles, endoplasmic reticulum and chloroplasts (Moore and Åkerman, 1984).

It is generally assumed that plant mitochondria are similar to their mammalian counterparts in that they possess an efficient mechanism for accumulating Ca^{2+} from the cytosol (see Moore and Åkerman, 1984). However, it is becoming increasingly evident that this is not the case for, although a respiration-dependent net accumulation of Ca^{2+} can be observed with some plant tissues (Chen and Lehninger, 1973; Day et al., 1978), it normally requires high exogenous Ca^{2+} concentrations and is accompanied by a passive binding to the mitochondrial membrane surface. None of the typical responses to Ca^{2+} (such as stimulation of respiration etc.) are observed with plant mitochondria (Moore and Bonner, 1977) except under conditions of massive loading (Chen and Lehninger, 1973). Obviously in order to determine whether or not mitochondria do have a regulatory function in maintaining cytosolic or matrix [Ca^{2+}] in plants, information is required on total mitochondria [Ca^{2+}] and the kinetic parameters of Ca^{2+} uptake into them. This information is also of importance with respect to herbicide action since it has been proposed that primary effect of antimicrotubular herbicides and fungicides is to inhibit Ca^{2+} uptake into mitochondria, thereby resulting in a deregulation of cytoplasmic Ca^{2+} (Hertel et al. 1983).

Relatively few studies have considered chloroplasts as a possible intracellular organelle for sequestering or regulating cytoslic Ca^{2+}. There are a number of reports to suggest that the stromal [Ca^{2+}] is low (see Moore and Åkerman, 1984) and furthermore that the choroplast envelope possesses a light-stimulated ATP dependent Ca^{2+} uptake mechanism which has kinetic parameters comparable to that of animal mitochrondria (Muto et al., 1982). Similarly, little information is available as to the Ca^{2+}-transporting systems residing in the tonoplast membrane of the vacuole even though these organelles occupy as much as 80% of the cell volume and appear to be the main site of accumulation of intracellular calcium (see Roux and Slocum, 1983). It is not known how calcium is accumulated, although the mechanism of uptake is dependent upon ATP and oxalate (Gross, 1982).

The best characterised Ca^{2+}-uptake system in plants are the calmodulin-activated (Ca^{2+}/Mg^{2+})-ATPase of the plasma membrane and the (Ca^{2+}/Mg^{2+})-ATPase of the endoplasmic reticulum and it is suggested that they both play a role in the regulation of cytosolic Ca^{2+} (Marme and Dieter, 1984).

The main aim of this article is to summarise our understanding of calcium accumulation by cytoplasmic organelles and in particular to consider evidence in favour of mitochondria being a major Ca^{2+} buffering organelle.

RESULTS AND DISCUSSION

(a) Mitochondria

In mammalian systems, mitochondria accumulate Ca^{2+}-electrophoretically by Ca^{2+} uniport which is driven by a membrane potential generated by electron transport activity (Åkerman and Nicholls, 1983). Uptake is specifically inhibited by ruthenium red and compounds that interfere with the generation or maintenance of $\Delta\psi$. Ca^{2+} is released on an antiporter in exchange with H$^+$ or Na$^+$ and this cycle is considered to be important in the regulation of intra- and extramitochondrial Ca^{2+}. The existence of such a cycle, or indeed, the presence of an active uniporter in plant mitochondria is still under discussion. Although respiration dependent net accumulation of Ca^{2+} can be observed in some tissues (Chen and Lehninger, 1973; Day et al. 1978; Hertel et al., 1983) its rate of accumulation is slow, it normally requires high exogenous Ca^{2+} concentrations and is accompanied by a passive binding to the membrane surface. With the possible exception of sweet potato mitochondria (Chen and Lehninger, 1973) Ca^{2+} uptake does appear to differ from that of animal cells in many respects.

For instance, as shown in Fig. 1 Ca^{2+} uptake by plant mitochondria when monitored using purified arsenazo III is substrate dependent, inhibited by respiratory chain inhibitors such as CN$^-$ or antimycin A, released upon anaerobiosis or the addition of an uncoupler, such as FCCP. Similar to mammalian systems there is a very fast release of the accumulated Ca^{2+} upon addition of the divalent cation ionophore A23187 suggesting that a Ca^{2+} gradient is formed across the inner membrane as a result of uptake. However, in contrast to the mammalian uniporter, Ca^{2+} uptake by plant mitochondria has an absolute requirement for phosphate. Neither acetate nor any other permeant anion can substitute for phosphate. The

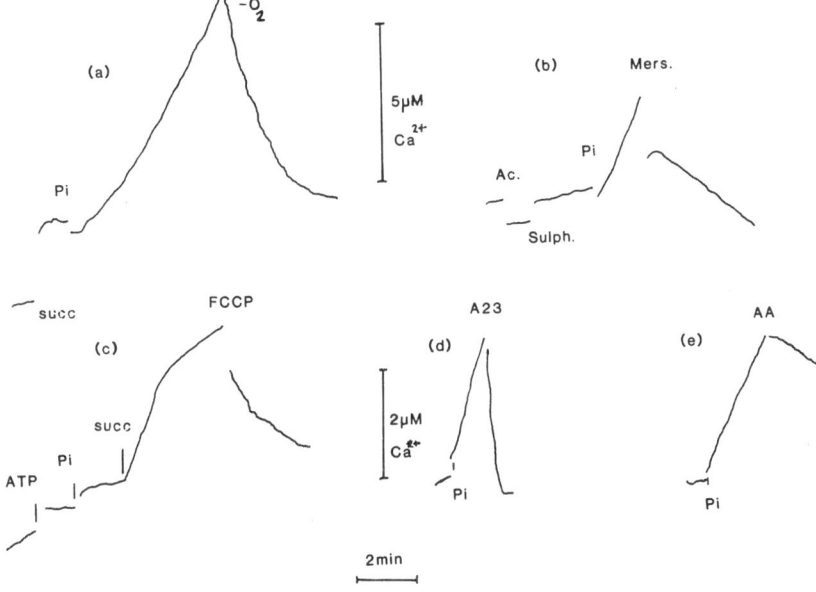

Fig. 1. Respiration dependent Ca^{2+} uptake.

Mung bean mitochondria (approx. 0.5mg) were suspended in basal medium containing Ca^+ and 100μM arsenazo III. Additions: 4mM succinate (succ), 0.4mM KH_2PO_4 (Pi), 2mM ATP, 0.4μM FCCP, 3.2μM A23181, 0.4mM K-Acetate (Ac), 0.4mM K-sulphate (sulph), 0.2mM Mersalyl (mers), 1μg/ml antimycin (AA). Upwards deflection-increase in Ca^{2+} uptake.

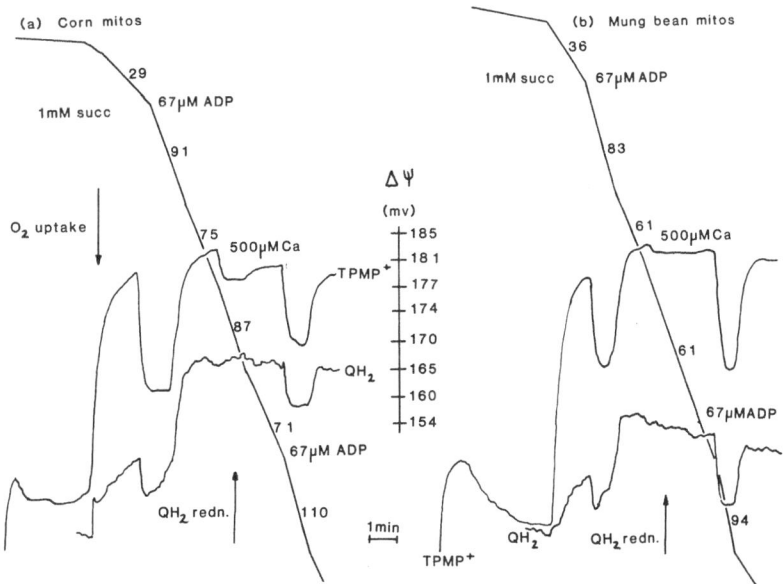

Fig. 2. Effect of Ca^{2+} on repiratory parameters.

O_2 consumption, $\Delta\psi$($TPMP^+$), and steady-state Q (QH_2red^n) were measured simultaneously in 1.5ml reaction vessel containing approx. 0.75mg protein. Upwards deflection - Q reduction and $TPMP^+$ uptake.

phosphate requirement is very probably due to a co-transport of phosphate by the phosphate translocator rather than a direct coupling to Ca^{2+} influx, since net Ca^{2+} uptake is inhibited if the phosphate translocator is blocked by mersalyl (Day et al., 1978; Åkerman and Moore, 1983). As other permeant ions such as acetate, cannot substitute for phosphate it is more likely that the Ca^{2+} is chelated or precipitated by phosphate within the mitochondrial matrix, thus aiding accumulation.

It is now clear that in the sense in which the term is used for mammalian mitochondria there is very little evidence for Ca^{2+} being a permeant cation in plant mitochondria. None of the typical responses to Ca^{2+} addition that are well documented for respiring animal mitochondria are observed with plant mitochondria (See Hanson, 1985; Moore and Bonner, 1977). As indicated in Fig. 2 the addition of 250µM or 500µM Ca^{2+} to mung bean mitochondria oxidising succinate causes a slight depolarisation of the membrane potential, but relatively little effect on steady state levels of ubiquinone or the respiratory rate. The addition of Ca^{2+} to corn seedling mitochondria, however, results in cyclic membrane potential depolarisations and slight respiratory stimulations. In either case, however, the effects on membrane potential or oxidation of the respiratory components is slight when compared to mammalian systems. Although direct measurements of Ca^{2+} uptake (as indicated in Fig. 1 and see Dieter and Marmé, 1980; Hertel et al., 1983; Day et al., 1978; Åkerman and Moore, 1983) are indicative of some form of metabolism-dependent uptake it is apparent that uptake is not by simple uniport and is sufficiently slow enough not to interfere with energy transduction processes. At 10µM external Ca^{2+} we observed (Åkerman and Moore, 1983) an initial rate of Ca uptake into mung bean mitochondria of approximately 5nmol/mg protein/min i.e. a rate which is 2 orders of magnitude lower than that reported for liver mitochondria). Note that this rate was calculated after taking into account endogenous Ca^{2+} which, with some preparations, is considerable. Such high binding Ca^{2+} capacity may account for the energy-independent accumulation observed by others (cf. Hertel et al., 1983).

It has been suggested that the photomorphogenic photoreceptor, phytochrome, can modulate Ca^{2+} levels in the cytosol by affecting mitochondrial Ca^{2+} accumulation both in oat (Roux et al., 1981) and corn seedlings (Yamaya et al., 1984). An apparent increase in efflux and decrease in uptake of mitochondrial Ca^{2+} was reported to occur upon irradiation with red light, which was reversed by far red light. However, in both cases Ca^{2+} accumultion as detected using murexide which is relatively insensitive to Ca^{2+} (Scarpa, 1979, and Dieter and Marmé (1981) were unable, to repeat these results when fluxes were measured using $^{45}Ca^{2+}$.

Data has also been reported to suggest that calmodulin may play a role in the regulation of both Ca^{2+} uptake, the external NADH dehydrogenase activity and ATPase in oat and apple mitochondria (Roux et al., 1981; Fukumoto and Nagai, 1982). This is based upon the observation that calmodulin antagonists, such as trifluoperazine, inhibit uptake and enzyme activity. It must be stressed, however, that such antagonists are powerful inhibitors of the mitochondrial respiratory chain even at micro-molar concentrations (Dunn et a., 1984) and furthermore, in the majority of cases (see Dieter and Marmé, 1980; Dunn et al., 1984; Yamaya et al., 1984) added calmodulin has relatively little effect on either Ca^{2+} uptake or ATPase activity. With respect to the external NADH dehydrogenase, although it is regulated by miromolar concentrations on Ca^{2+} (Moore and Åkerman, 1982) the isolated enzyme is insensitive to either calmodulin or its antagonists (Cottingham and Moore, 1984). Incidently, since the enzymes activity is regulated by Ca^{2+}, we are left with the apparent enigma that at low cytosolic Ca^{2+} levels (in the resting or dark condition) the enzyme is rate limiting. This conclusion obviously has considerable implications with respect to the regulation of electron flux through the external NADH dehydrogenase (see Palmer and Ward, 1985; Moore and Rich, 1985).

In conclusion it would appear that although plant mitochondria do accumulate Ca^{2+} in an energy-dependent fashion, it is not via a simple electrophoretic uniport and furthermore the low rates of transport cast doubt on the postulated role (Dieter and Marmé, 1980) of plant mitochondria in the regulation of cytosolic Ca^{2+} in the plant cell.

(b) Chloroplasts

Several investigators have presented results that suggest that the free Ca^{2+} concentration within the chloroplast stroma is low (for details see Moore and Åkerman 1984) and a net accumulation into the stroma occurs upon illumination (Muto et al., 1982) which is released in the dark. The light dependent uptake has been reported to require either exogenous or endogenous ATP and is inhibited when photophosphorylation is uncoupled by NH_4Cl.

Several lines of evidence suggest that NAD kinase, which is calmodulin activated, is localised in the chloroplast stroma (Jarrett et al., 1982; Muto et al., 1981) although in some species a considerable proportion is found in the cytosol (Simon et al., 1982). NAD kinase is an important regulatory component in the light-induced conversion of NAD to NADP, the terminal electron acceptor of photosystem I. The activation of this enzyme in situ by Ca^{2+} has been reported (Cormier et al., 1981; Jarrett et al., 1982) as indeed, has its sensitivity to calmodulin antagonists (Jarrett et al., 1982).

It may be concluded, on the basis of the aforementioned data, that chloroplasts have the capacity to maintain a low Ca^{2+} concentration within the stoma and that light dependent Ca^{2+} uptake results in an increase in stromal free Ca^{2+} which not only regulates NAD kinase but also may be of prime importance in regulating photosynthetic metabolism since it would trigger the production of NADP with the end result of providing both NADPH for the reductive pentose phosphate pathway and for the indirect transfer of reducing equivalents to the cytosol via metabolite shuttles.

(c) Vacuoles

To date, there is relatively little published information as to either the Ca^{2+} concentration, or the Ca^{2+}-transporting systems associated with the vacuole even though these organelles occupy as much as 80% of the cell volumne and appear to be the main site of accumulation of intracellular Ca^{2+} (see Macklon, 1984). Evidence has been presented by both Gross (1982) and Marin et al., (1982) to suggest that tonoplast vesicles take up Ca^{2+} in the presence of ATP and oxalate. The data of Marin et al. (1982) indicate the presence of a Ca^{2+}/H^+ antiporter in these vesicles, although no data concerning its characteristics or kinetics have been presented.

(d) Endoplasmic reticulum and plasma membrane vesicles

The best charactersed Ca^{2+} uptake systems reside in the endoplasmic reticulum and plasma membrane. It would appear from the literature that microsomal fractions isolated from plant tissues possess a calmodulin-activated $(Ca^{2+} + Mg^{2+})$ ATPase located in the plasma membrane and involved in Ca^{2+} extrusion from the cell as well as a $(Ca + Mg^{2+})$ ATPase involved in Ca^{2+} accumulation by the endoplasmic reticulum. Recent data (Buckout, 1983, 1984) on the characteristics of Ca^{2+} uptake into purified endoplasmic reticulum vesicles reveal that it is oxalate-dependent, specific for ATP and the enzyme has an apparent affinity for Ca^{2+} in the µM region. Furthermore there was no indication that calmodulin regulated Ca^{2+} uptake into these vesicles (Buckout, 1984). For a more exhasustive review of this topic readers are referred to other more specific articles in this book.

CONCLUSIONS

The role of cytoplasmic organelles in the maintenance of low cytosolic Ca^{2+} in higher plant cells is still controversial. The low Ca^{2+} transporting activity of mitochondria indicates that cytosolic Ca^{2+} Cannot be maintained by sequestration and extrusion by this organelle. A clearer picture emerges from the studies with chloroplasts, in so much as the presence of a light-mediated uptake system along with a Ca^{2+} stimulated NAD-kinase, suggests that Ca^{2+} activation is an important regulatory component in photosynthesis. Whether chloroplasts play a regulatory role in the maintenance of cell Ca^{2+} is uncertain. The role of endoplasmic reticulum in the reglation of cytosolic Ca^{2+} does appear more certain and in view of its high affinity for Ca^{2+} and large surface area it may be considered to be the principle cytoplasmic organelle involved in Ca^{2+} homeostasis.

ACKNOWLEDGEMENTS

The experimental work described was supported by the AERC (ALM), the Academy of Finland and the Sigrid Juselius Foundation (K.E.O.Å).

REFERENCES

Åkerman, K.E.O. 1982. Ca^{2+} transport and cell activation, Med.Biol., 60:168.

Åkerman, K.E.O. and Moore, A.L., 1983, Phosphate dependent, ruthenium red insensitive Ca^{2+} uptake in mung bean mitochondria. Biochem. Biophys. Res. Commun., 114:1176.

Åkerman, K.E.O. and Nicholls, D.G., 1983, Physiological and bioenergetic aspects of mitochondrial transport, Revs. Physiol. Biochem. Pharmacol., 95:150.

Åkerman, K.E.O., Proudlove, M.O., and Moore, A.L., 1983, Evidence for a Ca^{2+} gradient across the plasma membrane of wheat protoplasts, Biochim. Biophys. Res. Commun., 113:171.

Borle, A.B., 1981, Control, modulation and regulation of cell calcium, Revs. Physiol. Buichem. Pharmacol., 90:13.

Buckout, T.J., 1983, ATP-dependent Ca^{2+} transport in endoplasmic reticulum isolated from roots of Lepidium sativum, Planta, 159:84.

Buckout, T.J., 1984, Characterisation of Ca^{2+} transport in purified endoplasmic reticulum membrane vesicles from Lepidium sativum L. roots, Plant Physiol., 76:962.

Chen, C-H. and Lehninger, A.L., 1973, Ca^{2+} transport activity in mitochondria from some plant issues, Arch. Biochem. Biophys., 157:183.

Cormier, M.J., Charbonneau, H. and Jarrett, H.W., 1981, Plant and fungal calmodulin: Ca^{2+} dependent regulation of plant NAD kinase, Cell Calcium., 2:313.

Cottingham, I.R. and Moore, A.L. 1984, Partial purification and properties of the external NADH dehydrogenase from cuckoo-pint (Arum maculatum) mitochondria, Biochem. J., 224:171.

Day, D.A., Bertagnolli, B.L. and Hanson, J.B., 1978, The effect of Ca^{2+} on the respiratory responses of corn mitochondria, Biochim. Biophys. Acta, 502:289.

Dieter, P. and Marmé, D., 1980, Ca^{2+} transport in mitochrondrial and microsomal fractions from higher plants, Planta, 150:1.

Dieter, P. and Marmé, D., 1981, Far-red light irradiation of intact corn
    seedlings affects mitochondrial and calmodulin dependent microsomal
    Ca^{2+} transport, Biochem. Biophys. Res. Commun., 101:749

Dunn, P.P.J., Slabas, A.R., Cottingham, I.R. and Moore, A.L., 1984,
    Trifluoperazine inhibition of electron transport and adenosine
    triphosphatase in plant mitochondria, Arch.Biochem. Biophys., 229:287.

Fukomoto, M. and Nagai, K., 1982, Effect of calmodulin antagonists on the
    mitochondrial and microsomal Ca^{2+} uptake in apple fruit, Plant Cell
    Physiol., 23:179.

Gross, J., 1982, Oxalate-enhanced active calcium uptake in membrane
    fractions from zucchini squash, in: "Plasmalemma and Tonoplast", D.
    Marmé, E. Marré and R. Hertel, eds., Elsevier Biomedical Press,
    Amsterdam.

Hanson, J.B., 1985, Membrane transport systems of plant mitochondria,
    in: "Encyclopedia of Plant Physiol." Vol. 18, R. Douce and D.A. Day,
    eds, Springer-Verlag, Berlin.

Hertel, C., Affolter, H., Marmé, D. and Dierks-Ventling, C., 1983, Effects
    of amiprophosmethyl on massive and limited calcium loading of maize
    mitochondria. Biochem. J., 212:567.

Jarrett, A.W., Brown, C.J., Black, C.C. and Cormier, M.J., 1982, Evidence
    that calmodulin is in the chloroplast of peas and serves a regulatory
    role in photosynthesis, J. Biol. Chem., 257:13795.

Macklon, A.E.S., 1984, Calcium fluxes at plasmalemma and tonoplast,
    Plant, Cell Environ., 7:407.

Marmé, D. and Dieter, P., 1984, The role of Ca^{2+} and calmodulin in
    plants, in: "Calcium and Cell Function", Vol. IV, W.Y. Cheung, ed.
    Academic Press, New York.

Marin, B., Cretin, H. and d'Auzac, J., 1982, Energisation of solute
    transport and accumulation in the Hevea latex vacuole, in: Plasmalemma
    and Tonoplast;: their functions in the Plant Cell", E. Marme, E.
    Marre and R. Hertel, eds, Elsevier Biomedical Press, Amsterdam.

Moore, A.L. and Åkerman, K.E.O., 1982, Ca^{2+} stimulation of the
    external NADH dehydrogenase in Jerusalem artichoke (Helianthus
    tuberosus) mitochondria, Biochem. Biophys. Res. Commun., 109:513.

Moore, A.L. and Åkerman, K.E.O., 1984, Calcium and plant organelles,
    Plant Cell and Environ., 7:423.

Moore, A.L. and Bonner, W.D. Jr., 1977, The effect of calcium on the
    respiratory responses of mung bean mitochondria, Biochim. Biophys.
    Acta., 460:455.

Moore,A.L. and Rich, P.R., 1985, Organisation of the respiratory chain and
    oxidative phosphorylaton, in: "Encyclopedia of Plant Physiol." Vol.
    18, R. Douce and D.A. Day eds., Springer-Verlag, Berlin.

Muto, S., Hiyachi, S., Usuda, H., Edwards, G.E. and Bassham, J.A., 1981,
    Light-induced conversion of nicotinamide adenine dinucleotide to
    nicotinamide adenine dinucleotide phosphate in higher plant leaves,
    Plant Physiol., 68:324.

Muto, S., Izawa, S. and Hiyachi, S., 1982, Light-induced Ca^{2+} uptake by
    intact chloroplasts, FEBS Letts., 139:250.

Palmer, J.M. and Ward, J.A. 1985, The oxidation of NADH by plant
    mitochondria, in: "Encyclopedia of Plant Physiol." Vol. 18, R. Douce
    and D.A.Day, eds., Springer-Verlag, Berlin.

Roux, S.J. and Slocum, R.D., 1983, Role of calcium in mediating cellular
    functions important for growth and development in higher plants, in:
    "Calcium and Cell Function", Vol. III, W.Y. Cheung ed., Academic
    Press, New York.

Roux, S.J., McEntire, K., Slocum, R.D., Dedel, T.E. and Hale, C.C. 1981,
    Phytochrome induces photoreversible calcium fluxes in a purified
    mitochondrial fraction from oats, Proc. Natl. Acad. Sci. USA., 78:283.

Scarpa, A., 1979, Measurements of cation transport with metallochromic
    indicators, Meth. in Enzymol., 56, 301.

Simon, P., Dieter, P., Bonzon, M., Greppin, H. and Marmé, D. 1982, Calmodulin depndent and independent NAD kinase activity from cytoplasmic and chloroplastic fractins of spinach (Spinacia oleracea L.) Plant Cell Rep., 1:119.

Yamaya, T., Oaks, A. and Matsumoto, H. 1984, Stimulation of mitochondrial calcium uptake by light during growth or corn shoots, Plant Physiol., 75, 773.

CALCIUM INVOLVEMENT IN PLANT HORMONE ACTION

Daphne C. Elliott

School of Biological Sciences
The Flinders University of South Australia
Bedford Park, S.A. 5042

ABSTRACT

Many hormones and extracellular stimuli in animal systems regulate
cell function by inducing an increase in calcium concentration in the
cytoplasm. The evidence that plant hormones also work through calcium as
a second messenger is reviewed in this paper. Modulation of hormone responses by added calcium,by inhibitors of calcium transport or by calcium
ionophores indicates calcium involvement and there are many instances of
this phenomenon in plants. Hormone effects on intracellular calcium levels
have not been so widely studied, but some effects on calcium transport
have been reported. The current model of how the calcium messenger system
works is examined with particular reference as to where the gaps are in
applying this model to transmembrane signalling in plants.

INTRODUCTION

The importance of calcium in plants has long been recognised,
initially in the structural role related to membrane integrity and cell
wall mechanical properties (Jones and Lunt, 1967) and more recently in
processes such as ion transport, cell polarity and cytoskeletal protein
activities such as spindle function in cell division, cell plate formation
and cell wall biogenesis (Roux and Slocum, 1982). How such calcium related
events are ordered by environmental and other stimuli is largely unexplored.
In animals the regulation of a wide variety of cellular responses by the
use of calcium as an intracellular "second messenger" mediating extracellular signals is now accepted without reservation particularly for cell
functions such as contraction, ion transport and secretion (Rasmussen et
al., 1984). In plants, however, the application of this cellular control
system to the mode of action of plant hormones and other stimuli has
so far not had much experimental support. What is lacking in the plant
field is information at the molecular level on how calcium transport
from membrane, organelle and cell wall stores is controlled and what are
the initial events that trigger the biological processes which depend on
changes in cytoplasmic calcium concentration.

This paper reviews the evidence for modulation of hormone responses
by calcium and the evidence for hormone effects on calcium transport and/
or binding. A cytokinin-dependent response in Amaranthus seedlings, the
induction of betacyanin, is used as the model hormone system to illustrate

Table 1. Calcium-modulated Hormone Responses

Physiological Response	Calcium-mediated Process	References
AUXIN		
Elongation growth enhanced	H^+ secretion stimulated by low concentrations	Cohen and Nadler, 1976[a]
Elongation growth inhibited	H^+-dependent biochemical wall-loosening processes inhibited by high concentrations	Cleland and Rayle, 1977[a]
Tanada effect	Cell surface potential (hyperpolarization)	Tanada, 1968
Polar transport	Auxin-binding to membranes or hyperpolarization of plasma membrane	Dela Fuente and Leopold, 1973[a]; Poovaiah and Leopold, 1976[a], Higinbotham et al., 1967[a]
Elongation growth	Anticalmodulin compounds stimulate at low concentrations, inhibit at higher levels	Elliott et al., 1983
CYTOKININ		
Senescence retardation	Calcium enhances	Poovaiah and Leopold, 1973[b]
Cotyledon enlargement	Calcium stimulates	Leopold et al., 1974[b]
Cell expansion	Calcium stimulates at low concentration, additive with cytokinin	Ralph et al., 1976[b]
Ethylene production	Calcium stimulates	Lau and Yang, 1975[b]
Anthocyanin synthesis	Calcium inhibits, EDTA promotes	Bassim and Pecket, 1975[b]
Betacyanin synthesis	Calcium at low concentrations, EGTA, ruthenium red stimulate; calcium at higher levels, A23187 inhibit	Elliott, 1976, 1979; Table 2, this paper
Betacyanin synthesis, cell enlargement, cell growth	Anticalmodulin compounds stimulate at low concentrations, inhibit at higher levels	Elliott, 1983
GIBBERELLIC ACID		
Hypocotyl growth	Ruthenium red inhibition of Ca^{2+} uptake and efflux inhibits GA_3-induced growth; EDTA, EGTA stimulates	Moll and Jones, 1981
α-Amylase synthesis, secretion and release	Calcium necessary	Carbonell and Jones 1984
α-Amylase synthesis	Anticalmodulin compounds stimulate at low concentrations, inhibit at higher levels.	Elliott et al., 1983

[a]Ref. given in Roux and Slocum, 1982.
[b]Ref. given in Elliott, 1983.

calcium modulation and to explore some areas where there are gaps in the calcium second messenger system as applied to plants.

CALCIUM-MODULATED HORMONE RESPONSES

The overlapping of functions of cytokinins, auxins and gibberellic acid lends itself to the concept that their mode of action may be similar, depending for specificity on pre-determined target cells. For example, construction of ionic channels in membranes may involve the assembly of modular units so that activation of a controlling channel may depend on any one of a number of sensors, present in different cell types, and activatable by different effectors. The many indications of calcium interaction in plant hormone and phytochrome action and in plant tropic responses suggests that calcium may play a pivotal role in a common mode of action.

The actions of all plant hormones, in their regulation of germination, growth or senescence are modulated by the presence of Ca^{2+}. Table 1 gives examples of processes where calcium modulation has either been shown directly following supply of exogenous calcium, or inferred by effects of calcium ionophores or inhibitors. This subject has been reviewed previously by Bangerth (1979), Leopold (1977) and Van Steveninck (1976).

The difficulty of demonstrating an effect of added Ca^{2+} in many plant tissues is due to high levels of stored Ca^{2+} particularly in cell walls and vacuoles and to inhibitory effects if Ca^{2+} reaches high levels in the cytoplasm. For example, A23187, the calcium ionophore, inhibits benzyladenine-dependent betacyanin synthesis (Elliott, 1976) as does $CaCl_2$ at 1 mM and $Ca(NO_3)_2$ at 5 mM (Elliott, 1979), but if cell wall calcium is depleted by carrying out a heat/aging pretreatment in the presence of EGTA, then A23187 and added calcium can be shown to stimulate (Table 2). Experiments 1 and 2 show that 1 mM $Ca(NO_3)_2$ gives a greater effect after EGTA, while Experiment 2 compared with Experiment 1 shows that A23187 increases the % stimulation by calcium.

HORMONE EFFECTS ON CALCIUM TRANSPORT OR BINDING

Measurement of cytoplasmic calcium ($[Ca^{2+}]_{cyt}$) in plants has indicated a value close to 10^{-7} M (Williamson, 1981). Since this is low compared to external (apoplastic) and total cell concentration, large changes can take place following altered influx, sequestration or binding.

Recently there has been some work on the calcium transport systems present in plants and it appears that calcium entry into plant cells is passive and that cytoplasmic calcium levels are controlled both by active calcium transport out of the cell via plasma membrane Ca^{2+}-translocating ATPase and by Ca^{2+}-sequestering by the mitochondria, vacuoles and possibly other organelles (Marmé, 1983). Movements of calcium at the cellular and tissue levels have been shown in the case of phototropic and gravitropic stimuli but there is less evidence of similar movement in response to plant hormones (Roux and Slocum, 1982). Effects of hormones on calcium transport and intracellular distribution have, however, been demonstrated in both in vivo and in vitro systems. Table 3 gives some examples of these effects.

MODEL OF THE CALCIUM MESSENGER SYSTEM

The current model of how calcium works as a second messenger in animals (Rasmussen et al., 1984) involves interaction of a hormone with a plasma membrane receptor which results in phosphoinositide hydrolysis

Table 2. Effect of EGTA Pre-treatment on Betacyanin Synthesis in *Amaranthus* Seedlings

Pre-treatment with EGTA[a]	Addition to Induction Medium[b]	Betacyanin Content[c] nmol/seedling		BA-dependent Synthesis	
		−BA	+BA	% Control	Effect Ca^{2+} (%)
Experiment 1					
−	Nil	0.043 ± .006	0.200 ± .011	100	
−	+ $Ca(NO_3)_2$	0.029 ± .003	0.305 ± .020	108	+ 8
+	Nil	0.038 ± .009	0.266 ± .011	89	
+	+ $Ca(NO_3)_2$	0.048 ± .005	0.337 ± .023	113	+27
Experiment 2					
−	A23187	0.033 ± .001	0.258 ± .024	100	
−	A23187 + $(Ca(NO_3)_2$	0.029 ± .006	0.322 ± .008	130	+30
+	A23187	0.029 ± .001	0.268 ± .021	107	
+	A23187 + $Ca(NO_3)_2$	0.046 ± .002	0.380 ± .018	148	+39

[a]*Amaranthus tricolor* seeds were germinated at $25^{o}C$ for 88 hr in the dark and the cotyledons plus hypocotyls from 1g seeds cut into either 20 ml distilled H_2O or 20 ml 1 mM EGTA, pH 7.4, in a 15 cm Petri dish. Dishes were placed at $40^{o}C$ in the dark for 1.5 hr and further 1.5 hr with the lid removed at $25^{o}C$. This heat/shock aging has been shown to potentiate cytokinin action (Elliott, 1982). The half-seedlings were then drained, washed in 200 ml distilled water and replaced in 20 ml fresh distilled H_2O. Cotyledons plus the top 5 mm hypocotyl were pinched off into the induction medium with two pairs of forceps (40 half-seedlings/5 ml induction medium).

[b]The basic induction medium was 5 mM L-tyrosine (betacyanin precursor) in 10 mM K_2HPO_4-NaH_2PO_4 (pH 6.8) ± 0.5 µM BA and the incubation was at $25^{o}C$ for 24 hr in the dark. $Ca(NO_3)_2$ and A23187 were added at 1 mM and 1 µM respectively, where indicated.

[c]Betacyanin was extracted in 3.33 mM acetic acid and measured at 537 nm (Elliott, 1979). Values are means ± SE (n=3).

and an increase in diacylglycerol and a rise in $[Ca^{2+}]_{cyt}$ (Fig.1). While there is some disagreement about which phosphoinositide is broken down by the phospholipase C-activated hormone occupation of cell surface receptors (Berridge, 1984; Majerus et al., 1985) it seems clear that both second messengers (diacylglycerol and Ca^{2+}) are essential for full cellular response. Together they activate protein kinase C which catalyses the phosphorylation of a distinct sub-set of cellular proteins of which one may be a cellular tyrosine kinase that phosphorylates a protein (P42) which appears to be involved in the regulation of cellular multiplication (Sefton and Hunter, 1984). The rise in $[Ca^{2+}]_{cyt}$ also leads to the activation of calmodulin-dependent reactions either directly by activation of an enzyme or, if that enzyme is a protein kinase, then indirectly by phosphorylation of a different sub-set of cellular proteins.

Table 3. Hormone Effects on Calcium Transport/Binding

Plant Material	Effect	References
AUXIN		
Pea and bean plants	Auxin-transport inhibitors inhibit calcium translocation	Wieneke et al., 1971[a] Goswami and Audus, 1976[a]
Soybean protoplasts	Calcium uptake inhibited, efflux increased	Cohen and Lilly, 1984
Soybean hypocotyl	Auxin binding to isolated membranes promotes release of calcium, decreasing the thickness of membranes	Buckhout et al., 1981[a] Morré and Bracker, 1976[a]
Rice root	Ca^{2+} ATPase in vivo and in vitro stimulated by auxin	Erdei et al., 1979
Elongating zone of soybean hypocotyl	Pre-treatment with auxin increases ATP-dependent calcium transport in plasma membrane fractions	Kubowicz et al., 1982[b]
CYTOKININ		
Mung bean hypocotyl	Kinetin increases calcium uptake (and vice versa)	Lau and Yang, 1975[b]
Fungus _Achyla_	Increased calcium uptake; cytokinins allosterically regulate binding of calcium to cell surface glycoprotein	Le John et al., 1974[b]
Funaria	Localisation of membrane-associated calcium at presumptive bud site in target caulonema cells after cytokinin treatment	Saunders and Heppler, 1981[b]
Meristematic (or maturing) zones of soybean hypocotyl	Pre-treatment with cytokinin increases ATP-dependent calcium transport in plasma membrane fractions	Kubowicz et al., 1982[b]
Wheat root	BA-induced coupling between calmodulin and Ca^{2+}-ATPase in microsomal membranes	Olah et al., 1983
GIBBERELLIC ACID		
Lettuce hypocotyl	Release of Ca^{2+} from cell wall by GA_3 leads to wall loosening and growth	Moll and Jones, 1981

[a]Ref. given in Roux and Slocum, 1982.
[b]Ref. given in Elliott, 1983.

$[Ca^{2+}]_{cyt}$ increase is caused by release from intracellular stores (endoplasmic reticulum) by IP_3, another product of PIP_2 breakdown catalysed by phospholipase C. Transport of Ca^{2+} across plasma membrane is also altered by hormones but there is no evidence yet that IP_3 is involved.

If this model is applied to plants there are some clear parallels. Plant calmodulin is similar in many respects (e.g. calcium-binding properties) to that from other organisms (Cox et al., 1984) although with differences in electrophoretic mobility and in amino acid composition (Marmé, 1983). Some Ca^{2+}-calmodulin dependent enzymes have been found in plants (NAD kinase, Ca^{2+}-ATPase, see Roux and Slocum, 1982 and Marmé, 1983, for refs.; quinate : NAD^+ oxidoreductase, Ranjeva et al., 1983; protein kinases, see Elliott and Skinner, 1985, for refs.). Recently a phospho-lipid-stimulated protein kinase has been demonstrated in plant extracts which is similar to protein kinase C in its behaviour on a DE52 cellulose column, its substrate specificity and its calcium dependence (Elliott and Skinner, 1985). Moreover the identification of tyrosine phosphorylation in plant proteins and its apparent increase in cells dividing in response to auxin and cytokinin (Elliott and Geytenbeek, 1985) suggests that the postulated mechanism for protein kinase C phosphorylation of a protein necessary for cellular multiplication (Sefton and Hunter, 1984) may be operating in plants. Other effects of hormones on protein phosphorylation patterns have been reported (e.g. Morré et al., 1984; Elliott, 1985).

The major gaps in our knowledge in applying this model (Fig. 1) to plants are in the area of the cyclic turnover of inositol lipids. There

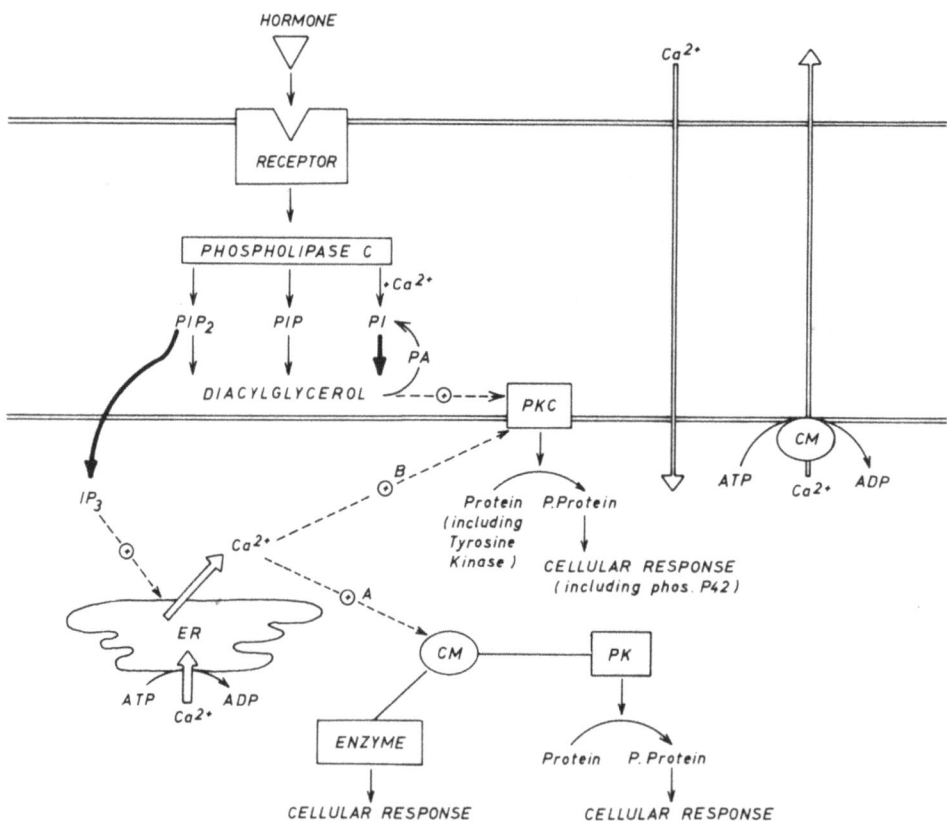

Fig. 1.   Model to illustrate the diacylglycerol and calcium signal pathways. A, amplitude modulation pathway, responsible to initiating; B, sensitivity modulation pathway, responsible for sustaining (Rasmussen et al., 1984); CM, calmodulin; PKC, protein kinase C; PA, phosphatidic acid; Pl, phosphatidylinositol; PIP, phosphatidyl-inositol 4-phosphate; PIP_2, phosphadidylinositol 4,5-bisphosphate; IP_3, inositol 1,4,5-trisphosphate.

Table 4. Effect of Diacylglycerol on Betacyanin Synthesis in <u>Amaranthus</u> Seedlings.

Pre-treatment with EGTA[a]	Addition to Induction Medium[b]	Betacyanin Contant[c] nmol/seedling		Synergism[d]	
		-BA	+BA	-BA	+BA
Experiment 1					
-	Nil	0.058	0.358		
-	DOG, 10 µg/ml	0.088	0.413		
-	DOG, 25 µg/ml	0.095	0.387		
-	DOG, 50 µg/ml	0.088	0.342		
Experiment 2					
+	Nil	0.022	0.097		
+	A23187 + Ca(NO$_3$)$_2$	0.037	0.267		
+	DOG, 10 µg/ml	0.028	0.122		
+	A23187 + Ca(NO$_3$)$_2$ + DOG, 10 µg/ml	0.045	0.323	+52	+14

[a] <u>A. tricolor</u> half-seedlings were pre-treated as in Table 2, except that seedcoats were removed and 20 seedlings/1 ml induction medium was used. Temperature in Experiment 2 was at 42°C for the heat pre-treatment.

[b] Induction medium was as in Table 2. Dioctanoylglycerol (DOG) was prepared by the action of phospholipase on dioctanoyl-L-α-phosphatidyl-choline (Sigma), A solution in chloroform was dried down under N$_2$ and sonicated for 2' at 4°C in buffer containing 1% DMSO just before use. In treatments + DOG, 10 µg was added at the beginning of the induction, and a further 10 µg after 2 hr to allow for the possibility that the DOG was being rapidly degraded by the seedlings.

[c] Betacyanin assay as in Table 2.

[d] Percentage increase due to synergism between Ca^{2+} and DOG is calculated after subtracting the appropriate control ($-Ca^{2+}$ -DOG) and is defined as:

$$\frac{(Ca^{2+} + DOG) - Ca^{2+} - DOG}{(Ca^{2+} + DOG)} \times 100$$

has been some evidence of increased inositol phospholipid turnover in response to auxin (Monroe et al., 1982) and to cytokinin (Batchelor, 1984; Elliott, unpublished work). The inositol analogue, gammaexane (Lindane, γ-hexachlorocyclohexane) which interferes with phosphatidylinositol metabolism and inhibits phospholipid turnover in membranes was found to prevent the potentiation of cytokinin-induction of betacyanin synthesis caused by a pre-treatment of heat shock and aging (Elliott, 1982). The cytokinin-dependent response itself however was not markedly inhibited by gammaexane.

While the areas of polyphosphoinositide identification and diacyl-glycerol production have been largely ignored, recent experiments have shown stimulatory effects of exogenous diacylglycerol in <u>Amaranthus</u> betacyanin system (Table 4). The synergism seen in Experiment 2 between added calcium and dioctanoylglycerol is characteristic of the dual signal pathway in the calcium messenger system (Rasmussen et al., 1984).

These recent results, together with earlier indications in the key area of control of cytosol calcium concentration summarized above (Table 3), have now mapped out a framework for applying the model of the calcium-messenger system to plants which justifies further work in the area.

Modulation of phytochrome action or plant tropisms by calcium, or changes in cellular calcium ion fluxes by these stimuli, have not been reviewed here (see Roux and Slocum, 1982; Roux, 1984). The many parallels with calcium and plant hormone action however make it clear that the model of calcium as a second messenger could apply to all these extracellular signals.

REFERENCES

Bangerth, F., 1979, Annu. Rev. Phytopathol., 17: 97.
Batchelor, S.M., 1984, Cytokinin induced responses in Amaranthus tricolor, Ph.D. Thesis, Flinders University of South Australia.
Berridge, M.J., 1984, Biochem. J., 220: 345
Carbonell, J. and Jones, R.L., 1984, Physiol. Plant., 63: 345.
Cohen, J.D. and Lilly, N., 1984, Plant Physiol., 75S: 109.
Cox, J.A., Comte, M., Malnoë, A., Burger, D. and Stein, E.A., 1984, Mode of action of the regulatory protein calmodulin, in: "Metal Ions in Biological Systems", Vol. 17, H. Sigel, ed., Marcell Dekker, Inc.
Elliott, D.C., 1976, Abstracts, 9th International Conference on Plant Growth Substances, Lausanne, pp. 86-87.
Elliott, D.C., 1979, Plant Physiol., 63: 264.
Elliott, D.C., 1982, Plant Physiol., 69: 1169.
Elliott, D.C., 1983, Plant Physiol., 72: 215.
Elliott, D.C., 1985, Protein phosphorylation in normal and transformed cells: effect of cytokinins on calcium regulation in vivo and in vitro, Abstracts, This Vol.
Elliott, D.C., Batchelor, S.M., Cassar, R.A. and Marinos, N.G., 1983, Plant Physiol., 72: 219.
Elliott, D.C. and Geytenbeek, M., 1985, Biochim. Biophys. Acta, 845: 317.
Elliott, D.C. and Skinner, J.D., 1985, Phytochemistry (in press).
Erdei, L., Toth, I. and Zsoldos, F., 1979, Physiol. Plant, 45: 448.
Jones, R.G.W. and Lunt, O.R., 1969, Bot. Rev., 33: 407.
Leopold, A.C., 1977, Adv. Chem. Ser., 159: 33.
Majerus, P.W., Wilson, D.B., Connolly, T.M., Bross, T.E. and Neufeld, E.J., 1985, Trends in Biochem. Sci., April: 168.
Marmé, D., 1983, Encyl. Plant Physiol., New. Ser. 15: 599.
Moll, C.M. and Jones, R.L. 1981, Planta, 152: 450.
Monroe, A., Gripshover, B. and Morré, D.J., 1982, Plant Physiol., 69S: 151.
Morré, D.J., Morré, J.T. and Varnold, R.L., 1984, Plant Physiol., 75: 265.
Olah, Z., Berczi, A. and Erdei, W., 1983, FEBS Lett., 154: 395.
Ranjeva, R., Refeno, G., Boudet, A.M. and Marmé, D., 1983, Proc. Nat. Acad. Sci., 80: 5222.
Rasmussen, H., Kojima, I., Kojima, K., Zawalich, W. and Apfeldorf, W., 1984, Adv. Cyc. Nuc. Prot. Phos. Res., 18: 159.
Roux, S.J., 1984, BioScience, 34: 25.
Roux, S.J. and Slocum, R.D., 1982, Role of calcium in mediating cellular functions important for growth and development in higher plants, in: "Calcium and Cell Function", Vol.111, W.Y. Cheung, ed., Academic Press, Inc.
Sefton, B.M. and Hunter, T., 1984, Adv. Cyc. Nuc. Prot. Phos. Res., 18: 195.
Tanada, T., 1968, Plant Physiol., 43: 2070.
Van Steveninck, R.F.M., 1976, Encycl. Plant Physiol., New Ser., 2: 307.
Williamson, R.E., 1981, What's New in Plant Physiology, 12: 45.

CALCIUM MODULATION OF AUXIN-MEMBRANE INTERACTIONS IN PLANT CELL ELONGATION

D. James Morré

Department of Medicinal Chemistry and Department of
Biological Sciences
Purdue University, West Lafayette, IN 47907 U.S.A.

INTRODUCTION

Regulatory molecules of the auxin type specifically accelerate elongation growth in plants and in excised plant parts. Despite more than 50 years of intensive investigation, very little is known concerning the molecular details of elongation growth in plants or how auxin regulation of elongation growth is achieved.

It is usual, however, that effector (hormone)-responsive systems include at least three essential elements:

1) A receptor site to recognize and bind the extracellular signal (hormone),

2) Some form of transduction mechanism that recognizes a change in the configuration or conformation of the receptor, and

3) An amplifier that translates the received message into normally an increase (or decrease) in the intracellular concentration of a chemical species capable of exerting control over intracellular processes. These second messengers can then diffuse away from the membrane into the cytosol and carry the receptor message to remote parts of the target cell usually with the initiation of a biochemical response cascade (Michell et al., 1977).

Calcium is well established as a second messenger, alone or in combination with calmodulin, in other systems (Marmé, 1985). In this report, a role for calcium as a second messenger important to modulation of auxin-responsive plant systems will be discussed.

MATERIALS AND METHODS

The plant material was etiolated hypocotyls of soybean (Glycine max [L] Herr, var. Wayne) germinated and grown 4 days in darkness. Segments 1-2 cm long, cut 5 mm below the cotyledons, were harvested under ordinary laboratory light and used for membrane isolations. The hypocotyl segments were homogenized for 60 to 90 s at 60 chops/s with a mechanized razor blade chopper (Morré, 1971) in 1 to 2 volumes of 0.25 to 0.3 M sucrose, buffered at pH 6.5 with a Tris buffer containing, normally, 10 mM KCl and 1 mM $MgCl_2$

or a similar medium prepared in pre-centrifuged coconut water as solvent. To remove cell walls, homogenates were filtered through a single layer of Miracloth (Chicopee Mills, New York). To remove starch, starch plastids and nuclei, an initial centrifugation for 5,000 to 10,000 X g for 10 min was used. This was followed by a second centrifugation of 45,000 X g for 20 min to yield a fraction enriched in surface (plasma membrane and tonoplast) membranes but containing also endoplasmic reticulum, Golgi apparatus and mitochondria. This pellet, freshly prepared and used immediately upon resuspension, was employed as the primary cell-free system to investigate the role of calcium in the auxin-membrane response cascade.

RESULTS

The first two steps in the stimulus-response cascade of auxin action are supported by available experimental evidence. Soybean membranes, including plasma membranes (Fig. 1) were shown to bind auxin (Williamson et al., 1977). Subsequently, soybean membranes were shown to undergo a change in conformation upon exposure to auxin (Morré and Bracker, 1976; Helgerson et al., 1976) and to release divalent ions, including calcium (Buckhout et al., 1981) with a dose dependency that paralleled the growth response of tissue sections to auxin. For both the "membrane thinning" response seen by electron microscopy of plasma membrane vesicles stained by phosphotungstic acid at low pH (Roland et al., 1972), and the corresponding increase in microviscosity measured by fluorescence polarization, as well as the release of divalent ions, we observed an optimum curve where auxin concentrations either higher or lower than the optimum resulted in a diminished response.

To account for the release of divalent ions as a result of the treatment of isolated vesicles, Scatchard analyses revealed a decrease in the number of sites in the membrane available to bind calcium rather than an overall change in the affinity of binding (Table 1).

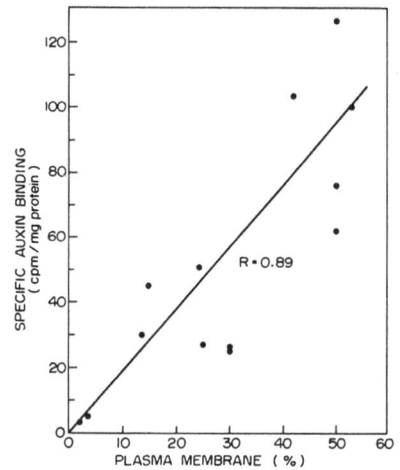

Fig. 1. Regression line showing the relationship between plasma membrane content and the specific acitivty of IAA binding of cell fractions from sucrose-coconut water gradient without added magnesium (From Williamson et al., 1977).

Table 1. Scatchard Analysis of Calcium Binding to Isolated Membranes from Etiolated Hypocotyls of Soybean and Effect of 1 μM 2,4-D. From Buckhout et al., 1981.

Treatment	Ca^{2+} Binding Sites (Nanomoles/Mg Protein)	
	Low Affinity	High Affinity
No auxin	226	47
Auxin (1 μM 2,4-D)	187	28
Change due to 2,4-D	-39	-19

Membrane constituents expected to bind calcium were analyzed and, among those, a major response may now be ascribed to the hydrolysis of the membrane phospholipid, phosphatidyl inositol.

When compositional changes of sufficient magnitude to account for the loss of anionic binding sites of soybean membranes resulting from auxin treatment (Table 1) were sought, a major finding was a reduction in the relative proportions of the anionic phospholipids phosphatidyl inositol (PI), phosphatidyl glycerol (PG) and phosphatidic acid (PA). Together with phosphatidyl choline (PC), these phospholipids decreased in proportion. In contrast, the amount of phosphatidyl ethanolamine (PE) of the inner membrane leaflet was apparently unchanged so that this phospholipid showed a corresponding increase in proportional amount (Table 2).

Table 2. Quantitation of Phospholipid Changes in Auxin-Treated Membranes. Memrane preparations were resuspended and divided. Half was incubated 10 min with buffer alone and half was incubated 10 min with buffer containing 1 µM 2,4-D. The change in phospholipid composition was derived from four such determinations. From Morré et al., 1984a.

| Phospholipid | Phospholipid Phosphorous (Mg/100 Mg) | | | |
| | Total | | Auxin Minus | % |
	Control	1 µM 2,4-D	Control	Decrease
Phosphatidic Acid (PA)	14.8	13.3	-1.5 ± 3.7	10
Phosphatidyl Choline (PC)	24.1	21.0	-3.1 ± 4.6	13
Phosphatidyl Inositol (PI)	9.7	6.2	-3.5 ± 0.9	36
Phosphatidyl Glycerol (PG)	25.9	23.1	-2.8 ± 5.8	11
Phosphatidyl Ethanolamine (PE)	25.5	36.4	$+10.9 \pm 5.7$	-

The auxin-stimulated hydrolysis of PI together with the subsequent hydrolyses of the PA produced would be sufficient to account for the release of divalent ions from the membranes reported by Buckhout et al. (1981).

In subsequent experiments, the ability of auxin to stimulate the degradation of PI was measured from the release of [3H]inositol from membranes prelabeled with [3H]inositol. The isolated membrane vesicles from soybean were incubated for periods of time varying from 1 to 4 h employing an exchange reaction that is unaffected or inhibited by Ca^{2+}, stimulated by Mn^{2+} and CMP, but not dependent upon added CTP, Mg^{2+} or diglyceride (Morré et al., 1984a; Sandelius and Morré, unpublished). The exchange activity used to prelabel the membranes was similar to the type of reaction described for castor bean endosperm by Sexton and Moore (1981).

Using membranes prelabeled by the exchange mechanism, PI hydrolysis was found to be rapid ($T_{1/2}$ < 5 min) and enhanced by auxin throughout the physiological range of growth promoting concentrations with an optimum of about 7×10^{-7} M (Morré et al., 1984a). The dose-response curve for PI breakdown in isolated membrane vesicles paralleled that of the growth response of excised hypocotyl sections such that, as with membrane thinning and the calcium release phenomenon, descending limbs were observed both in the supraoptimal and suboptimal dose ranges (Morré et al., 1984a). That the dependency of the membrane-located inositol exchange reaction on auxin concentration so closely paralleled the growth response to auxin in vivo, provided one line of evidence that the response observed may have physiological significance. Additionally, the growth inactive analog of 2,4-D, 2,3-dichlorophenoxyacetic acid (2,3-D), was without effect in promoting PI exchange.

In the absence of the action of phosphatidic acid phosphatase, PI hydrolysis would not of itself account for Ca^{2+} release. However, any phosphatidic acid formed from PI hydrolysis would be expected to be cleaved by the action of phosphatidic acid phosphatase present in the membrane to account both for the enhanced release of inorganic phosphate (Scherer and Morré, 1978a) and of calcium (Buckhout et al., 1981) observed previously from isolated membrane vesicles by incubation in the presence of auxins.

More recently, using membrane vesicles prelabeled by the exchange pathway, auxin-stimulated hydrolysis of PI was found to be promoted by both 1 to 10 μM Ca^{2+} and by calmodulin (Sandelius and Morré, this volume). In these experiments, Ca^{2+} alone had little or no effect on PI hydrolysis whereas the rate of hydrolysis was stimulated 2- to 4-fold by the combination of auxin plus Ca^{2+} and more than 10-fold by the combination of auxin plus Ca^{2+} together with calmodulin. These findings now point to the strong possibility that the auxin-stimulated degradative activity is distinct from the exchange activity used to prelabel the membranes (Table 3). Both appear to be stimulated by cytidine nucleotides (Sandelius and Morré, unpublished) but the exchange reaction has shown little or no stimulation by Ca^{2+} over a wide range of Ca^{2+} concentrations tested.

Table 3. Properties of Auxin-Stimulated Phosphatidyl Inositol Turnover Reactions of Soybean Membranes

Parameter	Exchange Reaction	Hydrolytic Reaction
Ion Requirement	Mn^{2+*}	Ca^{2+}
Duration	> 1 h	$T_{1/2}$ < 5 min
Response to Calmodulin	Not determined	Stimulated

* Also stimulated by cytidine nucleotides but calcium was without effect or, at high concentrations, inhibitory.

The exchange reaction which is stimulated perhaps 20 to 40% by 2,4-D is approximately linear for 1 h or longer while the auxin-stimulated breakdown reaction goes to completion in a much shorter time. The response of the exchange activity to calmodulin has not been determined but calmodulin has a marked accelerating effect on the hydrolytic reaction. The response of the latter to auxin then is ca. 2-fold.

The findings of Boss and Massel (1985) suggest the possibility that some of the polyphosphoinositides currently under investigation in mammalian cells (Nishizuka, 1984) also exist in plants. With the vesicles isolated from soybean homogenates labeled in the presence of tritiated inositol, 96 to 98% of the radioactivity of acid extracts is recovered as phosphatidyl inositol. Results are nearly identical when incubations are carried out in the presence of ATP and provide no certain indications of a hormone-responsive formation of polyphosphoinositides under these simple conditions of incubation in the cell-free system (Anna Stina Sandelius, unpublished).

While degradative changes of membranes isolated from homogenates of soybean were documented earlier (Scherer and Morré, 1978b), there was no detailed information on the nature of the membrane-associated phospho-lipases involved. While now it is becoming clear that the auxin-stimulated PI degradation reaction is distinct from the exchange reaction used to

prelabel the membranes (Table 3), there is still only fragmentary information on the characteristics of the hydrolytic activity.

If freshly isolated membrane vesicles are incubated with ^{14}C-choline
or -ethanolamine instead of ^{3}H-inositol under the same conditions as were
used to prelabel membranes with inositol, essentially no incorporation is
observed. With ^{14}C-choline, a weak Ca^{2+}-promoted activity of not more
than 0.1 nanomoles/h/mg protein is observed but without stimulation by
2,4-D. However, in vesicles prelabled in situ with ^{14}C-choline, an
efflux of radioactivity was observed when freshly and rapidly isolated
vesicles were incubated for 15 min at room temperature. This efflux was
observed both as a loss of radioactivity from the membrane and an appearance of radioactivity in the soluble supernatant (Table 4). The loss of

Table 4. Auxin Plus Calcium-Accelerated Loss of ^{14}C-Choline from Vesicles
Isolated from Etiolated Segments of Soybean Hypocotyls Prelabeled for 4 H
With 2 μCi Methyl-[^{14}C]Choline in 1 ml Total Volume/30 1 cm Segments With
or Without 1 μM 2,4-D. Unpublished Data of D. J. Morré, G. Scherer and
A. Sievers.

Treatment	Membrane	Supernatant
Membranes Prelabeled in Intact Hypocotyl Segments + 1 μM 2,4-D		
Initial	52,000 ± 3,000	-
Control 15 min	49,000 ± 2,500	4,130 ± 1,400
1 μM 2,4-D + 1 μM Ca^{2+} 15 min	46,300 ± 2,000	7,300 ± 2,750
Membranes Prelabeled in Intact Hypocotyl Segments + No 2,4-D		
Initial	40,000 ± 2,500	-
Control 15 min	35,500 ± 950	3,400 ± 1,900
1 μM 2,4-D + 1 μM Ca^{2+} 15 min	29,500 ± 2,700	7,300 ± 1,000

^{14}C-choline was accelerated approximately 2-fold by treatment with 1 μM
2,4-D plus 0.1 or 1 μM Ca^{2+}. With membranes prelabeled in situ in the
presence of auxin, only Ca^{2+} was required to promote ^{14}C-choline efflux
from the isolated vesicles. However, with vesicles isolated from tissue
sections prelabeled in the absence of 2,4-D, both Ca^{2+} and 2,4-D were
required to promote efflux (data not shown). The effect of calmodulin
has not been tested in this sytem.

The above findings (Table 4), taken together with the information on
PI hydrolysis (Morré et al., 1984a) would suggest that, with the apparent
exception of PE, there exist pools of phospholipid within the membranes
(ca. 10 to 15 % of the PC and ca. 30 to 40 % of the PI) that may be
available for rapid degradation in response to an auxin stimulus. The
reaction in combination with the action of phosphatidic acid phosphatase
would result not only in release of calcium and diglyceride but the
calcium released would now be expected to further accelerate phospholipid
breakdown as diagrammed in figure 2.

Breakdown of even 10% of the membrane phospholipid within 10 to 15
min in response to auxin could be significant in terms of the physiology
of the plant. It would readily account for the 10% membrane thinning
observed previously both in vivo and in vitro (Morré and Bracker, 1976)
in response to auxin, for the observed increase in microviscosity in

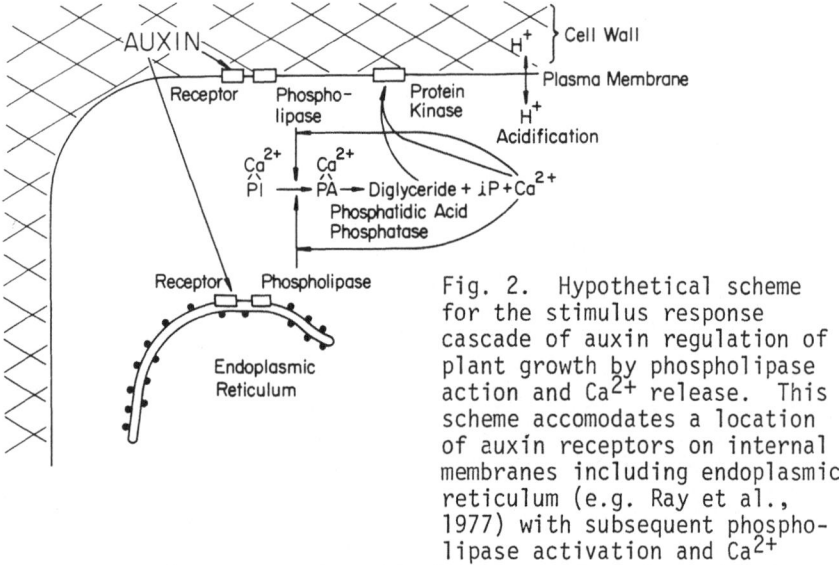

Fig. 2. Hypothetical scheme for the stimulus response cascade of auxin regulation of plant growth by phospholipase action and Ca^{2+} release. This scheme accomodates a location of auxin receptors on internal membranes including endoplasmic reticulum (e.g. Ray et al., 1977) with subsequent phospholipase activation and Ca^{2+} functioning as a second messenger to activate some series of reactions leading perhaps to acidification of the cytoplasm, the cell wall region or both.

A putative major product of the phospholipid hydrolysis, 1,2-diglyceride, is an activator of protein kinases of the C-type in mammalian cells (Nishizuka, 1984) and, together with Ca^{2+}, has been reported to stimulate protein phosphorylation when added to isolated membrane vesicles from soybean (Morré et al. 1984b). Also reported was a rapid (10 sec or less) response of protein phosphorylation of bands of ca. 20,000 and 40,000 apparent molecular weight in response to auxin. This latter response, however, appears to be too early to result indirectly from the diglyceride released from auxin-stimulated phcspholipid breakdown and is inhibited by Ca^{2+} (Varnold and Morré, 1985). In contrast, more recent findings using membranes prepared by two-phase partition and by the technique of free-flow electrophoresis (Sandelius et al., this volume), demonstrate an auxin-responsive protein phosphorylation reaction of plasma membranes that is stimulated by Ca^{2+} (Varnold et al., this volume). This could provide an example of a plasma membrane-located reaction potentially responsive to released Ca^{2+} (Fig. 2) but the role of diglyceride, if any, in this system remains to be investigated.

DISCUSSION

The roles of calcium ions as primary or secondary effectors of elongation growth in plants will doubtless be both complex and important. Formerly, Ca^{2+} at high concentrations, i.e. 1 mM, was well established as one of the few treatments known that rapidly and completely suppressed auxin effects of cell wall loosening (Morré and Eisinger, 1968). This effect on cell wall stiffening was not a result of the non-metabolic cross-bonding of carboxyl groups of, for example, pectic chains but required living cells for expression (Coartney and Morré, 1980). Also associated with millimolar concentrations of calcium ions was a change in the conformation of the plasma membrane (thickening) opposite in sign to that observed with auxin (Morré et al., 1974; Morré and Bracker, 1976), an effect accompanied by loss of auxin responsiveness in tissue (Coartney and Morré, 1980). The high-dose calcium effect may be responsible, as well, for protoplast stabilization (Morré et al. 1984c), membrane fusions (Morré et al., 1974), interruption of tip growth in pollen tubes (Morré and Van DerWoude, 1974; Reiss and Herth, 1979), and other membrane phenomena observed at high calcium concentrations. While offering often valuable insights into the growth process and various membrane-related phenomena, the high-dose calcium responses were not normally regarded as either physiological nor as evidence for a second messenger role for calcium.

While the present studies are based largely on experiments with membrane vesicles in a cell-free system, there is morphological evidence from membrane thinning that a corresponding loss of phospholipid in response to auxin could take place in incubated tissue segments (Morré and Bracker, 1976). More important, the findings provide the basis for a signal-response cascade of auxin action involving calcium as a second messenger. As such, the system is ammenable to much more detailed future investigations that would presently be possible with whole cells or intact tissues. Future investigations must, however, include experiments to determine what form the cascade may take when operating in vivo.

ACKNOWLEDGEMENT

Work supported in part by a grant from the National Science Foundation PCM 8206222.

REFERENCES

Boss, W. F., and Massel, M., 1985, Inositol phospholipid metabolism in fusogenic carrot cells and protoplasts, Plant Physiol. 77 (Suppl.): 147.

Buckhout, T. J., Young, K. A., Low, P. S., and Morré, D. J., 1981, In vitro promotion by auxins of divalent ion release from soybean membranes, Plant Physiol. 68:512.

Coartney, J. S., and Morré, D. J., 1980, Studies on the role of wall extensibility in the control of cell expansion, Bot. Gaz. 141:56.

Helgerson, S. L., Cramer, W. A., and Morré, D. J., 1976, Evidence for an increase in microviscosity of plasma membranes from soybean hypocotyls induced by the plant hormone indole-3-acetic acid. Plant Physiol. 58:548.

Marmé, D., 1985, Calcium, Naturwissenschaften, 72:113.

Michell, R. H., Jafferji, S. S., and Jones, L. M., 1977, The possible involvement of phosphatidyl inositol breakdown in the mechanism of stimulus-response coupling at receptors which control cell-surface calcium gates, Adv. Exp. Med. Biol., 83:447.

Morré, D. J., 1971, Isolation of Golgi apparatus, Methods Enzymol. 22:130.

Morré, D. J., and Bracker, C. E., 1976, Ultrastructural alteration of plasma membranes induced by auxins and calcium ions, Plant Physiol. 58:544.

Morré, D. J., and Eisinger, W. E., 1968, Cell wall extensibility: its control by auxin and relationship to cell elongation, in "Biochemistry and Physiology of Plant Growth Substances," F. Wightman and G. Setterfield, eds., Runge Press, Ottawa, Canada, p. 625.

Morré, D. J., and VanDerWoude, W. J., 1974, Origin and growth of cell surface components, in: "Macromolecules Regulating Growth and Development," E. D. Hay, T. J. King and J. Papaconstantionou, Academic Press, New York, p. 81.

Morré, D. J., Bracker, C. E., and VanDerWoude, W. J., 1974, Influence of calcium ions on the plant cell surface: Membrane fusions and conformational changes, Proc. Electron Microsc. Soc. Amer., 32:94.

Morré, D. J., Gripshover, B., Monroe, A., and Morré, J. T., 1984a, Phosphatidyl inositol turnover in isolated soybean membranes stimulated by the synthetic growth hormone 2,4-dichlorophenoxyacetic acid, J. Biol. Chem. 259:15364.

Morré, D. J., Morré, J. T., and Varnold, R. L., 1984b, Phosphorylation of membrane located proteins of soybean in vitro and response to auxin. Plant Physiol. 75:265.

Morré, D. J., Gripshover, B., Boss, W. F., and Tuite, P. J., 1984c, Membrane cell-free systems for probing growth mechanisms, in: "Annual Proceedings of the Phytochemical Society of Europe, Volume 24: Membranes and Compartmentation in the Regulation of Plant Functions," A. M. Boudet, G. Alibert, G. Marigo, and P. J. Lea, eds., Clarendon Press, Oxford, p. 247.

Nishizuka, Y., 1984, Turnover of inositol phospholipids and signal transduction, Science 225:1365.

Ray, P. M., Dohrman, U. C., and Hertel, R., 1977, Specificity of auxin-binding sites on maize coleoptile membranes as possible receptor sites for auxin action. Plant Physiol. 60:585.

Reiss, H.-D., and Herth, W., 1979, Calcium ionophore A 23187 affects localized wall secretion in the tip region of pollen tubes of Lilium longiflorum, Planta, 145:225.

Roland, J.-C., Lembi, C. A., and Morré, D. J., 1972, Phosphotungstic acid-chromic acid as a selective electron-dense stain for plasma membrane of plant cells, Stain Technology 47:195.

Scherer, G. F. E., and Morré, D. J., 1978a, In vitro stimulation by 2,4-dichlorophenoxyacetic acid of an ATPase and inhibition of phosphatidic acid phosphatase of plant membranes. Biochem. Biophys. Res. Commun. 84:238.

Scherer, G. F. E, and Morré, D. J., 1978b, Action and inhibition of endogenous phospholipases during isolation of plant membranes. Plant Physiol. 62:933.

Sexton, J. C., and Moore, T. S., 1981, Phosphatidyl inositol synthesis by an Mn^{2+}-dependent exchange enzyme in castor bean endosperm, Plant Physiol. 68:18.

Varnold, R. L., and Morré, D. J., 1985, Phosphorylation of membrane-located proteins of soybean hypocotyls. Inhibition by calcium in the presence of 2,4-dichlorophenoxyacetic acid, Bot. Gaz.

Williamson, F. A., Morré, D. J., and Hess, K., 1977, Auxin binding activities of subcellular fractions from soybean hypocotyls, Cytobiologie 16, 63.

EVIDENCE FOR A MECHANISM BY WHICH AUXINS AND FUSICOCCIN

MAY INDUCE ELONGATION GROWTH

Roger W. Parish[1], Hubert Felle[2] and Benno Brummer[3]

[1]Plant Biology Institute, University of Zürich
CH-8008 Zürich
[2]Botanical Institute, Justus Liebig University
D-6300 Giessen
[3]Department of Physiology, University of Yale
Conn. 06510

INTRODUCTION

Hyperpolarization of membrane potentials of coleoptile tissue and seedling stems is now generally accepted as a specific auxin effect (Bates and Goldsmith, 1983). Hyperpolarization is thought to result from stimulation of the outwardly-directed, electrogenic proton pump in the plasma membrane (Marré and Ballarin-Denti, 1985). Pump stimulation leads to acidification of the apoplast which is postulated to loosen the cell wall either by direct breakage of acid-labile cell wall bonds or by activation of wall-loosening enzymes (Cleland, 1980). This acid-growth theory had its origins in the discovery that acid solutions can stimulate plant cell elongation (Hager et al., 1971; Rayle and Cleland, 1970). Inconsistencies in the theory have become apparent over the years (see references in Brummer and Parish, 1983; Kutschera and Schopfer, 1985). In particular, no straightforward correlation between proton efflux and growth has been found.

We have sought an alternative to the wall-acidification theory of auxin-induced elongation growth (Brummer and Parish, 1983). This article presents a slightly modified version of our original hypothesis and reexamines its predictions in the light of new data. No attempt is made to review the field and, due to space limitations, most of the articles originally cited in support of the hypothesis are omitted. The references given, usually recent, should be considered representative and as sources for the earlier literature.

WORKING HYPOTHESIS

This is summarized in Fig. 1. When bound to their receptors on the plasma membrane and endoplasmic reticulum (ER) (and possibly even the tonoplast, see later), auxins induce an increase in the cytosolic levels of free Ca^{2+}. This Ca^{2+} may be taken up (sequestered) by the ER, mitochondria or vacuole, which secrete protons into the cytosol. Alternatively, occupancy of tonoplast receptors may induce the release (leak) of protons from the vacuole. Fusicoccin (FC) leads to cytosolic acidification via a mechanism different from auxins, directly stimulating an unknown H^+-producing reaction. The increase in the cytosolic proton concentration will stimulate the proton pump in the plasma membrane, resulting in acidification of the cell wall. Although this acidification may be involved in the initial growth response, the model proposes that continuous elongation growth depends on changes in membrane structure and transmembrane ion gradients resulting from pump stimulation. The transmembrane ion gradients could act as a second messenger and influence processes such as secretion, protein synthesis and gene transcription.

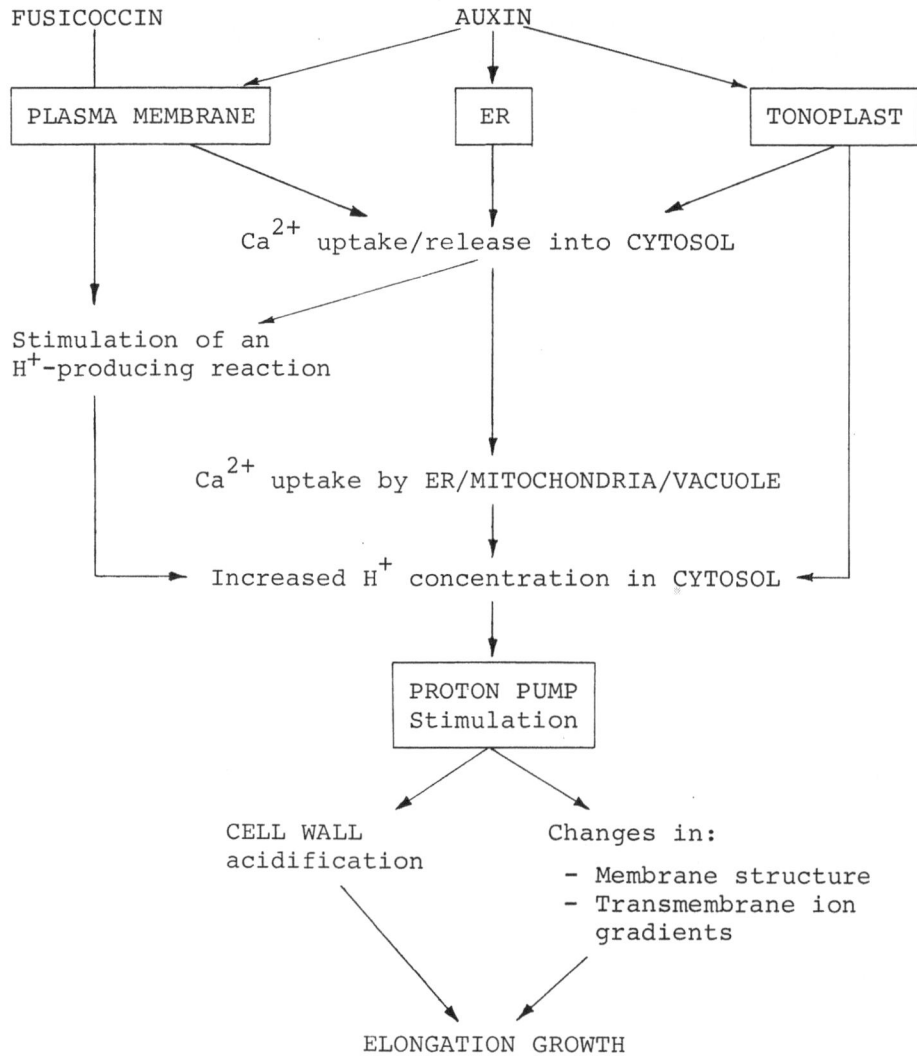

Fig. 1. Model of mechanisms by which auxin and fusicoccin induce plant cell elongation.

CYTOSOLIC ACIDIFICATION

The two most likely ways in which auxin could stimulate the H^+ pump are to increase its efficiency (eg. by configuration changes) or to provide more protons (cytosolic acidification) (see Göring and Bleiss, 1982; Brummer and Parish, 1983; Bates and Goldsmith, 1983). There are reports that auxin stimulates the ATPase activity in plasma membrane-enriched fractions (Scherer, 1984). However, there is a considerable delay (longer than auxin transport would be expected to take) before membrane hyperpolarization is induced by auxin in vivo. FC does not activate the ATPase in vitro (Hanson and Trewavas, 1982).

Experimental evidence

A number of techniques have been developed for the measurement of intracellular pH (Nuccitelli and Deamer, 1982). For a variety of reasons (eg. central vacuole, cell wall, turgor, autofluorescence) these methods are often not readily applicable to plant cells. The method of choice is probably pH microelectrodes since membrane potential changes can be measured simultaneously in the same cell (eg. Brummer et al., 1984a). However, application of this technique to tissues which elongate in response to auxins has proved diffucult. Nevertheless, root epidermal cells can be studied and a rapid cytosolic acidification can be detected following FC treatment (Brummer et al., 1985). This decrease commences within 30 sec and precedes the membrane hyperpolarization, indicating the pump stimulation is a consequence of the increased proton (substrate) concentration.

In coleoptile tissue we have used the dual wavelength absorbance technique to measure cytosolic pH. This method employs the indicators fluorescein diacetate or 6-carboxyfluorescein diacetate. These diffuse into the cytosol where an esterase releases the chromophores fluorescein or carboxyfluorescein which are charged and consequently trapped. A decrease in absorbance and a blue-wavelength shift occur with decreasing pH. The rapid decrease in cytosolic pH caused by FC in root cells could also be demonstrated in maize coleoptiles using this method (Brummer et al., 1985). Indole-3-acetic acid (IAA) also caused a decrease in cytosolic pH, beginning after a lag of 5 min. Membrane hyperpolarization and extracellular acidification were first detectable after 10 min and 30-60 min respectively. Hager et al. (1985), using the same method, also found FC acidified the cytosol of Avena coleoptiles. Using ^{32}P nuclear magnetic resonance signals, FC treatment of maize root tips resulted in some spectra indicating a slight decrease in cytosolic pH (Roberts et al., 1981).

Manipulating the cytosolic pH provided indirect evidence that pH_c is involved in elongation growth. Procaine, which gains protons in the cytosol and depolarized the membrane, inhibited IAA-induced growth (Brummer et al., 1985). Furthermore, 1-naphthyl acetate acidified the cytosol, hyperpolarized the membrane and stimulated coleoptile growth.

Our microelectrode measurements indicate the cytosolic pH drops about 0.1 pH units (to ca. pH 7.0) and remains fairly constant following FC treatment of root cells. Unfortunately, the absorbance technique has proved impossible to calibrate. However, FC appears to acidifiy the cytosol to a greater extent than IAA. (Extrapolation gives a factor of between 2 and 5, however different rates of spread and uptake in the tissue by the two compounds may contribute to the difference.) This raises the question of whether the reduction in cytosolic pH by auxin is sufficient to entirely account for the H^+ pump stimulation observed. Two points should be made. Firstly, the extracellular acidification detected following treatment of coleoptiles by FC or IAA at concentrations used for the cytosolic pH measurements also differs to a similar degree (i.e. 3-4 times greater with FC). Secondly, stimulation of the pump will remove excess H^+ and tend to raise the cytosolic pH. However, the pH remains lowered following IAA treatment, indicating an increased H^+ flow into the cytoplasmic pool is occurring.

Mechanisms of cytosolic pH regulation

Since FC and possibly auxins do lower the cytosolic pH, knowledge of the mechanisms of cytosolic pH regulation becomes relevant to their modes of action. The constancy of cytosolic pH reflects a steady state rather than an equilibrium, i.e. there is a steady flow of H^+ into and out of the cytosolic pool (Sanders and Slayman, 1982). However, pump activation and deactivation alone are theoretically inadequate to regulate cytosolic pH because the membranes are permeable to proton reentry. Hence, the extra H^+ pumped out following a drop in cytosolic pH would to a large extent be returned by the increased PMF (Slayman, 1985). A solution to this problem for the cell is to adjust the fraction of pump current returned by non-proton pathways (eg. by an increased leakage of organic anions). Evidence for such a mechanism has been obtained in Neurospora where initial hyperpolarization due to cytosolic acidification was followed by an increase in membrane leakage conductance leading to depolarization (Sanders et al., 1981). However, auxin- or FC-induced hyperpolarization is not followed by such a depolarization. Moreover, Sanders and Slayman (1982) found that H^+ entry into Neurospora is small and pH control probably results from a balance between H^+ production by metabolism and pump operation. Which of these is regulated to stabilize cytosolic pH is not known. Metabolism may actually consume H to stabilize pH. Finally, H^+ concentration is probably not the only error signal involved since the response to the acid load (regulation via pump activity or metabolism) depends on the form the load takes (Sanders and Slayman, 1982). These authors also invoke a feedback loop between metabolism and the pump, possibly related to the fact that the pump is a major consumer of ATP. When monensin is added to coleoptiles in low-pH medium, H^+ are transported into the cell and the pump stimulated (Brummer et al., 1984b). We expected this cycling of H^+ would not change wall pH, however net wall acidification was observed. At present one can only say that FC and IAA probably lower cytosolic pH by different mechanisms since the FC effect is rapid whereas the IAA effect follows a lag.

Ca^{2+} and cytosolic acidification

We originally suggested that, following the auxin-induced release of Ca from intracellular stores, this Ca^{2+} is sequestered by mitochondria with subsequent H^+ release (Brummer and Parish, 1983). However, plant mitochondria are now known to have a low Ca^{2+}-transporting activity and their role in the regulation of cytosolic Ca^{2+} is unclear (Moore and Akerman, 1984). Ca^{2+}-ATPases and Ca^{2+}/H^+ antiport systems have been found in membrane fractions, but their precise intracellular location is still not clear (Macklon, 1984). An increase in cytosolic Ca^{2+} may, ofcourse, stimulate H^+ producing metabolism. Hanson and Trewavas (1982) suggest reactivation of growth in excised tissues by auxin is due to release of the H^+ pump from inactivation. This inactivation may result from increased cytosolic Ca^{2+} levels and auxins function by indirectly stimulating a plasma membrane Ca^{2+}-ATPase. Auxin transport and Ca^{2+} influx and efflux appear to be related and in gravitropically stimulated organs opposite lateral movements of Ca^{2+} and IAA occur (see references in De Guzman and Dela Fuente, 1984). Unfortunately, the measurement of plant cytosolic Ca^{2+} with fluorescent indicators has not proved feasible, so hopefully the development of Ca^{2+} electrodes will end the speculation.

Phosphoinositide metabolism

Jumping onto the latest animal research bandwagon has always been a tempting but dangerous approach for plant biologists. Nevertheless, how elegant it would be if auxins stimulate phosphoinositide metabolism, modifying Ca^{2+} levels via inositol 1,4,5-triphosphate and H^+ transporters via diacylglycerol (Berridge, 1984). Phosphatidylinositol turnover in soybean membranes is stimulated by 2,4-dichlorophenoxyacetic acid at growth promoting concentrations (Morré et al., 1984). However, the reaction differs from that in animal cells and no increase in diglyceride was observed.

CONSEQUENCES OF PROTON PUMP STIMULATION

Wall acidification

The cell wall matrix is softened by acidification, however the extent to which this contributes to auxin-induced elongation growth is difficult to ascertain. Cleland (1983) suggests lack of correlation between proton flux and growth rate may be related to changes in the capacity of the wall to loosen in response to protons. Kutschera and Schopfer (1985a) have presented results incompatible with four predictions of the acid-growth theory. One difficulty with such experiments, which the authors recognise, is the assumed correlation between the pH changes at the outer surface of the plasma membrane and the measurements actually made.

Weak acids are known to penetrate the plasma membrane, acidify the cytosol and stimulate the proton pump (see Brummer et al., 1984a), also stimulating K^+ uptake in roots (Marré and Bellarin-Denti, 1985). Different acid solutions (at the same concentration) exhibit different pH optima for maximal growth induction (Hager et al., 1971). Using microelectrodes to measure membrane potential and cytosolic pH in the same cell, we found the degree of growth stimulation by citric and acetic acids was positively correlated with the extent of their cytosolic acidification and stimulation of the H^+ pump (Brummer et al. 1984a). Moreover, the carboxylic ionophore monensin lead to acidification of the cell wall when coleoptiles were incubated in alkaline buffer containing Na (Brummer et al., 1984b). Neither pump stimulation nor growth occurred even though IAA induced growth under similar conditions. These results suggest wall acidification alone is inadeqaute to induce growth.

Kutschera and Schopfer (1985b) suggest that FC, unlike auxin, does induce growth via wall acidification. This implies two quite independent mechanisms exist, which is open to doubt and difficult to reconcile with the results described above. The crucial difference between the two compounds may lie in the mechanisms for increasing the cytosolic H^+ concentration. If auxins, for example, raise Ca^{2+} levels, a variety of additional responses will be expected (eg. wall deposition, gene regulation).

Changes in membrane structure and transmembrane ion gradients

Changes in membrane potential are associated with the primary mode of action of animal glycoprotein hormones, certain toxins and interferon and might regulate cellulose synthesis in cotton fibers (see Brummer and Parish, 1983). John (1983) has suggested ethylene synthesis is coupled to a transmembrane, electrogenic H^+ flux. Membrane potential appears to modulate some processes, including cellulose synthesis, independently of any obvious function it serves as a driving force for transport or synthetic reactions (Delmer et al., 1982). Changes in bilayer fluidity, polarity and distribution of lipids as well as conformation and positioning of proteins in the bilayer may be involved.

The effects of FC and auxins are probably not due to hyperpolarization _per se_. Crucial would be the stimulation of the H^+ pump which could, for example, drive localized entry of ions required for growth. This implies, ofcourse, that the relevant carriers be already located at the presumptive growth site (Slayman, 1985). Current has been found to enter at the point of growth in many germinating or tip-growing cells (eg. root hairs, fungal hyphae, pollen tubes; see Brawley et al., 1984). In wild carrot embryos, current enters cotyledons and exits at the radicle (Brawley et al., 1984). A substantial portion of the current is carried by H^+ and exogenous IAA inhibits the current reversibly, presumably by stimulating H^+ secretion and interfering with endogenous gradients of H^+ and IAA. The electric field generated by current may serve as a driving force for electrophoresis of material toward the peak of inward current. Alternatively, the flow of specific ions could act as a signal conveying spatial information which localizes growth (see Kropf et al., 1983). An inward current precedes branch emergence in _Achyla_ whereas, during branching, the inward current in the original tip may decline or reverse transiently without the growth rate changing (Kropf et al., 1983). As well as localizing materials on the basis of their charge, endogenous currents may move proteins (eg. receptors and ion channels) in the membrane (Brawley et al., 1984). Slayman (1985) points out that the discovery of potassium-proton and chloride-proton symports suggests H^+ pumps may be important in energizing salt-dependent processes in plants. PMF may, for example, regulate the volume changes in the cells controlling transpiration and diurnal leaf movements (Raschke, 1979; Satter and Galston, 1981).

We have suggested that changes in transmembrane ion gradients induced by auxins can serve as a second messenger (Brummer and Parish, 1983). It has been known for many years that plant growth can be induced by electric fields. Brummer (1982) found that continuous or pulsed current could induce growth of coleoptiles placed at the anode but inhibited growth of those at the cathode. The most probable effect of weak current is to modify transmembrane ion gradients and it would be useful to measure directly membrane potential, cytosolic pH and ion transport during such treatments.

Finally, cytosolic acidification will affect metabolic patterns and it would be important to discover whether pump stimulation is vital for growth or simply reflects the cytosolic acidification. Unfortunately, these are difficult to uncouple, since inhibiting the pump with specific inhibitors will have pleiotropic effects on metabolism.

THE FUTURE

The development of microelectrodes to measure cytosolic pH, membrane potential, Ca^{2+} concentrations, and pH at the outer surface of the plasma membrane in single coleoptile cells will be important for the testing of the predications made by our hypothesis. Moreover, substances such as diacylglycerol may be directly injected into the cell where such measurements (including growth) are being made. The results of such experiments will still leave us a long way from understanding the actual mechanisms of growth, but will hopefully lead to identification of the signals involved.

B. Brummer was supported in part by the Roche Research Foundation.

REFERENCES

Bates, G.W., and Goldsmith, M.H.M., 1983, Rapid response of the plasma-membrane potential in oat coleoptiles to auxin and other weak acids, Planta, 159:231.

Berridge, M.J., 1984, Inositol triphosphate and diacylglycerol as second messengers, Biochem. J., 220:345.

Brawley, S.H., Wetherell, D.F., and Robinson, R.R., 1984, Electrical polarity in embryos of wild carrot precedes cotyledon differentiation, Proc. Natl. Acad. Sci. USA, 81:6064.

Brummer, B., 1982, Studies on auxin-induced elongation of Zea mays coleoptiles, Doctoral Thesis, University of Zürich.

Brummer B., and Parish, R.W., 1983, Mechanisms of auxin-induced plant cell elongation, FEBS Lett., 161:9.

Brummer, B., Felle, H., and Parish, R.W., 1984a, Evidence that acid solutions induce plant cell elongation by acidifying the cytosol and stimulating the proton pump, FEBS Lett., 174:223.

Brummer, B., Potrykus, I., and Parish, R.W., 1984b, The roles of cell-wall acidification and proton-pump stimulation in auxin-induced growth: studies using monensin, Planta, 162:345

Brummer, B., Bertl, B., Potrykus, I., Felle, H., and Parish, R.W., 1985, Evidence that fusicoccin and indole-3-acetic acid induce acidification in Zea mays cells, FEBS Lett., in press.

Cleland, R.E., 1980, Auxin and F excretion: the state of our knowledge, in: "Plant Growth Substances 1979," F. Skoog, ed., Springer, Berlin Heidelberg New York.

Cleland, R.E., 1983, The capacity for acid-induced wall loosening as a factor in the control of Avena coleoptile cell elongation, J. Exp. Bot., 34:676.

De Guzman, C.C., and Dela Fuenta, R.K., 1984, Polar calcium flux in sunflower hypocotyl segments, Plant Physiol., 76:347.

Delmer, D.P., Benziman, M., and Padan, E., 1982, Requirement for a membrane potential for cellulose synthesis in intact cells of Acetobacter xylinum. Proc. Natl. Acad. Sci. USA, 79:5282.

Göring, H., and Bleiss, W., 1982, Changes of growth rate and osmotic potential under the influence of decreased pH and IAA, in: "Plasmalemma and Tonoplast: Their Functions in the Plant Cell," D. Marmé, E. Marré, and R. Hertel, ed., Elsevier, Amsterdam New York.

Hager, A., Menzel, H., and Krauss, A., 1971, Versuche und Hypothese zur Primärwirkung des Auxins beim Streckungswachstum. Planta, 100:47.

Hager, A., and Moser, I., 1985, Active acid esters and permeable weak acids induce active H^+ excretion and extension growth of coleoptile segments by lowering the cytoplasmic pH.Planta, 163:391.

Hanson, J.B., and Trewavas, A.J., 1982, Regulation of plant cell growth: the changing perspective. New Phytol., 90:1.

John, P., 1983, The coupling of ethylene biosynthesis to a transmembrane, electrogenic proton flux. FEBS Lett., 152:141.

Kropf, D.L., Lupa,M.D.A., Caldwell, J.H., and Harold, F.M., 1983, Cell polarity: endogenous ion currents precede and predict branching in the water mold Achlya, Science, 220:1385.

Kutschera, U., and Schopfer, P., 1985a, Evidence against the acid-growth theory of auxin action, Planta, 163:483.

Kutschera, U. and Schopfer, P., 1985b, Evidence for the acid-growth theory of fusicoccin action, Planta, 163:494.

Macklon, A.E.S., 1984, Calcium fluxes at the plasmalemma and tonoplast. Plant, Cell and Environment, 7:407.

Marré, E., and Ballarin-Denti, A., 1985, The proton pumps of the plasmalemma and the tonoplast of higher plants, J. Bioenerg. Biomem., 17:1.

Moore, A.L., and Akerman, K.E.O., 1984, Calcium and plant organelles, Plant, Cell and Environment, 7:423.

Morré, D.J., Gripshover, B., Monroe, A., and Morré , J.T., 1984, Phosphatidylinositol turnover in isolated soybean membranes stimulated by the synthetic growth hormone 2,4-dichlorophenoxyacetic acid, J. Biol. Chem., 259:15364.

Nuccitelli, R., and Deamer, D.W., 1982, "Intracellular pH: its Measurement, Regulation and Utilization in Cellular Functions." Liss, New York.

Raschke, K., 1979, Movements of stomata, in "Physiology of Movements, W. Haupt and M.E. Feinleib, eds., Springer, Berlin.

Rayle, D.L., and Cleland, R.E., 1970, Enhancement of wall loosening and elongation by acid solutions, Plant Physiol., 46:250.

Roberts, J.K.M., Ray, P.M., Wade-Jardetsky, N., and Jardetsky, O., 1981, Extent of intracellular pH changes during H^+ extrusion by maize root-tip cells, Planta, 152:74.

Sanders, D., and Slayman, C.L., 1982, Control of intracellular pH, J. Gen. Physiol., 80:377.

Sanders, D., Hansen, U.-P., and Slayman, C.L., 1981, Role of the plasma membrane proton pump in pH regulation in non-animal cells, Proc. Natl. Acad. Sci. USA, 78:5903

Satter, R.L., and Galston, A.W., 1981, Mechanisms of control of leaf movements, Annu. Rev. Plant Physiol., 32:83.

Scherer, G.F.E., 1984, H^+-ATPase and auxin-stimulated ATPase in membrane fractions from zucchini and pumpkin hypocotyls, Z. Pflanzenphysiol., 114:233.

Slayman, C.L., 1985, Plasma membrane proton pumps in plants and fungi, Bio Science, 35:34.

POSTER ABSTRACTS

THE CONCENTRATION OF CALMODULIN ALTERS DURING EARLY STAGES OF CELL

DEVELOPMENT IN THE ROOT APEX OF PISUM SATIVUM

E.E.F. Allan and A.J. Trewavas

Botany Department
University of Edinburgh
Mayfield Road, Edinburgh EH9 3JH, U.K.

Quantitative changes in calmodulin were investigated during early stages of cell development in the root apex of Pisum sativum. Concentrations were estimated in two ways: highly purified samples of calmodulin extracted from serial developmental sections were estimated by NAD kinase activating ability, and whole cell extracts of individual tissue areas from serial sections were examined by two-dimensional polyacrylamide gel electrophoresis. Extraction conditions were designed to solubilise membrane-bound calmodulin, and to minimise interference from calmodulin-binding proteins which appeared to reduce the extractability of calmodulin from pea seedling tissues.

Large variations in concentration are found to occur within the apical few millimetres during early cell differentiation and root cap formation. Tissue-dependent differences are also detected from a very early stage, within the apical meristem. From analysis of serial developmental sections it is found that calmodulin falls rapidly in concentration from approximately 22uM in the root cap/apical meristem region to a minimum of 1-2uM at the base of the meristem. Thus calmodulin is at a minimum by 2mm from the apex, during early stages of cell expansion. A slight increase in concentration to reach about 3uM by 5mm from the apex is followed by a decline to the lower level of 1-2uM by 10mm. This pattern is also observed for changes in specific activity of calmodulin. More detailed analysis by gel electrophoresis of individual tissues reveals that calmodulin decreases rapidly in the cortex at the base of the meristem although remaining relatively high in stelar tissue. Calmodulin falls in stelar tissue after 2mm from the apex, and the increase in concentration 2.5-5mm from the apex that is observed in serial sections appears to occur only in cortical tissue.

Calmodulin therefore varies considerably during cell development. There is no simple correlation with growth, although high concentrations appear to coincide closely with the root cap and with areas of cell division such as the apical meristem, the stele at the base of the meristem, and the cortex in the area of localised meristematic activity in the inner cortex which is associated with initiation of lateral root primordia. Low concentrations appear to be associated with the onset of rapid cell elongation and with maturing cells.

The biological activity of calmodulin normally appears to be limited by levels and compartmentation of ionised calcium and calmodulin-binding proteins rather than by its concentration. However, the high concentrations in the root apex might reflect the presence of a large number of calmodulin-dependent proteins. It has also been observed that effects on polymerisation of cold-labile microtubules require relatively high levels of calmodulin, thus it is also possible that higher concentrations may be required for regulation of certain subpopulations of microtubules. It is probable that calmodulin regualtes several proteins in the root apex as a number of polypeptides from root tissue bind to a calmodulin affinity column in a calcium-dependent manner. These proteins will probably include isozymes of NAD kinase, Ca^{2+} ATPase and protein kinase, and several microtubule-associated proteins and tubulin kinase. Calmodulin may be involved in a variety of activities during division, expansion and differentiation in the primary root apex.

INTRACELLULAR LOCALIZATION OF CALMODULIN ON EMBRYONIC AXES OF

CICER ARIETINUM L.

Josefina Hernández-Nistal, Juan J. Aldasaro, Dolores
Rodriguez, Josefa Babiano and Gregorio Nicolás

Plant Physiology Department, Biology Faculty
University of Salamanca, Salamanca, Spain

INTRODUCTION

Calcium plays an important role in the regulation of plant metabolism
and in mediating the adaptation of plants to environmental changes. Some of
these functions are mediated through its binding to calcium binding regulatory proteins, the most important among these is calmodulin (CaM). We want
to know if CaM has a role in the early stages of seed germination. So we
have estimated its content in isolated subcellular fractions and in sections
of 36 and 72 hours old embryonic axes germinated in different media:
H_2O-25°C, H_2O-30°C and ABA-25°C, in order to prove if temperature and germination inhibitors have any effect on CaM levels.

MATERIALS AND METHODS

Seeds were germinated in H_2O (25-30°C) or ABA 25µM-25°C during 36-72h.
The starting material was the embryonic axes. To obtain protoplast, mitochondria and microsomal fractions Muto's (1982) method was followed. The
purification of the nuclear sap, chromatin and nuclear membranes enriched
fractions, as Matsumoto et al., (1984). To isolate cell wall proteins, LiCl
was used (Huber and Nevins, 1980). The localization of CaM on the embryonic
axes was done on 1.5 or 3 mm long radicle segments, both in 36 and 72h old.
CaM was assayed by radioimmunoassay using ^{125}I RIA kit (Amersham, England).

RESULTS

Table 1. Percentage CaM/total protein in cellular fractions of 36h old
embryonic axes.

Treatm.	Cell Wall	Proplast Fraction	Micros. Fract.	Mitoch. Fract.	Cytosol Fract.	Nucl. Membr.	Nucl. sap	Chromat. Fract.
H_2O-25°C	1.24	0.4	0.4	2.5	0.5	2.4	1.80	2.4
H_2O-30°C	2.31	0.6	0.1	9.3	1.1	0.9	0.86	0.04
ABA-25°C	0.32	1.3	0.8	4.1	1.7	0.4	0.44	0.1

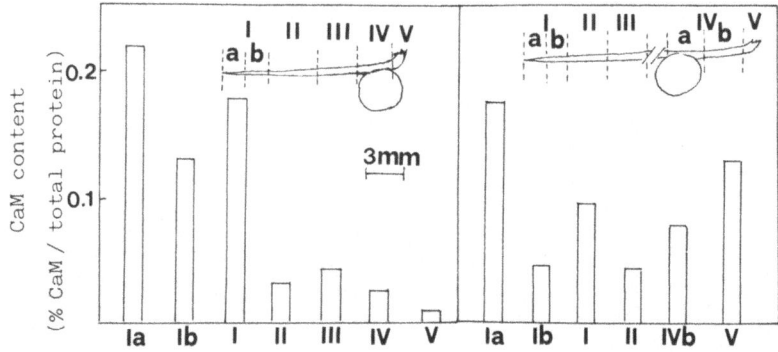

Fig. 1. CaM content in the different zones of A: 36h old; B: 72h old
   embryonic axes.

DISCUSSION

CaM level is greater in the intact embryonic axes from seeds germinated
at 30°C than in those germinated at 25°C. ABA reduces significantly the
amount of CaM in the embryonic axes as it does in radish seeds (Cocucci,
1984). In the cell wall, the main Ca^{++} store in plants, the ratio CaM/total
protein was the largest in H_2O-30°C. Among the subcellular fractions, the
mitochondria has more CaM than protoplast and microsomal fraction, as Muto
(1982) reported in wheat leaf cell, but in the cytosol, it appears at a low
level, probably due to non-specific binding of cytosolic CaM to membranes
during the experimental procedures. ABA increases CaM levels in cytosol,
protoplast and microsomal fractions. The known regulatory role of Ca^{++} in
mitosis and cytokinesis of plant cell are consistent with a nuclear locale
for CaM. In the fraction purified from nuclei the percentage CaM/total protein in H_2O-25°C is always the largest one, mainly in nuclear membranes and
in chromatin. Matsumoto et al., (1984) reported a CaM-like activity associated with chromatin of pea buds.

The involvement of Ca^{++} and CaM in the activation of growth in other
plant systems and in membrane functions suggests that the Ca-CaM complex
could be implicated in the early germination phases. To prove it CaM levels
were related to radicular growth ability as suggested by its relation to
segments of 36h old embryonic axes. Figure 1 shows that the largest percentage is in the Ia segment, the apical one, where the cells are going to
undergo mitosis. In the Ib segment, where the cells are in elongation, the
level of CaM is diminished. These results agree with those of Hausser et
al. (1984) who visualized increased CaM in the tip of growing algae, root
hairs and pollen tubes.

In order to complete this result, the percentage CaM/total protein was
analyzed in 72h old segments. Ia was also the highest value, but lesser
than in 36h. In Ib segments the level has decreased considerably. However,
the V segment (the bud), which had a very low level of CaM in the 36h old
embryonic axes, now has a higher rate. We can conclude that a high CaM level
allows cell elongation and division, and in this way could promote germination as suggested by Cocucci (1984).

REFERENCES

Cocucci, M., 1984, Plant Cell and Environ., 7:215.

Hausser, I., Herth, W. and Reiss, H.D., 1984, Planta, 162:33.

Huber, D.J. and Nevins, D.J., 1980, Plant Physiol. 65:768.

Matsumoto, M., Yamaha, T. and Tanigawa, M., 1984, Plant and Cell Physiol., 25:191.

Muto, S., 1982, FEBS Lett., 147:161.

ACTIVATION OF PLANT AND ANIMAL CALMODULIN-DEPENDENT ENZYMES BY PLANT,

ANIMAL, PROTIST, AND FUNGAL CALMODULINS

Alice C. Harmon, Harry W. Jarrett, and Milton J. Cormier

Department of Biochemistry
University of Georgia
Athens, GA   30602

The activities of the calmodulin-dependent enzymes, pea seedling NAD kinase, bovine brain cyclic nucleotide phosphodiesterase and human erythroctye Ca^{2+}-ATPase were determined in the presence of varying amounts of calmodulins from bovine brain, human erythrocyte, mung bean seed, sea pansy (a marine coelenterate), mushroom, and Tetrahymena. Pea seedling NAD kinase required 10 to 100-fold lower calmodulin concentration for 50% activation than did bovine brain cyclic nucleotide phosphodiesterase or human erythrocyte Ca^{2+}-ATPase. The activation of phosphodiesterase and Ca^{2+}-ATPase by all calmodulins tested was cooperative with Hill coefficients ranging from 1.3 to 2.0. The activation of NAD kinase by mushroom, Tetrahymena, sea pansy, bovine brain, and human erythrocyte calmodulin was not cooperative (Hill coefficients = 1.0) whereas activation of this enzyme by mung bean calmodulin was cooperative (Hill coefficient 1.3). The order of potency of the six calmodulins differed for each of the three enzymes. For Ca^{2+}-ATPase the order of potency was erythrocyte > brain = sea pansy = mung bean = mushroom = Tetrahymena. For phosphodiesterase the order was erythrocyte = brain > mung bean = sea pansy > Tetrahymena > mushroom. For NAD kinase the order was Tetrahymena > mung bean > erythrocyte > brain = sea pansy = mushroom. These data show that calmodulin-dependent enzymes recognize different features of calcium-calmodulin complexes.

The physical and chemical properties of mushroom calmodulin were determined and compared to the properties of plant and animal calmodulins. No differences in mushroom calmodulin were observed that could account for its low potency in activating phosphodiesterase.

ACKNOWLEDGMENT

This work was supported by grants from N.S.F. (PCM 8213177) and D.O.E. (DE-AS09-83ER13107).

THE CONTROL OF CALMODULIN SYNTHESIS AND LOCALISATION IN

PEA ROOT TISSUE

Russell Butcher and David E. Evans

The Botany School
University of Oxford
South Parks Road, Oxford, OX1 3RA

INTRODUCTION

Since the first demonstration of calmodulin as a protein activator of NADkinase in higher plants[1], a large volume of data has been accumulated indicating that calmodulin has a role in the control of a wide variety of processes in plant cells. In any given cell, it is apparent that calmodulin is involved in a variety of processes simultaneously and that some of these processes require high concentrations of calmodulin to accumulate at certain locations - for instance during cell division[2]. In addition, calmodulin has been shown to be present in a number of subcellular compartments in plant cells[3]. Regulation of cell function by calcium and calmodulin is thus likely to occur at three or more points; control of calcium concentration (both cytoplasmic and within organelles); control of the level and distribution of calmodulin (both within the cytoplasm and organelles) and finally the control of enzymic and other processes by calcium and calmodulin.

Control of calcium levels, both in the cytoplasm and organelles has been the subject of extensive investigation in various tissues. The nature of calmodulin acceptor proteins is also being increasingly studied. Regulation of calmodulin levels and distribution has, however, been comparatively little studied in plant systems.

Regulation of calmodulin levels may be the result of increased synthesis, turnover or changes in compartmentalisation. Dedman et al[4] using 3T3 fibroblast cells together with monospecific antibodies and fluorescent probes have demonstrated the presence of numerous occupied and free calmodulin acceptor proteins which may, in addition to other roles, be involved in compartmentalising calmodulin. Studies on the nucleotide sequence of genes for calmodulin from various species [5] have revealed that, although sequence homology is generally very high, homology is low in the 3' and 5' 'untranslated' regions, suggesting that these regions may be important in the control of transcription or translation, or may contain sequences involved in the control of transport. Studies on the expression of calmodulin in chicken have suggested that only a single calmodulin gene is present; however multiple species of calmodulin mRNA are present "which may contain information that would direct it to an intracellular area"[5].

We are therefore investigating a number of possible controls of calmodulin function in plant tissues and work is in progress using affinity

purified antibodies to higher plant calmodulin, immunocytochemistry and
in vitro protein synthesis: some of the preliminary results of this are
described.

ANTIBODY PREPARATION

Two methods of raising antibodies to higher plant calmodulin were
used successfully. Calmodulin was prepared from peanut seed by affinity
chromatography on phenothiazine sepharose and purity demonstrated by
electrophoretic techniques. The protein was then either injected into
rabbits without further treatment or derivatised with dinitrofluorobenzene
prior to injection. A significant increase in antibody titre was noted
after two to four injections (each of approximately 0.5mg). Serum was
collected and fractionated for IgG. In view of the problems associated
with raising antibodies to proteins closely related to the native ones,
antibodies to bovine calmodulin were obtained for comparative purposes and
antibodies to plant calmodulin were affinity purified on a calmodulin-
agarose column. Purified antibody was eluted from the column with KSCN
and its activity demonstrated.

IMMUNOCYTOCHEMISTRY

Root material of Pea was fixed in 3% paraformaldehyde and cryo-
sectioned. Sections were mounted in sucrose and prepared for immunofluorescence. Sections stained using pre-immune serum showed little or
no fluorescence, while those stained with antibodies to plant and bovine
calmodulin showed marked staining of bodies immediately surrounding the
nucleus of non-dividing cells. Electron microscopy showed these to be
protein body-like.

IN VITRO PROTEIN SYNTHESIS

A preliminary investigation of the in vitro translation of mRNA for
plant calmodulin has been undertaken. A total RNA fraction was prepared
from pea roots, translated in a wheatgerm system and immuno-precipitations
carried out using affinity purified antibody. A significant amount of
translation product was shown to be bound to the antibody, while controls
including message-free preparations showed little binding.

REFERENCES

1. J. M. Anderson and M. J. Cormier, Calcium dependent regulator of
   NADkinase, Biochem. Biophys. Res. Commun. 84:595 (1978).
2. S. W. Tiwari, S. M. Wick, R. E. Williamson and B. E. S. Gunning,
   Cytoskeleton and integration of cellular function in higher plants,
   J. Cell Biol. 99:63s (1984).
3. S. Muto, Distribution of calmodulin within wheat leaf cells, F.E.B.S.
   Letters 147:161 (1984).
4. J. R. Dedman, M. J. Welsh, M. A. Kaetzel, R. L. Pardue and
   B. R. Brinkley, Localisation of calmodulin in tissue culture cells,
   in: "Calcium and cell function" vol 3, W. Y. Cheung, ed., Academic
   Press, London (1982).
5. J. A. Putkey, K. F. Tsui, T. Tanaka, J. P. Stein, E. C. Lai and
   A. R. Means, Chicken calmodulin genes: a species comparison of
   cDNA sequences and isolation of a genomic clone, J. Biol. Chem.
   258:11864 (1983).

IMMUNOCYTOCHEMICAL LOCALIZATION OF CALMODULIN IN PLANT TISSUE

Bruce S. Serlin, Marianne Dauwalder, and Stanley J. Roux

Department of Botany, University of Texas at Austin

Austin, Texas 78713

It is becoming apparent that calmodulin is a key regulatory protein in plants. It has been shown to be involved in mediating gravitropic (Biro et al., 1984) and phytochrome (Wagner et al., 1984; Serlin and Roux, 1984) responses. It is involved in directed polar growth (Hausser et al., 1984) and in regulating the concentration of cytosolic calcium (Fukumoto and Nagai, 1982). Given the diversity and importance of calmodulin-mediated functions known thusfar, we became interested in developing a technique which allowed us to investigate the distribution of calmodulin in plant tissue. In this report, we describe a method which is applicable for both algae and higher plants.

Materials and Methods

Plummules harvested from 7 day-old etiolated pea seedlings, filaments of _Mougeotia_ and cells of _Mesotaenium_ were the study material. Each of these were fixed in 2% formaldehyde in phosphate buffer pH 7.6. The fixed material was then used for immunocytochemical localization of calmodulin. After fixation, the material was dehydrated and embedded in polyethylene glycol (PEG). The PEG found to be best suited for both light and electron microscopy was produced by mixing two molecular weights, 4000 and 1540. For immunofluorescent studies, 6 um sections of pea plummules and 1-2 um sections of the algae were mounted on albuminized glass slides and then divested of their embedding medium. The adherent tissue was exposed to affinity purified antibodies produced in sheep against calmodulin (Chafouleas et al., 1979). Following washing the material was then incubated with rhodamine-conjugated goat anti-sheep antibodies. Material was viewed with a Zeiss Universal fluorescent microscope.

Discussion

Both stem and leaf areas of the plummule were examined. In the stems, cells showed a general light staining pattern within both the cytoplasm and the nucleoplasm. There was no fluorescence within the nucleoli. Further, there appears to be a differential to the staining pattern across the tissue regions. The most intense staining was seen in the epidermal cells and in the vascular parenchyma. Within these cells, there is a high degree of vacuolar staining in addition to the cytoplasmic staining. The same staining patterns were repeated in the leaf tissue. Guard cells differed from adjacent epidermal cells by having no vacuolar staining.

Mougeotia and Mesotaenium exhibited strong fluorescence along the edges of the chloroplast. Additionally, in Mougeotia, an area of high staining is discernable adjacent to the plasma membrane. In cells in which chloroplast rotation had been induced prior to fixation, the leading turning edge of the chloroplast stained most intensely. Visable in these cells were also areas of cytoplasm extending from the chloroplast to the plasma membrane which showed a light staining. These initial promising results are now being extended to the E.M. level of observation with immunogold localization work.

References

Biro, RL, Hale, CC, Wiegand, OF, and Roux, SJ, 1982, Effects of chlorpromazine on gravitropism in Avena coleoptiles. Ann Bot. 50: 735-747.

Chafouleas, JG, Dedman, JR, Munjaal, RP, and Means, AR, 1979, Calmodulin: development and application of a sensitive radioimmunoassay. J. Biol. Chem. 254: 10262-10267.

Hausser, I, Herth, W, and Reiss, H-D, 1984, Calmodulin in tip-growing plant cells, visualized by flourescing calmodulin-binding phenothiazines. Planta 162: 33-39.

Fukumoto, M, and Koushiro, N, 1982, Effects of calmodulin antagonists on the mitochondrial and microsomal Ca^{2+} uptake in apple fruit. Plants & Cell Physiol. 23: 1435-1441.

Serlin, BS, and Roux, SJ, 1984, Modulation of chloroplast movement in the green alga Mougeotia by the Ca^{2+} ionophore A23187 and by calmodulin antagonists. Proc. Natl. Acad. Sci. 81: 6368-6372.

Wagner, G, Valentin, P, Dieter, P, and Marmé, D, 1984, Identification of calmodulin in the green alga Mougeotia and its possible function in chloroplast reorientational movement. Planta 162: 62-67.

ALUMINIUM-INDUCED DEFORMATIONS AND MALFUNCTIONS OF CALMODULIN

Alfred Haug and Christopher Weis

Department of Microbiology and Pesticide Research Center
Michigan State University, East Lansing, MI 48824 (U.S.A.)

Aluminium toxicity is recognized as a serious global problem because vast areas of the world, especially in the subtropic and tropic zones, in western Europe and in the eastern United States, suffer from soil acidity which mobilizes soil aluminium[1]. Because part of the toxic effects exerted by aluminium ions are known to interfere with calcium metabolism, and since calmodulin has been recognized as a multifunctional, calcium-dependent regulatory protein for a variety of cellular responses[2], Al-induced structural deformations of calmodulin would therefore have severe repercussions on a multitude of cellular functions. Consequently we have advanced the hypothesis that a complex formed between calmodulin and aluminium may be a key lesion in what is now the broadly defined syndrome of aluminium toxicity[3].

Upon application of stoichiometric quantities of solvated aluminium species to micromolar calmodulin solutions, the protein's helical content decreases as opposed to a calcium-induced increase[3]. Moreover, in the presence of aluminium, calmodulin's exposure of hydrophobic surface area is enhanced[3] compared with that in the presence of calcium[2]. Since these hydrophobic surface regions are potential sites for the interaction of calmodulin with target proteins, we used melittin (M=2900) as target model. In the presence of calcium, with aluminium absent, calmodulin forms a mononuclear, high-affinity complex with melittin, with a dissociation constant in the nanomolar range[4]. As evidenced by circular dichroism studies, in the presence of aluminium ions melittin's helix content is considerably reduced (Table I) compared to that formed upon association of melittin with calcium-calmodulin (4:1). Binding of aluminium to calmodulin thus results in structural changes which hamper melittin's association with the regulatory protein. Weaker hydrophobic interactions in the complex of melittin with aluminium-calmodulin (3:1) are also manifested by a smaller blue-shift in the fluorescence emission maximum of melittin's single tryptophanyl residue. It appears therefore that the existence of a more polar microenvironment around this fluorophore in the presence of aluminium ions is less favourable for helix formation when melittin associates with calmodulin. The notion that the tryptophanyl residue resides in a more polar microenvironment in the presence of aluminium is consistent with observations that aluminium-altered calmodulin has a more open structure as compared with that of calcium-calmodulin (4:1), with aluminium absent[3]. A more open structure is suggested by findings that application of aluminium ions

TABLE I. Molecular parameters of calmodulin-melittin interaction, as measured by optical techniques. The calmodulin-melittin(1:1) complex was studied in the absence or presence of 4 aluminium ions per calmodulin, pH 6.5, 25°C. Calcium was present at a molar ratio of 8:1 for [calcium]/[calmodulin]. ϕ is the rotational correlation time, and k_q the bimolecular quenching constant of melittin's tryptophanyl residue; θ_{222} is the observed ellipticity of the complex.

melittin-calmodulin	ϕ nsec	$k_q 10^{-9}$ $M^{-1}sec^{-1}$	θ_{222} millideg
Al absent	8.42 ± 0.1	1.20 ± 0.1	-65
Al present	3.59 ± 0.1	1.62 ± 0.1	-62

breaks hydrogen bonds and that a spin probe covalently attached to calmodulin is less immobilized in aluminium-altered calmodulin as opposed to calcium-calmodulin[5]. The existence of such an open structure was further demonstrated by quenching experiments, whereby acrylamide molecules were used to quench the fluorescence emission from the tryptophanyl residue of melittin in association with calmodulin. In the presence of micromolar aluminium concentrations, at pH 6.5, 25°C, the bimolecular quenching constant, k_q, is larger as opposed to that measured in the absence of aluminium ions. Fluorescence lifetime studies support the interpretation that the quenching process of melittin's tryptophanyl residue by acrylamide proceeds via collisions rather than by static interactions. From measurements of fluorescence anisotropy vs. tryptophan lifetime rotational correlation times have been obtained when melittin is bound to calmodulin.

In summary, aluminium-altered calmodulin has a solvation structure different from that of calcium-calmodulin (4:1). The increased presence of water in aluminium-altered calmodulin is less favourable for helix formation in melittin, a model for target proteins. Since structural, time-dependent fluctuations play a crucial role in protein dynamics, a proper fit of calmodulin with target proteins cannot take place in aluminium-altered calmodulin. This mismatch may thus be instrumental in malfunctions of calcium- and calmodulin-dependent processes when aluminium ions enter the plant cell.

References

1. C.D. Foy, R.L. Chaney, and M.C. White, The physiology of metal toxicity in plants, Ann. Rev. Plant Physiol. 29:511 (1978).
2. C.B. Klee and T.C. Vanaman, Calmodulin, Adv.Protein Chem.35:213 (1982).
3. A. Haug, Molecular aspects of aluminum toxicity, CRC Crit. Rev. Plant Sci. 1: 345 (1984).
4. J.A. Cox, Sequential events in calmodulin on binding with calcium and interaction with target enzymes, Fed. Proc. 43: 3000 (1984).
5. N. Siegel, R. Coughlin, and A. Haug, A thermodynamic and EPR study of structural changes in calmodulin induced by aluminum binding, Biochem. Biophys. Res. Comm. 115: 512 (1983).

CALCIUM AND CALCIUM-BINDING PROTEINS IN PHLOEM

Dinkar D. Sabnis and Alan R. McEuen

Department of Plant Science
University of Aberdeen
Aberdeen, Scotland

ABSTRACT

The sieve elements of most dicotyledenous, and some monocotyledenous, plants possess an elaborate arrangement of filamentous or tubular cytoskeletal elements. Since calcium and calcium-binding proteins are involved in the assembly and disassembly of other cytoskeletal proteins, we investigated the possibility that similar mechanisms might operate in phloem.

Exudate collected from the severed sieve tubes of several cucurbitaceous species contained total calcium in the range of 2-8 mM, considerably higher than cytoplasmic levels in other plant and animal cells. Using Chelex-100 or Murexide, it was determined that the amount of total calcium bound by all soluble ligands in exudate collected into buffer was between 2-18 per cent. Protein-bound calcium was a negligible fraction of this. A simple mathematical model which treated all low MW compounds as a single ligand, combined with the known formation constants of possible calcium complexes, permitted an estimate of the in vivo equilibrium between free and bound forms of calcium. The data indicate that 30-70 per cent calcium may be present in a bound state, and free calcium concentrations ranged from 57 μM to 330μM. The abnormally high levels of calcium in sieve tube exudate may be a consequence of the release of turgor pressure when the stems are cut.

A ^{45}Ca-binding molecule, with an apparent MW >100,000, was fractionated on Sephadex G-100, from exudate samples of Cucurbita maxima, C. pepo, Cucumis sativus and C. melo. The sieve tube origin of calcium-binding activity was confirmed. The protein nature of the calcium-binding moiety was indicated by its lability on storage, dependence on the concentration of thiol reagents, and sensitivity to protease digestion. Scatchard analysis suggested the presence of at least two binding sites.

A radioimmunoassay for calmodulin indicated relatively high levels of the protein in both sieve tube exudate as well as extracts of Cucurbita stem. Affinity chromatography and immunoblots were employed to relate the presence of calmodulin to the calcium-binding protein reported earlier.

CALCIUM NUTRITION, CALMODULIN CONTENT AND SUSCEPTIBILITY TO TMV IN TOBACCO (<u>NICOTIANA</u> <u>TABACUM</u> L. C.V. XANTHI)

Sylvie Ferrario, Alain Poupet and Denise Blanc

I.N.R.A., Stations d'Agronomie, et de Pathologie végétale
Bd. du Cap, B.P. 78
06602, Antibes, France

Studying the influence of plant nutrition on the development of para-sites in host plants is of great interest especially from an agronomic viewpoint. There are many reports that mineral deficiences affect plant susceptibility to viruses (Chessin and Scott, 1955; Seaker et al., 1982), but interpretations have been seldom proposed. In this view our choice of calcium nutrition has been made because of the recognized importance of calcium in the structure of membrane and cell wall where the primary inter-actions between the virus and the host cell do occur. Furthermore, the role of calcium as second messenger in many metabolic events is now accepted (Marmé and Dieter, 1983), and it would be interesting to investigate it in a virus infected cell host model. So we propose to study the susceptibility to TMV (tobacco mosaic virus) of tobacco plants (<u>Nicotiana</u> <u>tabacum</u> L. c.v. Xanthi) in relation with their calcium nutrition and their calmodulin content.

Before being inoculated with TMV, tobacco plants grown in controlled conditions were fed with eight different levels of calcium in nutrient solu-tions (from 0.5 to 10 meq/1 Ca), providing a linear variation (from 0.7 to 2.3 p. 100 of dry matter) of the total calcium contents of the whole plants and of their leaves (Ferrario et al., 1985). The variation of calcium con-centration in the mineral nutrient solutions is necessarily accompanied by other changes, and in our case the relative ratio of NO_3- to NH_4+ was affected. Also the pH of the culture medium (a mixture of peat and perlite) increased with calcium content. Therefore, in a complementary second experi-ment, calcium nutrition and pH substrate were maintained constant while the relative ratio of NO_3- to NH_4+ varied.

Considering the predominant action of calcium as a second messenger, and using a procedure of the activation of 3', 5' - AMPc - PDE method, the amount of the tobacco calmodulin has been determined in these plants in which the total calcium content varied. The tobacco calmodulin has also been characterized by its behaviour on phenothiazin affinity column and by electrophoresis.

The most important observation is that the calmodulin amount related to the total proteins seems independant from calcium nutrition. As total calcium content in tobacco plants increases from 0.7 to 2.3 p. 100 of dry matter, the relative ratio of calmodulin to total proteins remains constant (0.35 p. 100). In the first experiment the observed increase of total calmodulin and protein contents as total calcium decreases, could be related to the decrease of substrate pH. This latter could favour nitrate absorption (Rao and Rains, 1976) and its assimilation since there is no clear nitrate accumulation in plants. Besides, in the second experiment the calmodulin content of tobacco increases with NO_3- concentration in the nutrient solution.

The tobacco cultivar used (Nicotiana tabacum L. c.v. Xanthi) is characterized by its hypersensitive response to TMV: it reacts by the formation of necrotic lesions 48 h after its inoculation. The number of necrotic lesions appearing on the inoculated leaves is function of the number of TMV sensitive cells. This is accompanied by the activation of particular metabolisms such as synthesis of some phenolic compounds (Martin-Tanguy et al., 1976), and of the pathogenesis related proteins (PRs proteins) (Van Loon and Van Kammen, 1970). These events are correlated to the inhibition of TMV multiplication and its localization (Van Loon and Van Kammen, 1970).

In our first experiment, we have observed that the number of necrotic lesions varies as a function of calcium nutrition according to a bimodal curve with two maxima for tobacco susceptibility at two different levels of Ca (Ferrario et al., 1985). Furthermore, using HPLC methods, we have noted a similar variation of the contents of hydroxycinnamoyl putrescine and the PRs proteins the synthesis of which is recognized as a consequence of the necrosis. In a previous work (Ferrario et al., 1985) this bimodal curve has also been observed for the TMV content in the leaves and in the pistils of the sensitive cultivar Samsun.

For many reasons the interpretation of this observation is not clear because we cannot relate directly the total calcium content to the TMV multiplication rate (Ferrario et al., 1985).

Therefore the bimodal curve could result from the variation not only of calcium but also of nitrogen or pH in nutrient solutions. A concerted action is possible between the calcium and pH nutrition effects, which could be reflected down to an intracellular level. So the estimation of calcium and pH intracytoplasmic variations might be interesting in our model.

Moreover, no difference in calmodulin content has been detected between healthy and infected plants (or between calcium deficient and well-fed tobaccoes). Nevertheless the action of calmodulin in this virus infection of tobacco plants could be an intermediate in a series of events which we now try to uncover.

REFERENCES

Chessin, M. and Scott, H.A., 1955, Calcium deficiency and infection of Nicotiana glutinosa by tobacco mosaic virus, Phytopathology, 45:288-289.

Ferrario, S., Poupet, A. and Blanc, D., 1985, Influence de la nutrition calcique sur la multiplication du virus de la mosaique dy tabac chez le tabasc, Agronomie, in press.

Marmé, D. and Dieter, P., 1983, Role of Ca^{2+} and calmodulin in plants, in "Calcium and Cell Function" vol. IV, Academic Press, New York.

Martin-Tanguy, J., Martin, C., Gallet, M. and Vernoy, R., 1976, Sur de puissants inhibiteurs naturels de multiplication du virus de la mosaique du tabac, C. R. Acad. Sci., Ser. D, 282:2231-2234.

Rao, K.P. and Rains, D.W., 1976, Nitrate absorption by varley. I. Kinetics and energetics, Plant Physiol., 57:55-58.

Seaker, E.M., Bergman, E.L. and Romaine, C.P., 1982, Effects of magnesium on tobacco mosaic virus infected egg plants, J. Amer. Soc. Hort. Sci., 107(1):162-166.

Van Loon, L.C. and Van Kammen, A., 1970, Polyacrylamid disc electrophoresis of the soluble leaf proteins from Nicotiana tabacum var. Samsun and Samsun NN'. Changes in protein constitution after infection with tobacco mosaic virus. Virology, 40:199-211.

ROLE OF Ca^{2+} IN THE INDUCTION OF THE PHYTOALEXIN DEFENSE
RESPONSE IN SOYBEAN CELLS

Margarita R. Stäb and Jürgen Ebel

Biologisches Institut II
Universität Freiburg
D-78 Freiburg, FRG

One type of active defense response of plants to in-
vading microorganisms is the production of phytoalexins which
are low molecular weight antimicrobial compounds. Phytoalexin
synthesis can also be induced in various plant tissues by
molecules of microbial origin, called elicitors.

Cultured soybean (Glycine max) cells and a ß-glucan
elicitor of the fungus Phytophthora megasperma f.sp. glycinea
a pathogen of soybean, have been used as a model system for
studies of the induced synthesis of isoflavonoid phytoalexins
(Ebel, 1984). Phytoalexin accumulation is correlated with
transient increases in the activities of several enzymes of
phytoalexin biosynthesis. The transient increase in chalcone
synthase activity, the first enzyme specific for flavonoid/
isoflavonoid biosynthesis, was shown to be preceded by a
large and rapid enhancement in chalcone synthase mRNA activity and amount (Ebel et al., 1984; Schmelzer et al., 1984;
Grab et al., 1985). These results are consistent with the
hypothesis that phytoalexin synthesis in soybean cells is
regulated by temporary gene activation (Schmelzer et al.,
1984).

The mechanism by which the glucan elicitor causes the
rapid and drastic alterations in the metabolism of soybean
cells is largely unknown. We observed that changes in the
concentration of Ca^{2+} ions in the cell culture medium markedly affected the elicitor-mediated enzyme induction and phytoalexin accumulation. Removal of extracellular Ca^{2+} through
an EGTA treatment of the culture followed by a washing procedure almost completely abolished the elicitor-mediated
enzyme induction and phytoalexin accumulation, whereas readdition of Ca^{2+} to the culture medium restored the phytoalexin response. Transport studies in vitro demonstrated that
the partially purified elicitor did not inhibit active Ca^{2+}
uptake by the membrane vesicles. The results of studies with
antagonists of Ca^{2+} function in cells are presented and the
role of Ca^{2+} in signal transduction/transmission during the
elicitor-mediated phytoalexin response will be discussed.

REFERENCES

Ebel, J., 1984, Induction of phytoalexin synthesis in plants
following microbial infection or treatment with elicitors, in: "Bioregulators, Chemistry and Uses", R. L.
Ory and F. R. Rittig, eds., Amer. Chem. Soc. Symposium
Series, 257:257.

Ebel, J., Schmidt, W. E., and Loyal, R., 1984, Phytoalexin
synthesis in soybean cells: Elicitor induction of
phenylalanine ammonia-lyase and chalcone synthase
mRNAs and correlation with phytoalexin accumulation,
Arch. Biochem. Biophys., 232:240.

Grab, D., Loyal, R., and Ebel, J., 1985, Elicitor-induced
phytoalexin synthesis in soybean cells: Changes in the
activity of chalcone synthase mRNA and the total population of translatable mRNA, Arch. Biochem. Biophys.,
submitted.

Schmelzer, E., Börner, H., Grisebach, H., Ebel, J., and
Hahlbrock, K., 1984, Phytoalexin synthesis in soybean
(Glycine max). Similar time courses of mRNA induction
in hypocotyls infected with a fungal pathogen and in
cell cultures treated with fungal elicitor, FEBS Lett.,
172:59.

CA-STIMULATED α-AMYLASE SECRETION FROM BARLEY ALEURONE PROTOPLASTS

Douglas S. Bush, Maria-Jesus Cornejo,
Chunnong Huang and Russell L. Jones

Department of Botany
University of California
Berkeley, California 94720, USA

α-Amylase secretion by isolated barley aleurone layers is greatly enhanced in 1-10 mM extracellular Ca^{2+}. This effect of Ca^{2+} on α-amylase secretion is rapidly reversible (< 1 minute) and peculiar to a group of α-amylase isozymes with a high isoelectric point (high pI). Secretion of α-amylase isozymes with a low isoelectric point (low pI) is not stimulated by Ca^{2+}. It has been suggested that Ca^{2+} regulates secretion by altering either the cell wall or the plasma membrane. In order to determine whether the cell wall is obligatorily involved in Ca-stimulated secretion, we have prepared protoplasts from the aleurone and measured their amylase secretion in response to Ca^{2+} and gibberellic acid (GA_3).

Protoplasts prepared according to the method of Jacobsen et al. (<u>Planta</u>, in press) and incubated in 5 μM GA_3 show a 20-fold increase in α-amylase secretion as the external Ca^{2+} concentration is increased from 0 to 100 mM; half-maximal secretion occurs at 10 mM Ca^{2+}. Protoplasts, like layers, show a differential effect of Ca^{2+} on the secretion of high and low pI isozymes. There is a three-fold increase in the secretion of low-pI α-amylases between 1 and 100 mM Ca^{2+} compared to a 16-fold increase in high-pI isozyme secretion over the same range of Ca^{2+} concentration. Pulse-labelling of protoplasts with ^{35}S-methionine indicates that the secreted protein is made <u>de novo</u>. Unlike layers, however, Ca-stimulated secretion of amylase by protoplasts is not rapidly reversible; rather, the Ca^{2+} concentration does not slow the secretion of high-pI α-amylases for 3 to 6 hours.

Although Ca^{2+} influences the stability of the plasma membrane and the longevity of the protoplast, witholding Ca^{2+} does not alter the normal course of development which aleurone undergoes. Protoplasts isolated from untreated aleurone layers are densely packed with protein bodies and these become vacuolate even in the absence of GA and Ca^{2+}.

We conclude that the cell wall is not necessary for Ca^{2+}-stimulated amylase secretion and that Ca^{2+} exerts its regulatory effect at the plasma membrane or on secretory structures inside the cell.

CALCIUM UPTAKE AND EXCHANGE IN BARLEY ALEURONE LAYERS AND PROTOPLASTS DURING CA-STIMULATED AMYLASE SECRETION

Douglas S. Bush and Russell L. Jones

Department of Botany
University of California
Berkeley, California 94720, USA

Extracellular Ca^{2+} regulates α-amylase secretion in isolated barley aleurone layers. At high Ca^{2+} concentrations (1-10 mM) the secretion of isozymes of α-amylase with high isoelectric points (high pI) is greatly stimulated. This effect is rapidly reversed by withdrawing Ca^{2+} and is specific for Ca^{2+}. Extracellular Ca^{2+} may exert its effect either directly on the plasma membrane or by altering the Ca concentration of some compartment inside the cell. We have determined the influence of extracellular Ca^{2+} on intracellular Ca pools by measuring Ca uptake and exchange in aleurone layers and by direct measurement of cytoplasmic Ca^{2+} activity in aleurone protoplasts.

On an equivalent basis Ca is a minor constituent of the aleurone accounting for only .06% of the dry weight. Based on organelle fractionation less than 5% of aleurone Ca is associated with the endoplasmic reticulum and mitochondria. The intracellular Ca content of the aleurone declines during incubation as a result of Ca efflux. This efflux is not stimulated by gibberellic acid (which greatly increases amylase secretion) or by the presence of 20 mM extracellular Ca^{2+}. Over a period of 24 hours when layers actively secrete amylase, 20% of the intracellular Ca can be exchanged with extracellular Ca^{2+}. These results suggest that there is a small pool of metabolically active Ca whose size declines during incubation but which remains a constant fraction of total cell Ca. Using protoplasts from barley aleurone alyers and Quin-like dyes (provided by Dr. R. Tsien) we have confirmed that the level of cytoplasmic Ca is quite low (< 1 μM). Experiments to determine if there are transient or steady-state changes in cytoplasmic Ca^{2+} during Ca-stimulated secretion are in progress.

We conclude that intracellular Ca levels are influenced by extracellular Ca^{2+} and that the regulatory effect of Ca on amylase secretion is at the plasma membrane.

THE CONTROL OF α-AMYLASE SYNTHESIS BY CALCIUM IN THE BARLEY ALEURONE

Jill Deikman and Russell L. Jones

Department of Botany
University of California
Berkeley, California 94720, USA

Barley aleurone layers synthesize and secrete at least four isozymes of α-amylase which are separated by various criteria (including isoelectric points) into two groups: A (low pI; isozymes 1 and 2) and B (high pI; isozymes 3 and 4). Isozyme 2 is produced under all conditions of aleurone incubation but its rate of accumulation is enhanced by the presence of gibberellic acid (GA). Isozyme 1 requires GA for its production as do isozymes 3 and 4. Isozymes 3 and 4 also require Ca^{2+} for their accumulation in and secretion from aleurone layers; the production and secretion of isozymes 1 and 2 is completely independent of the level of Ca^{2+} in the external medium. The dependence of amylase secretion on Ca^{2+} has been investigated previously and shown to be specific to the B group (Jones, R. L. and Jacobsen, J. V., 1982, Planta 156:421-432), and the Ca^{2+} requirement was demonstrated to be at the level of the plasma membrane (Moll, B. A. and Jones, R. L., 1982, Plant Physiology 70:1149-1155). The work presented here concerns the effect of Ca^{2+} on amylase synthesis.

Pulse-labelling of barley aleurone layers incubated with 2.5 µM GA_3 in the presence or absence of 5 or 10 mM $CaCl_2$ shows that α-amylase isozymes 3 and 4 are not synthesized in vivo in the absence of Ca^{2+}. Aleurone layers were pulse-labelled with ^{35}S-methionine for 30 min and for 3 h, and isozymes in the aleurone extract were separated by agar gel electrophoresis and visualized by fluorography.

α-Amylase mRNA levels were measured by hybridization studies with α-amylase cDNA clones derived from group A and group B α-amylase mRNA. Clone 1-28 (group B) hybridized on RNA gel blots and dot blots equally to mRNA from aleurone layers incubated in GA with or without Ca^{2+}. Hybridization of the same blots with clone E (a group A clone provided by Dr. John C. Rogers) also resulted in equal hybridization to mRNA from aleurone layers incubated with GA in the presence or absence of Ca^{2+}. Clone E but not clone 1-28 hybridized to mRNA from layers incubated in H_2O, demonstrating our ability to distinguish between the A and B groups of α-amylase mRNA.

In vitro translation of aleurone mRNA produced by layers incubated in GA with and without Ca^{2+} and immunoprecipitation of translation products shows that the α-amylase mRNA produced in the absence of Ca^{2+} is functional, at least in vitro.

We conclude that Ca^{2+} is important in the synthesis of group B α-amylase at a step after mRNA accumulation and processing. Experiments designed to pinpoint this step are in progress.

TYROSINE SPECIFIC PROTEIN KINASES IN PLANT TISSUES

D. Blowers and A.J. Trewavas

Department of Botany, University of Edinburgh, King's Buildings, Mayfield Road, Edinburgh EH9 3JH

Until 1979 protein kinases were considered to use only serine and threonine as substrates, although some work had been carried out on the so called 'acid labile' histone phosphates thought to be phosphohistidine and phospholysine (Chen et al. 1974). The discovery of an ability to phosphorylate tyrosine in Polyoma tumor antigen immunoprecipitates (Eckhart et al. 1979) opened a door to the understanding of cell growth control. Numerous other tyrosine specific protein kinases have now been discovered in animals, these include the receptors of insulin, epidermal growth factor, platelet derived growth factor (Krebs 1983), 17-β-estradiol (Migliaccio et al. 1984) and numerous retroviral oncogene products. To date no tyrosine kinase activity has been reported for any plant material.

In some cases the overall level of phosphotyrosine in animal cells may be as low as 0.05% of the total phosphoamino acid content (Sefton et al. 1982). Clearly, techniques used to reveal its presence must either separate or exclude phosphoserine and phosphothreonine. Three basic methods have been employed to identify in vitro tyrosine kinase activity in plant material using $[\gamma-^{32}P]ATP$. Each approach will now be discussed in turn.

Differential Stability of Phosphoserine, Phosphothreonine and Phosphotyrosine in Hot Alkali

This method relies upon the higher stability of the phosphodiester bond to tyrosine as compared to serine and threonine in the presence of hot alkali (Plimmer 1941). However, the original data applies to the free phosphoamino acids and others have found that phosphothreonine also shows a degree of stability when present in certain proteins (Cooper and Hunter 1981). Thus, the persistence of labelled bands in alkali treated polyacrylamide gels, revealed by autoradiography, is not an entirely reliable means of assessing the presence of phosphotyrosine unless each resistant band is subjected to hydrolysis and phosphoamino acid analysis.

Hydrolysis of Labelled Total Protein and Two Dimensional Phosphoamino Acid Analysis

Both crude membrane and ammonium sulphate precipitated fractions were prepared from plant material and after labelling were subjected to hydrolysis in 6N HCl and two dimensional separation by thin layer electrophoresis. Autoradiography was used to locate spots comigrating with phosphoamino acid markers.

Experiments of this type have been carried out using pea bud, pea root, zucchini hypocotyl and crown gall callus as the starting material. Whilst large amounts of phosphoserine and in some cases smaller amounts of phosphothreonine were observed in none of these experiments was any phosphotyrosine detected.

Artificial Substrates as a Means to Amplify Tyrosine Kinase Activity

Low levels of phosphotyrosine could be the result of the presence of few sites suitable for phosphorylation. The use of artificial substrates should supply the enzyme(s) with unlimited, but not necessarily ideal, sites. Such substrates must be easily isolated/assayed and if peptides must not contain serine or threonine. Angiotensin II and tyramine are suitable and readily available substrates. Free tyrosine is not soluble in water to any great extent. The phosphorylated products are isolated after trichloroacetic acid (TCA) precipitation of protein and hydrolysis of remaining [$-^{32}$P]ATP (Braun et al. 1983).

Both angiotensin II and tyramine were found to reduce the amount of label isolated with TCA soluble material for pea membranes and soluble fraction proteins. However, this cannot be taken as evidence that tyrosine kinase activity is absent since products must also be identified. Phosphorylated angiotensin II could not be identified in high voltage paper electrophoretic separations of the products and similarly, two dimensional thin layer electrophoretic separations revealed no phosphotyramine.

References

Braun, S., Ghany, A.-M. & Racker, E. (1983). A rapid assay for protein kinases phosphorylating small polypeptides and other substrates. Anal. Biochem. 135, 369-378

Chen, C.-C., Smith, D.L., Bruegger, B.B., Halpern, R.M. & Smith, R.A. (1974). Occurrence and distribution of acid labile histone phosphates in regenerating rat liver. Biochemistry 13, 3785-3789

Cooper, J.A. & Hunter, T. (1981). Changes in protein phosphorylation in Rous Sarcoma Virus transformed chicken embryo cells. Mol. Cell. Biol. 1, 165-178

Eckhart, W., Hutchinson, M.A. & Hunter, T. (1979). An activity of phosphorylating tyrosine in Polyoma T antigen immunoprecipitates. Cell 18, 925-933

Krebs, E.G. (1983). Historical perspectives on protein phosphorylation and a classification system for protein kinases. Phil. Trans. R. Soc. Lond. 302, 3-11

Migliacchio, A., Rotandi, A. & Auricchio, F. (1984). Calmodulin stimulated phosphorylation of the 17-β-estradiol receptor on tyrosine. Proc. Nat. Acad. Sci. USA 81, 5921-5925

Plimmer, R.-H.A. (1941). Esters of phosphoric acid IV: Phosphoryl hydroxy amino acids. Biochem. J. 35, 461-469

Sefton, B.M., Hunter, T., Nigg, E.A., Singer, S.J. & Walter, G. (1982). Cytoskeletal targets for viral transforming proteins with tyrosine specific protein kinase activity. Cold Spring Harbor Symp. Quant. Biol. 46, 939-953

PROTEIN PHOSPHORYLATION IN NORMAL AND TRANSFORMED CELLS :

EFFECT OF CYTOKININ ON CALCIUM REGULATION *IN VIVO* AND *IN VITRO*

Daphne C. Elliott

School of Biological Sciences
Flinders University of South Australia
Bedford Park, S.A. 5042

ABSTRACT

The calcium stimulation of protein kinase activity in a crude 50,000 g supernatant from growing and non-growing normal callus and from crown gall material has been compared. There is a greater Ca^{2+} effect in non-growing (starved) soybean callus than in growing (i.e. plus cytokinin) callus and a very much lower effect in crown gall material. In the presence of added calmodulin cytokinin increases the stimulation by Ca^{2+} in all three tissues.

INTRODUCTION

In animals many hormones and extracellular stimuli regulate cell function by inducing an increase in calcium concentration in the cytoplasm (Rasmussen et al., 1984). There is increasing evidence that plant hormones work also through Ca^{2+} as a second messenger (Elliott, 1985). Based on a number of studies, including effects of calmodulin-binding drugs, a general hypothesis of cytokinin action has been formulated (Elliott, 1983) stating that 1. the hormone acts through membrane function (specifically K^+ and/or Ca^{2+} fluxes) and 2. formation of the participating membrane structures is stimulated by stress pre-treatment and by the hormone itself. When Ca^{2+} is the key regulator in animal systems, Ca^{2+} mobilisation, phosphatidyl-inositol turnover and protein phosphorylation are all aspects of the amplification cascade (Rasmussen et al., 1984). In this paper some effects of cytokinins on Ca^{2+}-dependent phosphorylation in plants are explored.

RESULTS AND DISCUSSION

Using a 50,000 g supernatant from soybean callus and T37-transformed crown gall the optimum conditions for demonstrating calcium stimulation of protein kinase activity were found (Table 1). These crude extracts both showed the best calcium-dependent response in 1.6 mM EDTA/EGTA plus 2.0 mM Ca^{2+}, adopted as standard in subsequent experiments. The titration of endogenous calcium-stimulated protein kinase with EGTA gives a different shaped curve for normal and transformed tissue.

The effect of cytokinin *in vivo* and *in vitro* on Ca^{2+}-calmodulin-stimulated protein kinase in crude extracts from plant tissue cultures is

Table 1. Titration of endogenous Ca^{2+}-stimulated protein kinase with EGTA

	Protein kinase (pmole/min/mg protein)					
EDTA/EGTA (mM)	0.6	1.6	2.6	0.6	1.6	1.6
Ca(NO$_3$)$_2$ (mM)	0	0	0	1.0	2.0	3.0
Soybean callus Ca^{2+}-dependent activity	22.99	10.71	5.92	31.03 8.04 +35%	24.44 13.73 128%	23.21 12.50 117%
Crown gall T37 Ca^{2+}-dependent activity	31.39	21.08	17.93	34.0 2.61 + 8%	30.26 9.18 +44%	26.56 5.48 +26

Table 2. Effect of cytokinin on calmodulin-stimulated protein kinase

	Effect of Calcium (%)			
	− Cytokinin		+ Cytokinin	
Tissue	−CBP	+CBP	−CBP	+CBP
Soybean callus (starved)	136	134	140	165
Soybean callus (growing)	81	82	76	128
Tobacco crown gall (T37)	40	60	40	80

shown in Table 2. Comparison of the calcium stimulation reveals a greater % effect in non-growing (starved) soybean callus than in growing (i.e. +BA) callus and a very much lower effect in crown gall material. This can be correlated with probable cytokinin content of the tissues. The hypothesis that cytokinins work via changes in internal Ca^{2+} concentrations (Elliott, 1983) gains some support if the % response to calcium is related to the calcium status of the extracted tissue i.e. the higher the $[Ca^{2+}]$, the less effect there is of Ca^{2+} added to the assay mixture. This correlation may be carried further with the crown gall response which has the lowest % stimulation of calcium. This tissue is now known to be able to grow without added cytokinin because one of the Ti genes supplies an enzyme in the cytokinin biosynthesis pathway with resulting high cytokinin content.

When cytokinin is added *in vitro*. a further pattern develops. Previous work on the effect of BA on protein kinase assays *in vitro* has drawn attention to the inhibiting effect of BA used at high concentrations (10^{-4}M) presumably because of competition with ATP (Elliott and Murray, 1974; Keates and Trewavas, 1974; Ralph and Wojcik, 1981). In the present work, at 1 μM, BA has no effect on basal protein phosphorylation, but increases the calcium effect in the presence of calmodulin in all tissues. This effect (Table 2) may be some kind of sensitivity modulation (as defined by Rasmussen et al., 1984), while the *in vivo* effect (Table 1,2) can be seen as an amptitude modulation.

REFERENCES

Elliott, D.C., 1983, Plant Physiol., 72 : 215.
Elliott, D.C., 1985, Calcium involvement in plant hormone action, This Vol.
Elliott, D.C. and Murray, A.W., 1974, "Plant Growth Substances, 1973".
        Hirokawa, Tokyo.
Keates, R.A.B. and Trewavas, A., 1974, Plant Physiol., 54 : 95.
Ralph, R.K. and Wojcik, S.T., 1981, Plant Sci. Lett., 22 : 127.
Rasmussen, H., Kojima, I., Kojima, K., Zawalich, W. and Apfeldorf, W.,
        1984, Adv. Cyc. Nuc. Prot. Phos. Res., 18 : 159.

PROPERTIES OF A CALCIUM-DEPENDENT PROTEIN KINASE FROM SILVER BEET LEAVES

Gideon M. Polya and Vito Micucci

Department of Biochemistry
La Trobe University
Bundoora, Victoria, Australia

Protein phosphorylation catalyzed by cyclic nucleotide- or Ca^{2+}-dependent protein kinases represents a major means by which external stimuli change animal cellular processes[1]. In higher plants, cyclic nucleotide-dependent protein kinase has not been unequivocally resolved[2,3] although many possible elements of a cyclic nucleotide-regulatory system have been found in plants[3-6]. Ca^{2+} clearly acts as a "second messenger" in plant as in animal cells[7]. Various agents have been demonstrated or inferred to change cytosolic free Ca^{2+} concentration in plant cells with consequent changes in plant cellular processes[7]. Plants contain calmodulin and various Ca^{2+}-calmodulin- or Ca^{2+}-activated enzymes, including Ca^{2+}-dependent protein kinases[7,8-11]. Soluble[8-10] as well as membrane-bound[7,11,12] Ca^{2+}-dependent protein kinase is present in plants. Two soluble Ca^{2+}-dependent protein kinases (protein kinases I and II) have been partially-purified from wheat embryo and characterized[8-10]. This abstract describes the properties of a Ca^{2+}-dependent protein kinase from leaf tissue.

A soluble protein kinase that is dependent on Ca^{2+} for activity was partially purified from silver beet (Beta vulgaris) leaves by a procedure involving chromatography on DEAE-Sephacel and gel filtration. The protein kinase catalyzes the Ca^{2+}-dependent phosphorylation of histones and casein. The Ca^{2+}-dependence of casein (but not of histone) phosphorylation is elicited after gradient elution of the preparations from DEAE-Sephacel i.e. this step removes an inhibitor of net Ca^{2+}-dependent casein phosphorylation. Ca^{2+}-dependent casein and histone phosphorylating activities exactly co-purify on gel filtration and ion exchange chromatography. The rate of casein phosphorylation is half-maximal at 1 μM free Ca^{2+} concentration and maximal at 10 μM; 0.2 mM and 1.0 mM free Ca^{2+} are required for half-maximal and maximal histone phosphorylation rates, respectively. Millimolar Mg^{2+} is required in addition to Ca^{2+} for maximal activity. Millimolar Mn^{2+} can substitute for the (Ca^{2+} + Mg^{2+}) requirement. The Km for ATP is 19 μM; other nucleoside-5'-triphosphates and ADP inhibit phosphoryl transfer from ATP to protein. Serine, rather than threonine, residues of casein are preferentially phosphorylated by the enzyme.

The leaf protein kinase is inhibited by the phenothiazine-derived calmodulin antagonists trifluoperazine, fluphenazine and chlorpromazine. Histone phosphorylation catalyzed by the protein kinase is stimulated by calmodulin and heparin. The leaf Ca^{2+}-dependent protein kinase is similar in many properties to the wheat embryo Ca^{2+}-dependent protein kinases I and

II. However the leaf protein kinase is distinct in molecular weight (49,000) from the wheat embryo Ca^{2+}-dependent protein kinases I and II (molecular weights 90,000 and 86,000, respectively)[9,10]. The leaf enzyme resembles wheat embryo protein kinase I in that phosphoryl transfer to protein from ATP catalyzed by both enzymes is inhibited by ITP, CTP and UTP; the reaction catalyzed by protein kinase II is not inhibited by these nucleoside 5'-triphosphates[10].

We propose that Ca^{2+}-dependent protein kinases of this kind are involved in Ca^{2+}-mediated stimulus-response coupling in photosynthetic higher plant cells. In animal systems, elements of the Ca^{2+}- and cyclic nucleotide-based regulatory systems interact[1]. We have resolved a high affinity cyclic nucleotide-binding phosphohydrolase from silver beet leaves (K_d values for 3',5'-cyclic AMP and 3',5'-cyclic GMP: 3 μM and 0.3 μM, respectively) and have demonstrated in the same tissue protein phosphorylation that is activated (up to _circa_ 2-fold) by 3',5'-cyclic GMP (10-50 μM) (Polya, Gantinas and Basiliadis, manuscript in preparation). The latter phenomenon may derive from inhibition of ATP hydrolysis by 3',5'-cyclic GMP and not to direct activation of a cyclic nucleotide-dependent protein kinase. The Ca^{2+}-dependent protein kinase and the high-affinity 3',5'-cyclic GMP-binding phosphohydrolase may represent sites of action of Ca^{2+} and 3',5'-cyclic GMP, respectively, acting as second messengers in higher plant leaves.

REFERENCES

1. P. Cohen, _Nature_ 296:613-620 (1982).
2. R. Kato, I. Uno, T. Ishikawa and T. Fujii, _Plant Cell Physiol_. 24:841-848 (1983).
3. E.G. Brown and R.P. Newton, _Phytochemistry_ 20:2453-2463 (1981).
4. G.M. Polya and J.A. Bowman, _Plant Physiol._ 68:577-584 (1981).
5. D.A. Francko, _Adv. Cyclic Nucleotide Res._ 15:97-117 (1983).
6. R.P. Newton, E.E. Kingston, D.E. Evans, L.M. Younis and E.G. Brown, _Phytochemistry_ 23:1367-1372 (1984).
7. P. Dieter, _Plant, Cell & Environment_ 7:371-380 (1984).
8. G.M. Polya and J.R. Davies, _FEBS Lett._ 150:167-171 (1982).
9. G.M. Polya, J.R. Davies and V. Micucci, _Biochim. Biophys. Acta_ 761:1-12 (1983).
10. G.M. Polya and V. Micucci, _Biochim. Biophys. Acta_ 785:68-74 (1984).
11. A.M. Hetherington and A. Trewavas, _FEBS Lett._ 145:67-71 (1982).
12. G.M. Polya, A. Schibeci and V. Micucci, _Plant Sci. Lett._ 36:51-57, (1984).

CALCIUM-DEPENDENT PROTEIN PHOSPHORYLATION IN GERMINATED POLLEN

Gideon M. Polya[a], Vito Micucci[a], Anne L. Rae[b],
Philip J. Harris[b] and Adrienne E. Clarke[b]

[a]Department of Biochemistry, La Trobe University, Bundoora
Victoria, Australia. [b]Plant Cell Biology Research Centre
University of Melbourne, Parkville, Victoria, Australia

Ca^{2+}-dependent protein phosphorylation represents a major mechanism for stimulus-response coupling in animal systems[1-3]. Soluble[4-6] and membrane-associated[7,8] Ca^{2+}-dependent protein kinases are present in higher plants and are likely to be involved in Ca^{2+}-mediated signal transduction in plant systems. We have initiated investigations into the possible involvement of Ca^{2+}-dependent protein phosphorylation in pollen tube growth.

Self-incompatibility genes (S-genes) are present in many families of flowering plants and operate to prevent inbreeding. Thus in incompatible pollinations the products of identical S-alleles in the pollen and pistil are presumed to interact to produce arrest of pollen tube growth[9]. The mechanism of action of the S-gene product is not known. However the involvement of Ca^{2+}-dependent protein phosphorylation in signal transduction in incompatible pollinations represents an attractive hypothesis given the evidence for a "second messenger" role of Ca^{2+} in plants[8], the presence of Ca^{2+}-dependent protein kinases in plants[4-8] and the involvement of Ca^{2+} in pollen germination and pollen tube growth[10-12].

Soluble and particulate fractions were prepared from extracts of the germinated pollen of _Nicotiana alata_ (self incompatibility genotype S_2S_3). These soluble and particulate fractions contain protein kinases that catalyze the phosphorylation of histones, casein and endogenous proteins. Histone phosphorylation is largely Ca^{2+}-dependent and is activated by high (0.5 mM) free Ca^{2+} concentration. The Ca^{2+}-dependent phosphorylation of particular casein-derived polypeptides is activated at low (0.01-0.04 mM) free Ca^{2+} concentration. No activation of exogenous protein phosphorylation by calmodulin was found, but phenothiazine-derived calmodulin antagonists markedly stimulate phosphorylation of exogenous and endogenous proteins catalyzed by the protein kinases of the germinated pollen fractions. A range of endogenous polypeptides in the soluble and particulate fractions are substrates for the Ca^{2+}-dependent and independent protein kinases. The Ca^{2+}-dependent protein kinase(s) may be involved in the normal growth of pollen tubes after germination _in vitro_. The presence of this type of enzyme in germinated pollen allows for the possible involvement of Ca^{2+}-dependent protein phosphorylation in signal transduction between style and pollen tubes during _in vivo_ pollen tube growth.

REFERENCES

1. Y. Nishizuka, Nature 308:693-698 (1984).
2. P. Cohen, Nature 296:613-620 (1982).
3. L.F. Reichardt and R.B. Kelly, Ann. Rev. Biochem. 52:871-926 (1983).
4. G.M. Polya and J.R. Davies, FEBS Lett. 150:167-171 (1982).
5. G.M. Polya, J.R. Davies and V. Micucci, Biochim. Biophys. Acta 761:1-12 (1983).
6. G.M. Polya and V. Micucci, Biochim. Biophys. Acta 785:68-74 (1984).
7. A.M. Hetherington and A. Trewavas, FEBS Lett. 145:67-71 (1982).
8. P. Dieter, Plant, Cell & Environment 7:371-380 (1984).
9. A.E. Clarke, M.A. Anderson, A. Bacic and P.J. Harris, J. Cell Sci. In press (1985).
10. J.L. Brewbaker and B.H. Kwak, Am. J. Bot. 50:859-865 (1963).
11. J.M. Picton and M.W. Steer, Protoplasma 115:11-17 (1983).
12. H-D. Reiss and W. Herth, Planta 145:225-232 (1979).

IN VITRO AND IN VIVO PROTEIN PHOSPHORYLATION IN OAT COLEOPTILES:

EFFECTS OF CALCIUM, CALMODULIN ANTAGONISTS AND AUXIN

K. Veluthambi and B. W. Poovaiah

Department of Horticulture
Washington State University
Pullman, WA 99164-6414, U.S.A.

The biochemical basis of the transduction of extracellular signals into intracellular events has long been a subject of great interest. There is now convincing evidence suggesting that a wide variety of regulatory agents, including both extracellular and intracellular messengers, produce diverse types of biological responses by regulating the state of phosphorylation of specific proteins (Cohen, 1982). The role of calcium and calmodulin in regulating protein phosphorylation in plants has been studied with great interest in recent years (Hetherington and Trewavas, 1982; Salimath and Marme, 1983; Polya and Micucci, 1984; Veluthambi and Poovaiah 1984a and 1984b). All these studies have been performed under in vitro conditions. Information on in vivo phosphorylation would be helpful to understand the effect of hormones and other stimuli in altering calcium levels in the cell and thereby affecting protein phosphorylation. If intracellular Ca^{2+} concentration increases in response to an external stimulus, then calmodulin and calmodulin-dependent protein kinases will be activated leading to the phosphorylation of several proteins.

In this study we selected the auxin-responsive oat (Avena sativa L.) coleoptile segments as the experimental material and first studied the effects of Ca^{2+} and calmodulin antagonists on in vitro protein phosphorylation. Subsequently, in vivo phosphorylation was performed by feeding ^{32}Pi to coleoptile segments and the effects of auxin and calmodulin antagonists were tested.

In vitro and in vivo protein phosphorylations in oat coleoptile segments were analyzed by sodium dodecyl sulfate-polyacrylamide gel electrophoresis and by two-dimensional gel electrophoresis. In vitro phosphorylation of several proteins was distinctly promoted at 1 to 15 µM free Ca^{2+} concentrations. Ca^{2+}-stimulated phosphorylation was markedly reduced by trifluoperazine, chlorpromazine and naphthalene sulfonamide (W-7). Two proteins were phosphorylated both under in vitro and in vivo conditions but the patterns of phosphorylation of several other proteins were different under the two conditions. Trifluoperazine, W-7 and ethylene glycol-bis-(β-aminoethyl ether)-N,N'-tetraacetic acid (EGTA) + calcium ionophore A23187 treatments resulted in reduced levels of in vivo protein phosphorylation of both control and auxin-treated coleoptile segments. Analysis of proteins by two-dimensional gel electrophoresis

347

following in vivo phosphorylation revealed auxin-dependent changes of certain proteins. However, a general inhibition of phosphorylation by calmodulin antagonists suggested that both control and auxin-treated coleoptiles exhibited Ca^{2+}, and calmodulin-dependent protein phosphorylation in vivo.

REFERENCES

Cohen, P., 1982, The role of protein phosphorylation in neural and hormonal control of cellular activity, Nature, 296:613.

Hetherington, A., and Trewavas, A., 1982, Calcium-dependent protein kinase in pea shoot membranes, FEBS Lett., 145:67.

Polya, G. M., and Micucci, V., 1984, Partial purification and characterization of a second calmodulin-activated Ca^{2+}-dependent protein kinase from wheat germ, Bichim Biophys Acta, 785:68.

Salimath, B. P., and Marme, D., 1983, Protein phosphorylation and its regulation by calcium and calmodulin in membrane fractions from zucchini hypocotyls, Planta, 158:560.

Veluthambi, K., and Poovaiah, B. W., 1984a, Calcium-promoted protein phosphorylation in plants, Science, 223:167.

Veluthambi, K., and Poovaiah, B. W., 1984b, Calcium- and calmodulin-regulated phosphorylation of soluble and membrane proteins from corn coleoptiles, Plant Physiol., 76:359.

PHOSPHORYLATION OF TONOPLAST PROTEINS IN Acer

pseudoplatanus CELL SUSPENSION CULTURES

Chantal Teulières, Gilbert Alibert and Raoul Ranjeva

Centre de Physiologie Végétale
Université Paul Sabatier, U.A. CNRS n° 241
118 route de Narbonne
31062 Toulouse Cédex

ABSTRACT

Highly purified tonoplast preparations obtained from Acer cells allow us to study the autophosphorylation processes of intrinsic proteins in this membrane.

When cell suspensions are incubated with $^{32}PO_4$, the total membranes isolated from protoplasts in presence of sodium fluoride (NaF) contained about 16 % Pi bound to proteins (15 nmoles/10^6 protoplasts), 1 % of the bound ^{32}P is localized in tonoplast proteins (0.17 nmoles).

The phosphorylated proteins of the tonoplast lost rapidly (in about 10 min) 2/3 of their Pi when incubated in tris-Mes buffer at pH 8 after elimination of the NaF. This result indicates that part of the incorporated Pi in the tonoplast proteins is easily removed by hydrolisis involving protoplast-bound protein phosphatase (s).

In highly purified tonoplast fractions, protein phosphorylation is dramatically enhanced in vitro by Ca^{2+} in the presence of calmodulin. This reaction is completely inhibited by fluphenazine (an antagonist of calmodulin). So tonoplast bears membrane-bound protein kinase (s) Ca^{2+}-calmodulin dependent.

The phosphorylation reaction is very rapid (within one min for maximum phosphorylation in our conditions) and followed by dephosphorylation (75 % loss of Pi in 10 min). The dephosphorylation process appears to be catalysed by tonoplast-bound protein phosphatase (s) not dependent on divalent ions. Moreover the Ca^{2+}-calmodulin dependent phosphorylation affects a restricted set of proteins as indicated by SDS PAGE analysis.

The reversible Ca^{2+}-calmodulin dependent phosphorylation of tonoplast proteins is discussed in relation with the role of vacuole in the sequestration of cellular metabolits and ions.

CALCIUM-CALMODULIN REQUIREMENTS OF PHOSPHATIDYL INOSITOL TURNOVER

STIMULATED BY AUXIN

Anna Stina Sandelius and D. James Morré

Department of Biological Sciences and Department of
Medicinal Chemistry
Purdue University, West Lafayette, IN 47907 U.S.A.

In contrast to animal cells (Michell et al., 1977), the metabolism of
phosphatidyl inositol (PI) in plant membranes has been little studied as a
potential component in the signal-response cascades important to growth and
development. PI is synthesized by pathways involving CDP-diglyceride and
inositol. The terminal enzyme required Mg^{2+} or Mn^{2+}, is inhibited by Ca^{2+},
and is not known to be hormone responsive. A second pathway of inositol
incorporation involves an exchange enzyme (Sexton and Moore, 1981), stimulated by cytidine nucleotides and Mn^{2+} and unaffected or slighly inhibited
by Ca^{2+} (Sandelius and Morré, unpublished). This activity is stimulated
or inhibited by auxin hormones in vitro with isolated membrane vesicles in
a dose-dependent fashion (Morré et al. 1984). PI hydrolysis, also stimulated by auxin in vitro with isolated membrane vesicles (Morré et al.,
1984), is less well characterized but appears to require calcium and/or
calmodulin for maximum activity (Table). The assay procedure developed is
as follows: Membrane vesicles from soybean are prelabeled for 10 to 30 min
by the exchange enzyme activity, and washed by centrifugation. The method
is rapid, facile, and requires only substrate amounts of radioactive inositol. It specifically labels only PI and the availability of prelabeled
membrane vesicles greatly facilitates the subsequent analyses of PI hydrolysis. Auxin (2,4-dichlorophenoxyacetic acid = 2,4-D or indole-3-acetic acid
= IAA) at near optimum concentrations of between 0.1 and 1μM stimulate both
the inositol incorporation and the subsequent release of inositol but auxin-
stimulated release is observed most clearly in the presence of calcium
(1 to 10 μM). Additionally, inositol release is enhanced by the addition
of calmodulin. These findings suggest the involvement of a specific hydrolytic activity in the breakdown of PI, amenable to hormone modulation as
well as to control by calcium and calmodulin.

The inositol incorporation is rapid, a maximum response to auxin is
observed within the first 10 min after auxin addition, and the dose-response
curve parallels that of the growth response to auxin in that both descending
limbs in the supraoptimal and suboptimal dose ranges are observed (Morré et
al., 1984). Thus, the dependence of the membrane-located inositol exchange
reaction on the concentration of auxin parallels that of the growth response
of the tissues to auxin. The latter provides one line of evidence for a
possible physiological significance of the reaction in the control of plant
growth.

Table. Calcium-Calmodulin Requirements of Phosphatidyl Inositol Turnover Stimulated by Auxin during a 10 Minute Incubation

Calcium	IAA	Calmodulin*	Loss of Phospatidyl Inositol** (Cpm/mg protein)
None	None	None	None
	1μM	None	6,300
1μM	None	None	6,600
		125 units/ml	14,400
	1μM	None	10,200
		125 units/ml	18,000
10μM	None	None	3,000
		125 units/ml	11,400
	1μM	None	11,700
		125 units/ml	19,200

* Bovine (Sigma)   ** Initial = 120,000 cpm/mg protein

The turnover (degradation) of phosphatidyl inositol by head group removal-exchange could contribute substantially to auxin-induced calcium release from isolated membranes. Phosphatidyl inositol is an acidic phospholipid capable of binding calcium. Its degradation to phosphatidic acid and the subsequent cleavage of the phosphatidic acid by the very active phosphatidic acid phosphatase present in soybean membranes (Scherer and Morré, 1978) could account for the calcium release observed previously in response to auxin (Buckhout et al., 1981) as well as an auxin stimulated release of inorganic phosphate (Scherer and Morré, 1978).

While the findings reported must be regarded as preliminary, the observations do support the concept that isolated plant membranes respond directly to auxins. In addition, isolated membranes provide a useful system to examine calcium requirements of auxin-responsive phenomena.

REFERENCES

Buckhout, T. J. Young, K. A., Low, P. S., and Morré, D. J., 1981, In vitro promotion by auxins of divalent ion release from soybean membranes, Pl. Physiol. 68:512.

Michell, R. H., Jafferji, S. S., and Jones, L. M., 1977, The possible involvement of phosphatidyl inositol breakdown in the mechanism of stimulus-response coupling at receptors which control cell-surface calcium gates, Adv. Exp. Med. Biol. 83:447.

Morré, D. J., Gripshover, B., Monroe, A., and Morré, J. T., 1984, Phosphatidyl inositol turnover in isolated soybean membrane stimulated by the synthetic growth hormone, 2,4-dichlorophenoxyacetic acid, J. Biol. Chem. 259:15364.

Sandelius, A. S., and Morré, D. J., Characteristics of phosphatidyl-inositol turnover of soybean microsomes involving a reversible inositol exchange, unpublished.

Scherer, G. F. E., and Morré, D. J., 1978, In vitro stimulation by 2,4-dichlorophenoxyacetic acid of the ATPase and inhibiton of phosphatidic acid phosphatase of plant membranes, Biochem. Biophys. Res. Commun. 84:238.

Sexton, J. C. and Moore, T. S., 1981, Phosphatidylinositol synthesis by an Mn^{2+}-dependent exchange enzyme in castor bean endosperm, Plant Physiol. 68: 18.

Work supported in part by National Science Foundation Grant PCM 8206222

ISOLATION OF PLASMA MEMBRANE AND TONOPLAST FROM THE SAME HOMOGENATES OF PLANT CELLS BY FREE-FLOW ELECTROPHORESIS

A. Sandelius, C. Penel, G. Auderset, A. Brightman,
K. Safranski, H. Greppin and D. James Morré

Department of Biological Sciences and Department of
Medicinal Chemistry, Purdue University, West Laf-
ayette, IN 47907 U.S.A. and Laboratory of Plant
Physiology, University of Geneva, Switzerland

Plasma membranes and tonoplast have proven difficult to separate from plant homogenates by conventional sucrose grad- ient and differential centrifugation (Quail,1979, Gripshover et al., 1984). Here we describe a procedure based on free flow elec- trophoresis that is applicable to both etiolated shoots as well as green leaves. Homogenates were centrifuged at $10,000 g_{max}$ for 10 min to remove mitochondria and the resulting supernatant was centrifuged for 20 min at $45,000 g_{max}$. The latter pellet was re- suspended as the starting material for free-flow electrophoresis.

A typical sep- aration is illus- trated in Figure 1 for membranes from etiolated hypocot- yls of soybean. It consisted of approx- imately 40 fractions with the bulk of the protein falling be- tween fractions 38 and 48. In addition, there were two shoul- ders. The one near- est the anode and most distant from the point of sample injection had char- acteristics of tono- plast (nitrate-inhi- bited ATPase) while the shoulder far- therest from the anode had character-

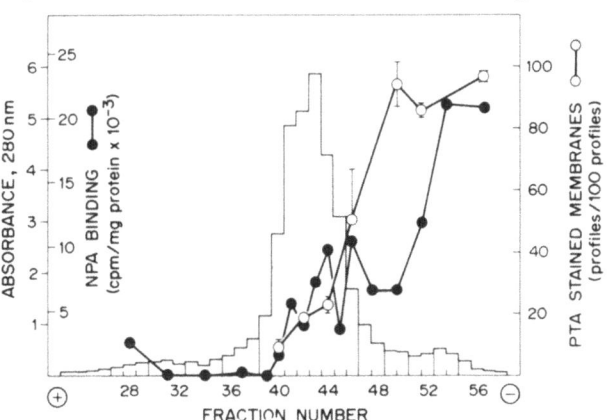

Fig. 1. Free-flow electrophoresis sep- aration correlating the distribution of N-l-naphthylphthalamic acid (NPA) binding with the distribution of vesicles reac- tive with phosphotungstic acid at low pH. The step function gives the distribution of total protein.

istics of plasma membranes (stained with phosphotungstic acid at low pH, bound N-l-naphthylphthalamic acid, NPA, exhibited glucan synthetase II and vanadate-inhibited ATPase activity). Marker characteristics for mitochondria (succinate-INT reductase), Golgi apparatus (fucosyl transferase/latent IDPase) and endoplasmic reticulum (NADPH-cytochrome c reductase) as well as carotenoids and chlorophyll, when present, were found primarily in the center of the separation between fractions 38 and 48.

Further differentiation of tonoplast and plasma membrane was aided by measurements of membrane thicknesses after simultaneous fixation with glutaraldehyde plus osmium tetroxide (Table 1). Class I membranes were 10 nm thick, showed the dark-light-dark pattern very clearly, reacted with phosphotungstic acid at low pH and consisted of plasma membrane and occasional elements of the Golgi apparatus derived from the mature pole. Class II membranes were 8-9 nm thick, showed the dark-light-dark pattern poorly and did not react with phosphotungstic acid at low pH. The prin-

Table 1. Membrane dimensions of soybean membranes fixed by simultaneous glutaraldehyde plus osmium tetroxide.

Cell component	Membrane Tickness, nm ± S.D.	Class
Plasma membrane	10.5 + 0.5	I
Tonoplast	8.5 + 0.7	II
Endoplasmic reticulum	6.0 + 0.3	III
Mitochondria		
Outer membrane	6.0 + 0.4	III
Inner membrane	6.3 + 0.5	III
Etioplast		
Outer envelope memb.	6.0 + 0.4	III
Inner envelope memb.	5.8 + 0.3	III
Prothylakoid memb.	4.8 + 0.2	III
Peroxisome	6.2 + 0.3	III
Golgi apparatus	6-9	Mixed
Nuclear envelope	6.0 + 0.3	III

ciple membrane in this class was tonoplast although some Golgi apparatus membranes would not be distinguished from the tonoplast on the basis of these criteria. Class III membranes were thinner than 8.5 nm, showed the dark-light-dark pattern little, if at all, and did not react with phosphotungstic acid at low pH. Membranes in this category included all of the remaining internal membranes of the cell. These results provide a basis for the tentative identification of plasma membrane and tonoplast membrane solely on morphological criteria independent of any biochemical assumptions (Table 2). A similar procedure was followed by Powell et al. (1982) to identify tonoplast membranes of the fungus Gilbertella persicaria. On this basis, tonoplast was 86% pure (fraction A) and plasma membrane was 97% pure (fraction E)(Table 2).

Table 2. Composition of free-flow electrophoresis fractions based on membrane morphology.

Free-flow fractions	% of Total membranes 9.5-11 nm	8-9.5 nm	5-7 nm	PTA-pos.
A (22-33)	0	86	14	0
B (34-38)	0	54	46	7
C (39-45)	14	11	75	18
D (46-50)	33	3	64	35
E (51-57)	98	0	2	97

REFERENCES

Gripshover, B., Morré, D. J., and Boss, W. F., 1984, Fractionation of suspension cultures of wild carrot and kinetics of membrane labeling. Protoplasma 123:213.
Quail, P. H., 1979, Plant cell fractionation. Ann. Rev. Plant Physiol. 30:425.
Powell, M. J., Bracker, C. E., and Morré, D. J., 1982, Isolation and ultrastructural identification of membranes from the fungus Gilbertella persicaria. Protoplasma 111:87.

Work supported in part by a grant from the National Science Foundation to D.J.M 8206222.

PHOSPHORYLATION OF MEMBRANE-LOCATED PROTEINS OF SOYBEAN: IN VITRO RESPONSE OF PURIFIED PLASMA MEMBRANES TO AUXIN AND CALCIUM

Robert L. Varnold, D. James Morré and Anna Stina Sandelius

Department of Medicinal Chemistry and Department of
Biological Sciences
Purdue University, West Lafayette, IN 47907 U.S.A.

Isolated membranes of soybean incorpated ^{32}P from γ-[^{32}P]ATP in vitro. The incorporation was rapid and labeling was achieved in 10 s or less. When displayed on 10% sodium dodecylsulfate-polyacrylamide gels, several labeled protein bands were revealed (Morré et al., 1984). Incorporation into protein bands with apparent molecular weights of 45,000 and 50,000 as well as into total material insoluble in trichloroacetic acid was observed in the presence of auxin (2,4-dichlorophenoxyacetic acid = 2,4-D). Also, incorporation into a low molecular weight constituent was stimulated 2- to 3-fold by auxin in some experiments. The activity may be the result of protein kinases of the C-type since stimulation also was given by diglyceride plus calcium (Morré et al., 1984), constituents known to augment C-type kinases in other systems (Nishizuka, 1984).

The response of phosphorylation in vitro of membrane proteins in response to calcium and auxin, alone and in combination, was investigated in more detail first with total microsomes (Varnold and Morré, 1985). While 2,4-D, at growth promoting concentrations, stimulated phosphorylation in the absence of calcium, the overall effect of auxin in the presence of calcium was to inhibit phosphorylation. The inhibition was observed over a wide range of calcium and auxin concentrations. We have now extended this investigation to purified plasma membrane fractions. These membranes exhibit a markedly different pattern of response to auxin and calcium than do other membranes.

Plasma membranes were prepared by the two-phase technique (Kjellbom and Larsson, 1984) modified for microsomal membranes isolated from hypocotyl segments of dark-grown soybeans. The two-phase system contained 6.4% (w/w) dextran T-500 (Pharmacia) and 6.4% polyethyleneglycol, PEG 3350 (Fischer) in 5 mM K_2HPO_4-KH_2PO_4, pH 6.8, 0.25 M sucrose and microsomal membranes from 20-25 g soybean tissue. After mixing and separation of the phases, the upper phase containing the plasma membranes was repurified two times using fresh two-phase systems. The lower phases containing other membranes also were collected. Final membrane pellets were prepared following dilution with distilled water and centrifugation at 90,000 X g for 45 min (Spinco SW28 rotor). The membrane pellets were suspended in 0.5 to 1 ml of 50 mM MES, pH 6.5 containing 50 mM $MgCl_2$ and were incubated for 0.2 min in an assay solution (total volume of 250 μl containing γ-[^{32}P]ATP (Amersham, ca. 3000 Ci/mmol) with and without 2,4-D and with and without calcium. The reaction was stopped with 1 ml 5% trichloroacetic acid.

The acid-insoluble pellet was washed twice with 1 ml 5% TCA and once with 1 ml absoute ethanol. It was then solubilized and specific radioactivity was determined. Based on content of marker enzymes for endoplasmic reticulum, Golgi apparatus and mitochondria together with morphometric analyses based on staining with phosphotungstic acid at low pH, the purified upper phase fractions were found to contain more than 90% plasma membranes.

The purified plasma membranes responded to auxins in a much more striking manner (optimum response at 3 X 10^{-7} M 2,4-D) than was observed previously for total microsomes (Fig. 1A) while other membranes obtained from the lower phase of the two-phase separation responded to auxin little if at all (Fig. 1B). Additionally, the protein kinase activity of the purified plasma membranes was stimulated by cacium ions over the range 10^{-8} to 10^{-4} M (Fig. 2A). ^{32}P incorporation into other membranes, like that into total microsomes (Varnold and Morré, 1985), was stimulated by added calcium only in the range of 1 to 10 mM (Fig. 2B).

ACKNOWLEDGEMENT

Work supported in part by a grant from the National Science Foundation PCM 8206222.

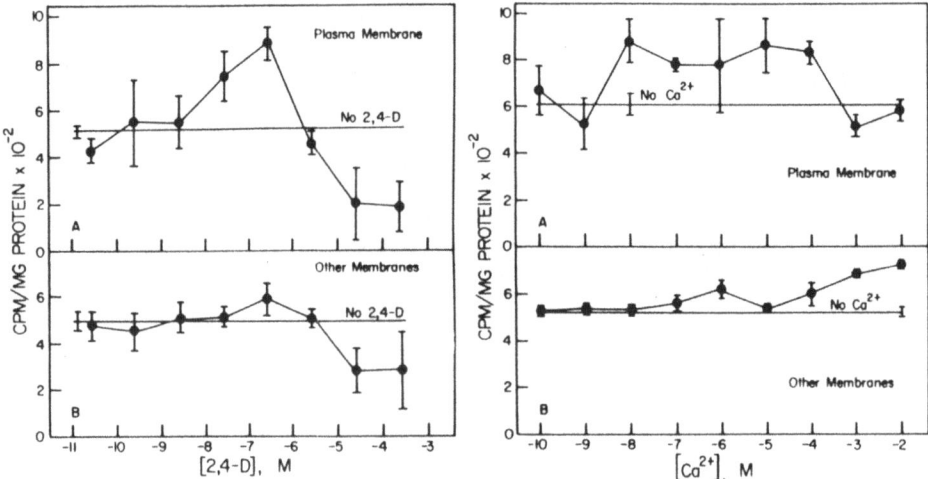

Fig. 1. Incorporation of ^{32}P from 5 μCi of γ-[^{32}P]ATP in the presence of varying concentrations of 2,4-D. Results are from 7 determinations ± standard deviations.

Fig. 2. Incorporation of ^{32}P from 5 μCi of γ-[^{32}P]ATP in the presence of varying concentrations of Ca^{2+}. Results are from 6 determinations ± standard deviations.

REFERENCES

Kjellbom, P., and C. Larsson, 1984, Preparation and polypeptide composition of chlorophyll-free plasma membranes from leaves of light-grown spinach and barley, Physiol. Plantarum 62:501.

Morré, D. J., Morré, J. T., and Varnold, R. L., 1984, Phosphorylation of membrane-located proteins of soybean in vitro and response to auxin. Plant Physiol. 75:265.

Nishizuka, Y., 1984, Turnover of inositol phospholipids and signal transduction. Science 225:1265.

Varnold, R. L., and Morré, D. J., 1985, Phosphorylation of membrane-located proteins of soybean hypocotyl. Inhibition by calcium in the presence of 2,4-dichlorophenoxyacetic acid. Bot. Gaz.

DISTRIBUTION OF CALMODULIN-DEPENDENT AND CALMODULIN-INDEPENDENT NAD

KINASE DURING EARLY CELL DEVELOPMENT IN THE ROOT APEX OF PISUM SATIVUM

E.E.F. Allan and A.J. Trewavas

Botany Department
University of Edinburgh
Mayfield Road, Edinburgh EH9 3JH, U.K.

Two forms of NAD kinase activity are detectable in shoot and root tissues of Pisum sativum. Detection and separation of the two forms are both dependent on extraction conditions. A Ca^{2+}-CaM-independent component can be extracted by gentle homogenisation conditions, whereas an additional Ca^{2+}-CaM-dependent component is dependent on the presence of high salt concentrations during homogenisation.

The two forms of activity can be separated by ion exchange chromatography or by calmodulin-affinity chromatography indicating that the calmodulin-independent activity does not represent basal enzyme activity in the absence of Ca^{2+} and calmodulin. Following separation, the calmodulin-dependent form is found to be totally dependent on Ca^{2+} and calmodulin for activity, whereas the calmodulin-independent form is completely independent of both. The calmodulin-independent activity is not inhibited by phenothiazines while heat-treated samples do not activate bovine brain phosphodiesterase indicating that the enzyme is not contaminated with calmodulin. Furthermore, it is unlikely that the calmodulin-independent form is extract-generated as the total activity eluting from ion exchange columns remains the same regardless of extraction procedures, although the calmodulin-dependent form requires harsher conditions for its extraction. The difference in proportions of the two forms of activity in root and shoot tissue also indicates that the calmodulin-independent form is unlikely to be extract-generated. It is therefore suggested that there are at least two isozymes of NAD kinase in pea seedlings. This is in agreement with the results of Simon et al., (1984) who found both calcium-calmodulin-dependent NAD kinase activity and EGTA-and trifluoperazine-resistant activity in pea seedlings. The differences in extraction requirements indicate that the calmodulin-independent isozyme(s) may be present largely in the soluble phase while the calmodulin-dependent form(s) may be present largely in membrane systems.

Pyidine nucleotides are required for both basic metabolic activities and more specialised cellular functions. The presence of differentially regulated NAD kinase isozymes clearly will permit finer and more flexible control over localised pyridine nucleotide levels than would be the case with a single NAD kinase that was entirely dependent on calcium and calmodulin for activity as originally suggested (Cormier et al., 1981).

357

Total and relative activities of the calmodulin-dependent and calmodulin-independent isozymes of NAD kinase were measured during early cell development in the root apex. The calmodulin-independent form was found to constitute a much higher proportion of total potential NAD kinase activity in root than in shoot tissue, and to increase only slightly on a cellular basis during the apical 30 millimetres of root development. There was a much larger increase in calmodulin-dependent activity over this region, activity increasing 17-fold/cell, 5-6-fold/mg protein, and 3-fold/segment within the apical 20 millimetres, with a lower rate of increase over the following 10 millimetres. The ratio of potential calmodulin-dependent to calmodulin-independent activity therefore rose throughout at least the apical 30 millimetres. This increase in total potential in vitro activity may be paralleled by an increase in NADP(H):NAD(H) ratio as observed in Vigna roots (Yamamoto, 1963). Thus, the apparent in vivo increase in NAD kinase activity may reflect an increase in calmodulin-dependent NAD kinase.

As pyridine nucleotides are present in limiting concentrations in the root apex (Yamamoto, 1963), any alterations in their concentrations or ratios may alter activities of a number of metabolic pathways. During early stages of cell development in the root apex, for example, an increasingly higher proportion of glucose is metabolised through the NADP-requiring pentose phosphate pathway rather than the NAD-requiring glycolytic pathway. As there are few changes in relative in vitro activities of enzymes from these pathways beyond about 6mm from the apex despite continued changes in relative activities of the pathways (Gibbs and Beevers, 1955; Fowler and Ap Rees, 1970), a major function of the increase in NADP:NAD ratio beyond 6mm may be regulation of glycolysis and the pentose phosphate pathway. Regulation of these pathways will have further implications for pathways dependent on their products such as the shikimic acid pathway, lignin biosynthesis, amino acid and fatty acid biosynthesis, which are also known to vary during root development. It is therefore suggested that the subcellular concentrations and distributions of calcium, calmodulin and isozymes of NAD kinase will be important in regulating directly and indirectly several aspects of root metabolism through compartmentation of NAD(H) and NADP(H) levels.

REFERENCES

Cormier, M.J., Jarrett, H.W. and Charbonneau, H., 1981, Role of Ca^{2+}-calmodulin in metabolic regulation in plants, in "Calmodulin and intracellular Ca^{2+} receptors". S. Kakiuchi, H. Hidaka and A.E. Means, eds., Plenum Press, New York, 125.

Fowler, M.W. and Ap Rees, T., 1970, Carbohydrate oxidation during differentiation in roots of Pisum sativum, Biochim. Biophys. Acta., 201:33.

Gibbs, M. and Beevers, H., 1975, Glucose dissimilation in the higher plant. Effect of age of tissue, Plant Physiol., 30:343.

Simon, P., Bonzon, M., Greppin, H. and Marme, D., 1984, Subchloroplastic localization of NAD kinase activity: evidence for a Ca^{2+}, calmodulin-dependent activity at the envelope and for a Ca^{2+}, calmodulin-independent activity in the stroma of pea chloroplasts, FEBS Letts., 167:332.

Yamamoto, Y., 1963, Pyridine nucleotide content in the higher plant. Effect of age of tissue, Plant Physiol., 38:45.

Ca^{2+}, CALMODULIN DEPENDENT NAD KINASE: REGULATION BY LIGHT

Peter Dieter[a] and Dieter Marmé[b]

[a]Biochemical Institute, Hermann Herder Strasse 7
[b]Institute of Biology III, Schänzlestrasse 1 and
Gödecke Research Institute, D-7800 Freiburg, FRG

The NAD kinase is the only enzyme which catalyzes the phosphorylation of NAD to NADP. This enzyme has therefore an important function in plant metabolism since many plant key enzymes involved in the synthesis and degradation of sugars, lipids and amino acids are dependent on pyridine nucleotides and have a clear preference for NAD or NADP.

We could show that the NAD kinase activity from dark grown corn coleoptiles is almost totally dependent on Ca^{2+} and calmodulin (Dieter and Marmé, 1984). Nearly all of the enzyme activity is associated with the outer membrane of mitochondria. This membrane-bound NAD kinase, associated with intact mitochondria can be activated by exogenously added Ca^{2+} and calmodulin supposing that in vivo changes of the free cytoplasmic Ca^{2+} concentration can be sensed by this enzyme.

Light irradiation of intact corn seedlings as well as an increase of the cellular Ca^{2+} concentration in segments of corn coleoptiles leads to an increase of NADP and to a decrease of NAD (Marmé and Dieter, 1982; Dieter et al., 1984) suggesting that the NAD kinase can be activated in vivo by light as well as by an enhancement of the cellular Ca^{2+} concentration.

We could show that light irradiation of segments of corn coleoptiles as well as an irradiation of an isolated mitochondrial fraction does not change the NAD kinase activity. We furthermore demonstrated that neither the cellular distribution nor the total activity of the Ca^{2+}, calmodulin-dependence of the NAD kinase is changed after light irradiation of intact corn seedlings (Dieter and Marmé, 1985). The enzyme is in both cases located in the outer mitochondrial membrane and its activity is totally dependent on the presence of both Ca^{2+} and calmodulin. In intact mitochondria and the presence of calmodulin the enzyme activity increases linearly from 10^{-7} to 10^{-3} M Ca^{2+}. At 10^{-4} M Ca^{2+} half maximal activation occurs at about 10^{-8} M calmodulin. After solubilization and purification by calmodulin-Sepharose affinity chromatography the Ca^{2+}-dependence of the enzyme has changed. The activation reaches a plateau at about 10^{-5} M Ca^{2+} and half maximal activation occurs at about 6×10^{-6} M Ca^{2+}.

The calmodulin content of corn coleoptiles is also not changed by light irradiation of corn seedlings. We determined a value of 3.3 ± 0.7 and 2.6 ± 0.5 mg calmodulin/kg fresh weight for coleoptiles of dark and far red light grown corn seedlings respectively. These values have been determined using

TABLE 1. EFFECT OF LIGHT ON THE NAD KINASE ACTIVITY OF CORN COLEOPTILES

Light irradiation	Fraction		
	Homogenate	Particulate	Soluble
Seedlings[a]			
-	48 ± 8[c]	43 ± 7	7 ± 1
+	44 ± 9	41 ± 8	4 ± 1
Mitochondria[b]			
-	-	42 ± 5	-
+	-	43 ± 4	-

[a]Corn seedlings were irradiated with far red light for 5.5 days (Dieter and Marmé, 1981).

[b]The mitochondrial fraction was irradiated at 0°C for one hour with red light (Dieter and Marmé, 1981).

[c]The NAD kinase activity is expressed as m units/g fresh weight.

TABLE 2. EFFECT OF LIGHT AND THE Ca^{2+}-IONOPHORE A23187 ON THE NICOTINAMIDE NUCLEOTIDE POOL OF CORN COLEOPTILES

	NAD	NADH	NADP	NADPH	$\dfrac{NAD/NADH}{NADP/NADPH}$
	(nmole/g fresh weight)				
Light irradiation					
Segments[a]					
-	11.3	2.4	6.2	1.2	1.9
+	11.6	2.9	6.1	1.1	2.0
Seedlings[b]					
-	12.0	1.2	7.2	1.1	1.6
+	9.6	1.3	12.1	1.2	0.8
A 23187					
Segments[c]					
+EGTA	8.0	0.7	1.5	0.5	4.4
+CaCl₂	6.5	0.9	2.7	0.9	2.1

[a]Segments of corn coleoptiles (length: 0.5 cm) were irradiated at 0°C for one hour with red light (Dieter and Marmé, 1981).

[b]Corn seedlings were irradiated with far red light for 5.5 days (Dieter and Marmé, 1981).

[c]Segments of corn coleopitles (length: 0.5 cm) were incubated at 25°C for 30 min in the presence of 10^{-5}M A23187 with 1 mM EGTA or with 1 mM $CaCl_2$.

a calmodulin-dependent cAMP phosphodiesterase and bovine brain calmodulin as a standard.

These results clearly show that there is no significant change in the total activity and the properties of the NAD kinase in dark and light grown corn tissue. Furthermore, it is shown that the NAD kinase cannot be activated by light irradiation in vitro of a mitochondrial fraction or by irradiation of intact corn segments. Otherwise the enzyme is activatable by irradiation of intact corn seedlings. Since it has been shown that light irradiation of intact seedlings may lead to an increase of the cellular Ca^{2+} concentration (Dieter and Marmé, 1981) we suggest that the activation of the NAD kinase in vivo is mediated by Ca^{2+} and calmodulin.

REFERENCES

Dieter, P. and Marmé, D., 1981, Far red light irradiation of intact corn seedlings affects mitochondrial and calmodulin-dependent microsomal Ca^{2+} transport. Biochem. Biophys. Res. Commun., 101:749-755.
Dieter, P. and Marmé, D., 1985, Comparison of the Ca^{2+}, calmodulin-dependent NAD kinase of coleoptiles from dark and far red light grown corn coleoptiles. Plant Physiol., submitted.
Dieter, P., Salimath, B.P. and Marmé, D., 1984, The role of calcium and calmodulin in higher plants, in "Annual Proceedings of the Phytochemical Society of Europe", Vol. 23, pp. 213-229, A. Boudet et al., eds, Oxford University Press.
Marmé, D. and Dieter, P., 1982, Calcium and calmodulin-dependent enzyme regulation in higher plants, in "Plasmalemma and Tonoplast: Their Functions in the Plant Cell", D. Marmé, E. Marré, R. Hertel, eds, pp. 111-118, Elsevier Biomedical Press, Amsterdam, New York, Oxford.

Ca^{2+}-DEPENDENT GENERATION OF A MEMBRANE-BORNE PROTEINASE INVOLVED IN

VOLUME REGULATION OF *POTERIOOCHROMONAS*

Heinrich Kauss

Dept. of Biology
University of Kaiserslautern
F.R.G.

POSTER ABSTRACT

Cells of the golden-brown alga *P. malhamensis* regain their volume after osmotic shrinkage mainly by production of isofloridoside (IF, α-galactosyl-1→1-glycerol)[1]. A few min after shrinkage a soluble key enzyme of the relevant pathway, the UDP-Gal:α-glycerol-3-phosphate 1α-galactosyltransferase has gained activity. This increase parallels the degree of shrinkage and also disappears shortly after an artificial swelling, indicating a physiological role of the activation process. Evidence is accumulating that this is due to a cascade of proteolytic events, which can be induced in homogenates from cells of standard volume by Ca^{2+}. The proteinase responsible for activating the galactosyltransferase has been purified recently and its involvement in the regulation process demonstrated[2]. The overall sequence of events appears to be clear for the right-hand part, but hypothetical for the membrane-part of the following scheme:

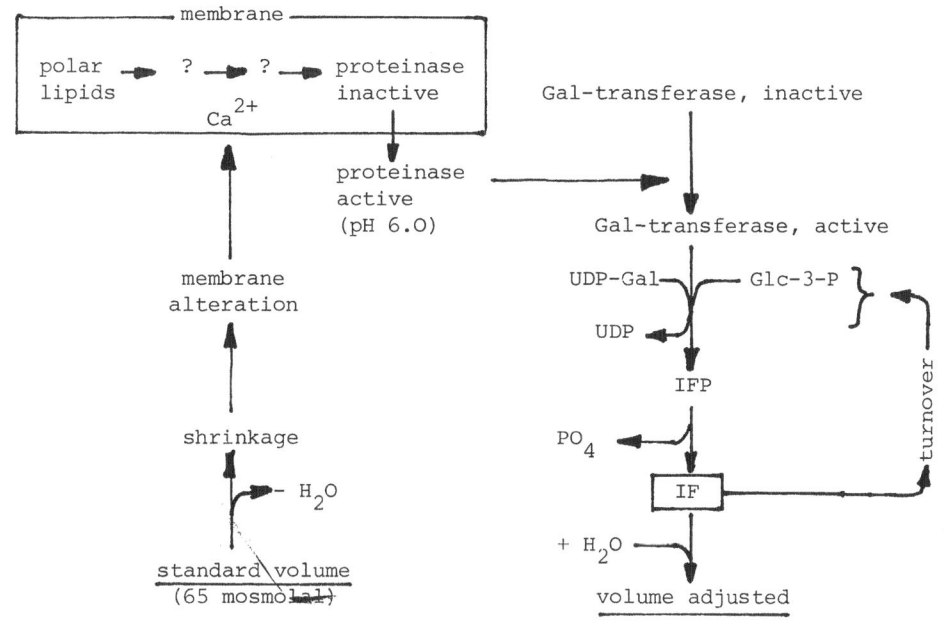

Although the involvement of membranes and the *in vitro* induction by Ca^{2+} has been shown, it remains unclear whether calmodulin, which has been isolated from the organism, is indead mediating the Ca^{2+}-effect. These doubts are due, in part, to the fact that induction *in vitro* can also be effected by some detergents and by the detergent-like action of phenothiazine drugs[3]. At the other hand, the sequence leading to the production of the proteinase can be inhibited by certain compounds known to interact with phospholipids, indicating their involvement in the unknown sequence of events occuring in the membrane[4]. Attempts to directly demonstrate a change of $[Ca^{2+}]_{cyt}$ shortly after cell shrinkage gave results which are in part contradictory[5]. The system will be used to discuss some difficulties in integrating the results of *in vitro* studies on the role of Ca^{2+} with *in vivo* experiments.

REFERENCES

1. H. Kauss, Volume regulation in *Poterioochromonas*. Involvement of calmodulin in the Ca^{2+}-stimulated activation of isofloridoside-phosphate synthase, Plant Physiol. 71:169 (1983).
2. D. Köhle and H. Kauss, Purification of a membrane-derived proteinase capable of activating a galactosyltransferase involved in volume regulation, Biochim. Biophys. Acta 799:59 (1984).
3. H. Kauss, Volume regulation: Activation of a membrane-associated cryptic enzyme system by detergent-like action of phenothiazine drugs, Plant Sci. Lett. 26:103 (1982).
4. H. Kauss, A membrane-derived proteinase capable of activating a galactosyl-transferase involved in volume regulation of *Poterioochromonas*, in: "Plant Proteolytic Enzymes," M. J. Dalling, ed., CRC Press, Boca Raton (1985).
5. H. Kauss and U. Rausch, Compartmentation of Ca^{2+} and its possible role in volume regulation of *Poterioochromonas*, in: "Compartments in Algal Cells and their Interaction," W. Wiessner, D. Robinson, and R. C. Starr, ed., Springer-Verlag, Berlin Heidelberg (1984).

THE REGULATION OF PLANT PEROXIDASES BY CALCIUM

Claude Penel, Federico J. Castillo, Stefanie Kiefer and
Hubert Greppin

Laboratoire de Physiologie végétale, Université de Genève
3 place de l'Université, 1211 Genève 4, Switzerland

Among the functions currently attributed to plant peroxidases is the
assembly of cell wall constituents, establishing the cell wall architecture.
This role results from their ability to form cross-links between sugars,
proteins, and phenols (Epstein and Lamport, 1984). It implies the transport
of peroxidase molecules outside the cells and, most likely, the control of
their activity in the wall. We present here experimental evidences showing
that calcium plays an important role in these processes. There are at least
three distinct modes of action of calcium on peroxidases, which are briefly
described below.

Peroxidase Secretion

The transport of peroxidases outside the cells is a calcium-mediated,
energy dependent mechanism. This was demonstrated in cell suspensions
(Sticher et al., 1981) and in whole leaves (Castillo et al., 1984). The
rate of enzyme secretion can be increased by the external addition of
calcium and lowered by EGTA. In some cases, the presence of a calcium
ionophore (A23187, bromolasalocid ethanolate) enhances the effect of calcium,
showing that the increase of the cellular calcium concentration is important (Castillo et al., 1984; Penel et al., 1984). Experiments with methionine labelled with ^{35}S have clearly shown that exogenously supplied calcium
induces the secretion of newly synthesized peroxidase molecules from leaf
cells and that this effect is increased by the ionophore A23187 (Castillo,
1985). This calcium-mediated peroxidase secretion is enhanced by an exposure of plants to ozone and, in that case, was inhibited by a pretreatment
with EGTA (Castillo et al., 1984). In suspension cultures, auxins increased
the effect of calcium, provided that cells were previously cultured in the
presence of an auxin (Gaspar et al., 1983).

Peroxidases binding to membranes

The most cathodic isoperoxidase of several plant species can be bound
in vitro to membranes in the presence of calcium and, in some cases, in
the presence of manganese. This binding only occurs when the carbohydrate
moiety of the isoperoxidase is present. It could be essential in the mechanism of transport of the peroxidases towards their final destination

(Kiefer et al., 1985). Preliminary data have shown that plasmalemma and tonoplast could bind the enzyme in the presence of calcium.

Peroxidase activation

Peroxidases contain one or two atom(s) of calcium, which are necessary for their enzyme activity (Haschke and Friedhoff, 1979). A fraction of the cathodic isoperoxidase ionically bound to cell walls of spinach leaves can be activated in vitro by addition of calcium. This calcium-activable iso-peroxidase, which cannot be distinguished from the non-activable form by isoelectric focusing, can be separated by affinity chromatography through a column of concanavalin A Sepharose. After separation, this isoperoxidase is activated by rather low calcium concentrations. By successive additions of calcium and oxalate (or EGTA), it is possible to increase and decrease its activity.

Peroxidases are related to the control of cell elongation since they can rigidify walls by their cross-linking activity and their ability to form lignin. The data presented here suggest that, for this role, they are under the control of cellular and apoplastic calcium levels, which re-gulate the rate of their transfer to walls and, maybe, their activity in situ.

References

Castillo, F. J., 1985, Activité Peroxydasique et Pollution Atmosphérique, thesis, Université de Genève.

Castillo, F. J., Penel, C. and Greppin, H., 1984, Peroxidase release induced by ozone in Sedum album leaves. Involvement of Ca^{2+}, Plant Physiol., 74:846.

Epstein, L., and Lamport, D. T. A., 1984, An intramolecular linkage involving isodityrosine in extensin, Phytochemistry, 23:1241.

Gaspar, Th., Kevers, C., Penel, C., and Greppin, H., 1983, Auxin control of calcium-mediated peroxidase secretion by auxin-dependent and auxin-independent sugarbeet cells, Phytochemistry, 22:2657.

Haschke, R. H., and Friedhoff, J. M., 1978, Calcium-related properties of horseradish peroxidase, Biochem. Biophys. Res. Commun., 80:1039.

Kiefer, S., Penel, C., and Greppin, H., 1985, Ca^{2+}- and Mn^{2+}-mediated binding of the glycoprotein peroxidase to membranes of Pharbitis cotyledons, Plant Sci. Lett., in press.

Penel, C., Sticher, L., Kevers, C., Gaspar, Th., and Greppin, H., 1984, Calcium controlled secretion by sugarbeet cells: effect of ionophores in relation to organogenesis, Biochem. Physiol. Pflanzen, 179:173.

Sticher, L., Penel, C., and Greppin, H., 1981, Calcium requirement for the secretion of peroxidases by plant cell suspensions, J. Cell Sci., 48:345.

THE IN VITRO REVERSIBLE ASSOCIATION OF THE REGULATORY AND CATALYTIC SUBUNITS OF QUINATE: NAD$^+$ OXIDOREDUCTASE

Annick Graziana, Marietta Dillenschneider,
Martine Charpenteau and Raoul Ranjeva

Centre de Physiologie Végétale, Université Paul Sabatier
U.A. CNRS n° 241, 118 route de Narbonne
31062 Toulouse Cédex

ABSTRACT

Quinate: NAD$^+$ oxidoreductase (QORase) occurs as a heterodimeric protein in dark-grown cell suspension cultures of carrot. The enzyme contains a catalytic (42 kDa) and a regulatory (65 kD) subunit that behaves as a calciprotein and renders QORase dependent upon calcium ions. Upon illumination, the enzyme dissociates and becomes independent upon calcium.

The work reported here has been done in order to get more information about:
- the forces that link the catalytic and regulatory subunits;
- the possible occurrence of photodependent messenger(s) that trigger(s) the dissociation process.

INVOLVEMENT OF HYDROPHOBIC FORCES

When QORase is incubated with increasing concentrations of protease(s), the enzyme loses both its activity and sensitivity to calcium. Such a result suggests that, at least in our conditions, it is not possible to dissociate the heterodimeric enzyme by limited proteolysis.

By varying the chaotropic parameters of the reaction medium, it appears that:
- increasing the salting-out effect makes QORase less accessible to calcium;
- increasing the chaotropic effects renders the enzyme partially independent of calcium;
- changing the ratio of the chaotropic/antichaotropic agents leads to the modulation of the calcium effects.

GTP AS A POSSIBLE PHOTODEPENDENT MESSENGER

Our previous results have shown that the association-dissociation process is independent upon protein synthesis, but also on potential endoprotease(s). Therefore, we have tried to look at a possible molecule that would trigger the dissociation, namely the insensitization to calcium.

The experiments were as follows:
- partially or highly purified fractions of QORase were prepared;
- the preparation was incubated for different times and with different concentrations of various nucleotides;
- the sensitivity of QORase to calcium was measured.

In these conditions, only GTP leads to a loss in activity and sensitivity to calcium, when partially purified preparations were used. GTP was without effects on pure QORase. The idea is that GTP may be, by some manner, degraded before reacting on QORase.

The ongoing work intends to test if GTP is the actual photodependent messenger that would connect the light effects with the calcium response.

PLANT ASPARTATE KINASE IS NOT ACTIVATED BY CALMODULIN OR CALCIUM[*]

P.L.R. Bonner, A. Hetherington and P.J. Lea

Dept. of Biological Sciences, University of Lancaster
Lancaster, LA1 4YQ, U.K.

Calmodulin extractable from plant tissue bears a striking similarity to the calmodulin from animal tissue, these similarities include: molecular weight, amino acid composition and sequence, the affinity towards calcium and the potency in activating animal and plant calmodulin-dependent enzymes. These enzymes include protein kinases, NAD kinase and calcium ATPase (Dieter, 1984). Recently, Sane et al., (1984) reported that the enzyme aspartate kinase (ATP:aspartate 4-phospho-transferase, EC.2.7.2.4) is activated by calcium and a calmodulin like factor from plants.

Aspartate kinase is present in both plants and microbes and catalyses the phosphorylation of aspartate to aspartyl phosphate. The reaction is the first step in the biosynthesis of the amino acids isoleucine, lysine, threonine and methionine. The enzymes activity is regulated by feedback inhibition in plants by the end products of the pathway (Bright et al., 1984).

Aspartate kinase was purified from carrot (Daucus carota. L.DC.3) cell suspension culture. 100g of material (stored in liquid nitrogen) was homogenised in 100ml of 50mM K_2PO_4/NaOH pH 7.4 containing 1mM DTT, 50mM KCl and 20% (v/v) ethanediol. After centrifugation at 20,000g for 30 minutes the supernatant was brought to 60% (Sat.) with ammonium sulphate and centrifuged at 20,000g for 30 minutes. The precipitate was dissolved in the extraction buffer and applied to a Sephacryl S-200 column (PHARMACIA.) (2.6x65.0cm) equilibrated with 25mM TRIS/HCl pH 7.1 containing 50mM NaCl, 1mM DTT and 10% (v/v) ethanediol. Active fractions were pooled and applied to a DEAE-Sephacel (PHARMACIA.) column (2.5x8.0cm) equilibrated with 25mM TRIS/HCl pH 7.1 containing 50mM NaCl, 1mM DTT and 10% (v/v) ethanediol, bound material was eluted with a linear gradient of NaCl 50mM-300mM, active fractions were pooled and used in future experiments. The activity was measured by the reaction of acidic ferric chloride with hydroxamate formation, 0.1ml of enzyme was incubated at 30°C with 0.1ml of 250mM aspartate pH 7.4, 0.1ml 50mM TRIS/HCl pH 7.4 containing 1mM DTT and 10% (v/v) ethanediol, 0.05ml 4M hydroxylamine pH 7.4, 0.05ml 200mM ATP, 0.05ml 250mM $MgSO_4$, 0.05ml H_2O. terminated by the addition of 0.5ml ferric chloride reagent (0.37M $FeCl_3$, 0.67M HCl, 0.2M TCA) the absorbance 505nm was then measured. Additional plant material from barley and pea seedling leaves as purified to the ammmonium sulphate

[*] We are indebted to ICI, Plant Protection Division for financial support.

stage and desalted on Sephadex G-25. One unit of activity is defined as nanamoles of apartyl-hydroxamate formed min^{-1} mg^{-1} protein. All reagents tested in Table 1 were purchased from Sigma Chemical Co.Ltd.

Table 1. The Effect of inhibitors of Calmodulin on the activity of Aspartate Kinase purified from D.carota cells.

Sample	Specific Activity(units)	% Inhibition
Enzyme	329	Zero
Enzyme + 5mM lysine	29	91
Enzyme + 5mM Threonine	316	4
Enzyme − ATP	Zero	100
Enzyme + 0.1mM EGTA	334	Zero
Enzyme + 1mM EGTA	329	Zero
Enzyme + 0.05mM Ca^{2+}	334	Zero
Enzyme + 0.5mM Ca^{2+}	310	6
Enzyme + 0.4µM Calmodazolium	329	Zero
Enzyme + 2µM Calmodazolium	329	Zero
Enzyme + 25µg ml^{-1} Compound 48/80	278	16
Enzyme + 50µg ml^{-1} Compound 48/80	291	12
Enzyme + 100µM Trifluoperazine	214	35
Enzyme + 500µM Trifluoperazine	94	72
Enzyme + 0.5mM Ca^{2+} + 12µg ml^{-1} Calmodulin	308	6
Enzyme + 0.5mM Ca^{2+} + 12µg ml^{-1} Calmodulin + 25µg ml^{-1} Compound 48/80.	278	16

Table. 1 shows the effect of a number of activators or inhibitors of calmodulin dependent enzymes on lysine sensitive aspartate kinase purified from D. carota cells. Essentially similar data was obtained with relatively impure preparations of H. vulgare and P. sativum leaves.

The only significant inhibition of aspartate kinase activity was observed with trifluoperazine at the same concentrations used by Sane, et al. (1984). Trifluoperazine has been cited as a non specific antagonist of calmodulin dependent enzymes and Roufogalis (1982) has questioned the relevance of studies using concentrations above 100µM. There was no inhibition of enzyme activity by the more specific antagonist Calmodazolium and compound 48/80 (Gietzen, et al., 1983) and no increase in activity observed at the addition of bovine calmodulin and calcium.

We conclude that we are unable to find any evidence at the present time to indicate a role for calmodulin in activating plant lysine sensitive aspartate kinase activity.

Bright, S.W.J., Lea, P.J., Aruda,P., Hall, N.P., Kendall, A.C., Keys, A.J., Kueh, J.S.H., Parker, M.L., Rognes, S.E., Turner, J.C., Wallgrove, R.M. and Miflin, B.J. 1984. Manipulation of Key Pathways in photorespiration and Amino Acid Metabolism by mutation and selection in "The Genetic manipulation of Plants and its Application to Agriculture." P.J.Lea and G.R.Stewart,(eds.) Oxford Univ.Press. Oxford, U.K.
Dieter, P. 1984. Calmodulin and calmodulin mediated processes in plants. Plant Cell Environ. 7:371.
Gietzen, K., Sanchez-Delgado, E. and Bader, H. 1983. Compound 48/80: A powerful and specific inhibitor of Calmodulin-dependent Ca^{2+}-Tranport ATPase. IRCS Med.Sci. 11:12.
Roufogalis, B.D. 1982. Specificity of trifluoperazine and related Phenothiazines for calcium-binding proteins, in Calcium and cell function III. Cheung, W.Y., ed. Academic Press, N.Y., London.
Sane, P.V., Kochhar, S., Kumar, N. and Kochhar, V.K. 1984. Activation of plant Aspartate Kinase by calcium and Calmodulin-like factor from plants. Febs Letts. 175:238.

AN ASSESSMENT OF THE USEFULNESS OF QUIN-2-AM FOR CYTOPLASMIC CALCIUM

MEASUREMENTS IN PLANT CELLS

R. J. Cork*

Department of Developmental Biology
RSBS, Australian National University
Canberra, ACT, Australia

INTRODUCTION

There are only a few techniques for measuring cytoplasmic free calcium and in general these are technically difficult and may only have limited applicability. In the past few years a new generation of fluorescent indicators have been developed (Tsien 1980) which promise to be of great value. The first of these indicators to be commercially available is quin-2. While there are some inherent disadvantages in the detection of quin-2 in cells, the main advantage of the method lies in the use of the membrane-permeable acetoxymethyl ester (quin-2-AM) which should be hydrolysed by cytoplasmic esterases to trap the fluorescent indicator in the cell (Tsien 1981). This method of loading cells is much less disruptive than micro-injection and should be applicable to a whole range of cells which cannot be microinjected or have microelectrodes inserted in them. It has been used on a variety of animal cells but as yet there are no published reports of its use in plant cells. I describe here some attempts to use quin-2-AM on a number of plant cells, outline some of the problems found and suggest some possible answers.

METHODS AND RESULTS

Most experiments were done with the tip growing cells of Funaria protonema and Tradescantia pollen tubes as calcium gradients are thought to be important in the development of these cells, but cells from a wide variety of other plants were also tested for accumulation of quin-2 via incubation in the acetoxymethyl ester. Incubation solutions contained 50µM quin-2-AM and cells were incubated at room temperature or 25°C for periods ranging from 30 mins to 10 days. Intracellular fluorescence was measured either in a spectrofluorimeter (Tsein et al 1982) for cell suspensions or with a fluorescence microscope with an attached photomultiplier tube for single cells (Rogers et al 1983).

When cells were examined with the fluorescence microscope no intra-cellular fluorescence, above the cells autofluorescence, could be measured. Using the spectrofluorimeter, hydrolysis of the ester could be followed as a shift in the fluorescence emission peak from 430nm to 496nm. All of the

*Present Address: Biological Sciences, Purdue University, Lafayette, IN
                   47907, USA

371

cell suspensions tested were able to hydrolyse the acetoxymethyl ester though the rate of hydrolysis was quite slow for some cells. However, the rapid quenching of all the fluorescence on the addition of $MnCl_2$ showed that this hydrolysis was extracellular (Hesketh et al 1983).

To see if longer incubations would allow cells to slowly accumulate sufficient quin-2 to give measurable intracellular fluorescence, Funaria protonema were incubated for periods of up to 10 days from germination in nutrient solution containing quin-2-AM. Although there was still no measurable intracellular fluorescence there was significant inhibition of the growth of these cells. It was initially thought that this inhibition might be the result of the calcium buffering effect of quin-2 in the cells, however, incubations with the control ester APDA-AM (Hesketh et al 1983) or formaldehyde produced similar effects showing that it was the formaldehyde produced by the extracellular hydrolysis of the ester which was responsible. Although this effect was only apparent after about 24 hrs in Funaria protonema it did not seem to be solely the result of long incubations as similar inhibition of growth was observed in Tradescantia pollen tubes within 90 mins.

DISCUSSION

The exact reasons for the failure of plant cells to accumulate quin-2 is unknown. Accumulation of fluorescence may be slow due to the lower temperatures used for incubations (Tsien et al 1984), however, this could probably be overcome if extracellular hydrolysis was not occurring. The cell wall is probably a major site of the extracellular hydrolysis as it contains various esterases but a number of wall-less cells tested (eg. Phytopthora zoospores, Pteridium spermatazoids) also hydrolysed the ester extracellularly. There seems to be some specificity for the acetoxymethyl ester as the tetraethyl ester was not hydrolysed at all by any of the cell suspensions tested. Also, most of the cells tested were able to rapidly accumulate fluorescien if incubated in solutions containing the ester fluorescien diacetate indicating that loading plant cells using membrane permeable esters is possible. It seems that another ester of quin-2 may be required which will be more specific for cytoplasmic esterases and which will not produce toxic by-products when hydrolysed. The alternative to other esters would most likely have to be microinjection of the free acid, though this would greatly reduce the applicability of the method.

REFERENCES

Hesketh, T.R., Smith, G.A., Moore, J.P., Taylor, M.V. and Metcalfe, J.C., 1983, Free cytoplasmic calcium concentration and the mitogenic stimulation of lymphocytes, J. Biol. Chem., 258:4876.
Rogers, J., Hesketh, T.R., Smith, G.A., Beaven, M.A., Metcalfe, J.C., Johnson, P. and Garland, P.B., 1983, Intracellular pH and free calcium changes in single cells using quene 1 and quin 2 probes and fluorescence microscopy. FEBS Letts., 161:21.
Tsien, R.Y., 1980, New calcium indicators and buffers with high selectivity against magnesium and protons. Design, synthesis and properties of prototype structures, Biochemistry, 19:2396.
Tsien, R.Y., 1981, A non-disruptive technique for loading calcium buffers and indicators into cells, Nature, 290:527.
Tsien, R.Y., Pozzan, T. and Rink, T.J., 1982, Calcium homeostasis in intact lymphocytes: cytoplasmic free calcium measured with a new, intracellularly trapped fluorescent indication, J. Cell Biol., 94:325.
Tsien, R.Y., Pozzan, T. and Rink, T.J., 1984, Measuring and manipulating cytosolic Ca^{2+} with trapped indicators, TIBS, 9(6):263.

SOME PRACTICAL ASPECTS OF THE APPLICATION OF QUIN-2 TO PLANT SYSTEMS

S. Gilroy, W. Hughes and A.J. Trewavas

Department of Botany, University of Edinburgh, The King's Buildings, Mayfield Road, Edinburgh

The fluorescent and ^{19}F NMR calcium indicators Quin-2 (Tsien et al., 1982) and 5FBAPTA (Smith et al., 1983) have been successfully applied to the measurement of free, cytosolic calcium concentration in a range of animal cells. The indicator is 'loaded' into the cytosol by the intra-cellular hydrolysis of a permeant tetraacetoxymethyl ester (Tsien et al., 1982). Thus ease of intracellular localization and their high calcium sensitivity and selectivity (Tsien, 1980) would make such indicators ideally suited for use in plant systems. However, Quin-2 ester is hydrolysed extracellularly by plant cells (unpublished data), perhaps due to esterase activity in the cell wall. So the use of these indicators with the wall-less, mesophyll protoplasts of Nicotiana tabacum L. Cv. wisconsin 38 was investigated.

Methods

Protoplast suspensions, 10^5-10^6 ml^{-1}, were incubated with 40μM Quin-2 ester under conditions as in Tsien et al. (1982). Hydrolysis was completed in 70-90 minutes, as judged by the change in fluorescence emission spectrum from a maximum at 430_{nm}, characteristic of the ester, to 520-540$_{nm}$, characteristic of Quin-2 in the optical environment of the protoplast suspension.

Results and Discussion

Ester hydrolysis was essentially extracellular with 3-5% of the Quin-2 produced being protoplast associated, compared with 30-60% in animal systems (Tsien et al., 1982). Protoplast pretreatment at pH 1-4 with trypsin, 0.1-0.5% for 5 to 25 minutes was ineffective at reducing this activity. The protoplast associated Quin-2 was insensitive to extracellular EGTA (30mM) or Mn^{2+} (100μM), a Quin-2 fluorescence quenching ion (Hesketh et al., 1983); perhaps indicating minimal uptake of Quin-2 to an intracellular location. An autoradiographic study has been undertaken to resolve this question. Protoplasts loaded with 5FBAPTA, a Quin-2 analogue of superior cation discrimination, for use with ^{19}F NMR (Smith et al., 1983), under similar conditions to those for Quin-2, did not reveal a ^{19}F NMR signal. This reinforces the view of the failure of protoplasts to accumulate indicator to NMR detectable (millimolar) levels. The use of an alternative permeant ester with the required characteristics of intracellular hydrolysis, as exhibited by the diacetate of fluorescein, or of liposome mediated indicator uptake may overcome this problem and are currently under investigation.

References

Hesketh, T. R., Smith, G. A., Moore, J. P., Taylor, M.V. and Metcalfe, J. C., 1983, Free cytoplasmic calcium concentration and the mitogenic stimulation of lymphocytes, J. Biol. Chem., 258:4876

Tsien, R. Y., 1980, New calcium indicators and buffers with high selectivity against magnesium and protons. Design, synthesis and properties of prototype structures, Biochemistry, 19:2396

Tsien, R. Y., Pozzan, T. and Rink, T. J., 1982, Calcium homeostasis in intact lymphocytes: cytoplasmic free calcium measured with a new, intracellularly trapped fluorescent indicator, J. Cell Biol., 94:325

Smith, G. A., Hesketh, R. T., Metcalfe, J. C., Feeney, J. and Morris, P. G., 1983, Intracellular calcium measurements by ^{19}F NMR of fluorine-labelled chelators, Proc. Natl. Acad. Sci. USA, 80:7178

CALCIUM IONS AND THE DYNAMICS OF THE MICROTUBULAR CYTOSKELETON DURING

THE INITIATION AND THE PROGRESSION OF MITOSIS IN ENDOSPERM CELLS

M. Vantard°, H. Stoeckel°, P. Picquot°, L. Van Eldik+,
and A.M. Lambert°

°Laboratoire de Biologie Cellulaire Végétale, Université
Louis Pasteur, 67083 Strasbourg-cedex, France
+Howard Hughes Medical Institute, Vanderbilt University
Nashville, TN 37232, USA

Experiments have been carried out on endosperm cells of Amaryllidaceae
(Haemanthus and Clivia) in order to investigate the role of Ca^{2+} as a
second messenger in the regulation of the mitotic process.

Effects of modulations of the Ca^{2+} concentration in the surrounding
of the microtubular cytoskeleton, as well as the specific colocalization
of Calmodulin (CaM) and Ca^{2+} rich microdomains in the mitotic spindle point
to a particular regulation implicating the Ca^{2+}/CaM complex for the control
of chromosome movements.

Various Ca^{2+} concentrations have been tested:

(a) on whole cells, maintained in temporary culture (Mole-Bajer and Bajer,
1968), using the ionophore A 23187 to promote the Ca^{2+} entry through
the plasmalemma;
(b) on lyzed cell models obtained in Ca^{2+}/EGTA buffers (Cande, 1980), where
microtubules (MTs) polymerized in vivo are accessible to in vitro
experimentation with controlled free Ca^{2+} concentrations.

Analysis of the microtubular cytoskeleton "in toto" was conducted with
immunocytochemistry, using antibodies against mammalian brain Tubulin. These
are revealed either by indirect fluorescence (FITC) or by immunogold
staining (IGS: De Mey et al., 1982).

Although the effects observed in whole cells show a very large varia-
bility, probably due to the intervention of cytoplasmic factors and of
intracellular Ca^{2+} sequestering systems, the results obtained in the two
experimental systems converge.

In cell models, during interphase, MTs are depolymerized by Ca^{2+} in
the range of micromolar concentrations. Disassembly proceeds from their
distal end, near the cytoplasmic membrane, towards the nucleus. Short fragments, close to the nuclear envelope, persist. The question remains concerning the role of microtubular polarity and anchorage at the nuclear
envelope. During division, MTs involved in the mitotic spindle resist higher
concentrations of free Ca^{2+}, and sometimes kinetochore attached MTs show a
striking resistance, up to millimolar concentrations of Ca^{2+}. This difference

in sensitivity to Ca^{2+} ions shown by the kinetochore attached MTs, in regard to free spindle MTs and to interphasic MTs, can be related to similar differences appearing during cold treatments (Lambert and Schmit-Benner, 1980). They suggest differences in the regulation of the polymerization/depolymerization in these three populations of MTs.

Additional arguments are brought by the immunocytochemical detection of CaM in whole cells (Vantard et al., 1985), using antibodies against brain CaM (De Mey et al., 1980) or spinach CaM (Watterson et al., 1980). During interphase, the CaM pattern is rather scattered. As soon as the transformation of the interphasic microtubular network into the mitotic spindle begins, specific localizations of CaM become visible. In the early prophase, during the first stages characterized by the formation of converging MTs centers (Schmit-Benner et al., 1983), CaM appears to be linked to these aster-like regions. Later on, during metaphase, the distribution of CaM in the two half-spindles is comparable with the pattern of kinetochore attached MTs. This parallelism between the localization of CaM and kinetochore attached MTs is more obvious during anaphase, where CaM is restricted to the regions between kinetochores and poles. Checking the distribution of the anti-CaM antibodies with IGS in electron microscopy confirms their localization in the close vicinity of MTs. Not all the MTs are labelled: some of them are free of gold particles. This suggests that labelled MTs correspond to kinetochore attached ones.

The distribution of membrane associated Ca^{2+}, as revealed by Chlorotetracycline (CTC) fluorescence, shows specific localizations in the mitotic endosperm cells (Wolniak et al., 1980). Further development of this technique was carried out in order to compare CTC fluorescence and CaM patterns. Previous results are confirmed, and the data support a preferential accumulation of Ca^{2+} rich microdomains in the regions where CaM is also detected, i.e. in the half spindles, between kinetochores and the polar region.

This colocalization of CaM and Ca^{2+} rich microdomains in the vicinity of the kinetochore attached MTs, specifically involved in the chromosome movements, is consistent with the hypothesis of a functional Ca^{2+}/CaM dependent regulation, which could act either directly on the GTP dependent assembly/disassembly of Tubulin, or in relation with mechanisms involving MTs associated proteins.

REFERENCES

Cande, W.Z., 1980, A permeabilized cell model for studying cytokinesis using mammalian tissue culture cells, J. Cell Biol., 87:326.
De Mey, J., Moeremans, M., Geuens, G., Nuydens, R., Van Belle, H. and Dr Brabander, M., 1980, Immunocytochemical evidence for the association of Calmodulin with microtubules of the mitotic apparatus, in: "Microtubules and Microtubule Inhibitors", M. Borgers and M. De Brabander, eds, North Holland Publishing Co., Amsterdam, 228.
De Mey, J., Lambert, A.M., Bajer, A.S., Moeremans, M. and De Brabander, M., 1982, Visualization of microtubules in interphase and mitotic plant cells of Haemanthus endosperm with the immuno gold staining (IGS) method, Proc. Natl Acad. Sci. USA, 79:1898.
Lambert, A.M. and Schmit-Benner, A.C., 1980, Inhibition of microtubule assembly in vivo by colchicine, vinblastine and cold treatments during anaphase. Chromosome and spindle dynamics, J. Cell Biol., 87:234a.
Mole-Bajer, J. and Bajer, A., 1968, Studies of selected endosperm cells with the light and electron microscope. The technique, Cellule 67:257.
Schmit, A.C., Vantard, M., De Mey, J. and Lambert, A.M., 1983, Aster-like microtubule centers establish spindle polarity during interphase-mitosis transition in higher plant cells, Plant Sci. Rep., 2:285.

Vantard, M., Lambert, A.M., De Mey, J., Picquot, P. and Van Eldik, L.J., 1985, Characterization and immunocytochemical distribution of Calmodulin in higher plant endosperm cells: localization in the mitotic apparatus, J. Cell Biol., submitted.

Watterson, D.M., Iverson, D.B. and Van Eldik, L.J., 1980, Spinach Calmodulin: isolation, characterization, and comparison with vertebrate Calmodulins, Biochemistry, 19:5762.

Wolniak, S.M., Hepler, P.K. and Jackson, W.T., 1980, Detection of membrane-calcium distribution during mitosis in haemanthus endosperm with Chlorotetracycline, J. Cell Biol., 87:23.

MITOSIS AND CYTOKINESIS IN FUNARIA ARE ACCOMPANIED

BY INWARD CURRENT LOCALIZED AT THE NUCLEAR ZONE

Mary Jane Saunders

Botany Department
Louisiana State University
Baton Rouge, LA 70803 U.S.A.

A trigger for cell division in a wide variety of systems has been postulated to result from ionic fluxes (particularly Ca^{2+}) across the plasma membrane. Furthermore, it has been proposed that control of the different stages of mitosis (in particular the metaphase/anaphase transition) may be regulated by spatially localized increases in intracellular Ca^{2+} concentration ($[Ca^{2+}]_i$) that are controlled by an endomembrane system (Wolniak and Hepler, 1983). Calcium has also been implicated in the regulation of the cytokinetic apparatus in plant cells (see Gunning, 1982). In an attempt to elucidate the role of ion currents in mitosis and cytokinesis I investigated temporal and spatial localization of ion fluxes across the plasma membrane of dividing tip-growing filamentous Funaria hygrometrica cells.

Transcellular ion currents were measured with a non-intrusive vibrating probe (Jaffe and Nuccitelli, 1974). In brief, the sensing element of this instrument consists of a microelectrode tipped with a ball of platinum black. The electrode was vibrated (200 cps) with an excursion of 20 um perpendicular to the long axis of Funaria cells and measured the small potential differences between the extremes of its sweep. These potential differences are converted to current densities by using Ohm's law.

The major zone of inward current was not at the growing tip, but at the nuclear zone (0.5 uA/cm^2). In tip cells the nucleus moves forward at the rate of 50 um/hr remaining 120 um distant from the growing tip; the zone of maximal inward current also moves. In every Funaria cell measured (₵ 300), the major zone of inward current was at the nucleus; this includes cells located up to seven cells back from the growing tip cell. In no cell was any outward current detected. It is possible that outward current may be located either on upright branches or in the center of the mat of protonemal filaments where it was impossible to position the probe.

A tremendous change was detected in the magnitude of the inward current at the nuclear zone during mitosis and cytokinesis of tip cells. The current increased 3 fold just before the initiation of mitosis and remained high but pulsated during the stages of mitosis. The zone of increased inward current is detected only at the nuclear region and falls off to resting levels when the probe is moved along the length of the cell. The next major change detected in the currents was during cell plate formation. The maximal inward current shifted slightly from the center of the spindle axis to that region at which the now obliquely-oriented cell plate will fuse with the parental wall. As soon as the cross wall fuses with the parent wall the current falls to zero. After division of tip cells the two daughter nuclei migrate to positions approximately halfway along the length of the daughter cell. No change in current was associated with either the migrating nucleus or the site where the nucleus finally comes to rest. These observations differ from hormonally triggered asymmetrical division in Funaria where an inward current predicts the site to which the nucleus will migrate before dividing.

An attempt was made to initially characterize the ions responsible for the current. When gadolidum nitrate (a competitive Ca^{2+} uptake inhibitor) was added (1 um final concentration) the current associated with mitosis and cytokinesis immediately fell to zero.

These data suggest that Funaria cells have an asymmetrical distribution of Ca^{2+} channels concentrated over the nuclear region in quiescent cells. Upon the initiation of mitosis, these channels open to admit a spatially localized increase in Ca^{2+} ions. This correlates well with the data of Mineyuki et al. (1983) who described a local change in cytoplasmic viscosity at the nuclear region just before and concurrent with mitosis in filamentous Adiantum cells. In addition, Schmeidel et al. (1980) found that premitotic and mitotic nuclei in Funaria could not be displaced by centrifugation. This argues for a change in viscosity or anchoring of nuclei just prior to and during mitosis. This increased viscosity may also be the driving force behind the creation of the phragmosome, a raft of cytoplasm seen in highly vacuolate plant cells that positions the nucleus before division.

These results in Funaria may indicate a cytoplasmic microdomain of increased $[Ca^{2+}]$ around the nucleus. This $[Ca^{2+}]$ increase may be postulated to sequentially trigger nuclear envelope breakdown, mitosis and cytokinesis. Thus, Ca^{2+}_i as a trigger for mitosis would not only be temporally controlled to a specific point in the cell cycle, but also spatially controlled by the plasma membrane. One might hypothesize that a similar mechanism would operate in cells in which the exact placement of the cytokinetic apparatus is essential to establish resulting developmental morphology. This is especially true in plant systems in which asymmetric division can lead daughter cells on differing developmental pathways.

ACKNOWLEDGMENT

This work supported in part by NSF grant PCM-84-08496.

REFERENCES

Gunning, B. E. S., 1982, The cytokinetic apparatus: Its development and spatial regulation in: "The Cytoskeleton in Plant Growth and Development," C. L. Lloyd, ed., Academic Press, London.

Jaffe, L. H. and Nuccitelli, R., 1974, An ultrasensitive vibrating probe for measuring steady external currents, J. Cell Biol., 63:614.

Mineyuki, Y., Yamada M., Takagi, M., Wada, M., Furuya, M., 1983, A digital image processing technique for the analysis of particle movements: Its application to organelle movements during mitosis in Adiantum protonemata, Plant and Cell Physiol., 24:225.

Schmiedel, G., Schnepf, E., 1980, Polarity and growth of caulonema tip cells of the moss Funaria hygrometrica, Planta, 147:405.

Wolniak, S. N., Hepler, P. K. and Jackson, W. T., 1983, Ionic changes in the mitotic apparatus at the metaphase/anaphase transition, J. Cell Biol., 96:598-605.

QUANTIFICATION OF <u>MESOTAENIUM</u> CALMODULIN BY IMPROVED CYCLIC NUCLEOTIDE

PHOSPHODIESTERASE TEST

Sigrid Jacobshagen, Franz Grolig and Gottfried Wagner

Botanisches Institut I der Justus-Liebig-Universität
Senckenbergstrasse 17 - 21
D-6300 Giessen, FR Germany

Calmodulin alike actin and a few other eukaryotic polypeptides is a highly conserved protein expressed in animal and plant cells (Cheung, 1970; Marmè and Dieter, 1983). Thus, despite its plant source, calmodulin from the green algae <u>Chlamydomonas</u> and <u>Mougeotia</u>, from fungi including <u>Dictyostelium</u> and <u>Neurospora</u>, from the moss <u>Funaria</u> and from a large set of higher plants (Marmè and Dieter, 1983) was found to active bovine heart cyclic nucleotide phosphodiesterase in a calcium-dependent manner much the same as bovine calmodulin does. Other remarkable properties of calmodulin, although of minor distinctive strength, are heat stability, Ca^{2+}-dependent high affinity binding to antipsychotic drugs including chlorpromazine and fluphenazine, and Ca^{2+}-dependent electrophoretic mobility in sodium-dodecyl-sulfate polyacrylamide gels. Thus, the cyclic nucleotide phosphodiesterase test has turned out the major functional tool of calmodulin identification. The cyclic nucleotide phosphodiesterase test, however, as run so far (Cheung, 1969), appears lengthy and tedious with more than 15 individual steps of sample processing including multiple transfers of sample aliquots, protein denaturation and sample centrifugation.

Here, the test was improved in three major ways and finally allowed, <u>e. g</u>. from a chlorpromazine affinity column eluate, determination of up to 45 calmodulin samples within 3 h at fourfold increased sensitivity.

The improvements include:
- Phosphodiesterase activity was stopped instantaneously at given times by the specific inhibitor 3-Isobutyl-1-methylxanthine (IBMX) (Ashcroft et al., 1972) rather than by 5 min heat denaturation. The IBMX inhibitor in concentrations up to 1 mmol l^{-1} was found not to interfere with the alkaline phosphatase activity in the coupled enzymatic reaction.
- Alternatively, phosphodiesterase activity and alkaline phosphatase activity were stopped simultaneously by enzyme dissolution in sodium-dodecyl-sulfate (SDS), avoiding protein precipitation in trichloric acid (TCA).
- The clear SDS-sample, free of TCA, facilitated the colorimetric determination of released inorganic phosphate with fourfold increased sensitivity at the peak wavelength of 820 nm.

Growth of <u>Mesotaenium</u> cells, cell homogenization and chlor-promazine affinity column chromatography followed the methods as described for <u>Mougeotia</u> before (Wagner et al., 1984). 0.81 ml of reaction medium contain-

ed 100 mmol l^{-1} imidazole at pH 7.5, 4 mmol l^{-1} $MgSO_4$, 2 mmol l^{-1} adenosine 3':5'-cyclic monophosphate, 0.1 mmol l^{-1} $CaCl_2$, 1 milliunit of calmodulin-deficient phosphodiesterase (PDE) from bovine heart (Nr. 709 883,
Boehringer, Mannheim FRG) and different amounts of bovine brain calmodulin
or Mesotaenium protein. The mixture was incubated at 37° C for 30 min. The
reaction was stopped by the addition of 20 µl of IBMX resulting in an inhibitor concentration of 1 mmol l^{-1}. 5 units of alkaline phosphatase, added
in a volume of 10 µl, catalyzed release of inorganic phosphate (P_i) at
37° C for 10 min. The reaction was stopped by the addition of 0.16 ml of
phosphate complex medium yielding a final concentration of 0.2 mmol l^{-1}
sulfuric acid, 0.8 mmol l^{-1} ammonium-molybdate, 4.8 mmol l^{-1} ascorbic acid
and 0.25 % (w/v) SDS. Alternatively, 0.84 ml of reaction medium included
both enzymes, i. e. 1 milliunit of PDE and 5 units of alkaline phosphatase,
at 37° C for 30 min. The coupled enzymatic reaction was stopped by the
addition of 0.16 ml of phosphate complex medium as given above. The final
sample volume of 1 ml, devoid of TCA, was incubated at 60° C for 30 to
60 min. This allowed the colorimetrically useful bluish ascorbic acid-reduced complex of molybdophosphate to reach steady state in its fromation,
(Fig. 1), in contrast to the conditions as often used before (Marmè and
Dieter, 1983). Furthermore, shift to the peak absorption of the ascorbic
acid-reduced molybdophosphate complex at 820 nm wavelength improved total
sensitivity of the abbreviated test by a factor of four, a precondition to
detect calmodulin in plant material of limited quantities as is true for
slowly growing cells of Mesotaenium, with a cytoplasmic calmodulin concentration close to 10^{-5} mol l^{-1} (data not shown).

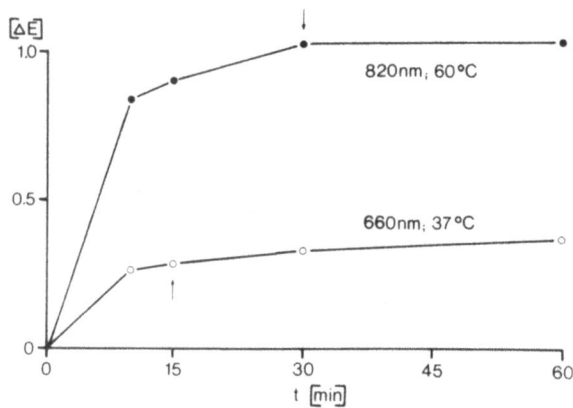

Fig. 1.
Formation of the bluish coloured ascorbic acid-reduced complex of molybdophosphate as a function of time and temperature. The two conditions were
37° C with wavelength of detection at 660 nm (o - o) and 60° C with wavelength of detection at 820 nm, respectively (● - ●). Time of routine data
reading is indicated by arrows and shows fourfold increased sensitivity of
the test.

REFERENCES

Ashcroft, S.J.H., Randle, P.J., and L.-B. Täljedal, 1972, FEBS Lett. 20:263.
Cheung, W.Y., 1969, Biochim. Biophys. Acta 191:303.
Cheung, W.Y., 1970, Biochem. Biophys. Res. Comm. 38:533.
Marmè, D. and Dieter, P., 1983, in: "Calcium and Cell Function", Vol. 4,
        W.Y. Cheung, ed., Academic Press, New York, London.
Wagner, G., Valentin, P., Dieter, P., and D. Marmè, 1984, Planta 162:62.

Financial support from the Deutsche Forschungsgemeinschaft is thankfully
acknowledged.

CALCIUM EFFECTS ON STOMATAL GUARD CELLS

E.A.C. MacRobbie

Botany School, University of Cambridge

Downing Street, Cambridge CB2 3EA, U.K.

One of the peculiarities of guard cell behaviour is their extreme sensitivity to the presence of Ca^{2+}. The striking effects of even low concentrations of Ca^{2+} on stomatal aperture have been known since the early studies of Fujino (1967), but the mechanism of the effect remains unknown. It is therefore important to establish the effects of Ca^{2+} on the levels of ion accumulation in guard cells in a range of conditions, and on the ion fluxes, influx and efflux, by which the steady state levels are maintained.

For the measurement of ion fluxes in guard cells it is necessary to use 'isolated' guard cells, in epidermal strips in which all cells other than guard cells have been killed; treatment at low pH is the most convenient way of achieving this, and methods for flux measurements in such isolated guard cells of Commelina communis L. are described in MacRobbie (1983, 1984). It is necessary to describe the steady state conditions at different apertures (steady state ion content, and steady state ion fluxes), but it is also necessary to establish the changes in ion fluxes, either influx or efflux, responsible for the transition between two steady states, induced by a change of conditions, as for example the addition of Ca^{2+}. It is important to measure the influx during the early stages of opening, and its sensitivity to Ca^{2+} and other factors, as well as the two-way fluxes in the steady state achieved after overnight incubation in given conditions.

Effect of Ca^{2+} on steady state ion content

The sensitivity to Ca^{2+} is retained in isolated guard cells. The addition of 2 mM Ca^{2+} to the 10-75 mM RbCl bathing the guard cells results in smaller apertures and lower ion contents, but the relation between aperture and ion content is not changed; the regression lines of ion content on aperture are not significantly different in the presence or absence of Ca^{2+}. The effect of Ca^{2+} is therefore on the ion transport properties of the tissue, and not on the mechanical properties of the cell wall.

Effect of Ca^{2+} on influx

Guard cells were isolated, starting from the closed state, to investigate the effects of Ca^{2+} on the influx during opening. Influxes were measured during the opening induced by incubation on 30 mM RbCl, with either low Ca^{2+} (10μm) and Mg^{2+} (5 mM), or high Ca^{2+} (5 mM), at pH 3.9. The influx of ^{86}Rb was inhibited by Ca^{2+}; the ratio of the influx in 5 mM Ca^{2+} to that in (5 mM Mg^{2+} + 10μM Ca^{2+}) was 0.71 $\pm$ 0.09, 0.71 $\pm$ 0.09, and 0.83 $\pm$ 0.10, in three separate experiments. There was also a marked inhibition of influx in the dark in the presence of 5 mM Ca^{2+}, but not in 5 mM Mg^{2+}. In the presence of 5 mM Ca^{2+} the ratio of the influx in the dark to that in the light was, in three different experiments, 0.66 $\pm$ 0.10, 0.70 $\pm$ 0.09, and 0.58 $\pm$ 0.06; in the presence of 5 mM Mg^{2+} and only 10 μM Ca^{2+} the ratio of the influx in the dark to that in the light was 1.15 $\pm$ 0.10 and 1.16 $\pm$ 0.07. Thus the effect of dark in inhibiting ^{86}Rb influx is a Ca^{2+}-dependent process. Similar, though smaller, inhibition of ^{86}Rb influx in the dark was also observed at pH 7 in the presence of 2 mM Ca^{2+}.

The dark-induced inhibition of ^{86}Rb influx in the presence of Ca^{2+} is abolished by the addition also of 2mM $LaCl_3$ in the incubation medium, which is consistent with the hypothesis that the effect involves Ca^{2+} influx, and that this is blocked by La^{3+}.

It has been established that in steady state open guard cells (after overnight incubation in light), transfer to the dark has no significant effect on ^{86}Rb influx (MacRobbie, 1983, 1984). In these conditions control of aperture is dependent on regulation of ion efflux, not influx. In the only experiment yet done on steady state tissue in the presence of Ca^{2+}, no significant difference in influx was found between light and dark, but this needs to be confirmed by further measurements.

Effects of Ca^{2+} on ^{86}Rb efflux

As yet the effects of Ca^{2+} on ^{86}Rb efflux in steady state tissue have been measured only in partly open guard cells, and not in wide open. In 30 mM RbCl, at pH 6, the ^{86}Rb efflux was reduced by 24-37% by the addition of 1 mM Ca^{2+}; the addition of 2 mM EGTA in a parallel experiment increased the efflux by 21-32%. This is similar to effects in other plant cells, where membrane permeability is increased in Ca^{2+}-free conditions, and reduced in the presence of Ca^{2+}. It remains to be tested whether there is any effect of Ca^{2+} on the ion efflux in wide open guard cells, where the maintenance of opening is dependent on the control of ion efflux.

References

Fujino, M., 1967, Role of adenosinetriphosphate and adenosine-triphosphatase in stomatal movement, Sci. Bull. Fac. Educ. Nagasaki Univ., 18:1.
MacRobbie, E.A.C., 1983, Effects of light/dark on cation fluxes in guard cells of Commelina communis L., J. Exp. Bot., 34:1695.
MacRobbie, E.A.C., 1984, Effects of light/dark on anion fluxes in isolated guard cells of Commelina communis L., J. Exp. Bot., 35:707.

CALCIUM AND THE CURRENT-VOLTAGE RELATIONS OF STOMATAL GUARD CELLS

Michael R. Blatt

Botany School, University of Cambridge, Cambridge, England

The last 10 years have marked a resurgence of interest in the electrical properties of guard cells as they bear on the ionic relations of the cells and on stomatal movements. The effects of light (Moody and Zeiger, 1978), extracellular K^+ and Ca^+ (Saftner and Raschke, 1981), and CO_2 (Edwards and Bowling, 1985) among other conditions, have been examined. These studies have focused almost exclusively on changes in membrane potential. The question of membrane conductance and its relation to voltage has been wholly ignored.

Early results of an ongoing study of guard cell current-voltage (I-V) relations are summarized below (data shown is for <u>Vicia</u>). Measurements were carried out on epidermal strips by two-electrode voltage clamp using double-barrelled microelectrodes (see Blatt and Slayman, this volume).

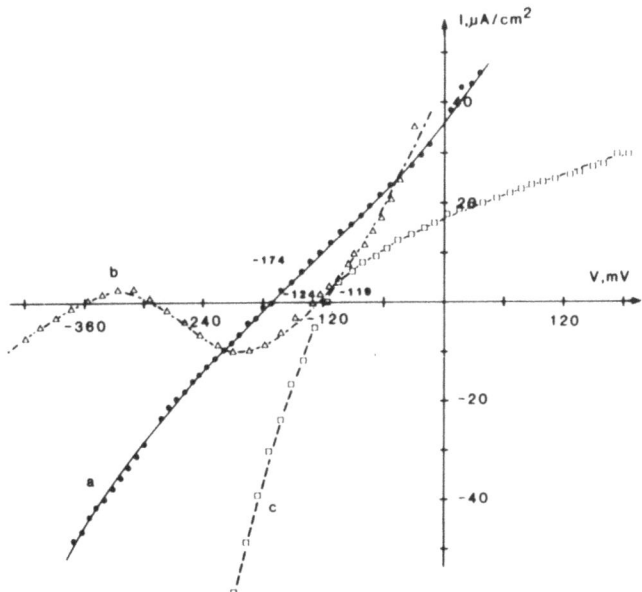

Fig. 1. I-V relations for 3 <u>Vicia</u> guard cells. Measurements in Ca-MES buffer, pH 6.1 with (a) 0.25mM K_2SO_4 [1mM Ca^{2+}], (b) and (c) 0.005mM K_2SO_4 [0.2mM Ca^{2+}]. The current scale for (b) has been compressed 4-fold.

The I-V relations for the guard cells fell into two broad categories, as illustrated in Fig. 1: (1) a low conductance (1-2 S m^{-2}) high potential (c. -180 mV) characteristic with a roughly linear profile (curve $\underline{a}$), and (2) a high conductance (10-50 S m^{-2} near the "resting" potential) low potential (generally positive of -100 mV) characteristic with an N-shaped profile (curve $\underline{b}$) or marked inward rectification (curve $\underline{c}$).

The I-V curves in the first category are reminiscent of those from Neurospora (Gradmann, et al., 1978) and Chara (Beilby, 1984) which reflect significant contributions of electrogenic pumping to the membrane characteristic. I-V relations in the second category show features of voltage-dependent conductances previously described in plant (cf. M. Beilby, this volume) and animal (Chenoy-Marchais, 1982) cells. There may be a link between salt loading from the impaling electrode and the "gated" conductance(s). This link is currently under study.

Calcium and its withdrawal (with/without 1mM EGTA) had no qualitative effect on the I-V profile (Fig. 2) over periods of 2-15 min. Both Mg^{2+} and La^{3+} could substitute for Ca^{2+}. Prolonged exposure to Ca^{2+}-free medium did lead to 10- to 100-fold increases in conductance. Thus, the primary effect of the cation, with reference to the membrane potential changes observed by Saftner and Raschke (1981), is "charge masking."

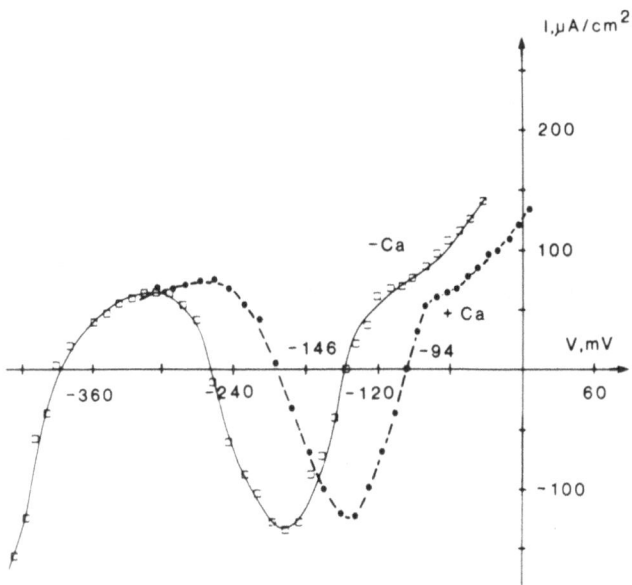

Fig. 2. Displacement of the I-V characteristic with 1 mM Ca^{2+}. +Ca^{2+} medium: as in Fig. 1, curve $\underline{a}$. -Ca^{2+} medium: 1mM Na-MES, pH 6.1. Both media contained 0.01 mM K$^+$ (as the sulphate).

REFERENCES

Beilby, M (1984) Current-voltage characteristics of the proton pump at Chara plasmalemma I. pH dependence. J. Membr. Biol. 81, 113.
Chenoy-Marchais, D. (1982) A Cl$^-$ conductance activated by hyperpolarization in Aplesia neurones. Nature 299, 359.
Edwards, A, Bowling, D.J.F. (1985) Evidence for a CO$_2$ inhibited proton extrusion pump in the stomatal cells of Tradescantia virginiana. J. Exptl. Bot. 36, 91.
Moody, W., Zeiger, E. (1978) Electrophysiological properties of onion guard cells. Planta 139, 159.
Saftner, R., Raschke, K. (1981) Electrical potentials in stomatal complexes. Plant Physiol. 67, 1124.

ABSCISIC ACID, CALCIUM IONS AND STOMATAL FUNCTION

A.M. Hetherington, D.L.R. De Silva, R.C. Cox, and T.A. Mansfield

Dept. of Biological Sciences, University of Lancaster
Bailrigg, Lancaster, LA1 4YQ, U.K.

One of the most clearly defined roles for abscisic acid (ABA) is the induction of stomatal closure when water supply is suboptimal (Mansfield & Davies, 1981). Ca^{2+} is also known to induce stomatal closure but its role as a regulator of guard cell turgor has received little attention. As Elliot and coworkers (1983) reported that calmodulin (CaM) antagonists influence growth regulator action it seemed logical to investigate whether Ca^{2+} and ABA interacted in influencing stomatal aperture.

Using isolated epidermis (Weyers & Travis, 1981) we confirmed that Ca^{2+} inhibited the opening of stomata from <u>Commelina communis L</u>. When epidermal strips were incubated in the presence of 10^{-6} mol m^{-3} ABA there was little inhibition of stomatal opening but when the same concentration of ABA was applied in combination with 0.10 mol m^{-3} $CaCl_2$, the opening response was inhibited. This suggested a synergistic response between ABA and Ca^{2-} (De Silva, Hetherington & Mansfield, 1985). We have also found that ABA action can be blocked by $LaCl_3$ (Fig. 1A) and the CaM antagonist W-7 (Fig. 1C) and that ABA action can be mimicked by the Ca^{2+} ionophore A23187 (Fig. 1B). These data are consistent with the hypothesis that ABA increases the permeability of the guard cells to Ca^{2+}. Ca^{2+} might then operate as a 2nd messenger to regulate the ionic fluxes that determine guard cell turgor. We have found no synergism between the synthetic auxin napth-1-ylacetic acid (which inhibits stomatal opening (Snaith & Mansfield, 1984)) and Ca^{2+} (Fig. 1D), suggesting that this mechanism relates specially to ABA.

ACKNOWLEDGEMENT

The authors wish to acknowledge the support of the AFRC and the ACU.

REFERENCES

De Silva, D.L.R. Hetherington, A.M. and Mansfield, T.A. (1985). Synergism between Ca^{2+} and ABA in preventing stomatal opening. New Phytol. (in press).

Elliot, D.C., Batchelor, S.M., Cassar, R.A. and Marinos, N.G. (1983). Calmodulin-binding drugs affect responses to cytokinin, auxin and gibberellic acid. Plant Physiol. 72, 219-224.

Mansfield, T.A. & Davies, W.J. (1981). Stomata and stomatal mechanisms. In: The Physiology and Biochemistry of Drought Resistance in Plants (Ed. by L.G. Paleg & D. Aspinall) p. 315-346. Academic Press, Sydney.

Snaith, P.J. & Mansfield, T.A. (1984). Studies of the inhibition of stomatal opening by napth-1-ylacetic acid and abscisic acid. J. Expt. Bot. 35, 1410-1418.

Weyers, J.D.B. & Travis, A.J. (1981). Selection and preparation of leaf epidermis for experiments on stomatal physiology, J. Expt. Bot. 32, 837-850.

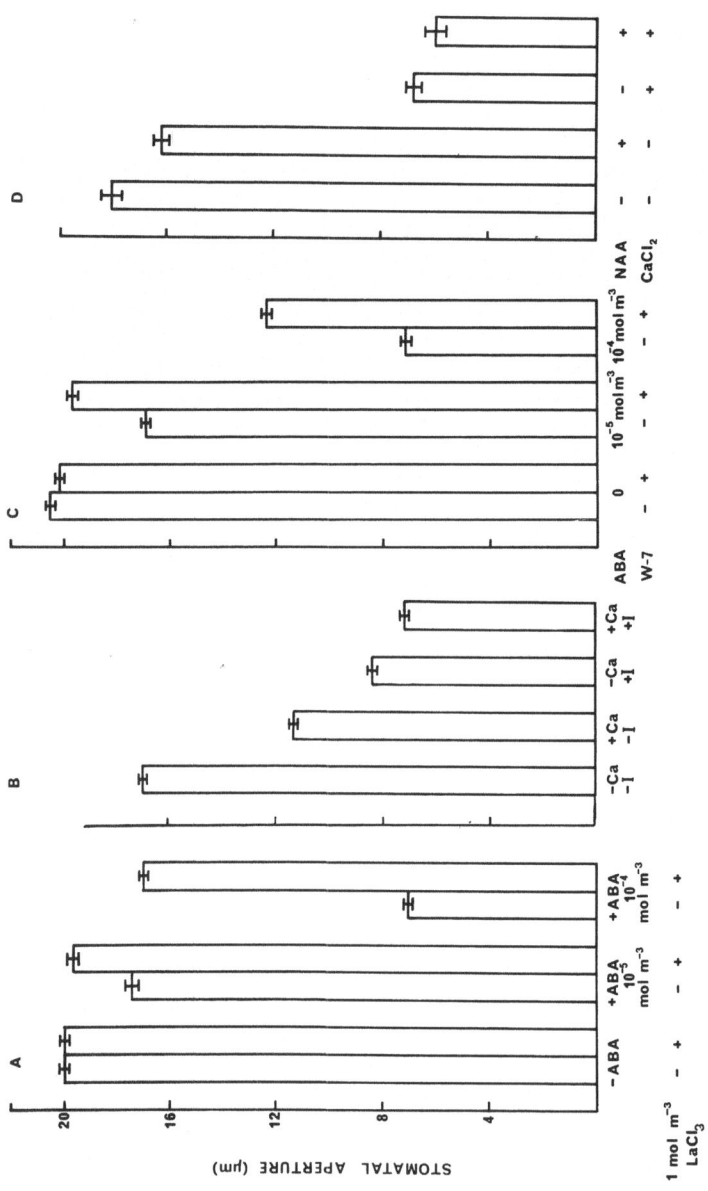

Fig.1. Abaxial epidermal strips from <u>Commelina communis</u> L. were incubated for 3h at 25 ± 1°C under a photon flux density of 160 μmol m^{-2} s^{-1} in 10 mol m^{-3} MES, 50 mol m^{-3} KCl pH 6.15 in which the stated concentrations of the various additives were dissolved. At the end of the incubation period stomatal apertures were measured using a Watson Hilux 70 microscope fitted with a projection eyepiece. Measurements were made of 90 stomata, located at random, 10 from each of 9 replicate strips. The strips were incubated in (A) the presence or absence of LaCl$_3$ with various concentrations of ABA, (B) the presence or absence of combinations of CaCl$_2$ (0.1 mol m^{-3}) and an ionphore A23187 (0.05 mol m^{-3}) (C) in various concentrations of ABA in the presence or absence of W7 (0.05 mol m^{-3}) (D) in the presence or absence of combinations of CaCl (0.1 mol m^{-3}) and NAA (10^{-4} mol m^{-3}).

CALCIUM REGULATION IN APPLE FRUIT

Masashi Fukumoto* and Michael A Venis

East Malling Research Station
Maidstone
Kent, ME19 6BJ

Several physiological disorders in apple fruit (e.g. bitter pit) are associated with calcium deficiency. We have begun to explore the hypothesis that these disorders arise from lesions in the normal cytoplasmic calcium regulation machinery.

As a first step, Cox apples were removed from cold store and vacuum infiltrated at 0.2 atmospheres with water (control) or with the calmodulin antagonists chlorpromazine or fluphenazine at 10^{-4}M. They were then returned to store at either 20°C or 0°C. At 20°C, symptoms resembling bitter pit or blotch pit were first visible after 2 days following fluphenazine treatment and after 7 days with chlorpromazine treatment. The symptoms increased in severity with time (Fig. 1) and several of the treated apples developed rot and senescent breakdown, again in the fluphenazine fruit first. At 4°C, pit-like spots first appeared about 3 weeks after fluphenazine treatment.

Fig. 1. Apple fruit 14 days after infiltration with water (control) or calmodulin antagonists.

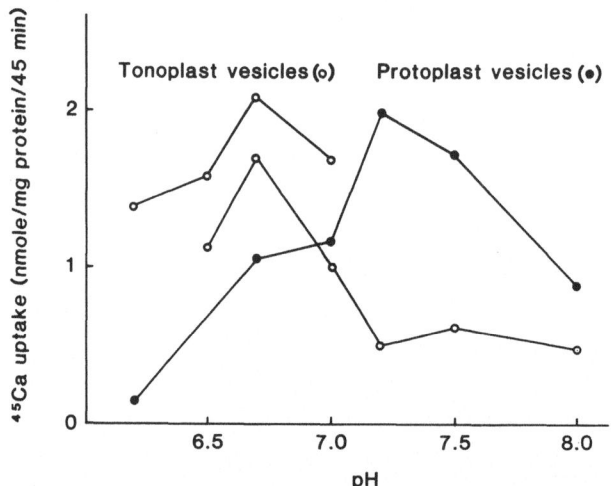

Fig. 2. ATP-dependent $^{45}Ca^{2+}$ transport as a function of pH in vesicles prepared from either tonoplasts or protoplasts of apple fruit.

Good protoplast yields (5×10^5/g tissue) were obtained from immature (30–50 g) Golden Delicious fruit using 2% Cellulase Onozuka RS, 0.5% Macerozyme R10 and 0.03% Pectolyase Y23, 16 h at 20^oC. Attempts to measure cytoplasmic $[Ca^{2+}]$ with quin 2-AM were unsuccessful under the loading conditions used (25–100 μM, 2 h at 21^oC). Higher temperatures led to protoplast rupture.

Tonoplast vesicles prepared from these protoplasts show ATP-dependent $^{45}Ca^{2+}$ uptake, stimulated by calmodulin and inhibited by fluphenazine. The optimum pH for this tonoplastic Ca^{2+} transport (pH 6.7) is distinct from that of vesicles prepared from intact protoplasts, pH 7.2–7.5 (Fig. 2). Further characterisation is in progress.

---

* Permanent address: Fruit Tree Research Station, Morioka, Japan.

Ca^{2+} CONTRIBUTES TO THE SIGNAL TRANSDUCTION CHAIN IN PHYTOCHROME-MEDIATED

SPORE GERMINATION

Randy Wayne and Peter K. Hepler

Department of Botany
University of Massachusetts
Amherst, MA 01003   USA

Phytochrome controls spore germination in the fern Onoclea. A major stumbling block in uncovering the calcium requirement for germination has been the presence of large quantities of the ion in the spore wall. Even if one sows the spores in calcium-free media the response to the ion can not be detected unless the wall bound calcium has been previously removed through washing in EGTA. When the wall-bound calcium is removed the spores display a remarkable sensitivity to the concentration of ion added back. The threshold concentration for calcium ions for germination is submicromolar while 3 μM supports a half-maximal response. Although strontium and barium can substitute for calcium, magnesium is completely ineffective.

The relationship between phytochrome and its presumptive second messenger is revealed more persuasively by the results showing that red light stimulates net calcium uptake in Onoclea spores while far-red light inhibits this process. Measurements made by atomic absorption spectroscopy indicate that enough calcium is taken up during the red light irradiation to raise the internal concentration by 500 μM. A large excess may be needed to saturate cellular binding sites so that sufficient calcium remains free to act as a stimulus. While these data do not directly show that the free calcium concentration has increased they are entirely consistent with that conclusion.

Further evidence that calcium couples the red light stimulus to the response comes from the observations showing that lanthanum blocks calcium uptake and also inhibits germination. Lanthanum applied prior to irradiation completely inhibits germination whereas it progressively loses its effectiveness if given short times after the onset of irradiation. After 5 minutes of red light lanthanum is completely ineffective indicating that uptake of external calcium needed for germination, occurs rapidly following the initiation of irradiation.

One of the most exciting and unexpected results is that in calcium-free media, the plasma membrane remains maximally poised to accept and transport calcium for at least 4 hours following red light irradiation. Subsequent irradiation with far-red light 0.5-4 hours after the red irradiation fails to prevent germination indicating that calcium transport has been uncoupled from transformed phytochrome (P_{fr}). These observations provided evidence for the occurrence of intermediary steps in the transduction chain between phytochrome and calcium influx. Thus the conversion of P_r to P_{fr} may alter the conformation of specific calcium channel proteins on the plasma membrane and lock them into a relatively stable open configuration. Even though P_{fr} may quickly decay to an inactive state the channels remain open for several hours during which time calcium enters the spore from the extracellular space and stimulates germination.

The final evidence showing that calcium couples the stimulus to the response in Onoclea comes from studies in which an artificially generated influx of calcium causes germination in the absence of red light irradiation. By applying the ionophore A 23187 to dark sown spores it is possible to partially mimic the effect of red light. Once calcium has entered the cell it seems likely that it binds calmodulin. Compounds that interfere with the function of the calcicalmodulin complex, e.g. CPZ, TFP, calmidazolium (R 24571) inhibit germination. Although the drug concentrations used are high, raising concern over their specificity, the effectiveness of the drugs corresponds to their affinity in binding to calcicalmodulin. Moreover, at high concentrations (300 µM) the effects of CPZ and TFP are reversible. These data indicate that calcium contributes to the signal transduction chain in phytochrome-mediated fern spore germination.

A CHANGE IN INTRACELLULAR pH DOES NOT CONTRIBUTE TO THE SIGNAL TRANSDUCTION

CHAIN IN PHYTOCHROME-MEDIATED SPORE GERMINATION

Randy Wayne and Peter K. Hepler

Department of Botany
University of Massachusetts
Amherst, MA 01003 USA

Calcium ions act as a "second messenger" in phytochrome-stimulated spore germination in <u>Onoclea</u> <u>sensibilis</u> L. In this study we determine whether changes in intracellular pH also contribute to the signal transduction chain. Using ^{31}P-nuclear magnetic resonance spectrometry, in which it is possible to resolve changes as small as 0.2 pH units, we fail to observe any change in intracellular pH associated with red light irradiation. In addition artificially inducing an intracellular change in pH of greater than 1 pH unit (5.8-7.2) has no effect on germination. These data indicate that a sustained increase in intracellular pH does not contribute to the signal transduction chain in phytochrome-mediated fern spore germination.

LOCALIZATION OF POTENTIAL CA^{2+}-BINDING SITES IN LILY POLLEN TUBES AND MAIZE CALYPTRA CELLS

W. Herth, H.-D. Reiss, B. Hertler, R. Bauer[+], and K. Traxel

Zellenlehre,and Physikalisches Institut, University of Heidelberg, D-6900 Heidelberg
+ Applikationslabor Elektronenmikroskopie, Carl Zeiss D-7082 Oberkochen

A sucessful localization of potential Ca^{2+}-binding sites would be of high value for interpreting the possible involvement of cell organelles in calcium regulated phenomena. We therefore tested the method of Belitser (1,2) on pollen tubes of <u>Lilium longiflorum</u>, where Ca^{2+} seems to be involved in maintenance of oriented growth, vesicle fusion, and transport, cytoplasmic streaming, regulation of the cortical microfilament network, and cell wall stiffening (3-6), and on maize calyptra cells, where Ca^{2+} should be involved in the regulation of the slime secretion process, and in transducing the gravitropism signal (7,8).

Pollen tubes of <u>Lilium longiflorum</u> show good ultrastructural preservation after fixation in the presence of 10 mM Ca^{2+}. In comparison to conventionally fixed pollen tubes, the cell wall, the subapical endoplasmic reticulum, mitochondria and the plasma membrane are heavily contrasted. An intensified staining reaction of the maize calyptra cells is recognized immediately in the survey electron micrograph. Cortical endoplasmic reticulum, circumnuclear endoplasmic reticulum cisternae, nuclear envelope and the nucleolus appear almost black. The plasma membrane is also stained. At higher magnifications this staining reaction consists of globular deposits. The electron deposits can be removed by washing in EDTA-buffer.

PIXE (<u>P</u>roton-<u>i</u>nduced <u>X</u>-ray <u>e</u>mission) analyses shows the low amount of K and Ca present in a control-fixed pollen tube of <u>Lilium longiflorum</u>. After fixation in the presence of 10 mM Ca^{2+}, the pollen tube shows a markedly enhanced signal for Ca. In the maize control, a slightly higher amount of K and Ca is detected. For K, this may be due to the embedding resin epon/araldite. After fixation in the presence of Ca^{2+}, the Ca-level is markedly enhanced in comparison to the control.

For electron spectroscopic imaging (ESI), ultrathin sections without supporting foil were used. Due to the contrast enhancement possible with this method (9), the pollen tube fixed in the presence of 10 mM Ca^{2+} clearly reveals electron dense globular structures close to the plasma membrane. The globular structures as well as the ribosomes and the inner layer of the cell wall are relatively rich in P. The electron dense globules are enriched in Ca. The mitochondria also shows some enrichment in Ca. For the maize calyptra cells fixed in the presence of Ca comparable results were obtained.

The Ca^{2+}-binding structures possibly represent all kinds of intracellular components capable of binding Ca^{2+} when Ca is provided, including specific Ca^{2+}-binding sites. It is hoped to detect the specific Ca^{2+}-binding sites with improved preparative conditions and ESI.

REFERENCES

1. Belitser, N.V., and Zaalishvili, G.V. (1983). Protoplasma 115, 222-227.

2. Belitser, N.V., Zaalishvilli, G.V., and Sytnianskaja, N.P. (1982). Protoplasma 111, 63-78.

3. Morré, D.J., and Van Der Woude, W.J. (1974). In: Macromolecules regulating growth and development, 30th Symp. Soc. Dev. Biol., pp. 81, Academic Press, New York

4. Picton, J.M., and Steer, M.W. (1983). Protoplasma 115, 11-17.

5. Reiss, H.-D., and Herth, W. (1985). J. Cell Sci., in press.

6. Reiss, H,-D., Herth, W., Schnepf, E., and Nobiling, R. (1983). Protoplasma 115, 153-159.

7. Behrens, H.M., Weisenseel, M.H., and Sievers, A. (1982). Plant Physiol. 70, 1079-1083.

8. Sievers, A., and Volkmann, D. (1972). Planta 102, 160-172.

9. Egle, W., Kurz, D., and Rilk, A. (1984). Magazine for Electron Microscopists (Zeiss Information) 3, 4-9.

Ca^{2+} REGULATION OF MYOSIN SLIDING ALONG CHARACEAE ACTIN BUNDLES

Teruo Shimmen[1], Masafumi Yano[2], and Kazuhiro Kohama[3]

[1] Department of Botany, Faculty of Science, [2] Faculty of Pharmaceutical Sciences, [3] Department of Pharmacolgy Faculty of Medicine, University of Tokyo, Hongo, Tokyo 113 Japan

Internodal cells of Characeae have bundles of actin filaments at the inner surface of the chloroplast layer which are anchored to the gel ectoplasm. It is believed that the motive force for cytoplasmic streaming is generated by the sliding of myosin in the flowing sol endoplasm along these fixed actin bundles. Since heavy meromyosin isolated from skeletal muscle can bind to Characeae actin bundles (Williamson 1974), it was thought that myosin from other organisms may be able to move along the actin bundles. In the present study, we reconstituted the sliding movement of myosin extracted from skeletal muscle, Physarum and scallop and studied Ca^{2+} control of the reconstituted sliding.

METHODS

The vacuole of the Chara internodal cell was perfused with an EGTA containing medium to remove the tonoplast. Endoplasm was completely removed by perfusing the tonoplast-free cell with EDTA-containing medium. Latex beads coated with myosin were suspended in Mg-ATP medium and introduced into the tonoplast-free cells.

RESULTS

1) Skeletal muscle myosin: The sliding velocity was 1-5 um/sec. The sliding direction depended upon the polarity of the actin bundles (Shimmen & Yano 1984). The sliding was insensitive to Ca^{2+}. When FITC-labeled native tropomyosin of rabbit skeletal muscle was introduced into tonoplast-free Chara cells, strong fluorescence appeared on the actin bundles, indicating that skeletal native tropomyosin can bind to Chara actin bundles. Latex beads coated with skeletal muscle myosin were applied to actin bundles treated with native tropomyosin. Movement did not occur in the absence of Ca^{2+} but was activated by Ca^{2+} (actin-linked regulation, Shimmen & Yano 1985).

2) Scallop myosin: Sliding did not occur in the absence of Ca^{2+} but was activated to a velocity of 1.2 um/sec by 10^{-4} M Ca^{2+} (Shimmen &

Kohama 1984). This is consistent with a previous report that in scallop, Ca^{2+} activates actomyosin ATPase via myosin (stimulative myosin-linked regulation, Kendrick-Jones et al. 1976).

3) <u>Physarum</u> myosin: The sliding velocity was 1.6 um/sec in the absence of Ca^{2+} and was inhibited to 0.6 um/sec by 10^{-4} M Ca^{2+} (Shimmen & Kohama 1984). This result supports the report of Kohama and Kendrick-Jones (1982) that in <u>Physarum</u>, Ca^{2+} inhibits actomyosin ATPase via myosin (inhibitory myosin-linked regulation).

Since the Ca^{2+} sensitivity of reconstituted sliding was completely dependent on myosin or native tropomyosin, it is concluded that actin bundles of <u>Chara</u> have no Ca^{2+} sensitivity. In Characeae, cytoplasmic streaming is inhibited by Ca^{2+} (Tominaga et al. 1983). Absence of Ca^{2+} sensitivity in <u>Chara</u> actin bundles suggests that Ca^{2+} regulation of cytoplasmic streaming is myosin-linked. Success in reconstituting inhibitory myosin-linked regulation using <u>Physarum</u> myosin also supports this hypothesis.

REFERENCES

1. Kendrick-Jones, J., Szentkiralyi, E.M., Szent-Gyorgyi, A.G. 1976. Regulatory light chains in myosins. J. Mol. Biol. <u>104</u>, 747.
2. Kohama, K., Kendrick-Jones, J. 1982. Negative Ca^{2+}-sensitivity of actin-activated Mg-ATPase of myosin from <u>Physarum polycepharum</u>. J. Muscle Research & Cell Motility 3, 491.
3. Shimmen, T., Kohama, K. 1984. Ca^{2+}-sensitive sliding of latex beads coated with <u>Physarum</u> or scallop myosin along actin bundles in Characeae cell. Third International Congress on Cell Biology (Abstract) p.504.
4. Shimmen, T., Yano, M. 1984. Active sliding movement of latex beads coated with skeletal muscle myosin or <u>Chara</u> actin bundles. Protoplasma <u>121</u>, 132.
5. Shimmen, T., Yano, M. 1985. Ca^{2+} regulation of myosin sliding along <u>Chara</u> actin bundles mediated by native tropomyosin. Proc. Japan Acad. Ser. B. <u>61</u>, 86.
6. Tominaga, Y., Shimmen, T., Tazawa, M. 1983. Control of cytoplasmic streaming by extracellular Ca^{2+} in permeabilized <u>Nitella</u> cells. Protoplasma <u>116</u>, 75.
7. Williamson, R.E. 1974. Actin in the alga <u>Chara corallina</u>. Nature <u>248</u>, 801.

MECHANISM OF Ca^{2+}-CONTROL OF CYTOPLASMIC STREAMING IN CHARACEAE

Yoshito Tominaga[1] and Masashi Tazawa[2]

[1] St. Agnes' Junior College for Women, Kamikyo-ku, Kyoto 602, Japan, [2] Department of Botany, Faculty of Science University of Tokyo, Hongo, Bunkyo-ku, Tokyo 113, Japan

The motile system of Characean cells is believed to be an acto-myosin-ATP system[1]. Streaming stops the moment an action potential is generated at the plasmalemma. It has now been established that the cessation of streaming is caused by a transient increase in cytoplasmic Ca^{2+} activity. The following facts support this hypothesis.

(1) During excitation Ca^{2+} influx increases greatly[2]. (2) Injection of Ca^{2+} into endoplasmic droplets transiently stops rotation of chloroplasts[3] which is also caused by an actomyosin system. (3) Injection of Ca^{2+} into the flowing cytoplasm of intact cells causes transient inhibition of streaming[4]. (4) When the Ca^{2+}-sensitive photoprotein, aequorin, that is introduced into the cell, it emits light upon excitation[5,6].

However, there is a fact which is not consistent with the Ca^{2+}-hypothesis. The streaming of cells which are made tonoplast-free by vacuolar perfusion with a medium containing the Ca^{2+}-chelating agent EGTA, is almost insensitive to intracellular Ca^{2+} up to 10^{-4} M[7]. But in cells whose plasmalemma is permeabilized by plasmolyzing in a slightly hypertonic medium containing EGTA, streaming is quite sensitive to Ca^{2+} and is stopped by 1 μM Ca^{2+}[8]. In tonoplast-free cells the streaming endoplasm is dispersed in the cell due to the absence of the tonoplast, while in plasmalemma-permeabilized cells it is confined between cell wall and tonoplast. The conclusion we draw from these facts is that a Ca^{2+}-sensitizing component mediating the cessation of streaming may exist in the streaming endoplasm.

The next problem is whether or not calmodulin (CaM) is involved in the Ca^{2+} mediated cessation of streaming. CaM was proved to be present in Chara at 400 ng/ml homogenate by the ^{121}I-radioimmunoassay method[9]. Trifluoperazine (TFP) at 25 μM, fluphenazine (FPH) at 1 μM and W-7 at 10 μM did not inhibit either the streaming or the Ca^{2+}-induced cessation of streaming in plasmalemma-permeabilized Nitella cells, but inhibited the recovery of streaming[9]. In tonoplast-free cells these CaM-binding drugs did not interfere with the streaming in the presence or absence of 10 μM Ca^{2+}. Thus, in Characean cells there are two kinds of Ca^{2+}-sensitizing components in the streaming sol endoplasm, one responsible for the cessation of streaming and the other, probably CaM, for the recovery of streaming.

To approach the mechanism of Ca^{2+} action we assumed that phosphorylation of some protein may be involved in the cessation or recovery of streaming. The presence of ATP-γ-S with ATP did not inhibit the Ca^{2+}-mediated cessation of streaming in plasmalemma-permeabilized cells of Nitella, but the recovery of streaming was sometimes strongly inhibited.

Although the molecular mechanism is completely unclear, we tentatively assume the following sequence of events. First, some component of the streaming system is phosphorylated by the action of Ca^{2+} and an unknown Ca^{2+}-sensitizing component, and the streaming stops. During application of Ca^{2+} Ca^{2+}-CaM is formed which on removal of Ca^{2+} causes dephosphorylation of some component. In view of the fact that myosin from Physarum plasmodium can be phosphorylated[10, 11], it must be examined whether the Characean myosin that is assumed to be present in the endoplasm is phosphorylated or not.

REFERENCES

1. N. Kamiya, Physical and chemical basis of cytoplasmic streaming, Ann. Rev. Pl. Physiol. 32: 205 (1981)
2. T. Hayama, T. Shimmen and M. Tazawa, Participation of Ca^{2+} in cessation of cytoplasmic streaming induced by membrane excitation in Characeae internodal cells, Protoplasma 99: 305 (1979)
3. T. Hayama and M. Tazawa, Ca^{2+} reversibly inhibits active rotation of chloroplasts in isolated cytoplasmic droplets of Chara, Protoplasma 102: 1 (1980)
4. M. Kikuyama and M. Tazawa, Transient increase of intracellular Ca^{2+} during excitation of tonoplast-free Chara cells, Protoplasma 117: 62 (1983)
5. R.E. Williamson and C.C. Ashley, Free Ca^{2+} and cytoplasmic streaming in the alga Chara, Nature 296: 647 (1982)
6. M. Kikuyama and M. Tazawa, Transient increase of intracellular Ca^{2+} during excitation of tonoplast-free Chara cells, Protoplasma 117: 62 (1983)
7. Y. Tominaga and M. Tazawa, Reversible inhibition of cytoplasmic streaming by intracellular Ca^{2+} in tonoplast-free cells of Chara australis, Protoplasma 109: 103 (1981)
8. Y. Tominaga, T. Shimmen and M. Tazawa, Control of cytoplasmic streaming by extracellular Ca^{2+} in permeabilized Nitella cells, Protoplasma 116: 75 (1983)
9. Y. Tominaga, T. Shimmen, S. Muto and M. Tazawa, Calmodulin and Ca^{2+}-controlled cytoplasmic streaming in Characean cells (submitted to Cell Structure Function)
10. S. Ogihara, M. Ikebe, K. Takahashi and Y. Tonomura, Requirement phosphorylation of Physarum myosin heavy chain for thick filament actin activation of Mg^{2+}-ATPase activity, and Ca^{2+}-inhibitory super-precipitation, J. Biochem. 93: 205 (1983)
11. K. Kohama, R. Craig, T. Kohama and J. Kendrich-Jones, Characterization of Ca^{2+}-sensitive Physarum myosin, Europ. J. Cell Biol., suppl. 1: 25 (1983)

THE EFFECT OF SEGMENT ORIENTATION AND CELL GROWTH ON THE ACROPETAL FLUX OF CALCIUM

R.K. dela Fuente and C.C. de Guzman

Department of Biological Sciences, Kent State University
Kent, Ohio 44242, USA

Currently, the basipetal and lateral. polarity of auxin transport and the ensuing cellular growth that occurs is the only system that can explain the relatively rapid and effective means by which plants are able to respond to such stimuli as unilateral light or gravity. Very little information is known about how the polarity of IAA transport is accomplished. Dela Fuente and Leopold (1973) and dela Fuente (1984) showed that the basipetal transport of IAA is diminished in plant tissues deprived of Ca^{2+}; a short incubation of such tissues in Ca^{2+} solution restored normal auxin transport.

It was hypothesized that the basipetal transport of IAA, which can be considered as a secretion of IAA at the basal end of individual cells, is probably a manifestation of the mechanism known in animal cells. The secretion of many kinds of substances in animal cells from neurotransmitters to proteins and ions is thought to occur by the mechanism of stimulus-secretion coupling (Rubin (1982), Rasmussen (1981), Campbell (1984). A stimulus caused the passive influx of Ca^{2+} which then causes the perturbation that leads to the secretion process.

If a similar mechanism is responsible for the basal secretion of IAA, it would look more likely that Ca^{2+} would enter the basal end, and if such were the case, then there could be an acropetal polarity of Ca^{2+} movement in plant axes. Indeed, de Guzman and dela Fuente (1984) found a higher rate of Ca^{2+} efflux at the apical end, compared to the basal end of 20 mm sunflower hypocotyl segments. However, the method we used to gather these data was such that basal Ca^{2+} efflux was performed with segments oriented in the normal position while in the case of apical Ca^{2+} efflux, the apical end of the segment was placed down in contact with the aqueous receiver solution.

To overcome the criticism that our earlier report could have been an artefact of segment orientation, we now present a new technique where apical and basal Ca^{2+} efflux can be determined in segments oriented in the normal position. The data gathered from this new technique showed the normal or reverse vertical orientation of the hypocotyl segments do not have an effect on the observed acropetal polarity of Ca^{2+} movement.

It was also found that the IAA-induced cell elongation correlates positively with the IAA-induced Ca^{2+} efflux. IAA in the presence of its transport inhibitor TIBA does not promote growth or Ca^{2+} efflux. The non-auxin β-NAA does not promote either growth of the hypocotyl segments or its calcium efflux, but the auxin α-NAA increased both parameters. Thus the question arose, is the IAA-induced Ca^{2+} efflux the result of

IAA-induced relaxation of cell wall pressure? We found these two processes can be separated from each other. By adding increasing amounts of mannitol to the aqueous receiver solution, growth of the segments were inhibited. This osmotic inhibition of growth however, did not result in an inhibition of the IAA-induced Ca^{2+} efflux.

In conclusion, the observed acropetal flux of Ca^{2+} in plant axes, which is promoted by exogenous IAA was found to be independent of the vertical orientation of the segments as well as the reactions accompanying cell wall relaxation.

REFERENCES

Campbell, A. K. 1983. Intracellular calcium – its universal role as regulator. John Wiley and Sons Ltd., New York.

De Guzman, C. C., and R. K. dela Fuente. 1984. Polar calcium flux in sunflower hypocotyl segments. I. The effect of auxin, Plant Physiol. 76: 348-353.

Dela Fuente, R. K. 1984. Role of calcium in the polar secretion of indoleacetic acid. Plant Physiol. 76: 342-346.

Dela Fuente, R. K. and A. C. Leopold. 1973. A role for calcium in auxin transport. Plant Physiol. 51: 845-847.

Rasmussen, H. 1981. Calcium and cAMP as synarchic messengers. Wiley Interscience. New York.

Rubin, R. P. 1982. Calcium and cellular secretion. Plenum Press. New York.

THE ROLE OF CALCIUM IN THE PHOTOTACTIC RESPONSE

OF Chlamydomonas reinhardtii

Nicole Morel-Laurens

Biology Department, Tufts University

Medford, Massachusetts 02155 USA

INTRODUCTION

Chlamydomonas reinhardtii, a unicellular green flagellate, displays
two kinds of photomovements in response to a blue-green stimulus light:
an oriented swimming toward or away from the light source (phototaxis), and
a non-oriented stop response (photophobic response) at the onset of high-
intensity illumination.

While there is very good evidence that the photophobic response is
the result of an influx of calcium (Ca) through the flagellar membrane
(Schmidt and Eckert, 1976; Hyams and Borisy, 1978), such evidence is lacking
for the phototactic response. There are, however, indirect indications that
Ca is involved in phototaxis. Stavis (1974) and Stavis and Hirschberg
(1973) have shown that photaccumulation, which may result from a photophobic response and/or phototaxis, requires Ca. Nultsch (1979) found that
the net phototactic response of a population of cells decreases to zero
(Nultsch, 1979) when Ca is excluded from the growth medium. However, Ca-
deprived medium is not completely free of Ca; it still contains Ca impurities from the other salts making up the medium which can add enough Ca to
the medium to make the final concentration as high as 2×10^{-6}M. Ca^{2+} in the
medium used by Nultsch can be calculated to be @ 7.3×10^{-8}M at pH 5.85 if
the Ca impurities are @ 2×10^{-6}M. Since at this extracellular Ca^{2+} concentration the electrochemical gradient for Ca must still be large (assuming
a negative transmembrane potential of about -50mv similar to that of
Haematococcus pluvialis, a close relative of Chlamydomonas; Litvin et al.,
1978), one would expect that Ca^{2+} would still move inside the cell and thus
elicit a response.

The purpose of this work was to (1) establish unequivocally if phototactic behavior persists at very low Ca^{2+}, i.e. in Ca-EGTA buffered medium
containing 7.4×10^{-10}M Ca^{2+}, (2) if the sign of phototactic response is positive or negative as Ca^{2+} is increased, and (3) if Ca is specifically responsible for phototaxis.

MATERIAL AND METHODS

Phototactic behavior was studied in individual cells rather than cell
populations. Synchronous cultures of Chlamydomonas reinhardtii CC 125 mt+,
also called 137c, were grown directly in modified Minimum Medium (Levine,
1971) without added Ca and with the trace metal mixture of Aquil (Morel,
et al., 1979), an algal growth medium specially devised for trace metal

control. The same medium was used for the behavioral assays after addition of EGTA and concentration of $CaCl_2$ computed to yield the desired Ca^{2+}. $CaCl_2$ was replaced by $MgCl_2$ to test whether Ca^{2+} is required for phototactic response. In all experiments, phototactic behavior was assayed when the cultures were in the exponential phase of growth.

The stimulus source was an argon ion laser (Spectra Physics Model 162), with a continuous output at 488nm. In these experiments two stimulus fluence rates were assayed: 1 and 10 mW/cm^2.

The behavior of individual cells was recorded with a videomicroscope system modified from that used by Morel-Laurens and Feinleib (1983). Swimming paths were tracked and analyzed using a system based on an IBM-Personal Computer as developed in my laboratory. Tracking began 20s after the onset of stimulation. The computer generated swimming paths of the individual cells were then analyzed using unit-vector analysis developed for circular distributions (Batschelet, 1966, 1972; Morel-Laurens and Feinleib, 1983). From each cell's swimming path a mean vector was determined. For each sample of cells a unit mean vector was then obtained by averaging all the mean vectors. The angle m of this mean vector indicates the "preferred" direction taken by the cells. It has a length r which varies between one and zero. The V test (Tabschelet, 1972) was used to estimate the statistical significance of the directedness of the cell sample.

RESULTS AND DISCUSSION

Even at the lowest Ca^{2+} concentration, 7.4×10^{-10}M, a statistically significant (P <0.05), phototactic response was observed. Furthermore, the sign of the response was a function of the stimulus fluence rate: at 1mw/cm^2 the response was negative (m=211°. r=0.16. n=94), at 10 mW/cm^2 it was positive (m=42°. r=0.15. n=98).

As the stimulus fluence rate was kept constant at 10 mW/cm^2 the sign of the response was a function of the extracellular Ca^{2+}. At 4.3×10^{-7}M Ca^{2+} the phototactic response was positive (m=18°. r=0.40. n=48. P<<0.0005). At a higher concentration of 9.2×10^{-4}M, Ca^{2+} elicited a strong negative response (m=177°. r=0.96. n=44. P<<0.0005).

Ca was specifically required for this response. Addition of 10^{-3}M $MgCl_2$, yielding a concentration of Mg^{2+} of 1.1×10^{-3}M, did not produce a strong phototactic response. There was however, a very weak but significant negative response (m=194°. r=0.17. n=57. P<0.05) which could be due to Mg^{2+} or to the small concentration of Ca^{2+} released from the EGTA-Ca complex.

A study of Ca^{2+} fluxes will be necessary to assess whether the behaviors described above are the results of increase or decrease in intracellular Ca^{2+}.

REFERENCES

Batschelet, E. 1965. An A.I.B.S. Monograph. Washington, D.C. Batschelet, E. 1972. In "Symposium on Animal Orientation and Navigation". NASA (Sci. and Tech. Info. Office, Washington, D.C.) Doc:NAS 1.21 252:62-91. Hyams, J.S. and G.G. Borisy. 1978. J. Cell Science. 33, 235-253. Levine, R.P. 1971. Chapter 10 in "Methods in Enzymology". Litvin, F.F., O.A. Sinescheckov, and V.A. Sineshchekov. 1978. Nature. 271, 476-478. Morel, F.M.M., J.G. Rueter, D.M. Anderson and R.R. L. Guillard. 1979. J. Phycol. 15, 135-141. Morel-Laurens, N.M.L. and M.E. Feinleib. 1983. Photochem. Photobiol. 37, 189-194. Nultsch, W. 1979. Arch. Microbiol. 123, 93-99. Nultsch, W. 1983. In "The Biology of Photoreception" pp. 521-541. Cambridge Univ. Press. Schmidt, J.A. and R. Eckert. 1976. Nature 262, 713-715. Stavis, R. and R. Hirschberg. 1973. J. Cell Biol. 59, 367-377.

CALMODULIN ANTAGONISTS INHIBIT ADVENTITIOUS ROOT GROWTH OF TRADESCANTIA

Shoshi Muto and Takayasu Hirosawa*

Institute of Applied Microbiology, University of Tokyo
Tokyo 113, Japan
*Present address: Plant Research Center, Kirin Brewery Co. Ltd
Kituregawa-cho, Tochigi-ken 329-14

INTRODUCTION

Calmodulin has been localized in the mitotic apparatus of cultured animal cells and is thought to sensitize microtubules in vitro to the depolymerization action of Ca^{2+} (Means and Tash, 1982). Recently, we investigated the distribution of calmodulin relative to that of tubulin at all stages of the cell cycle in onion and pea root meristematic cells, using a double immunofluorescence labelling with anti-spinach calmodulin and anti-tubulin antibodies (Wick et al., 1985). Calmodulin was associated with the mitotic spindle and with the phragmoplast at cytokinesis, but patterns of calmodulin that coincided with those of interphase microtubule arrays were not demonstrated. These results suggest that calmodulin plays an important role in mitosis and cytokinesis in plant cells. This led us to test the effect of calmodulin antagonists on the growth of adventitious root of Tradescantia albiflora, a spiderwort plant.

MATERIALS AND METHODS

Tradescantia albiflora was planted in a pot containing soil. The plants were grown hanging from the pot so that the adventitious roots did not grow. Stem cuttings with two leaves were prepared from plants with more than 20 nodes. The lower leaves were detached from the cuttings and their stems were immersed in distilled water in 100-ml Erlenmyer flasks which were wrapped with aluminium foil. The plants were incubated at 23°C, illuminating with daylight fluorescent lamps at 6000 lux the tops of the plants.

Trifluoperazine (a generous gift from Yoshitomi Pharmaceutical Industries Ltd) and compound 48/80 (Sigma) were added as aqueous solutions. Chlorpromazine (Sigma) and calmidazolium (Boehringer) were dissolved in ethanol and dimethylsulfoxide, respectively.

RESULTS AND DISCUSSION

When cuttings of Tradescantia stem were incubated in distilled water for one week, the buds located at the node grew into adventitious roots, 2 to 5 cm long. The number of roots was 2 to 5 per node. The root growth was inhibited by trifluoperazine in a concentration dependent manner. At 10 ppm (ca. 1.6 µM), it was completely inhibited. Since the trifluoperazine used was a dimaleic acid salt, the effect of the same molar concentration of maleic acid was tested. The acid had essentially no effect, thus the inhibition of root growth appeared to be the effect of trifluoperazine itself. When the drug was washed out with distilled water, the adventitious roots started to grow. This indicates that the inhibition is reversible. Prolonged incubation (2 weeks) with 5 µM trifluoperazine resulted in a delayed growth of roots. The decrease of the drug concentration by spontaneous degradation and/or by metabolism in the plant might allow the delayed growth. Similar results were obtained with another phenothiazine compound chlorpromazine. When the control plants were treated with the drug after one week's incubation, the growth of root was inhibited thereafter, indicating that the drug affected not only the buds but also the growing roots. These results suggest that calmodulin is involved in the growth of adventitious roots. To test this possibility, the effect of different types of calmodulin antagonists, compound 48/80 and calmidazolium were examined. Both drugs inhibited root growth in a concentration dependent manner. Root growth was completely inhibited by either 10 µg/ml (10 ppm) of compound 48/80 or 10 µM (ca. 7 ppm) calmidazolium. The solvent used to dissolve calmidazolium had essentially no effect. The inhibition by both drugs was reversible when treated with relatively low concentrations, but the high concentration of calmidazolium irreversibly inhibited root growth. The concentrations of drugs used here were comparable to the concentrations which inhibited in vitro calmodulin dependent pea NAD kinase. The K_i value for trifluoperazine was reported to be 2.8 µM (Muto, 1983). The enzyme was completely inhibited by 1 µM calmidazolium.

Similar growth inhibition of adventitious roots by the calmodulin antagonists used above was observed in another spiderwort plant Commelina communis.

Western blots (Towbin et al., 1979) indicated the presence of calmodulin in Tradescantia adventitious roots. The roots were extracted with the sample buffer for SDS-polyacrylamide gel electrophoresis (Laemli, 1970), and heated for 2 minutes in a boiling water bath. To the supernatant after centrifugation at 15,000 x g for 20 minutes, either $CaCl_2$ or EGTA were added to make a final concentration of 1 mM. After electrophoresis in 15% gel and blot onto a nitrocellulose sheet, the single band reacting with anti-calmodulin antibody (Muto and Miyachi, 1984) was observed irrespective of the presence of either Ca^{2+} or EGTA. The mobility of the protein band in the presence of EGTA was less than its mobility in the presence of Ca^{2+}. This indicates that the protein band is calmodulin.

We have shown that the cellular concentration of calmodulin in pea seedlings is high in young dividing cells and low in mature cells (Laemli, 1970), and that calmodulin is associated with the mitotic spindle and with the phragmoplast in mersistematic cells of pea and onion roots (Wick et al., 1985). These facts suggest that the inhibition of adventitious root growth by the calmodulin antagonists results from the inhibition of cell division. The drugs may bind to calmodulin and cause the disorder of microtubule depolymerization which is essential for continuing mitosis and cytokinesis.

REFERENCES

Laemli, U.K., 1970, Nature, 227:680-685.
Means, A.R., Tash, J.S. and Chafouleas, G.J., 1982, Physiol. Rev., 62:1-39.
Muto, S., 1983, Z. Pflanzenphysiol., 109:385-393.
Muto, S. and Miyachi, S., 1984, Z. Pflanzenphysiol., 114:421-431.
Towbin, H., Staehelin, T. and Gordon, J., 1979, Proc. Natl Acad. Sci. USA, 76:4350-4354.
Wick, S.M., Muto, S. and Duniec, J., 1985, Protoplasma, 126:198-206.

DIRECT ESTIMATION OF PLASMODESMATAL CONDUCTIVITY

B.R. Terry and A.W. Robards

Department of Biology
University of York
York, YO1 5DD. U.K.

Plasmodesmata, being direct cytoplasmic links between adjacent cells, are potential channels for the transport of molecules from one cell to the next. The conductivity of these channels to molecules of different types is clearly under some form of control. The aim of this poster is to describe one method by which the molecular conductivity of plasmodesmata can be estimated directly.

The method uses fluorescent dyes conjugated to amino acids and oligo-peptides as molecular probes of known molecular weight and properties. In this work such dyes have been injected into the nectary trichomes of Abutilon flowers by iontophoresis. The nectary trichomes are a good model system for the study of plasmodesmata since each trichome is made up of a column of cells through which, when functional, there is a considerable symplastic flux of nectar constituents. The plasmodesmata of secreting trichomes are known to be conductive. Furthermore the cells of the trichomes are easily accessible for microinjection and observation.

Results have shown that the plasmodesmata in Abutilon trichomes have characteristics common to plasmodesmata in other systems; for instance they have molecular exclusion limits close to 750 daltons for dyes conjugated to oligopeptides with neutral or polar aliphatic side groups. The plasmodesmata are very much less conductive to dyes coupled to aromatic amino acids. A family of polyglycine oligopeptides has been adopted as a standard for estimating the conductivity of plasmodesmata in this work.

Plasmodesmata in the proximal wall of the basal cell to each trichome do not allow even the smallest of the molecular probes to pass out of the trichomes. These same plasmodesmata are the only possible route, other than across the plasmalemma, for the nectar constituents to enter the trichomes. The proximal wall of the basal cell would seem to be the most likely point for the control of entry of nectar constituents.

The nectary trichomes of Abutilon have been shown to be an amenable experimental system. Work is underway to investigate the effects of various treatments on the conductivity of the plasmodesmata. Of primary interest are the effects of divalent cations, calmodulin inhibitors, intracellular calcium releasers and chelating agents – all of which can be injected into specific cells. The molecular conductivity of the plasmodesmata under these conditions will be correlated with simultaneous measurement of their electrical conductivity. The results from this work should expand our knowledge of the properties of plasmodesmata and how they contribute to the control of intercellular communication in plants.

CURRENT-VOLTAGE ANALYSIS AS A MEANS TO IN VIVO "SEPARATION" OF PRIMARY ELECTROGENIC AND COUPLED SECONDARY TRANSPORT

Michael R. Blatt and Clifford L. Slayman

Botany School, University of Cambridge, Cambridge England, and Physiology Dept., Yale Medical School, New Haven, CT, USA

Two distinct strategies of energy-dependent, extramitochondrial calcium transport have been identified to date in plant and fungal cells: direct coupling to ATP hydrolysis (Ca^{2+}-ATPase; cf. Dieter and Marme, 1980) and secondary coupling to the electrochemical driving force for protons (Ca^{2+}/H^+ antiport; cf. Stroobant and Scarborough, 1979). Both modes of transport have been identified in in vitro microsomal or plasma membrane fractions. It remains, however, to identify the transport systems as they operate in vivo (presumably to export Ca^{2+}), and to characterize the physiological conditions in which each functions.

Characterizing a single transporter in vivo is generally complicated by the close interaction(s) of the ensemble of transport processes in biological membranes. However, the problem is greatly simplified if the transporter of interest carries charge. To illustrate how electrophysiological methods -- in particular, current-voltage (I-V) measurements -- can be used to "isolate" and characterize coupled secondary transport independent of primary electrogenic pumping we draw on the example of proton-coupled K^+ transport in Neurospora.

Potassium transport in Neurospora was shown previously to exhibit characteristics similar to those reported in yeast, Chlorella and higher plants, for which K^+ pumps or K^+/H^+-ATPases have been proposed (Rodriguez-Navarro, Blatt, Slayman, manuscript in preparation). For Neurospora potassium-proton cotransport was postulated to account for the observed charge to K^+ stoichiometry (2:1) and for the movement of K^+ against its electrochemical gradient (data obtained in part through I-V, and simultaneous H^+- and K^+-selective electrode measurements). The current series of experiments demonstrate that the apparent coupling of K^+ transport and ATP hydrolysis arises form the particular I-V profile of the porter and not from any ATP-dependent step in the transport cycle.

The I-V relations of spherical Neurospora cells (Blatt and Slayman, 1983) were examined by voltage clamping between -400 and 100 mV as described previously (Gradmann, et al., 1978). Measurements were carried out in the presence and absence of K^+ (0-200 μM) before, during and after treatments of the cells with cyanide or vanadate. The K^+-transport I-V characteristic was dissected from the whole membrane I-V relation by subtraction of the controls across the voltage spectrum. Pump (H^+ pump) and leak components of the controls (0 K^+) were identified by joint fittings, to a 3-state pump model and parallel non-linear leak, of I-V data gathered during inhibitor treatments. Cyanide was added to the bathing media. Vanadate was injected directly via intracellular iontophoresis with the advantage that inhibition was rapid and not uptake limited.

Potassium currents obtained at any one K^+ concentration showed a marked voltage dependence between -300 and -100 mV. Potassium currents became saturated with respect to voltage at potentials negative of -300 mV (close to the mean stable potential recorded in 0 K^+), and were decreased by 80-90% at -100 mV. At the prevailing membrane potential in K^+, the currents were balanced quantitatively by the H^+ pump, thus accounting for the apparent K^+/H^+ exchange observed in flux experiments.

Cyanide and vanadate inhibited the H^+ pump and depolarized the membrane potential by 126-249 mV over the voltage range between c. -300 and -40 mV without any immediate effect on the K^+ I-V characteristic. In the case of vanadate, in three separate experiments the membrane potential depolarized by 184-249 mV and the pump short circuit current was reduced 4- to 15-fold within 20-60 s and c. 100-fold within 4-5 min. K^+ currents at any one clamp potential decreased measurably only after 4-6 min. Similar results were obtained with cyanide, although the drop in pump activity was less dramatic (7- to 8-fold).

Using the H^+ pump as an assay for the efficacy of the inhibitor treatments, it is clear that ATP withdrawal by means either of metabolic blockade (CN^-) or of ATPase inhibition ($VO_4^=$) does not retard current passage through the porter (eg. when the cells are clamped at -300 mV). Hence, inhibition of K^+ transport apparent at the prevailing membrane potential reflects the intrinsic kinetic dependence of the porter on voltage only, and not a direct coupling to ATP hydrolysis.

References

Blatt, M., Slayman, C.L. (1983) KCl leakage from microelectrodes and its impact on the membrane parameters of a non-excitable cell. J. Membr. Biol. 72, 223.

Dieter, P., Marme, D. (1980) Ca^{2+} transport in mitochondrial and microsomal fractions from higher plants. Planta 150, 1.

Gradmann, D., Hansen, U.-P., Long, W.S., Slayman, C.L., Warnke, J. (1978) Current-voltage relationships for the plasma membrane and its principal electrogenic pump in Neurospora crassa I: Steady-state conditions. J. Membr. Biol. 59, 333.

Stroobant, P., Scarborough, G. (1979) Active transport of Ca^{2+} in Neurospora plasma membrane vesicles. Proc. Nat. Acad. Sci. (US) 76, 3102.

THE ROLE OF Ca^{++} IN CHARA K^+ STATE

Mary Jane Beilby

Botany School, Cambridge University, Downing Street

Cambridge, CB2 3EA, England

Introduction

In the last decade it has been established that Chara plasmalemma presents us with several electrophysiological states. For pH_o less than ~10.0, $[K^+]_o$ less than 1.0 mM and in presence of Ca^{++}, the transmembrane p.d. is large and the membrane conductance is dominated by the proton pump[1,2]. Above pH_o 10.0 Chara enters a proton-permeable state[3]. Upon a suitable stimulus, Chara displays action potentials analogous to those in the nerve. The transient currents are carried by Cl^- and probably Ca^{++} [4]. If the $[K^+]_o$ rises above ~2.0 mM, the transmembrane p.d. becomes small and K^+-sensitive. When the pump is inhibited and the K^+ channels closed, linear current-voltage (I/V) characteristics suggest a passive leak. Ca^{++} plays a part in most of these states, but in this communication we shall concentrate on its effect on the K^+ channels.

Materials and Methods

Very young cells of Chara corallina were space-clamped by the insertion of the Pt/Ir wire along the axis of the cell. The p.d.-measuring electrode was inserted into cytoplasm, so that only the plasmalemma membrane was voltage-clamped. To ascertain the position of this microelectrode, the unique characteristics of short circuit current across plasmalemma were exploited[2]. The voltage clamp was controlled by the MINC 11 computer and the data-logging provided a time resolution of 2 msec. To obtain an I/V curve, the computer generated bipolar staircase voltage clamp command. The cell p.d. was clamped at resting level with fast excursions in depolarizing and hyperpolarizing direction. This approach minimized disturbance of the ionic profiles across the cell wall/membrane interface. Detailed description of the apparatus and methods has been given previously[2,5,7]. In this study it was important to select cells without calcifications, as Ca^{++} from these would leach out and contaminate the low Ca^{++} solutions.

Results and Discussion

The I/V characteristics of the K^+ state have been described previously[6,7]. The regions of negative conductance (see Fig. 1) are due to time and voltage dependence of the K^+ channels. As $[K^+]_o$ changes, the resting p.d. follows E_K and the populations of open K^+ channels vary.

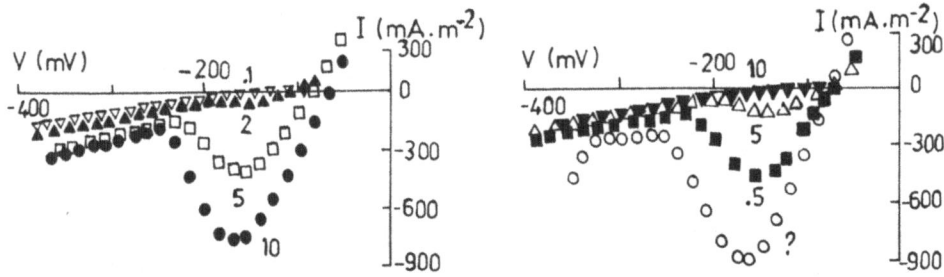

Fig.1 The I/V charasteristics of Chara K^+ state as function $[K^+]_o$(given in mM next to each curve). $[Ca^{++}]_o$ was 5.0 mM.

Fig.2 The I/V characteristics of Chara K^+ state as function of function of $[Ca^{++}]_o$(given in mM to each curve). The question mark indicates no Ca^{++} APW, see text. $[K^+]_o$ was 0.5 mM.

At 0.1 K^+ and at levels more negative than threshold p.d.(between -200 and -250 mV), most of the K^+ channels close leaving an unspecific leak conductance. Thus the I/V curves can be resolved into two essentially linear profiles with slope conductances of a leak and the K^+ channels.

Keifer and Lucas[8] suggested, that Ca^{++} closes the K^+ channels and the I/V data, indeed, supports this proposal. Fig. 2 shows that the general time and voltage dependence of the K^+ state is unchanged by increase of $[Ca^{++}]_o$ - it is very similar to decrease in $[K^+]_o$. It is difficult to reduce $[Ca^{++}]_o$ to 0 becouse of the strong Ca^{++} binding to the cell wall. Attempts to remove this by high concentrations of NaCl yield even larger conductances than in Fig. 2, which are often detrimental to the cell. At 10.0 mM Ca^{++} the channels are totally closed and the I/V profile coincides with that at 0.1 mM K^+ (Fig. 1) giving the leak or null state. Thus it seems that the presence of Ca^{++} antagonizes the K^+ state by closing the K^+ channels and the population of open channels depends on the ratio of $[K^+]_o$ and $[Ca^{++}]_o$.

References

1. D. W. Keifer and R. M. Spanswick, Activity of the electrogenic pump in Chara corallina as inferred from measurements of the membrane potential, conductance and potassium permeability. Plant Physiol. 62:653 (1978).
2. M. J. Beilby, Current-voltage characteristics of the proton pump at Chara plasmalemma. J. Membrane Biol. 81:113 (1984).
3. M. A. Bisson and N. A. Walker, The Chara plasmalemma at high pH. Electrical measurements show rapid selective passive uniport of H$^+$ or OH$^-$. J. Membrane Biol. 56:1 (1980).
4. M. J. Beilby and H. G. L. Coster, The action potential in Chara corallina III. The Hodgkin-Huxley parameters for the plasmalemma. Aust. J. Plant Physiol. 6:337 (1979).
5. M. J. Beilby and B. N. Beilby, Potential dependence of the admittance of Chara plasmalemma. J. Membrane Biol. 74:229 (1983).
6. A. I. Sokolik and V. M. Yurin, Transport properties of potassium channels of the plasmalemma in Nitella cells at rest. Soviet Plant Physiology 28:206 (1981).
7. M. J. Beilby, Potassium channels at Chara plasmalemma. J. Exp. Bot. 36:228 (1985).
8. D. W. Keifer and W. J. Lucas, Potassium channels in Chara corallina. Control and interaction with the electrogenic H$^+$ pump. Plant Physiol. 69;781 (1982).

THE ROLE OF Ca^{++} IN THE EXCITATION OF THE SINGLE MEMBRANE SAMPLES OF

CHARA

Mary Jane Beilby

Botany School, Cambridge University, Downing Street

Cambridge CB2 3EA, England

Introduction

The Hodgkin-Huxley equations[1] of the nerve action potential were fitted to Chara excitation[2]. Modifications were necessary to accommodate two activation-inactivation transients[3]. It was clear from radioactive tracer experiments[4] that one of these currents is carried by Cl^-. The other current appears as a second negative peak near the excitation threshold, diminishes at ∼-50 mV and reverses direction at more positive potentials, becoming a prompt positive spike. This transient responds to changes in $[Ca^{++}]_o$ and was thought to be carried by Ca^{++} ions[2]. Main problem with this hypothesis is the E_{Ca} given by the reversal potential, which makes $[Ca^{++}]_c$ ∼20 mM. In light of recent measurements with aequorin[4], such concentration seems clearly impossible.

The method of single membrane samples (SMS)[6] is hoped to provide the insight into the current "X".

Materials and Methods

Long (5-10 cm) internodal cells of Chara corallina were used for preparation of the SMS. Young cells without calcifications were selected and centrifuged slowly for ∼30 min. After the cells had been allowed to loose turgor in the air, the cytoplasm-rich end was tied off with a silk thread, making SMS 2 - 4 mm long. The SMS were retained for several days in APW (artificial pond water) to recover and then space-clamped (inserting the logitudinal electrode through the untied end) using same methods and apparatus as for intact leaf cells[7].

Results and Discussion

The SMS showed good streaming, which did not stop as the p.d.-measuring microelectrode was inserted. The resting p.d. rose immediately upon insertion, as if the cytoplasm of an intact cell was impaled. However, the resting p.d.s were less negative than in intact leaf cells (-170 mV as compared to -230 mV). Similarly to intact leaf cells, the voltage-current characteristics showed pump state, proton-permeable state or K^+ state in appropriate conditions. Both the steady state and excitation conductances were rather low compared to intact cells, despite the fact that the node was not tied off[6]. The excitation clamp currents showed the sharp spike

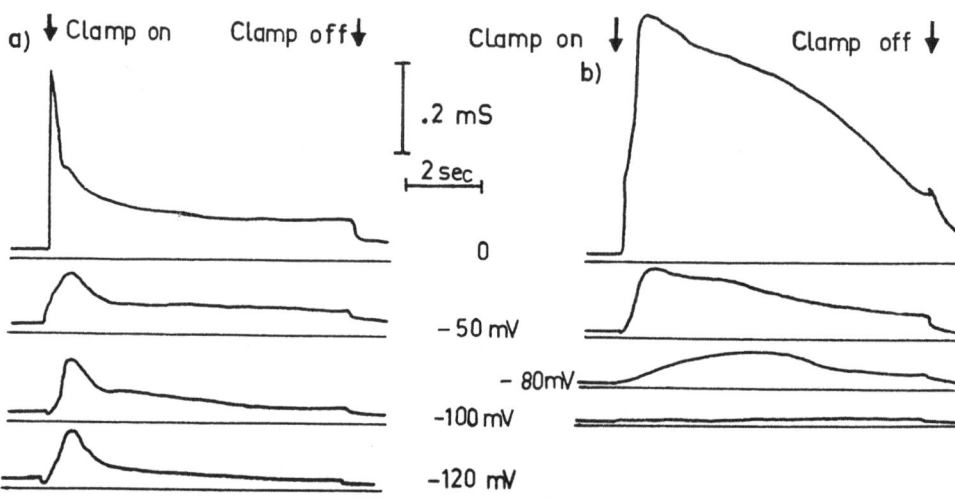

Fig. 1 Conductance profiles at p.d. levels given next to each curve a) in normal APW, Ca^{++} 0.5 mM, b) Ca^{++} 10.0 mM. The cell became inexcitable at -100 mV. Cell area .07 cm^2.

near positive potentials, but not the second negative peak near threshold. Further, preliminary experiments revealed that increasing the $[Ca^{++}]_o$ to 10.0 mM produced rather different currents (reflected by the conductances in Fig.1b) than in intact cells. The short circuit conductance increased, rather than declined and no second peak was visible at -50 mV[8]. The action potential peak did not become more negative[8].

While more experiments are necessary, it appears that SMS show a pattern of excitation currents unlike that of the intact cells. The differences might arise from the fact that there is no longer the two-compartment system of cytoplasm and vacuole. (Small vacuoles probably form soon after the SMS is prepared, but these are scattered through the cytoplasmic matter.) Damage in the centrifuging and ligation process must also be considered.

References

1. A. L. Hodgkin and A. F. Huxley, A quantitative description of membrane current and its application to conduction and excitation in nerve. J. Physiol. 117:500 (1952).
2. M. J. Beilby and H. G. L. Coster, Action potential in Chara corallina III. The Hodgkin-Huxley parameters for the plasmalemma. Aust. J. Physiol. 6:337 (1979).
3. M. J. Beilby and H. G. L. Coster, Action potential in Chara corallina II. Two activation-inactivation transients in voltage clamp of the plasmalemma. Aust. J. Physiol. 6:323 (1979).
4. A. B. Hope and G. P. Findlay, The action potential in Chara. Plant and Cell Physiol. 5:337 (1964).
5. R. E. Williamson and C. C. Ashley, Free Ca^{++} and cytoplasmic streaming in the alga Chara. Nature 296:647 (1982).
6. C. Hirono and T. Mitsui, The course of activation in plasmalemma of Nitella axilliformis., in: "Nerve Membrane", G. Matsumoto and M. Kotani, ed., University of Tokyo Press, Tokyo (1981).
7. M. J. Beilby and B. N. Beilby, Potential dependence of the admittance of Chara plasmalemma. J. Membrane Biol. 74:229 (1983).
8. M. J. Beilby, Calcium and plant action potentials. Plant, Cell and Environment 7:415 (1984).

CONTROL OF PLASMA MEMBRANE PERMEABILITY

OF <u>CHARA</u> BY CYTOPLASMIC CALCIUM

Dale Sanders, Ulf-Peter Hansen and Dietrich Gradmann

Pflanzenphysiologishes Institut der Universität Göttingen
Untere Karspüle 2
D-3400 Göttingen, Federal Republic of Germany

INTRODUCTION

Many hypotheses invoking cytoplasmic calcium as a second messenger in stimulus-response coupling in plant cells rely on indirect evidence: the correlation of a rise in cytoplasmic calcium activity with a given physiological stimulus or response; the effects of extracellular calcium, of calcium ionophores, of calcium channel blockers or of calmodulin inhibitors; <u>in vitro</u> effects of calcium on the activity of an enzyme suspected of playing a key intermediate role in coupling. We have taken advantage of an intracellular perfusion system—which disrupts the tonoplast and facilitates direct contact between the perfusion solution and the inside of the plasma membrane (Tazawa et al., 1976)—to perform a **direct** investigation into the effects of cytoplasmic calcium activity on the plasma membrane of the giant alga <u>Chara corallina</u>.

Normally, in "resting" internodal cells of <u>Chara</u>, the cytoplasmic pCa (pCa_c) is between 6.6 and 7.0, but during an action potential, pCa_c falls transiently but dramatically to about 5.2 (Williamson and Ashley, 1982). This change in pCa_c is probably due partly to increased calcium influx across the plasma membrane, and partly to release of calcium from intracellular stores (Beilby, 1984). Our investigation was directed at establishing whether the rise in cytoplasmic calcium activity is likely to be responsible for any of the changes in membrane electrical parameters during the action potential.

MATERIALS AND METHODS

Intracellular perfusion was carried out as follows. Chambers were placed over the ends of an internodal cell and filled with perfusion medium. A third chamber containing isotonic pond water was positioned over the centre of the cell. Perfusion was initiated by removing the cell ends and creating a small pressure difference between the two end chambers.

Electrical recordings were performed by an external electrode method (Smith and Walker, 1981). Current was passed between Ag/AgCl electrodes placed in one of the end chambers and the central chamber. Transmembrane electrical potential was measured as the difference in potential between two KCl/agar bridges positioned in the other end chamber and the central chamber. Current-voltage curves were obtained by clamping the plasma membrane at a series of discrete potentials with current pulses which were delivered as a bipolar staircase (Gradmann et al., 1978).

RESULTS

Lowering pCa_c from the control value of >7 to about 6 had relatively little effect on membrane electrical properties. However, re-perfusion of cells with medium of pCa 5.5 introduced a significant and reversible depolarization, which was even more marked at pCa 5.0. This depolarization occurred independently of the presence of ATP, and therefore appears to involve transport systems other than the electrogenic proton pump. Similar findings have been reported by Mimura and Tazawa (1983) for the alga Nitellopsis.

Since involvement of the proton pump could be ruled out, it seems reasonable to suppose that the depolarization involves opening of a channel which allows the dissipative flux of an ion whose equilibrium potential is more positive than the resting potential. This interpretation is supported by membrane current-voltage data from perfused cells. The current-voltage curves show a general calcium-dependent increase in conductance across a wide voltage range. The effect is not dependent on the presence of calmodulin (bovine brain) in the perfusion medium. The membrane depolarization occurs even at potentials more positive than the potassium equilibrium potential, and therefore it appears cytoplasmic calcium may well cause opening of a chloride channel.

DISCUSSION

Chloride channels are known to open during the action potential, since chloride efflux is responsible for the bulk of the inward current which flows (Beilby, 1984). The period of significantly depressed pCa_c during an action potential (Willamson and Ashley, 1982) coincides with the duration of current flow through the chloride channels. Our results show that experimental lowering of pCa_c to levels which occur spontaneously during an action potential results in reversible channel opening. Thus, one possible implication of the results is that the fall in pCa_c during an action potential is responsible for initiation of chloride efflux. An interesting corollary of this interpretation is that, once the initial fall in pCa_c has been triggered, the remainder of the action potential is **chemically,** and not electrically, controlled.

REFERENCES

Beilby, M.J., 1984, Calcium and plant action potentials, Plant Cell Environ., 7:415.
Gradmann, D., Hansen, U.-P., Long, W.S., Slayman, C.L., and Warncke, J., 1978, Current-voltage relationships for the plasma membrane and its principal electrogenic pump in Neurospora crassa: I. Steady-state conditions, J. Membr. Biol., 39:333.
Mimura, T., and Tazawa, M., 1983, Effect of intracellular Ca^{2+} on membrane potential and membrane resistance in tonoplast-free cells of Nitellopsis obtusa, Protoplasma, 118:49.
Smith, P.T., and Walker, N.A., 1981, Studies on the perfused plasmalemma of Chara corallina: I. Current-voltage curves: ATP and potassium dependence, J. Membr. Biol., 60:223.
Tazawa, M., Kikuyama, K., and Shimmen, T., 1976, Electrical characteristics and cytoplasmic streaming of characeae cells lacking tonoplast, Cell Struct. Funct., 1:165.
Williamson, R.E., and Ashley, C.C., 1982, Free Ca^{2+} and cytoplasmic streaming in the alga Chara, Nature, 296:647.

PROPERTIES AND DISTRIBUTION OF ATPases IN OAT AND WHEAT WITH SPECIAL REFERENCE TO THE PLASMALEMMA

Marianne Sommarin[1], Tomas Lundborg[2] and Anders Kylin[1]

1. Dept. of Plant Physiol., Univ. of Lund, S-220 07 LUND
Sweden   2. Dept. of Crop Genetics and Breeding, SLU
S-268 00 SVALÖV, Sweden

The transport properties of membrane systems are important for the maintenance of stable conditions for the biochemical reactions that take place in the cell and its organelles. The regulation of transport, particularly of transport through the plasmalemma, must be sensitive for the short-time variations that obviously occur to a greater extent in the surroundings of plant cells than around animal cells. In order to study functions related to transport and regulation, we have prepared plasmalemma vesicles from the 10,000 – 30,000 g microsomal fraction of root homogenates from oat and wheat, using the aqueous polymer two-phase separation technique (Lundborg et al. 1981, Widell et al. 1982).

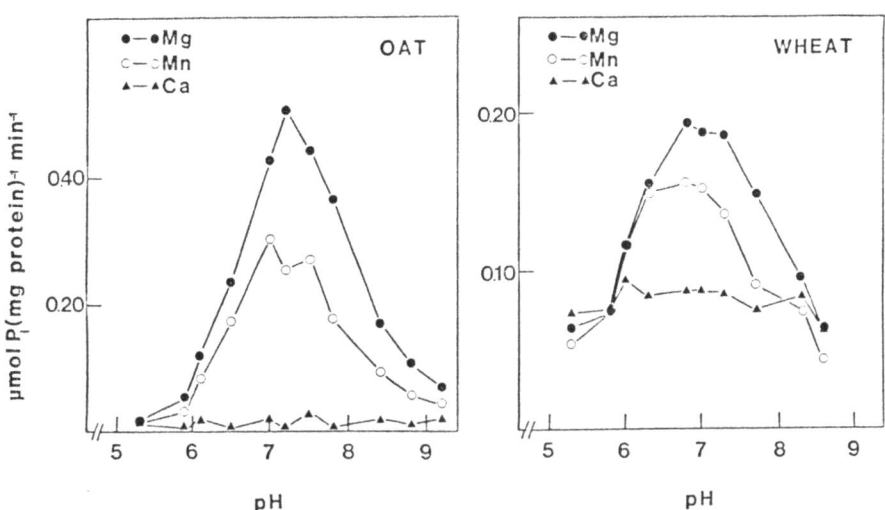

Figure 1. pH-dependence of ATPase activities in plasmalemma from roots of oat and wheat. Assay at 38°C for 15 min. Assay medium: 1 mM Tris-ATP, 1 mM MgCl$_2$ or 1 mM MnCl$_2$ or 2 mM CaCl$_2$, 40 mM Tris-Mes of desired pH and 8-10 µg of enzyme protein in a total volume of 0.5 ml.

Table 1. Substrate specificity of ATPase activities in plasmalemma from roots of oat and wheat. Assay at pH 6.8 under conditions stated in legend to figure 1.

Substrate	Oat			Wheat		
	Mg^{2+}	Mn^{2+}	Ca^{2+}	Mg^{2+}	Mn^{2+}	Ca^{2+}
		%			%	
ATP	100	100	6	100	100	100
CTP	6	7	28	32	63	62
GTP	4	6	28	34	30	67
ITP	6	9	100	31	39	67
UTP	3	6	10	26	54	74
ADP	2	2	0	26	26	67
CDP	1	0	0	19	32	61
GDP	4	4	16	25	25	62
IDP	4	4	42	20	14	61
UDP	4	6	21	24	29	68
AMP	1	2	12	0	0	0
PNPP	1	1	14	2	0	2
PP_i	2	2	6	1	4	0
100% activity corresponds to: (μmol/mg min)	0.49	0.280	0.032	0.303	0.277	0.155

We have studied the properties of ion-activated ATPases that may be part of the system for transport and regulation, and their distribution between different fractions during the purification procedure. In relation to ecological adaptation, it may be of interest that there is some Ca^{2+}-ITPase but at most traces of Ca^{2+}-ATPase present in plasmalemma from oat, which gives a good crop on acid soils low in Ca^{2+}, whereas activation of ATPase by Ca^{2+} occurs in plasmalemma from the Ca^{2+}-dependent wheat (Fig. 1, Table 1). ATPases activated by Mg^{2+} and Mn^{2+} are present in both species and show good specificity for ATP (Table 1); in contrast to the Ca^{2+}-activation in wheat, which appears almost as a general NTPase with certain preference for ATP. In both species the Mg^{2+}- as well as the Mn^{2+}-ATPase are further activated by K^+ in a way that in the Mg^{2+} case is often regarded as associated with transport of K^+. As for the Ca^{2+} activities, it is as yet unclear how they relate to the common picture of an out-pump for Ca^{2+} through the plasmalemma.

References

Lundborg, T., Widell, S. & Larsson, C. 1981, Distribution of ATPases in wheat root membranes, Physiol. Plant., 52: 89-95.

Widell, S., Lundborg, T. & Larsson, C. 1982, Plasma membrane from oats prepared by partition in an aqueous polymer two-phase system, Plant Physiol., 70: 1429-1435.

A CALCIUM ANTAGONIST BINDING SITE IN PLANTS

Elisabeth Andrejauskas, Rainer Hertel and Dieter Marmé

Institute of Biology III, University of Freiburg
Schänzlestr. 1
7800 Freiburg, West Germany

In animal cells the cytoplasmic free Ca^{2+} concentration is mainly regulated by opening and closing plasma membrane located Ca^{2+} channels. These Ca^{2+} channels are identified by specific binding of Ca^{2+} antagonists (i.e. diltiazem, nifedipine, verapamil) which are known to block Ca^{2+}-influx.

There is evidence that Ca^{2+} channels exist also in plants (e.g. Hayama et al. (1979) Protoplasma 99, 305-321). We have investigated specific Ca^{2+} antagonist binding to a particulate fraction from zucchini using [^{3}H]verapamil. Binding to zucchini membranes was saturable and reversible. The apparent equilibrium dissociation constant is K_D = 102 nM and the maximum number of binding sites is B_{max} = 60 pmol/mg of protein. [^{3}H]verapamil binding to zucchini membranes could not be inhibited by the Ca^{2+} antagonists nifedipine and diltiazem. Sucrose density fractionation of zucchini membrane preparations revealed that [^{3}H]verapamil binding sites are located primarily at the plasma membrane. However, binding activity could be detected also at lower density fractions, indicating the existence of specific binding sites at the endoplasmic reticulum or at the vacuoles.

A crude membrane preparation obtained from zucchini hypocotyls was used to solubilize the [^{3}H]verapamil binding site with the detergent CHAPS (3-((3-cholamidopropyl)dimethyl-ammonio)-1-propanesulfonate). Solubilization was carried out with 5 mM CHAPS at a detergent:protein ratio of 1:1 (w/w). After centrifugation for 1h at 100.000 x g 25-30% of the protein and up to 20% of the binding sites were recovered in the supernatant. Stability and yields of the solubilized receptor were increased when glycerol [10-20% (v/v)] or sucrose [10-20% (w/v)] were included in the solubilization media. The same rapid filtration technique, as for the membrane-bound receptor, was used for the CHAPS solubilized verapamil binding sites. Retention of the soluble receptor-ligand complex on glass-fiber filters was achieved by pretreating the filters with the cationic polymer polyethylenimine (PEI). The verapamil concentration that gave a half-maximal inhibition of [^{3}H]verapamil binding to the solubilized receptor was about 10^{-7} M, which was comparable to the IC_{50} obtained with membrane fractions.

CHARACTERIZATION OF CALCIUM CHANNELS IN CARROT CELLS

Marie-Jose Gillery and Raoul Ranjeva

Centre de Physiologie Végétale
Université Paul Sabatier, U.A. CNRS n° 241
118 Route de Narbonne, 31062 Toulouse Cédex

ABSTRACT

Plant cells may accumulate high levels of calcium but the mechanisms by which the cation enters the cell remain largely unknown.

Recently, Hetherington and Trewavas (1984) have shown that membranes isolated from pea seedlings bind calcium-channel antagonists in a specific manner but with a lower affinity constant compared to animal systems.

Such data suggest the occurrence of calcium channels in higher plant cells.

In our laboratory, we have prepared carrot protoplasts and membrane fractions (microsomes) from protoplasts and looked at their ability to bind calcium antagonists.

REVERSIBLE ASSOCIATION OF CALCIUM ANTAGONISTS

We have estimated the capacity of microsomes to bind reversible antagonists PN 200-110. The data obtained are as follows:
- microsomes may bind labelled channel antagonist that may be displaced by excess of unlabelled drug (nifedipine);
- high backgrounds due to non-specific adsorption prevent the accurate determination of the affinity constant.

PHOTOAFFINITY LABELLING OF CALCIUM CHANNELS

We have used a method that would allow the irreversible but specific binding of the antagonist, namely photoaffinity labelling with tritiated or iodinated antagonists.

Such a procedure allows us to work with intact protoplasts and microsomes as well. The data obtained were essentially the same as the above-mentioned results dealing with the reversible ligand. However, since the photoaffinity probe is irreversibly bound, it is possible to study the distribution of the label when the membranes are centrifuged through a sucrose gradient.

In these conditions, it clearly appears that most of the radioactivity is associated with "light fractions" but is bound in a non-specific manner. Such non-specific binding is responsible for the high backgrounds observed when the membranes are not purified.

In contrast only a very low amount of antagonists is specifically bound to two types of membranes with respective specific gravity of 1.1816 (M_1) and 1.13 (M_2).

M_1 may correspond to plasmalemma and M_2 to endomembranes.

Both types of membranes were labelled when either intact protoplasts or microsomes are used. However, the ratio of radiolabelled M_1/M_2 was higher when intact protoplasts are used suggesting that even if the affinity probe enters the cell, the plasmalemma is preferentially labelled on protoplast. This is not the case with crude microsome preparations.

From these data it is shown that both external and internal membranes from plants bind calcium-channel antagonists. However, channels occur in very low concentrations.

The minimal molecular size of the target protein will be determined.

REFERENCE

Hetherington, A.M. and Trewavas, A.L., 1984, Plant Sci. Lett. 35:109–113.

BINDING OF VERAPAMIL TO MAIZE ROOT MEMBRANES

D. Drakeford and A.J. Trewavas

Botany Department, University of Edinburgh, Mayfield Road
Edinburgh EH9 3JH

Calcium channels are membrane bound proteins which control entry of calcium to the cytosol. They have been extensively studied in animal tissues mainly by neurophysiologists because of the relevance of the calcium current to action potentials and heart beat. There seem to be at least 2 types of channel, those whose functioning is voltage dependent and those whose gating activity is modified by hormones. The activity of these channels can be modified by certain drugs notably the dehydro-pyridines and verapamil and these have been used to identify and purify calcium channel proteins.

We have recently reported (Hetherington and Trewavas 1984) that plant membranes bind nitrendipine with slightly weaker affinity than animal membranes and a very much lower density of binding sites. This poster reports on extension of these studies using labelled verapamil now that it is commercially available. Verapamil, unlike the dehydro-pyridines, is light stable and binding to maize root membranes is easily detectable. Verapamil rapidly inhibits maize root growth. The poster will illustrate aspects of the binding and describe the attempted solubilisation and further purification of the binding site. Such studies should hopefully advance our understanding of the control of calcium channels in plant cells.

Reference

Hetherington, A. and Trewavas, A.J. (1984). Plant Sci. Lett. 35, 109-113.

THE INFLUENCE OF CALCIUM ON MITOCHONDRIAL MEMBRANE TRANSPORT AND FLUIDITY

A. Cooke[1], D. Collison[2], F.E. Mabbs[2] and M.J. Earnshaw[3]

[1]Biology Department, Q.E.C., University of London
London WA8 7AH (U.K.). [2]Chemistry and [3]Botany Departments
University of Manchester, Manchester M13 9PL (U.K.)

INTRODUCTION

Calcium stabilises and maintains the semipermeability of plant cell membranes although the mechanism of action is still uncertain (Simon, 1978). Isolated plant mitochondria, which swell passively upon addition of a K-transporting ionophore, provide a possible model system for investigating the influence of Ca on inner membrane dynamics using electron paramagnetic resonance (e.p.r.) spectroscopy on nitroxide spin labels.

MATERIALS AND METHODS

Ionophore-induced swelling of uncoupled corn mitochondria (ca. 0.17 mg protein/ml) was measured at 520 nm in a medium containing 200mM KCl. Motion of the following nitroxide spin levels in the membranes of non-energised mitochondria (ca. 21.3 mg protein/ml) was determined by e.p.r. spectroscopy: 2-(3-carboxypropyl)- 4, 4-dimethyl-2-tridecyl-3-oxazolidinyloxyl [I(12,3), a membrane surface label] and 2-(14-carboxytetradecyl)-2-ethyl-4, 4-dimethyl-3-oxazolidinyloxyl [I(1,14), a membrane core label].

RESULTS

Half-maximal inhibition of passive K transport occurs at 0.35mM Ca for valinomycin, a mobile ionophore, and at 1.0mM Ca for gramicidin, a pore-forming ionophore (Fig. 1). Similar effects are produced by a range of inorganic divalent cations (Cooke and Earnshaw, 1985). The action of Ca is non-competitive implying that Ca does not interfere with K-ionophore complex formation and could act by slowing down ionophore penetration through the inner membrane. Fig. 1 shows that Ca also increases both $2T_{||}$ of I(12,3)-labelled mitochondria and τ of I(1,14)-labelled mitchondria which implies a decrease in membrane fluidity. Half-maximal effects occur at 1.0mM and 2.6mM Ca for I(12,3) and I(1,14) respectively. A similar decrease in the mobility of the labels is caused by decreasing temperature. Comparison of temperature-induced restrictions of label mobility with those induced by Ca implies that the membrane core is more sensitive to Ca than is the

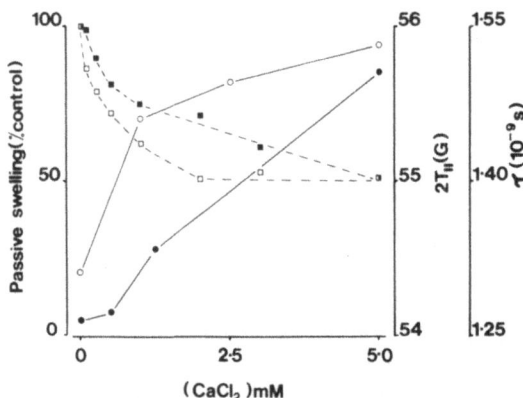

Fig. 1. Effect of Ca concentration on
passive swelling [valinomycin,
□ ; gramicidin, ■] and spin
label motion [$2T_{||}$ of $I(12,3)$,
○ ; τ of $I(1,14)$, ●].

surface. A survey of a range of multivalent cations further suggests
that this decrease in membrane fluidity is largely non-specific and due
to cation binding to membrane components.

DISCUSSION

Direct comparison of the Ca concentrations necessary to inhibit
ionophore-driven K transport and reduce membrane fluidity (Fig. 1) is not
possible due to the different mitochondrial protein concentrations
demanded by the two techniques. Nevertheless, it is reasonable to assume
that restriction of inner membrane motion by Ca could both decrease
valinomycin mobility and inhibit penetration of gramicidin monomers
required to initiate pore formation. The Ca concentrations used in these
experiments are much higher than those occurring in the cytosol.
However, mitochondrial matrix Ca concentrations *in vivo* are unknown and
furthermore Mg, which exerts similar effects to Ca, is found in mM
concentrations in the cytosol (Clarkson and Hanson, 1980). Previous work
has correlated membrane transport ATPase activity with membrane fluidity
(Caldwell and Haug, 1981) which, together with data in the present study,
suggests that the influence of cations on membrane fluidity could control
the rate of cellular ion and metabolite transporters.

REFERENCES

Caldwell, C.R., and Haug, A., 1981, Temperature dependence of the barley
    root plasma membrane-bound Ca- and Mg- dependent ATPase,
    Physiol. Plant., 53: 117.
Clarkson, D.T., and Hanson, J.B., 1980, The mineral nutrition of plants,
    Annu. Rev. Plant Physiol., 31: 239.
Cooke, A., and Earnshaw, M.J., 1985, Passive interactions of Ca and
    other multivalent cations with the membranes of isolated corn
    mitochondria, J. Exp. Bot., in press.
Simon, E.W., 1978, The symptoms of calcium deficiency in plants, New
    Phytol., 80: 1.

EFFECT OF PHOTOPERIOD ON THE CALCIUM UPTAKE BY MICROSOMES AND MITOCHONDRIA FROM GREEN LEAVES

Vladimir Stosic, Claude Penel and Hubert Greppin

Laboratoire de Physiologie végétale, Université de Genève
3 place de l'Université, 1211 Genève 4, Switzerland

The photoperiodic induction of flowering is a physiological process initiated in leaves of plants submitted to the appropriate photoperiod. This mechanism implies that leaves can perceive an environmental stimulus (i. e. length of day or night) and transmit a signal which induces the floral differentiation of meristems. As it is known that calcium is often involved in coupling stimulus to response in a wide variety of cells, it was tempting to look for a possible messenger function of this ion during floral induction, either in leaves or in meristems.

As a first attempt, membrane vesicles and mitochondria were isolated from leaves of spinach (a long day plant) at various times of a short day or after a transfer to continuous light, which promotes flowering. The uptake of labelled calcium by these vesicles and mitochondria was then measured in standardized conditions (Dieter and Marmé, 1981; Stosic et al., 1983). As shown in the figure, the maximum calcium uptake by microsomes from short-day grown plants occurs in the middle of the dark period, and the minimum during the 8-hour photoperiod. Apparently , the ATP-dependent calcium uptake follows a rhythmic activity during the photoperiod. A transfer of plants to continuous light does not change this rhythm, at least during the first 24 hours, but the activity is higher. The addition of bovine brain calmodulin activates the calcium uptake by a maximum of about 20 per cent. The highest activation was observed at the end of the short photoperiod, the lowest one occurring at the middle of the night. Transferring plants to continuous light at the end of the photoperiod -16 hours- somewhat extends the period of maximum activability by calmodulin, but then, the sensitivity to calmodulin falls to the level observed in microsomes from short day-grown plants and remains weak during the following hours. Interestingly, the lowest activation by calmodulin corresponds to the highest calcium uptake capacity without calmodulin. The transfer of spinach to continuous light, which induces flowering, seems to enhance the calcium transport capacity of some unknown membrane from leaf cells, but after several hours of light, the activation of this transport by calmodulin is inhibited.

Similar experiments were performed with mitochondria isolated from spinach at various times of the photoperiod. In the presence of appropriate substrates, the mitochondria accumulate calcium with apparent Km and Vmax greater than those of microsomes, but this accumulation does not change during the photoperiod.

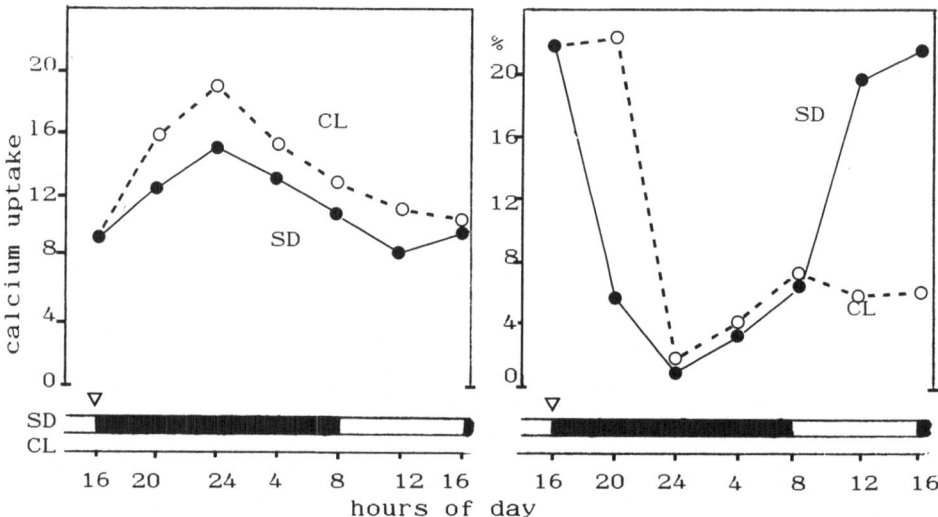

Fig. Calcium uptake by microsomes prepared from spinach leaves taken
at various times during a short day-long night cycle (SD) or after
the transfer (∇) of short day-grown plants to continuous light
(CL). Left: calcium uptake in presence of 1 mM ATP and 5 mM MgCl$_2$
(nmoles calcium/mg prot.30 min). Right: activation of the calcium
uptake by the addition of calmodulin (5 µg/ml).

REFERENCES

Dieter, P., and Marmé, D., 1981, Far red light irradiation of intact corn
    seedlings affects mitochondrial and calmodulin-dependent microsomal
    Ca^{2+}-transport, Biochem. Biophys. Res. Commun., 101:749.
Stosic, V., Penel, C., Marmé, D., and Greppin, H., 1983, Distribution of
    calmodulin-stimulated Ca^{2+} transport into membrane vesicles from
    green spinach leaves, Plant Physiol., 72:1136.

THE CALCIUM CONTENT OF CHLOROPLAST, MITOCHONDRIAL AND CYTOSOLIC FRACTIONS

OF PEA LEAF CELLS

Michael O. Proudlove and Anthony L. Moore

School of Biological Sciences
University of Sussex
Falmer, Brighton BN1 9QG, U.K.

INTRODUCTION

Evidence that small changes in the cytosolic pCa (from approximately 7 to 6) of animal cells may act as a secondary messenger of hormone action is now widely accepted (Åkerman, 1982). For higher plants, however, whilst there are reports that similar, small changes in $[Ca^{2+}]$ can affect the activity of several enzyme-mediated reactions (Moore and Åkerman, 1984), there are few data on the Ca^{2+} levels in the mature cell and the changes which may occur in response to effectors such as light and/or growth regulators (Åkerman et al., 1983; Moore and Åkerman, 1984). Using isolated protoplasts, which show high rates of CO_2-dependent O_2 evolution, we have measured both total intra-cellular Ca^{2+} content and that associated with chloroplastic, mitochondrial and cytosolic/vacuolar fractions of pea leaf cells. Isolating these constituents from protoplasts not only gives maximum yield of intact, active organelles, but also minimizes the time for gross changes in the steady-state $[Ca^{2+}]$ in such cell compartments by using conventional differential centrifugation techniques.

MATERIALS AND METHODS

Pea (Pisum sativum cv. Feltham First) seeds were soaked in running tap water for 12h at 20°C, planted in John Innes No.3 and grown under a 12h light/dark cycle at 22°C for 12-14 days. After gently abrading the lower epidermis whole, primary leaves were immersed in digestion medium (0.5M sorbitol; 1mM $CaCl_2$; 5mM MES-KOH, pH 6.0; 0.2% (w/v) pectolyase; 2% (w/v) cellulase) for 3h. Protoplasts were collected from gradients (Leegood and Walker, 1983), diluted 3-fold with 0.5M sorbitol; 40mM KCl; 10mM HEPES-KOH, pH 7.6 (HKM), centrifuged at 150g for 5 min, resuspended in $5cm^3$ of the same medium and again centrifuged at 150g for 5 min. Pellets were resuspended in HKM at a concentration of $500\mu g.cm^{-3}$ (3.35×10^6 cells.cm^{-3}). Protoplasts were diluted with $1cm^3$ HKM, passed once through $20\mu M$ mesh and centrifuged at 1500g for 2 min (chloroplast pellet). The supernatant was centrifuged for 3 min at 10,000g (mitochondrial pellet) and the final supernatant quickly decanted and taken as the cytosol/vacuole fraction. Each fraction was immediately quenched with 100mM HNO_3, containing 0.1% (w/v) $LaCl_3$, and $10\mu l$ samples were injected into a Pye Unicam SP9, using flame emission spectroscopy. Standards were made up in HKM and treated as for samples. Each measurement is the mean of at least five replicates per sample.

RESULTS AND DISCUSSION

Illuminated protoplasts, isolated from mature pea leaves, evolved oxygen at rates between 40-90 μmoles mg^{-1} chlorophyll h^{-1} in the presence of 10mM $KHCO_3$. As found for spinach, chloroplasts (Leegood and Walker, 1983) and mitochondria (Nishimura et al., 1982) were also rapidly prepared from pea protoplasts. The chloroplast pellet contained >90% of the chlorophyll associated with the protoplasts and these organelles were >98% intact, as measured by ferricyanide permeability and phase-contrast microscopy, and sustained PGA-dependent O_2 evolution at rates between 25-50 μmoles mg^{-1} chlorophyll h^{-1}. The oxygen evolved from chloroplasts was never as high as seen for the protoplasts from which they were isolated, presumably due to chloroplast dilution in the reaction medium and the lack of a complete balance of substrates and co-factors found in vivo. Mitochondria isolated from protoplasts were shown to be coupled but, because they constitute such a small fraction of the cell volume (Valles et al., 1984), the amount of mitochondrial protein pelleted was insufficient to obtain reasonable traces. After obtaining the various fractions from illuminated and non-illuminated pea leaf protoplasts the Ca^{2+} levels in each was measured. The results are presented in Table 1.

TABLE 1. Effect of light on Ca^{2+} distribution in pea leaf protoplasts.

Protoplasts (166μg chlorophyll) were suspended in 400μl HKM plus 10mM $KHCO_3$ and illuminated for 30 min at 25°C; dark controls were wrapped in foil. Results are expressed as nmoles (total) for each fraction and the numbers in brackets are the percentage of Ca^{2+} (total fractions) in each fraction.

Fraction	Dark	Light
Protoplasts	8350	6835
Chloroplasts	1440 (18)	1280 (20)
Mitochondria	80 (1)	71 (1)
Cytosol/vacuole	6250 (81)	5120 (79)

These show that Ca^{2+} can easily be measured in each fraction and that there is a small percentage increase in the chloroplast Ca^{2+} on illumination, consistent with its activation of stromal NAD^+ kinase (Moore and Åkerman, 1984). There were no measurable changes in mitochondrial Ca^{2+} but, again, this may reflect their small fractional cell volume. The majority of cell Ca^{2+} is associated with the cytosol/vacuole fraction and further details on their respective Ca^{2+} levels will be presented in the poster. The present technique takes several minutes from disruption to quench, possibly allowing some change in Ca^{2+} levels. We are presently developing a method for fractioning in seconds.

ACKNOWLEDGEMENTS: This work was supported by a grant from the S.E.R.C.

REFERENCES

Åkerman, K.E.O., 1982, Ca^{2+} transport and cell activation, Med.Biol., 60:168.
Åkerman, K.E.O., Proudlove, M.O. and Moore, A.L., 1983, Evidence for a Ca^{2+} gradient across the plasma membrane of wheat protoplasts, Biochem. Biophys.Res.Commun., 113:171.
Leegood, R. and Walker, D.A., 1983, Chloroplasts, in: "Isolation of membranes and organelles from plant cells", J.L. Hall and A.L. Moore, eds., Academic Press, London.
Moore, A.L. and Åkerman, K.E.O., 1984, Calcium and plant organelles, Plant, Cell Environ., 7:423.
Nishimura, M., Douce, R. and Akazawa, T., 1982, Isolation and characterization of metabolically competent mitochondria from spinach leaf protoplasts, Plant Physiol., 69:916.
Valles, K.L.M., Proudlove, M.O., Williamson, F.A., Beechey, R.B. and Moore, A.L., 1984, Intracellular-volume measurements of wheat-leaf mesophyll cells and protoplasts, Biochem.Soc.Trans., 12:850.

LIMITS ON A ROLE FOR CALMODULIN IN CHLOROPLAST METABOLISM

Anthony R. Ashton

Lehrstuhl für Pflanzenphysiologie
Universität Bayreuth
D-8580 Bayreuth, West Germany

Chloroplasts have been reported to contain a relatively small fraction (1-2%) of the total cellular calmodulin (Jarrett et al., 1982; Muto, 1982). The total concentration of calmodulin in the chloroplast stroma was estimated to be about 0.1 μM. Although low, this concentration of calmodulin (when free) would be quite sufficient to saturate known calmodulin-regulated enzymes. Thus it would seem plausible that calmodulin could play a role in chloroplast metabolism and that it might be worthwhile to screen chloroplast enzymes for calcium-dependent regulation. However, as discussed below, there are other severe constraints on a role for calmodulin in chloroplast metabolism.

THEORY

Apart from the need to have calmodulin in excess of the K_d for any calmodulin-enzyme interaction it is also necessary to have at least comparable molar concentrations of calmodulin and the target enzyme. For example, the concentration of RuBP carboxylase (Rubisco) active sites in the chloroplast stroma is about 3-5 mM - more than a 10,000-fold excess over calmodulin (Ashton, 1982). Thus calmodulin could potentially regulate 0.01% of the carboxylase molecules. Although Rubisco is the extreme example it is possible to estimate the concentration of other chloroplast enzymes and compare them with chloroplast calmodulin concentrations. The basis of the calculation is that the volume of the chloroplast stroma is about 25 μl/mg chlorophyll. By obtaining the enzyme activity (units/mg chlorophyll), the specific activity of the purified enzyme (units/mg protein) and the molecular weight from published work (e.g. Robinson and Walker, 1981) one can calculate the molar concentration of enzyme. Although approximate it is unlikely that the results will be in error by more than ten-fold.

RESULTS AND DISCUSSION

As can be seen from Table 1, all the enzymes of the Calvin cycle are in vast excess (>100-fold) relative to calmodulin.

The concentrations of photosystem I, photosystem II and the cytochrome b/f complex have been estimated to be 1 per 600 chlorophylls (Whitmarsh and Ort, 1984). This is equivalent to a concentration accessible to the stromal

Table 1. Concentrations of Calvin Cycle Enzymes

Enzyme	Active Site Concentration (μM)
Rubisco	3500
Phosphoglycerate kinase	90
Glyceraldehyde phosphate dehydrogenase	150
Triose phosphate isomerase	15
Aldolase	330
Fructose bisphosphatase	23
Transketolase	80
Sedoheptulose bisphosphatase	40
Ribose 5 phosphate epimerase	17
Ribose 5 phosphate isomerase	35
Ribulose 5 phosphate kinase	110

space of 67 μM. Ferredoxin and ferredoxin-NADP reductase are also more abundant than the above electron transport components (Böhme, 1978). These calculations suggest that calmodulin would be incapable of regulating photosynthesis directly without some sort of amplifying mechanism.

I have calculated the concentrations of a number of other chloroplast enzymes (but not yet the calmodulin-independent chloroplast NAD-kinase) and find all to be above 1 μM and usually above 10 μM. It seems that most metabolic enzymes catalyzing significant fluxes will be similarly abundant. I suggest that the only possible role remaining for calmodulin in chloroplasts would be at the beginning of an amplifying cascade.

While cytoplasmic calmodulin levels are higher than those in the chloroplast it seems reasonable to ask if there is enough calmodulin to saturate all cytoplasmic calmodulin-binding proteins. In a similar vein it may be that in animal cells the high affinity of calmodulin for its target proteins (nanomolar K_ds) compared with the total concentration of calmodulin in tissues of up to 10 μM, (Klee et al., 1980) may be a consequence of the high concentration of calmodulin-binding proteins in some tissues. In such cases much of the cellular calmodulin may be bound to these target proteins and the free calmodulin concentrations, for which the target proteins must compete, correspondingly low.

REFERENCES

Ashton, A.R., 1982, A role for ribulose-I, 5-biphosphate carboxylase as a metabolite buffer, FEBS Lett., 145:1-7.
Böhme, H., 1978, Quantitative determination of ferredoxin, ferredoxin-NADP reductase and plastocyanin in spinach chloroplasts, Eur. J. Biochem., 83:137-141.
Jarrett, H.W., Brown, C.J., Black, C.C. and Cormier, M.J., 1982, Evidence that calmodulin is in the chloroplast of peas and serves a regulatory role in photosynthesis, J. Biol. Chem., 257:13795-13804.
Klee, C.B., Crouch, T.H. and Richman, P.G., 1980, Calmodulin, Ann. Review Biochem., 49:489-515.
Muto, S., 1982, Distribution of calmodulin within wheat leaf cells. FEBS Lett., 147:161-164.
Robinson, S.P. and Walker, D.A., 1981, Photosynthetic carbon reduction cycle, in "The Biochemistry of Plants", Vol. 8, pp. 193-236, M.D. Hatch and N.K. Boardman, eds, Academic Press, New York.
Whitmarsh, J. and Ort, D.R., 1984, Stoichiometries of electron transport complexes in spinach chloroplasts, Arch. Biochem. Biophys. 231:378-389.

EFFECTS OF CALMODULIN ANTAGONISTS ON TRANSMEMBRANE AUXIN TRANSPORT IN

CUCURBITA PEPO L. HYPOCOTYL SEGMENTS

Mary C. Astle

University of Cambridge, Department of Biochemistry
Tennis Court Road, Cambridge CB2 1QW
United Kingdom

Several reports have related movement of calcium within plants to auxin transport. These have been considered by Hertel (1983), who postulates that auxin anion carriers, located on both plasma membrane and tonoplast, act as calcium gates. Cytoplasmic calcium concentration would increase concomitantly with auxin efflux into the extracellular free space or vacuole, and may act as a coupling messenger for auxin. I report here studies undertaken to examine the hypothesis that calcium and auxin transport are linked.

I have not yet been able to demonstrate any effect of IAA on calcium uptake by a membrane fraction prepared from Cucurbita pepo L. hypocotyls (modified method of Gross and Marmé, 1978). Possible regulation of the auxin anion efflux carrier by calmodulin in Cucurbita pepo L. hypocotyl segments has also been examined. The method used was based on Depta and Rubery (1984). Routinely, 2 mm segments were incubated in 0.3 μM [1-^{14}C] IAA at pH 5.0 for 15 min uptakes. Effects of calmodulin antagonists and other additives were compared to those of naphthylphthalamic acid (NPA), which inhibits the IAA efflux carrier by binding at a postulated regulatory site (Depta and Rubery, 1984).

The hydrophobic calmodulin antagonists chlorpromazine (CP) and trifluoperazine at 100 μM, and the calcium entry blocker flunarizine (but not verapamil) at 50 μM, increase IAA uptake, though not in the presence of 2 μM NPA. The extent of stimulation of net uptake was less than that observed for NPA. The effects of CP were studied in more detail. Over the concentration range 10 to 1000 μM, maximum stimulation of uptake was observed at 300 μM CP at external pH values of 4.0, 5.0 and 6.0. Similar concentration dependence of CP stimulations were observed for uptakes of ABA and the pH probe, DMO. This implies that cytoplasmic alkalinification, increasing the driving force for diffusive uptake of weak acids, may be involved. However, 300 μM CP stimulated IAA uptake to a greater extent than dimethylbenzylamine, employed as a basic analogue of CP, which stimulated IAA uptake to a constant level from 0.1 to 1000 μM. CP was observed to increase binding of IAA, though not of ABA or DMO, to frozen-and-thawed segments. This effect was shown not to be due to a stoichiometric ion pair effect by including excess nonradioactive IAA in the incubation media.

It appeared that NPA and CP could be acting at a similar site. CP can stimulate net IAA uptake at submaximal (0.02 to 0.1 μM) NPA. At maximal NPA, CP either had no effect or decreased net uptake. The extent of the

NPA stimulation was the more variable factor: where NPA had a large effect, CP generally decreased uptake at maximal NPA. Inclusion of nonradioactive IAA in the incubation media allowed NPA and CP effects to be separated. Uptake of radioactive IAA was decreased at 10 μM nonradioactive IAA, due to saturation of the uptake carrier. At higher nonradioactive IAA concentrations (100 μM) a slight stimulation of net uptake was observed, due to saturation of the NPA-sensitive efflux carrier. At 2 μM NPA, when the efflux carrier is inhibited at all IAA concentrations, low nonradioactive IAA had a greater inhibitory effect on net uptake, and there was no increase in uptake at higher IAA concentrations. However, nonradioactive IAA had no effect on the stimulation of uptake observed at 300 μM CP: neither uptake nor efflux carrier activities could be demonstrated kinetically in the presence of CP.

The ability of CP to increase membrane permeability to uncharged species and to interfere with carrier action may be due to hydrophobic action at the phospholipid bilayer. Direct hydrophobic interactions with the carrier molecules may also occur. Local anaesthetics are known to interact hydrophobically with membranes (e.g. Buckhout, 1984). Dibucaine and tetracaine, over the range 0.2 to 2 mM, stimulated IAA uptake, though not in the presence of maximal NPA, in a comparable manner to CP. They also stimulated uptakes of ABA and DMO. The effects of dibucaine and CP on IAA uptake were not additive, implying a common site of action, though dibucaine, in contrast to CP, did not affect binding of IAA to dead segments.

In conclusion, I have not yet demonstrated any direct link between calcium and auxin transport. CP has been shown to have non-specific effects on IAA uptake that may be related to its hydrophobic nature. Data proposing a role for calmodulin in a physiological response depending solely on modification of that response by calmodulin antagonists should be treated with caution.

ACKNOWLEDGMENT

This work was financed by SERC.

REFERENCES

Buckhout, T.J., 1984, Characterization of Ca^{2+} transport in purified endoplasmic reticulum membrane vesicles from Lepidium sativum L. roots, Plant Physiol., 76:962.

Depta, H. and Rubery, P.H., 1984, A comparative study of carrier participation in the transport of 2,3,5-triiodobenzoic acid, indole-3-acetic acid, and 2,4-dichlorophenoxyacetic acid by Cucurbita pepo L. Hypocotyl segments, J. Plant Physiol., 115:371.

Gross, G. and Marmé, D., 1978, ATP-dependent Ca^{2+} uptake into plant membrane vesicles, Proc. Natl Acad. Sci. USA, 75:1232.

Hertel, R., 1983, The mechanism of auxin transport as a model for auxin action, Z. Pflanzenphysiol., 112:53.

GROWTH CONTROL BY TRANSPLASMALEMMA REDOX: EVIDENCE FOR A CALCIUM ROLE

F. L. Crane, R. Barr and T. A. Craig

Department of Biological Sciences
Purdue University
West Lafayette, IN 47907   U.S.A.

We have considered the following evidence:

1. Auxin increases non-mitochondrial respiration.[1]
2. For carrot cells auxin increases cyanide and salicylhydroxamic acid insensitive (residual) respiration.[2]
3. Ferricyanide inhibits auxin induced oxygen uptake.[3]
4. Ferricyanide inhibits plant cell growth.[4]
5. Ferricyanide accepts electrons at the plasmalemma by trans-membrane electron transport.[5]
6. Highly purified plasma membranes have NAD(P)H ferricyanide reductase and cyanide, SHAM insensitive NAD(P)H oxidase.[6,7]
7. Auxin stimulates proton release and hyperpolarizes the membrane; electrons reducing oxygen do not accompany protons out of cell.[8]
8. Transplasmalemma ferricyanide reduction depolarizes the membrane, parallel electron and proton flow short circuits oxygen reduction.[9,10]
9. External ferricyanide stimulates proton release,[11,13] and oxidizes internal NAD(P)H.[12]
10. Auxin stimulates transmembrane ferricyanide reduction at low concentration and inhibits at high.[4,14]
11. Calmodulin antagonists inhibit auxin growth response.[15]
12. Calmodulin antagonists and internal calcium chelators inhibit transplasmalemma redox.[16]
13. Fusicoccin stimulated proton release is inhibited by anaerobiosis and restored by ferricyanide.[13]

From this evidence, we propose that a calcium-calmodulin dependent, auxin controlled, proton pumping, transplasmalemma NAD(P)H oxidase is involved in auxin control of plant cell growth and this NAD(P)H oxidase contributes to residual respiration of plant cells. Supported by NSF.

REFERENCES

1. L. J. Audus, 1960, p. 360, Vol. 12/2, in: "Encyclopedia of Plant Physiology," W. Ruhland, ed., Springer, Berlin.
2. P. C. Misra and F. L. Crane, unpublished.
3. V. V. Polevoy and T. Salamatova, 1977, p. 209, in: "Regulation of Cell Membrane Activity in Plants," E. Morré and O. Ciferri, eds., Elsevier, Amsterdam.

4. F. L. Crane, R. Barr, T. A. Craig, and P. C. Misra, Proceed. Plant Growth Regulator Soc. Amer. 11:87 (1984).
5. R. Barr, T. A. Craig, and F. L. Crane, Biochim. Biophys. Acta 812:49 (1985).
6. J. M. Ramirez, G. G. Gallego, and R. Serrano, Plant Sci. Lett. 34:103 (1984).
7. R. Barr, A. S. Sandelius, D. J. Morré, and F. L. Crane, Plant Physiol. in press (1985).
8. R. Cleland, Annu. Rev. Plant Physiol. 22:197 (1971).
9. V. A. Novak and N. G. Ivankina, Fisiologia Rast. 30:1107 (1983).
10. P. C. Sijmons, F. C. Lanfermeijer, A. H. deBoer, H. B. A. Prins, and H. F. Bienfait, Plant Physiol. 76:943 (1984).
11. J. D. C. Chalmers, J. O. D. Coleman, and N. J. Walton, Plant Cell Reports 3:243 (1984).
12. P. C. Sijmons, W. van den Briel, and H. F. Bienfait, Plant Physiol. 75:219 (1984).
13. R. Federico and C. E. Giartosio, Plant Physiol. 73:182 (1983).
14. R. Barr, F. L. Crane, and T. A. Craig, J. Plant Growth. Regul. 2:243 (1984).
15. D. C. Elliott, S. M. Batchelor, R. A. Cassar, and N. G. Marinos, Plant Physiol. 72:219 (1983).
16. R. Barr, B. Stone, T. A. Craig, and F. L. Crane, Biochem. Biophys. Res. Communs. 126:262 (1985).

THE EFFECTS OF FUSICOCCIN AND INDOLE-3-ACETIC ACID ON THE

CYTOSOLIC pH OF *ZEA MAYS* CELLS

Benno Brummer[1], Adam Bertl[2], Ingo Portrykus[3],

Hubert Felle[2] and Roger W. Parish[1]

[1]Institut für Pflanzenbiologie der Universität Zürich
Cytologie, Zollikerstrasse 107, CH-8008 Zürich
Switzerland

[2]Botanisches Institut, Justus Liebig Universität
Senckenbergstrasse 17-21, D-6300 Giessen, FRG

[3]Friedrich-Miescher-Institut, CH-4002 Basel, Switzerland

Microelectrodes were used to measure simultaneously the effects of fusicoccin on cytosolic pH and membrane potential in a maize root cell. Cytosolic pH began to fall within seconds of adding fusicoccin, whereas membrane hyperpolarization commenced after a lag of 2 min. The pH-microelectrode could not be used with coleoptile cells for technical reasons. However, the dual-wavelength absorbance technique showed that fusicoccin also induced a rapid cytosolic acidification in coleoptile cells. Indole-3-acetic acid lowered the cytosolic pH of these cells. The effect was less pronounced than with fusicoccin and began after 5 min, well before membrane hyperpolarization (10 min) and extracellular acidification (30 min) were detectable. Procaine, which penetrates the plasma membrane and gains protons in the cytosol, was shown to depolarize root cells and inhibit indole-3-acetic acid-induced growth of coleoptiles. 1-Naphthyl acetate, which acidifies the cytosol, hyperpolarized root cells and stimulated coleoptile growth. The results support the concept that fusicoccin and auxins induce elongation growth by lowering the cytosolic pH.

CALCIUM IN THE AGRICULTURE OF TURKEY

Serpil Terzioglu

Lecturer, Hacettepe University, Faculty of Science
Department of Biology, Botany Section
Beytepe, Ankara, Turkey

Calcium carbonate contents are quite high in the soil of Turkey. The country, Turkey, is divided into eight agricultural regions. The values of pH and contents of $CaCO_3$ have been shown in Figure 1.

Calcium causes agricultural problems only in Akdeniz, Karadeniz and Marmara regions of Turkey.

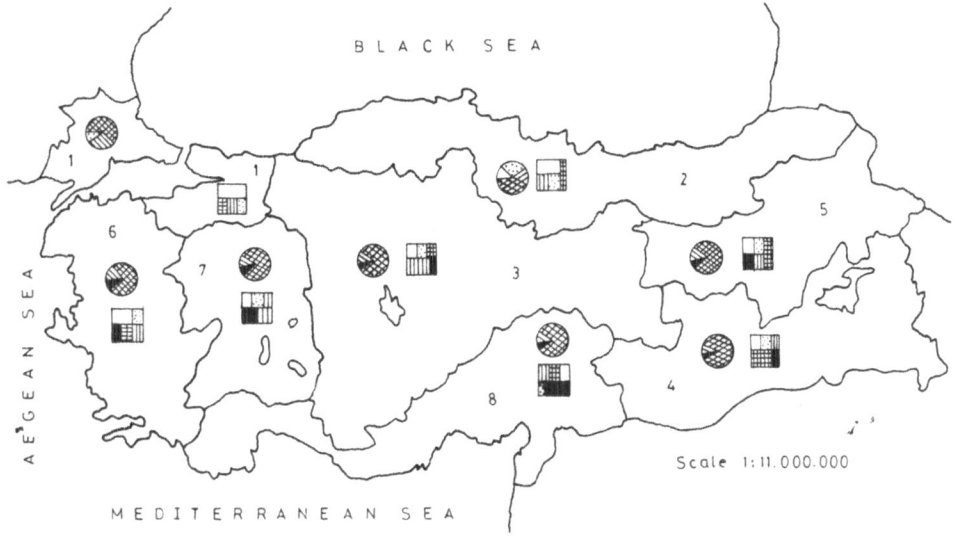

Agricultural Regions

1 Trakya ve Marmara	pH	$CaCO_3$
2 Karadeniz	○ 4-4.9 Highly acidic	□ <1 Very low
3 Orta Anadolu	◔ 5-5.9 Fairly acidic	▨ 1-5 Low
4 Güney Doğu Anadolu	◑ 6-6.9 Slightly acidic	▦ 5-15 Medium
5 Doğu Anadolu	⊕ 7-7.9 Slightly alkaline	▤ 15-25 High
6 Ege	● 8-8.9 Highly alkaline	■ >25 Very high
7 Göller		
8 Akdeniz		

Fig. 1. Average pH and $CaCO_3$ values for agricultural regions of Turkey (%).

439

The researches concerning the calcium in plant breeding are concentrated in two main groups.

The first group researches concern for the deficiencies of some plant mineral nutrients which appear in the alkali-reactioned soil of Akdeniz region. In this region the most important agricultural problem that is confronted is the lack of phosphorous and some micro-elements such as Fe, Zn, Cu, etc.

The second one concerns determining the sufficient amount of lime and fertilizer in order to increase the productivity of acid-reactioned soil of the Karadeniz and Marmara regions. In these regions, there are also the same problems confronted as in the Akdeniz region.

CALCIUM COMPLEXING LIGANDS DERIVED FROM CYSTINE : EXPLORATORY WORK

Louis Cazaux, Nadine Leygue, Claude Picard and Pierre Tisnès

Laboratoire de Synthese et Physicochimie organique
Unité Associée CNRS N° 471, Université Paul Sabatier
118 Route de Narbonne, 31062 Toulouse Cedex, France

INTRODUCTION

Two kinds of ligands derived from cystine are presented: macrocyclic dilactones and acyclic tetraester analogues of EGTA and BAPTA.

MACROCYCLIC LIGANDS

These compounds are synthethized by a new methodology: the key step is a tin template-driven cyclisation involving a bis (alkoxy triphenyl-tin) compound and the difluoride of the N-carbobenzoxy protected cystine.

Their complexing properties are measured in a first approach by the extraction of metallic picrates (Na^+, K^+, Mg^{2+}, Ca^{2+}) from an aqueous solution to a chloroformic solution, measured by ultraviolet spectro-photometry. Calcium ion is the unique ion extracted by the 12-membered ring (extraction ratio E = 19.3%) while the larger 21-membered ring, a better ligand for calcium (E = 36%), is less selective (Mg^{2+} : E = 7%).

A potential application of these lipophilic compounds is to use them as calcium carriers.

ACYCLIC LIGANDS

These compounds are derived from "modified" cystine where the disulfide linkage is replaced by a $-S-(CH_2)_2-S-$ arrangement. Their synthesis and the fixation of a physicochemical probe for the measurement of intra-cellular calcium is in progress.

FUTURE DEVELOPMENT OF CALCIUM STUDIES;

EVIDENCE IN THE SUPPORT OF CONCEPTS

David T. Clarkson

Plant Sciences Division
Long Ashton Research Station
Bristol BS18 9AF

A decade ago the plant physiologist or biochemist electing to work on the role of calcium in metabolism and development would have found many professional niches in which there would have been little competition. Today one casts a rueful glance back to such untroubled times realising that now there is great activity in nearly every niche and the professional survival pressure is intense. This change has accompanied the tardy realisation that plant cells may respond to environmental stimulae, growth factors and hormones in ways similar to animal cells. The literature relating to the triggering of biochemical reaction sequences by changes in cell calcium and calcium-calmodulin in animal cells is vast and has been increasing rapidly for thirty years. Plant scientists have a great deal of catching up to do but they have the advantage over the real pioneers in animal biology because techniques, concepts and drugs have all been developed which aid new work. Perhaps we have been in a position to make progress for longer than we realised. The research reports given at this meeting indicate that progress has indeed been rapid. So rapid in some directions that several subjects are being outstripped by others and that there is a danger of a new orthodoxy, in the way we think about plant cell development, becoming entrenched. Many participants at the meeting commented that changes in cell-free Ca^{2+} were very much more frequently spoken of than measured. The 'new orthodoxy' would be better founded on fact than on faith that plant cells are necessarily like their mammalian counterparts.

The preceding papers in this volume have given an impression of the depth and vigour of Ca^{2+} research. The aim of this postscript is to give some indication of where research is likely to develop, what sort of information is most urgently needed and where people had uneasiness about present positions.

One evening the participants were divided into nine groups to prepare statements about 'where do we go from here?' - each group centred around some relatively specialised interest. It is most instructive that several themes cropped up in nearly all of the group reports.

Some rules of evidence have been proposed by Jaffe (1980) in presenting a case for a calcium-mediated process: 1) that cellular response should be preceded or accompanied by an increase in the concentration of free Ca^{2+} in the cytoplasm; 2) that factors which block this increase in Ca^{2+} should also block the developmental response; 3) that experimental generation of an increase in free Ca^{2+} should stimulate the response. Most contributors were agreed that evidence on the first of these points is difficult to obtain and is far from satisfactory; all groups identified the solution of the technical problems involved as a major priority.

The problems experienced by plant biologists who had attempted to load tissues with esters of Quin 2 and Fura 2 dyes were due to extracellular esterases; these might be overcome by suppression of the enzyme activity or redesigning the ester groups so as to make them less effective substrates. In suitable cells, where there is sufficient thickness of cytoplasm to locate the micro-pipette tip unequivocally, direct injection of Fura 2 or Arsenazo dyes would seem to be the most attractive methods available as long as one can be certain that movement of the dye into the vacuole is negligible. Glass micro-electrodes containing liquid calcium ion-exchanger offer good prospects for measuring steady state Ca^{2+} activities but the response times of such electrodes are probably too slow to measure transients of physiological interest. Other physical averaging techniques, X-ray and proton microprobes, NMR etc., are unable to distinguish between free and sequestered or bound Ca^{2+}. Against this rather baffling technical background, Peter Hepler's suggestion of a physiological assay for cytosol Ca^{2+} activity has obvious attractions to those who work on cell systems where cytoplasmic streaming rates can be measured. Such a technique has the time resolution required for physiological experiments and it is rather comforting for a crypto-Darwinian, in an age of high technology, to contemplate a return to a method based on careful visual examination. There is no doubt, however, that the dependence of cytoplasmic streaming on cytosol Ca^{2+} activity is one of the best characterized calcium responses. It is highly probable that assays of this kind will be used in the future.

An area of related uncertainty concerns the transporters of calcium across membranes. Most workers accept, thermodynamics being what they are, that a cell with finite internal sequestering capacity must, in the long term, discharge Ca^{2+} back to the external medium via some energy linked process across the plasma membrane. The characterization of the molecular mechanism for this transport is long overdue. Its relationship with a di-valent ion stimulated ATPase in plasma membranes is far from clear. We will soon learn whether this is because current membrane isolation techniques are particularly disruptive of the plasma membrane, or whether we are looking for the wrong thing. It is significant, perhaps, that the calcium pump in rat liver plasma membrane has been shown to be a completely different pro-tein (MW 118,000, vanadate-inhibited phosphoenzyme) from the previously reported $(Ca^{2+}-Mg^{2+})$-ATPase in this tissue (Lin, 1985). The entry of Ca^{2+} into cells down its steep free energy gradient can occur sufficiently rapidly and with a specificity which makes it certain that some channel or carrier is involved. The inhibition of Ca^{2+}-requiring responses by drugs, developed to block Ca^{2+} channels in heart muscle, is an encouraging devel-opment; some of these are now available radioactively labelled, e.g. ^3H-desmethoxyverapamil (Amersham International UK) and will provide excel-lent tools for channel isolation from membranes. But which membranes? The question was raised, but could not be answered, whether the transient increases in cytosol Ca^{2+} depended on opening of channels to the outside solution or to sequestered stores of Ca^{2+} (more than ample to raise the activity into the 10^{-6} to 10^{-5} range) in the vacuole and endoplasmic reti-culum. If fine control of cytosol Ca^{2+} in plants depends, as it appears to in animal cells, on the activity of calcium pumps in the ER (Somlyo, 1984), the search for calcium channel proteins in the ER using labelled ligands might be fruitful. Participants were keenly interested in the development of patch clamping techniques to study the effects of pH, membrane potential and plant growth regulators on the gating of the calcium channels. In sum-mary, it seems that techniques are available for us to obtain a far more precise description of both the active and passive transport of calcium than we have at the moment.

444

After the technical problems of measuring cytosol Ca^{2+} have been resolved, we will be left with trying to understand what it is that the cell perceives about its calcium status so as to maintain a steady state activity in the sub-micromolar range. This constancy was much discussed but less consideration was given to another ion which is also closely regulated in the cytosol in the sub-micromolar range, viz. H^+. A personal prediction is that the cell pH and the cytosol Ca^{2+} people will be getting their acts together soon.

It has become evident that plant growth regulators can influence the cytosol free Ca^{2+} at an early stage in their effects on development. It is hard to accept that this is all that they do, since, within the same cell, different PGRs may produce different effects on development. Clearly the complex relationships between gibberellins and phosphatidyl inositol metabolism and calcium are likely to be different from those between, say, abscisic acid and the conductance of the plasma membrane to Ca^{2+}. The group considering this matter stressed the need for a model system in which there was a demonstrable receptor of the PGR and a way of measuring calcium fluxes across the membrane. The presence of auxin-binding sites in microsomes from maize coleoptiles and their possible association with the proton pump (Thompson, Krull and Venis, 1983) was suggested as a promising starting point. It is, however, a daunting task to reconstitute a meaningful system from cells in which intricate sequences of reactions may lead to changes in calcium fluxes. A further cautionary reminder to those who seek a role for calcium in all message carrying circumstances comes from a report of the Dahlem Conference on molecular mechanism of photoreception. In vision research it had long been accepted that calcium was the second messenger which controls the gating of Na^+-channels in the outer membrane of the rods. Recent evidence has shown, however, that this is done directly by cGMP rather than through stimulation of kinases and phosphatases by Ca-CaM (Vines, 1985). All of the sorts of changes in cytosol Ca^{2+} which we seek in plant cells were seen in this sytem but incorrect inferences seem to have been drawn.

Assuming that some environmental or hormonal signal does raise the free Ca^{2+} in the cytosol the next group of problems relate to the effects of the ion itself as opposed to those exerted by Ca-calmodulin (Ca-CaM) on cytoskeletal elements and enzyme activities. Both the ion and the protein are ubiquitous in cells, and in some respects CaM may mirror the distribution of calcium itself. It is not known how rapidly CaM levels change or how the concentration in a cell is regulated. The separate roles of CaM and Ca^{2+} might be distinguished on the basis of the sensitivity of a given structural rearrangement or enzyme activity to drugs, e.g. the phenothiazines, which are CaM antagonists, but the specificity of these substances is not very great. One approach to broadening knowledge of CaM-binding proteins would be to develop an overlay technique which could be used to label two-dimensional protein separations; the subsequent steps in sequencing and characterizing these proteins would utilize existing techniques of molecular biology. Concerning CaM itself, there are important details to be gathered about the significance of structural differences in animal, higher plant and algal CaMs and about the structural changes that accompany the binding of successive Ca^{2+} ions.

While it is clear that protein kinases can be activated by Ca-CaM and that a number of proteins in membranes and in the cytoplasm may become phosphorylated, participants confessed to much uncertainty about the physiological significance of these observations. Without doubt, this is due to the fact that the exact identity of the proteins which are phosphorylated is, for the most part, unknown. This is a matter of high priority for future research.

Papers delivered at the meeting showed convincingly that phytochrome conversion for Pr to Pfr can be accompanied by rapid changes in the Ca^{2+} flux across the plasma membrane. It is also established that light, via the phytochrome system, regulates gene expression in chloroplasts and that there are receptors for phytochrome in the nucleus. These findings raise the question as to whether Ca^{2+} or Ca-CaM have any direct role in regulating gene expression in plant cells. The group that considered this matter conveyed some skepticism and suggested that experimental evidence about the regulation of Ca^{2+} within the nucleus should be sought as well as the localization using immunocytochemical methods, of Ca-CaM.

Discussion of the structural role of Ca^{2+} centred on cytoskeletal proteins with additional references to its role in modifying the stability and surface charge properties of membranes. With respect to the cytoskeleton, some participants were dismayed to hear cell biologists, who had earlier presented such elegant work, declare that the actual role of Ca^{2+} and/or Ca-CaM, in microtubule assembly and in the formation of the mitotic spindle, was uncertain. The group stressed that efforts should be made to isolate intact microtubules and microfibrils and analyse all their components. Further refinements of Ca^{2+} and CaM localization were urgently needed since it is evident that, in dividing and tip-growing cells, there may be great heterogeneity within the cytoplasm. Although the quantitative relationships between Ca^{2+} and cytoplasmic streaming are established more information was needed on the molecular properties of plant/algal myosin and of the significance of its phosphorylation.

It is frequently stated that calcium has separate roles to play in the extracellular spaces (apoplast) and within the cytoplasm. These roles are quantitatively different in the amount of calcium required, the former utilizing concentrations around $10^{-3}M$. There is, therefore, a considerable gradient of Ca^{2+} across the plasma membrane as indeed there may be across the tonoplast and the ER. This gradient may have a considerable bearing on secretion of materials via membrane fusion. This is an area where experimental manipulation is very hard to achieve, but results like those described earlier by Jones (see p. 49) suggest that exocytosis of some amylase isozymes depends on the calcium concentration at the membrane surface. The group considering the role of calcium in secretion concluded that several complications could be removed by studying secretion of wall materials by protoplasts.

Throughout this session there was a conflict between the reductionists and those who believed that so much is lost when cells are disrupted that we would learn a great deal more if we watched them closely. The conflict was, in keeping with everything else at this meeting, good-humoured and even-tempered. When cornered the main protagonists would admit to the legitimacy of their opponents' claims and agree that the only way forward must combine both approaches. The battery of specialized techniques required does, however, make a multi-disciplinary team essential and suggests that the greatest progress will be made where funding can bring such groups together. What we have seen happen in the last few years represents a dramatic change of role for calcium, from a mere mineral nutrient whose properties and impact on the cell could be dealt with in a few lines, to a position centre stage in cell development and metabolic regulation (see the excellent review of Heppler and Wayne, 1985). How gratified the few who held a torch for this element (e.g. Burstrom, 1968) would be by this limelight. It is this which encourages further examination of other small players on the stage; boron is already clamouring for attention and there may be roles for other elements which have never been considered.

REFERENCES

Burstrom, H.G., 1968, Calcium and plant growth, Biol. Revs, 43:287-316.

Heppler, P.K. and Wayne, R.O, 1985, Calcium and plant development, Ann. Rev. Plant Physiol. 36:397-439.

Jaffe, L.F., 1980, Calcium explosions as triggers of development, Ann. N.Y. Acad. Sci., 339:86-101.

Lin, S.-H, 1985, Novel ATP-dependent calcium transport component from rat liver plasma membranes, J. Biol. Chem., 260:7850-7856.

Somlyo, A.P., 1984, Cellular sites of calcium regulation, Nature, 309: 516-517.

Thompson, M., Krull, U.J. and Venis, M.A., 1983, A chemoreceptive bilayer lipid membrane based on an auxin receptor ATPase electrogenic pump, Biochem. Biophys. Res. Comm. 110:300-304.

Vines, G., 1985, News Report, Dahlem Conference "The molecular mechanism of photoreception", New Scientist No. 1467, 40-43.

INDEX

Abscisic acid, 387
Acentriolar cells, 175
Acer, 349
Actin, 397
Adventitious root, 405
Agriculture, 439
Aleurone, 49
  layers, 335
  protoplasts, 333, 335
Aluminium, 19, 323
Amaranthus, 285
Amylase, 49, 333, 337
Anaphase, 167
Apple, 387
Aspartate kinase, 369
ATPase, 1, 417
Auxin, 84, 286, 293, 301, 347, 351, 355, 435, 437
  transport, 401, 433

Barley, 49, 333, 335
Bitter pit, 389
Bud formation, 190

Calcicole, 65, 262
Calcifuge, 65, 262
Calcium
  antagonist, 419
  ATPase, 261, 317
  binding proteins, 33, 41, 325
  channels, 381, 419, 421, 423
  dependent protein kinase, 75
  flux, 27, 242, 401
  measurement, 141, 149, 157, 371, 373
  signal, 81
  transport, 3, 241, 261
  uptake, 335
  wave, 28
Callose, 131
Calmidazolium, 180
Calmodulin, 2, 11, 19, 27, 83, 91, 107, 115, 123, 175,
        204, 269, 311, 313, 317, 319, 321, 323, 327
  antagonist, 274, 347, 405, 433
  dependant enzymes, 5
  synthesis, 319
Carrot, 421
Cell development, 311

Cell division, 185
Cell wall, 60
Cellulose, 57
Chara, 397, 399, 411, 413, 415
Chitosan, 131
Chlamydomonas, 403
Chloride channels, 416
Chloroplast, 201, 269, 281, 429, 431
Cicer arietinum, 313
Cold shock, 257
Commelina, 383, 385
Cress, 219
Crown gall, 341
Cucurbita, 325, 433
Cytokinin, 84, 185, 287, 341
Cytoplasmic calcium, 144, 149, 415
Cytoplasmic streaming, 399
Cytoskeleton, 375
Cytosol pH, 437

Dibucaine, 434
Dionaea, 233

Embryo, 76
Embryonic axis, 313
Endoplasmic reticulum, 219
Endosperm cells, 375

Free flow electrophoresis, 353
Funaria, 371, 379
Fusicoccin, 255, 301, 435, 437

Gadolinium, 380
Gibberellic acid, 286, 333, 335
Glucan synthase, 85, 131
Gravitropism, 219, 225
Gravity, 219
Growth, 293, 301

Hormones, 83, 284, 285
Hydrogen ion pump, 410

Intracellular calcium, 141, 277
Intracellular pH, 393
Iontophoresis, 407

Kinetochore, 175

Laser micromass analysis, 233
Leaf protein kinase, 77
Leaves, 343
Lepidium, 225
Light activation, 359
Lily pollen, 395
Lipid, 65
Lupin, 65, 262

Maize calyptra, 395
Mass flow, 243

Membrane
  -bound protein kinase, 79, 123
  excitation, 413
  proteins, 355
  voltage, 227
Mesotaenium, 193, 321, 381
Metaphase, 167
Microelectrodes, 149
Microsomes, 427
Microtubules, 180, 375
Mitochondria, 279, 425, 427, 429
Mitosis, 167, 175, 375, 379
Monensin, 54
Mougetia, 193, 201, 321
Mushroom, 317
Myosin, 397
  light chain kinase, 16, 207

NAD kinase, 1, 14, 91, 107, 317, 357, 359
Neurospora, 409
Nifedipine, 215, 421
Nitrendipine, 423
NMR, 157, 373
Nuclear
  ATPase, 116
  protein phosphorylation, 117
Nuclei, 115

Oat, 417
  coleoptile, 347
Oil palm, 160
Onoclea, 391
Oocystis, 57
Organelles, 277

Pea, 115, 319, 321
  leaf, 429
  nuclei, 119
Peroxidase, 99, 365
Phloem, 33, 326
Phosphatidyl-inositol, 292, 351
Phosphodiesterase, 381
Phospholipid, 65
Photoperiod, 427
Photosystems, 270
Phototaxis, 403
Phytoalexin, 331
Phytochrome, 115, 194, 201, 207, 391, 393
Pisum sativum, 311, 357
PIXE, 395
Plasmodesmata, 407
Plasmalemma, 435
Plasmamembrane, 170, 263, 353, 415, 417
  injury, 253
Polarity, 211
Pollen, 345
Polyamines, 134
Poterioochromonas, 363
Procaine, 437

Protein
    kinase, 1, 14, 75, 101, 343
    phosphorylation, 83, 99, 115, 206,
        341, 347, 349, 355
Proteinase, 363
Proton microprobe, 212
Proton pump, 305

Quin 2, 169, 371, 373
Quinate oxidoreductase, 42, 367

Root, 225, 319, 357
    apex, 311

Secretion, 333, 335, 365
Sensory cells, 233
Silver beet, 75, 343
Soybean, 99, 131, 331, 341, 355
Spermine, 133
Spinach leaves, 427
Spore germination, 391, 393
Statocytes, 219
Stomata, 383, 385, 387

Tetracaine, 434
Tobacco, 327
    mosaic virus, 327
Tonoplast, 349, 353, 390
Tradescantia pollen, 371, 405
Transcellular currents, 379
Trichomes, 407
Trifluoperazine, 112
Trigger hair, 233
Tuberisation, 84
Tyrosine kinase, 339

Vacuoles, 281
Verapamil, 170
Vibrating probe, 229, 379
Vicia, 65, 262
Volume regulation, 363

W 7, 195
Wall acidification, 301
Western blotting, 127
Wheat, 417

X-ray probe, 157

Zea mays, 437
Zeta potential, 198